Encyclopedia of Entomology

Volume I

Encyclopedia of Entomology
Volume I

Edited by **Christopher Fleming**

R CALLISTO REFERENCE

New York

Published by Callisto Reference,
106 Park Avenue, Suite 200,
New York, NY 10016, USA
www.callistoreference.com

Encyclopedia of Entomology: Volume I
Edited by Christopher Fleming

International Standard Book Number: 978-1-63239-240-4 (Hardback)

Printed in the United States of America.

Contents

Preface

Entomology is a field of study that might give nightmares to certain people with phobias. But what is it? Entomology is the discipline that focuses on the study of insects and their relationship to humans, the environment, and other organisms. It is an ancient field of study and dates back to the establishment of biology as a formal field of study by Aristotle. The insect world is a vast and incredibly diverse world and it is not easy to study it as a whole. Thus most people in this focus on a specific order, species or family of insects. There are many options for a career in entomology and range from forensic entomology to agricultural entomology. Entomology as a scientific discipline makes big contributions to such diverse fields such as biology, agriculture, chemistry, molecular science, human/animal health, criminology, and forensics. The study and research on insects often serves as the foundation for advancements in food and fibre production and storage, biological and chemical pest control, biological diversity, pharmaceuticals epidemiology, and numerous other fields. Entomologists also helpful in contributing towards the betterment of humankind by studying the role of insects in the spread of disease and discovering ways of protecting food and fibre crops and livestock from being damaged. Thus there is constant research to study the way insects could contribute beneficially for the advantage of humankind.

This book is an attempt to compile and understand new and ongoing research data in the field of entomology. I am thankful to those who put their effort and hard work into this field. I am also grateful to my family and friends who supported me in this endeavour. I also wish to acknowledge all the contributing authors who shared their expertise and took out the time for revisions irrespective of the part of world they were in, your efforts are deeply appreciated.

Editor

Accidental Fire in the Cerrado: Its Impact on Communities of Caterpillars on Two Species of *Erythroxylum*

Cintia Lepesqueur,[1] **Helena C. Morais,**[2] **and Ivone Rezende Diniz**[3]

[1] *Programa de Pós-Graduação em Ecologia, Instituto de Ciências Biológicas, Universidade de Brasília, 70910-900 Brasília, DF, Brazil*
[2] *Departamento de Ecologia, Instituto de Ciências Biológicas, Universidade de Brasília, 70910-900 Brasília, DF, Brazil*
[3] *Departamento de Zoologia, Instituto de Ciências Biológicas, Universidade de Brasília, 70910-900 Brasília, DF, Brazil*

Correspondence should be addressed to Cintia Lepesqueur, bioclg@gmail.com

Academic Editor: Helena Maura Torezan-Silingardi

Among the mechanisms that influence herbivorous insects, fires, a very frequent historical phenomenon in the cerrado, appear to be an important modifying influence on lepidopteran communities. The purpose of this study was to compare the richness, abundance, frequency, and composition of species of caterpillars in two adjacent areas of cerrado *sensu stricto*, one recently burned and one unburned since 1994, on the experimental farm "Fazenda Água Limpa" (FAL) (15°55′S and 47°55′W), DF, Brazil. Caterpillars were surveyed on two plant species, genus *Erythroxylum*: *E. deciduum* A. St.-Hil. and *E. tortuosum* Mart. (Erythroxylaceae). We inspected a total of 4,196 plants in both areas, and 972 caterpillars were found on 13.3% of these plants. The number of plants with caterpillars (frequency) differed significantly between the areas. The results indicate that recent and accidental fires have a positive effect on the abundance of caterpillars up to one year postfire, increase the frequency of caterpillars associated with *Erythroxylum* species in the cerrado and do not affect the richness of caterpillars on these plants. Moreover, the fires change the species composition of caterpillars by promoting an increase in rare or opportunistic species.

1. Introduction

Systems represented by the associations of plants and insects include more than one-half of the world's multicellular species. The impacts of disturbances, anthropogenic or otherwise, affect the characteristics of communities of herbivorous insects in any biome worldwide [1]. There is strong evidence that these disturbances result in complex changes in the interactions between plants and herbivores [2]. Fires affect communities of herbivorous insects and provide opportunities for changes in species richness, abundance and species composition in space and time [3]. Among herbivores, Lepidoptera can serve as good indicators of environmental changes caused by these disturbances in certain habitats [4].

Fires in the cerrado are a natural phenomenon of recognized ecological importance [5] and occur during the dry season, from May to September [6, 7]. The effects of fire on the structure, composition and diversity of plants in the cerrado are far more extensively documented [8–12] than the effects on the fauna [13–15]. The knowledge of the effects of fire on insect herbivores and their natural enemies is even more limited [3, 16, 17].

The general literature on the responses of insects to fire in comparison with the responses to other forms of management in open habitats indicates that a significant decrease of insects occurs soon after a fire. The magnitude of the decrease is related to the degree of exposure to flames and to the mobility of the insect [18]. In cerrado, a very rapid and vigorous regrowth of vegetation occurs [19] and this regrowth may favor an increase in the abundance of herbivores. The caterpillar community in the cerrado is species rich and the abundance of most species is low but is highly variable throughout the year [20, 21], due primarily to the climate variability that characterizes the two seasons (dry and wet) in the region. This pattern has also been observed for herbivorous insects in New Guinea. It is characteristic of the herbivorous insect communities in general and is also typical of tropical regions [22]. Among the mechanisms that influence these herbivorous insect community patterns, fires, a very frequent historical phenomenon in the cerrado, appear to be an important modifying influence on lepidopteran communities.

The objective of this study was to compare the richness, relative abundance, frequency, and species composition of caterpillars between two cerrado areas, one recently burned and one unburned since 1994. The study hypotheses that the richness, relative abundance, frequency, and species composition of the caterpillars on the host plants vary between recently burned areas and areas without recent burning (used as a control). We predict that the abundance and species richness of caterpillars will increase significantly in a recently burned area as a result of the intense regrowth of vegetation in the postfire environment [19]. The postfire environment differs greatly from the prefire environment because of the higher phenological synchrony of plants and because of changes in microclimate result from to increased exposure to the sun.

2. Methodology

External folivorous caterpillars were surveyed on two plant species, *Erythroxylum deciduum* A. St.-Hil. and *E. tortuosum* Mart. (Erythroxylaceae), in two adjacent areas of cerrado *sensu stricto*, on the experimental farm "Fazenda Água Limpa" (FAL) (15°55′S and 47°55′W), DF, Brazil. Both plant species were abundant and had similar size in the burned and unburned areas. This system, including only two plant species in the genus and their caterpillars, was chosen for study due to the need for simplification in the analysis and reduction of variables. This choice also reflected the ease of collection and identification and the prior knowledge of the system in the protected areas of the cerrado. The two plant species occur at high densities in the cerrado region and their lepidopteran fauna is known from previous studies in unburned areas [20, 23]. An accidental fire affected the entire area in 1994, and the area suffered another accidental fire in August 31, 2005. The area burned in 1994 was viewed as a control, and the area burned in 2005 was considered recently burned. Data were collected from September 2005 through August 2006.

In both study areas (recently burned and control), external folivorous caterpillars were collected weekly from foliage of 50 individuals of each of the two species of plants. All caterpillars were collected, photographed, numbered as morphospecies, and individually reared in the laboratory in plastic pots (except for gregarious caterpillars), with leaves of the host plant as a food. The adults obtained from laboratory rearing were, as far as possible, identified and deposited in the Entomological Collection, Departamento de Zoologia, Universidade de Brasilia.

A binomial test of two proportions was applied with a significance level of 0.05 to evaluate the occurrence of a consistent difference in the proportion of plants with caterpillars (relative abundance and species richness) between the areas [24]. Species rarefaction curves were constructed to analyze the species richness of caterpillars in each area [25]. EcoSim

Table 1: Number of plants with caterpillars, abundance, and richness of caterpillars on two species of *Erythroxylum*, in two areas of cerrado *sensu stricto* in the FAL (burned and control areas) from September 2005 to August 2006.

Variables	Areas		Total
	Control (%)	Burned (%)	
Inspected plants	2,065 (49.2)	2,131 (50.8)	4,196
Plants with caterpillars	226 (10.9)	333 (15.6)	559
Abundance of caterpillars	346 (35.6)	626 (64.4)	972
Richness of caterpillars	29 (59.0)	36 (74.0)	47*

*Species richness is not the sum total of the richness of the two areas because some species occur in both areas.

7.0 software was used to construct these curves based on 1000 replications [26].

The Shannon-Wiener index (H'), Simpson index (D) and Berger-Parker index (D_{bp}) were used to compare the diversity and dominance of the community of caterpillars on *Erythoxylum* in the two study areas. The indices were obtained with DivEs 2.0 software [27]. The Jaccard similarity index was also applied to evaluate the degree of similarity of the species composition of two communities. If the Jaccard index is equal to one ($B = 0$ and $C = 0$), all species are shared between the two communities. If the Jaccard index is near 0, few if any species are shared.

3. Results

We inspected a total of 4,196 plants, with similar numbers in both areas (Table 1). A total of 972 caterpillars were found on 13.3% of the plants inspected. The number of plants with caterpillars (frequency) differed significantly between areas ($p1 = 0.11$; $p2 = 0.16$; $Z = -4.46$; $P < 0.001$). The probability of finding a plant with a caterpillar in the control area (one out of nine plants inspected) was smaller than in the burned area (one to six plants). The relative abundance of caterpillars also differed significantly ($p1 = 0.17$, $p2 = 0.30$, $Z = -9.69$, $P < 0.001$) between areas. Almost twice as many caterpillars were found in the burned area as found in the control area (Table 1).

Forty-seven species or morphospecies (hereafter treated as species) of caterpillars were recorded, belonging to at least 15 families (two species belonged to unidentified families). The burned area had 36 species, compared with 29 species in the control area (Table 1). However, this difference in species richness between the areas was not significant ($p1 = 0.08$; $p2 = 0.06$; $Z = 1.57$; $P > 0.05$). Even after adjustment by the rarefaction method to a common basis of an equal number of caterpillars in both areas ($n = 346$) the species richness did not differ, and the estimated number of species varied between 24 and 32 (Table 2; Figure 1).

The value of dominance was higher in the burned area (34.5%) than in the control area (29.8%) (Table 2). Likewise, the dominance for the burned area, estimated by rarefaction, was between 31.2% and 37.9%, significantly higher than the value estimated for the control area on a common basis of 346 caterpillars in both areas (Table 2). These results are

TABLE 2: Diversity of caterpillars on two species of *Erythroxylum* in two areas of cerrado *sensu stricto* in the FAL (recently burned and control) from September 2005 to August 2006: number of caterpillars, species richness, estimated species richness through rarefaction in the control area ($n = 346$, 95% confidence interval), dominant species and dominance observed in both areas, estimated dominance by rarefaction in the control area ($n = 346$, 95% confidence interval), diversity index (H'), and dominance (D and D_{bp}).

	Control area	Burned area
Number of caterpillars	346	626
Observed species richness	29	36
Estimated richness (Rarefaction, $n = 346$)	—	24–32
Dominant species	*Antaeotricha sp.*	*Antaeotricha sp.*
Observed dominance	29.8%	34.5%
Expected dominance	—	31.2–37.9%
Diversity of Shanon-Wiener (H')	1.01	0.89
Dominance of Simpson (D)	0.16	0.21
Dominance of Berger-Parker (D_{bp})	0.30	0.35

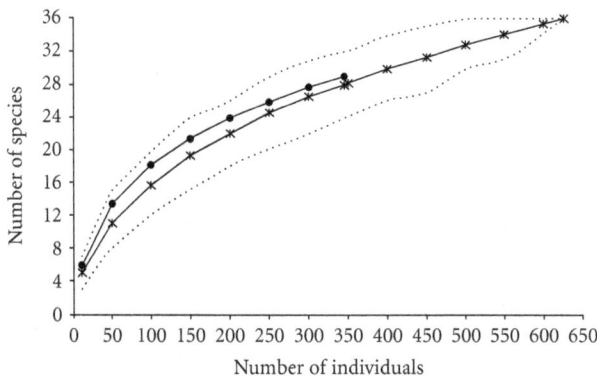

FIGURE 1: Rarefaction curves of caterpillar species of the control area (line with circle) and the burned area (line with star) in relative to the number of individuals estimated from randomizations of the order of 1000 samples in cerrado *sensu stricto* in the FAL from September 2005 to August 2006. The dotted line indicates 95% confidence intervals.

also consistent with the dominance index values D and D_{bp}, which were higher in the burned area. The diversity index H' was higher in the control area (Table 2).

An unidentified species of *Antaeotricha* (Elachistidae) was dominant, with 29.7% and 34.5% of the individuals found in the control and burned areas, respectively. Ten species recorded in the control area showed intermediate dominance, between 1.2 and 7.5%, whereas six species showed intermediate dominance in the burned area, with values between 1.1 and 8.0%. The proportion of rare species, those represented by less than 1% of all caterpillars, was significantly higher ($p1 = 0.55$, $p2 = 0.75$, $Z = -1.68$,

TABLE 3: Abundance of caterpillars and Jaccard similarity index between the two areas of cerrado *sensu stricto* in the FAL (recently burned and control) from September 2005 to August 2006 based on caterpillars found on two species of *Erythroxylum*.

Months	Abundance		Jaccard index
	Control area	Burned area	
Sep	16	0	0.00
Oct	12	17	0.22
Nov	3	7	0.20
Dec	12	30	0.29
Jan	31	40	0.70
Feb	16	27	0.50
Mar	12	33	0.27
Apr	26	42	0.25
May	132	242	0.33
Jun	51	144	0.62
Jul	26	37	0.29
Aug	9	7	0.00
Total	346	626	0.38

$P < 0.05$) in the burned area ($n = 27$) than in the control area ($n = 16$).

The similarity between the study areas was low (Sj = 0.38), even on a monthly basis, with January (Sj = 0.70) and June (Sj = 0.62) being the sole exceptions (Table 3). Of the 47 species recorded, 38.3% ($n = 18$ species) were common to the two areas (Table 4), and 25.5% of the species ($n = 12$) were restricted to the control area. The species restricted to the control area included the gregarious moth *Hylesia shuessleri* Strand, 1934 (Saturniidae) and the solitary *Dalcerina tijucana* (Schaus, 1892) (Dalceridae), both dietary generalists (Table 4). Approximately 40% of the species ($n = 18$) were found only in the burned area. These species included *Fregela semiluna* (Walker, 1854) (Arctiidae), a generalist species, and *Eloria subapicalis* (Walker, 1855) (Noctuidae) a dietary specialist. The effects of the fire appear to be more evident for Limacodidae as five of the eight species of this family found in the survey occurred exclusively in the control area. Certain species, however, appear to benefit from the effects of fire, for example, three species of Noctuidae found exclusively in the burned area: *Cydosia mimica* (Walker, 1866), *Cydosia punctistriga* (Schauss, 1904) and Noctuidae sp. The five most abundant species (more than 15 individuals per area) were found in both areas and are apparently restricted to the *Erythroxylaceae* in the region (Table 4).

No caterpillars were found on species of *Erythroxylum* until one month after the fire (Table 3). However, the relative abundance of caterpillars was higher in the burned area in all of the following months. Until 12 months after the occurrence of the fire, the caterpillar relative abundance in the burned remained higher than the abundance found in the control area (Figure 2). The temporal occupation of the species of *Erythroxylum* by caterpillars resulted in a pattern whose abundance and richness gradually increased with

TABLE 4: Families and species of caterpillars found on two species of *Erythroxylum* in burned and control areas of cerrado in the FAL from September 2005 to August 2006 (NI = no information about diet breadth; polyphagous = feeds on species from two or more families of plants; restricted = feeds only on species of Erythroxylaceae).

Family	Species	Control area	Burned area	Diet breadth
Arctiidae	*Fregela semiluna* (Walker, 1854)	0	4	Polyphagous
	Paracles sp.	6	2	Polyphagous
Dalceridae	*Acraga infusa* (Schauss, 1905)	4	2	Polyphagous
	Acraga sp. 1	0	1	NI
	Acraga sp. 2	0	2	NI
	Dalceridae sp.	0	1	NI
	Dalcerina tijucana (Schauss, 1892)	1	0	Polyphagous
Elachistidae	*Antaeotricha* sp.*	103	216	Restricted
	Timocratica melanocosta (Becker, 1982)	2	3	Polyphagous
Gelechiidae	*Dichomeris* sp. 1	1	10	Restricted
	Dichomeris sp. 2	22	6	Polyphagous
	Dichomeris sp. 3*	26	160	Restricted
	Dichomeris sp. 4	3	8	Polyphagous
	Dichomeris spp. (duas espécies)*	68	84	Restricted
	Gelechiidae sp.*	44	50	Restricted
Geometridae	*Cyclomia mopsaria* (Guenée, 1857)*	16	24	Restricted
	Geometridae sp. 1	3	0	Restricted
	Geometridae sp. 2	0	1	Restricted
	Stenalcidia sp. 1	0	5	NI
	Stenalcidia sp. 2	1	0	Restricted
Limacodidae	*Limacodidae* sp. 1	0	1	Polyphagous
	Limacodidae sp. 2	0	1	NI
	Limacodidae sp. 3	1	0	NI
	Limacodidae sp. 4	2	0	NI
	Limacodidae sp. 5	2	0	NI
	Miresa clarissa (Stoll, 1790)	0	1	Polyphagous
	Platyprosterna perpectinata (Dyar, 1905)	5	0	Polyphagous
	Semyra incisa (Walker, 1855)	2	1	Polyphagous
Megalopigydae	*Megalopyge albicollis* (Schauss, 1900)	0	1	Polyphagous
	Megalopyge braulio Schauss, 1924	0	1	Polyphagous
	Norape sp.	4	3	Polyphagous
	Podalia annulipes (Boisduval, 1833)	0	1	Polyphagous
Noctuidae	*Cydosia mimica* (Walker 1866)	0	1	Restricted
	Cydosia punctistriga (Schauss, 1904)	0	1	NI
	Eloria subapicalis (Walker, 1855)	0	7	Restricted
	Noctuidae sp.	0	1	Restricted
Notodontidae	*Heterocampa* sp.	7	12	Polyphagous
Oecophoridae	*Inga haemataula* (Meyrick, 1911)	6	1	Polyphagous
	Inga phaeocrossa (Meyrick, 1912)	1	0	Polyphagous
Pyralidae	*Carthara abrupta* (Zeller, 1881)	12	3	Polyphagous
Riodinidae	*Emesis* sp.	1	0	Polyphagous
	Hallonympha paucipuncta (Spitz, 1930)	0	1	Polyphagous
Saturniidae	*Hylesia schuessleri Strand*, 1934	1	0	Polyphagous
Tortricidae	*Platynota rostrana* (Walker, 1863)	0	3	Polyphagous
Urodidae	*Urodus* sp.	0	5	Restricted
Unidentified	sp. 1	1	0	NI
	sp. 2	1	1	NI

* Indicates the five commonest species.

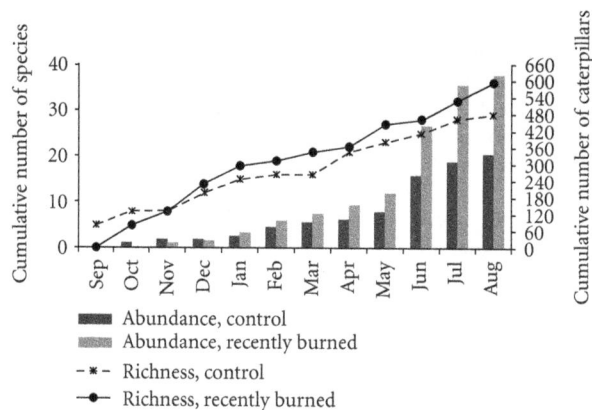

FIGURE 2: Cumulative number of caterpillars (bars) and species (rows) in two areas of cerrado *sensu stricto* in the FAL (recently burned and control) from September 2005 to August 2006.

sampling effort and showed a greater increase during the dry season, specifically during May and June (Figure 2).

4. Discussion

The sporadic and accidental fires in restricted areas of the cerrado may act to renew the vegetation [19], allowing the reoccupation of sites more rapidly by plant species. Several studies in tropical forests and in the cerrado have shown the importance of sprouting as a mechanism of postfire regeneration of shrub and tree species [28–32]. The new foliage that results from sprouting attracts a variety of herbivores.

In the cerrado, a low frequency of caterpillars on host plants is a common feature [20, 33–35]. However, recent fire in the cerrado study area produced as 4.7% increase in the frequency of caterpillars on plants of *Erythroxylum*. The reason for this increase may be that fire may benefit herbivores by increasing the availability of resources. This high availability of resources results from the regrowth of plants because many new leaves are synchronously produced.

Although species richness did not differ between areas, the higher dominance observed in the burned area suggests a higher diversity in the control area. The most interesting feature of this system is the increase of rare species in the burned area. This increase may result from intense regrowth, which may produce new oviposition sites and new environments for these species. At the same time, nearby areas were available to act as a source for re-colonization [17]. However, the rarefaction curves did not reach an asymptote. In fact, previous studies [23, 36] indicate that species caterpillars not found in our surveys occur on the two species of *Erythroxylum* that were examined. These additional species include *Erynnis funeralis* (Scudder & Burgess, 1870) (Hesperiidae), *Phobetron hipparchia* (Cramer, [1777]) (Limacodidae) and *Automeris bilinea* Walker, 1855 (Saturniidae). These species are all polyphagous and could be present on other species of host plants.

The variation in the abundance of insects in the cerrado occurs regardless of the passage of fire and remains seasonal [37]. However, the mortality caused by fire produces an immediate reduction in population size. Even, one month after the fire, caterpillars were not found on the plants surveyed. Moreover, the caterpillar abundance on both species of plants during all the subsequent months was higher in the area disturbed by the recent fire. Similar results have been found for adults of certain insect orders, such as Coleoptera, Hemiptera, Hymenoptera and Lepidoptera, in the cerrado of Brasilia [37]. The return to the previous levels of abundance depends on the order to which the insect belongs and ranges from two to more than thirteen months after the occurrence of the fire [3]. Up to 12 months after the occurrence of fire, the abundance of caterpillars associated with the *Erythroxylum* species studied here had not returned to a level comparable with that observed in the control area.

Research conducted in the same region with the community of caterpillars associated with *Byrsonima* (Malpighiaceae), showed that if the fire in the cerrado is recurrent every two years during the dry season, the results are quite different [38] from those previously discussed. In this case, the abundance and species richness of caterpillars in areas with frequent fires were markedly less than the abundance and species richness of caterpillars in areas protected from fire for more than 30 years. These results are consistent with other previous reports that fire reduces the populations of caterpillars [39], and may cause local extinction of some species [40]. However, these results from areas with frequent fires are in contrast to the results found if the fires are accidental and sporadic, as in the case of this study.

Even with smaller losses than those caused by recurrent fires, the recent accidental fire dramatically increased the abundance of caterpillars and as result, the attacks on plants in the postfire period, just at the time at which most synchronous leaf production in the cerrado occurs. For this reason, this process may produce extensive damage to vegetation and may harm biodiversity conservation in the region. Furthermore, a scheme of recurrent burns during several years in the same area results in the biological and physicochemical degradation of the soil and thus in the reduction of aerial biomass [41].

Although we did not replicate each treatment, our results reflect the effect of fire, as we have followed the changes in communities of caterpillars on various plant species for several years in protected areas from fire [21, 23, 38, 42], and in addition, we have surveyed caterpillars on other plant species in postfire conditions, with similar results (unpublished data). Furthermore, some studies suggest the impossibility of replication treatments when it comes from natural phenomena occurring on a large scale, as in the case of burning [43]. Thus, the results of this study indicate that the recent accidental fire had the following effects on the external folivorous caterpillars: (a) killed eggs and larvae at first but had a positive effect on the relative abundance of caterpillars up to one year postfire, (b) increased the frequency of caterpillars associated with two *Erythroxylum* species in the cerrado, (c) did not affect the richness of caterpillars on these plants and (d) changed the caterpillar species composition because the effects of the fire promoted increases of rare or opportunistic species.

Acknowledgments

The authors are grateful to Vitor O. Becker for the identification of the Lepidoptera and to the Administrators of the Experimental Station (Fazenda Água Limpa-FAL) of the University of Brasilia for facilitating our field work. C. Lepesqueur was granted a scholarship by CAPES, and I. R. Diniz was granted an award from CNPq (Research Productivity Grant). The project was financially supported by CNPq and FAPDF.

References

[1] L. M. Schoonhoven, J. J. A. Van Loon, and M. Dicke, Eds., *Insect-Plant Biology*, Oxford University Press, Oxford, UK, 2005.

[2] T. J. Massad and L. A. Dyer, "A meta-analysis of the effects of global environmental change on plant-herbivore interactions," *Arthropod-Plant Interactions*, vol. 4, no. 3, pp. 181–188, 2010.

[3] I. R. Diniz and H. C. Morais, "Efeitos do fogo sobre os insetos herbívoros do Cerrado: consensos e controvérsias," in *Efeitos do Regime do Fogo Sobre a Estrutura de Comunidades de Cerrado: Resultados do Projeto Fogo*, H. S. Miranda, Ed., pp. 121–131, Ibama, Brasília, Brazil, 2010.

[4] R. Panzer and M. W. Schwartz, "Effectiveness of a vegetation-based approach to insect conservation," *Conservation Biology*, vol. 12, no. 3, pp. 693–702, 1998.

[5] M. F. Simon, R. Grether, L. P. De Queiroz, C. Skemae, R. T. Pennington, and C. E. Hughes, "Recent assembly of the Cerrado, a neotropical plant diversity hotspot, by in situ evolution of adaptations to fire," *Proceedings of the National Academy of Sciences of the United States of America*, vol. 106, no. 48, pp. 20359–20364, 2009.

[6] L. M. Coutinho, "Fire in the ecology of the Brazilian cerrado," in *Fire in the Tropical Biota—Ecosystem Processes and Global Challenges*, J. G. Goldammer, Ed., pp. 82–105, Springer, Berlin, Germany, 1990.

[7] H. S. Miranda, M. M. C. Bustamante, and A. C. Miranda, "The fire factor," in *The Cerrados of Brazil: Ecology and Natural History of a Neotropical Savanna*, P. S. Oliveira and R. J. Marquis, Eds., Columbia University Press, New York, NY, USA, 2002.

[8] M. I. Miranda and C. A. Klink, "Colonização de campo sujo de cerrado com diferentes regimes de queima pela gramínea *Echinolaena inflexa* (Poaceae)," in *Impactos de Queimadas em Áreas de Cerrado e Restinga*, H. S. Miranda, C. H. Saito, and B. F. S. Dias, Eds., pp. 46–52, UnB, Brasilia, Brazil, 1996.

[9] E. A. De Castro and J. B. Kauffman, "Ecosystem structure in the Brazilian Cerrado: a vegetation gradient of aboveground biomass, root mass and consumption by fire," *Journal of Tropical Ecology*, vol. 14, no. 3, pp. 263–283, 1998.

[10] W. A. Hoffmann, "Post-burn reproduction of woody plants in a neotropical savanna: the relative importance of sexual and vegetative reproduction," *Journal of Applied Ecology*, vol. 35, no. 3, pp. 422–433, 1998.

[11] E. P. Rocha e Silva, *Efeito do regime de queima na taxa de mortalidade e estrutura da vegetação lenhosa de campo sujo de cerrado [M.S. thesis]*, Universidade de Brasília, Brasília, Brazil, 1999.

[12] H. S. Miranda, *Efeitos do Regime de Fogo Sobre a Estrutura de Comunidaaes de Cerrado: Resultados de Projeto Fogo*, Ibama, Brasília, Brazil, 2010.

[13] H. C. Morais and W. W. Benson, "Recolonização de vegetação de Cerrado após queimada, por formigas arborícolas," *Revista Brasileira de Biologia*, vol. 48, pp. 459–466, 1988.

[14] M. Prada, O. J. Marini-Filho, and P. W. Price, "Insect in lower heads of *Aspilia foliacea* (Asteraceae) after a fire in a central Brazilian savanna: evidence for the plant vigor hypothesis," *Biotropica*, vol. 27, pp. 513–518, 1995.

[15] R. P. B. Henriques, M. X. A. Bizerril, and A. R. T. Palma, "Changes in small mammal populations after fire in a patch of unburned cerrado in Central Brazil," *Mammalia*, vol. 64, no. 2, pp. 173–185, 2000.

[16] M. A. Marini and R. B. Cavalcanti, "Influência do fogo na avifauna do sub-bosque de uma mata de galeria do Brasil central," *Revista Brasileira de Biologia*, vol. 56, pp. 749–754, 1996.

[17] O. J. Marini-Filho, "Distance-limited recolonization of burned cerrado by leaf-miners and gallers in central Brazil," *Environmental Entomology*, vol. 29, no. 5, pp. 901–906, 2000.

[18] A. B. Swengel, "A literature review of insect responses to fire, compared to other conservation managements of open habitat," *Biodiversity and Conservation*, vol. 10, no. 7, pp. 1141–1169, 2001.

[19] M. B. Medeiros and H. S. Miranda, "Mortalidade pós-fogo em espécies lenhosas de campo sujo submetido a três queimadas prescritas anuais," *Acta Botânica Brasílica*, vol. 19, pp. 493–500, 2005.

[20] P. W. Price, I. R. Diniz, H. C. Morais, and E. S. A. Marques, "The abundance of insect herbivore species in the tropics: the high local richness of rare species," *Biotropica*, vol. 27, no. 4, pp. 468–478, 1995.

[21] I. R. Diniz and H. C. Morais, "Lepidopteran caterpillar fauna of cerrado host plants," *Biodiversity and Conservation*, vol. 6, no. 6, pp. 817–836, 1997.

[22] V. Novotný and Y. Basset, "Rare species in communities of tropical insect herbivores: pondering the mystery of singletons," *Oikos*, vol. 89, no. 3, pp. 564–572, 2000.

[23] M. S. Milhomem, H. C. Morais, I. R. Diniz, and J. D. Hay, "Espécies de lagartas em Erythroxylum spp. (Erythroxylaceae) em um cerrado de Brasília," in *Contribuição ao Conhecimento Ecológico do Cerrado*, L. L. Leite and C. H. Saito, Eds., pp. 107–111, UnB, Brasília, Brazil, 1997.

[24] D. L. Ayres and A. S. Santos, *Bioestat 5.0: Aplicações Estatísticas nas Áreas das Ciências Biológicas e Médicas: Desenvolvimento Estatístico*, Mamirauá & MCT/CNPq, Brasília, Brazil, 2005.

[25] H. L. Sanders, "Marine benthic diversity: a comparative study," *The American Naturalist*, vol. 102, pp. 243–282, 1968.

[26] N. J. Gotelli and G. L. Entsminger, *EcoSim: Null Models Software for Ecology*, Acquired Intelligence Inc. & Kesey-Bear, Jericho, Vt, USA, 2011.

[27] W. C. Rodrigues, "Dives—diversidade de espécies," 2005, http://www.ebras.bio.br/dives/.

[28] C. Uhl, K. Clark, and H. Clark, "Successional patterns associated with slash-and-burn agriculture in the upper Rio Negro region of the Amazon Basin," *Biotropica*, vol. 14, no. 4, pp. 248–254, 1982.

[29] J. B. Kauffman, "Survival by sprouting following fire in tropical forests of the eastern Amazon," *Biotropica*, vol. 23, no. 3, pp. 219–224, 1991.

[30] T. T. Castellani and W. H. Stubblebine, "Sucessão secundária inicial em mata tropical mesófila, após perturbação por fogo," *Revista Brasileira de Botânica*, vol. 16, pp. 181–203, 1993.

[31] A. S. Penha, *Propagação vegetativa de espécies arbóreas a partir de raízes gemíferas: representatividade na estrutura fitossociológica e descrição dos padrões de rebrota de uma comunidade florestal, Campinas, São Paulo [M.S. thesis]*, Universidade Estadual de Campinas, Campinas, Brazil, 1998.

[32] S. V. Martins, G. A. Ribeiro, W. M. Silva Junior, and M. E. Nappo, "Regeneração pós-fogo em um fragmento de floresta estacional semidecidual no município de Viçosa, MG," *Ciência Florestal*, vol. 12, pp. 11–19, 2002.

[33] I. Andrade, I. R. Diniz, and H. C. Morais, "A lagarta de *Cerconota achatina* (Zeller) (Lepidoptera, Oecophoridae, Stenomatinae): biologia e ocorrência em plantas hospedeiras do gênero *Byrsonima* Rich (Malpighiaceae)," *Revista Brasileira de Zoologia*, vol. 12, pp. 735–741, 1995.

[34] H. C. Morais, I. R. Diniz, and J. R. Silva, "Larvas de *Siderone marthesia nemesis* (Illiger) (Lepidoptera, Nynphalidae, Charaxinae) em um cerrado de Brasília, Distrito Federal, Brasil," *Revista Brasileira de Zoologia*, vol. 13, pp. 351–356, 1996.

[35] S. Scherrer, I. R. Diniz, and H. C. Morais, "Climate and host plant characteristics effects on lepidopteran caterpillar abundance on miconia ferruginata DC. and miconia pohliana Cogn (Melastomataceae)," *Brazilian Journal of Biology*, vol. 70, no. 1, pp. 103–109, 2010.

[36] I. R. Diniz, H. C. Morais, and A. J. A. Camargo, "Host plants of lepidopteran caterpillars in the Cerrado of the Distrito Federal," *Revista Brasileira de Entomologia*, vol. 45, pp. 107–122, 2001.

[37] I. R. Diniz, *Variação na abundância de insetos no cerrado: efeitos das mudanças climáticas e do fogo [Ph.D. thesis]*, Universidade de Brasília, Brasília, Brazil, 1997.

[38] I. R. Diniz, B. Higgins, and H. C. Morais, "How do frequent fires in the Cerrado alter the lepidopteran community?" *Biodiversity and Conservation*, vol. 20, no. 7, pp. 1415–1426, 2011.

[39] C. S. Crawford and R. F. Harwood, "Bionomics and control of insects affecting Washington grass seed fields," *Technical Bulletin of the Agricultural Experimental Station*, vol. 44, pp. 1–25, 1964.

[40] S. R. Swengel and A. B. Swengel, "Relative effects of litter and management on grassland bird abundance in Missouri, USA," *Bird Conservation International*, vol. 11, no. 2, pp. 113–128, 2001.

[41] E. L. Cardoso, S. M. A. Crispim, C. A. G. Rodrigues, and W. Barioni Júnior, "Efeitos da queima na dinâmica da biomassa aérea de um campo nativo do Pantanal," *Pesquisa Agropecuária Brasileira*, vol. 38, pp. 747–752, 2003.

[42] H. C. Morais, J. D. V. Hay, and I. R. Diniz, "Brazilian cerrado folivore and florivore caterpillars: how different are they?" *Biotropica*, vol. 41, no. 4, pp. 401–405, 2009.

[43] P. Van Mantgem, M. Schwartz, and M. Keifer, "Monitoring fire effects for managed burns and wildfires: coming to terms with pseudoreplication," *Natural Areas Journal*, vol. 21, no. 3, pp. 266–273, 2001.

Predation of Fruit Fly Larvae *Anastrepha* (Diptera: Tephritidae) by Ants in Grove

W. D. Fernandes,[1] M. V. Sant'Ana,[1] J. Raizer,[1] and D. Lange[2]

[1] *Faculdade de Ciências Biológicas e Ambientais, Universidade Federal da Grande Dourados (UFGD), MS 162, Km 12, 79804-970 Dourados, MS, Brazil*
[2] *Laboratório de Ecologia Comportamental e de Interações, Universidade Federal de Uberlândia (UFU), P.O. Box 593, 38400-902 Uberlândia, MG, Brazil*

Correspondence should be addressed to W. D. Fernandes, wedson@ufgd.edu.br

Academic Editor: Kleber Del-Claro

Based on evidence that ants are population regulatory agents, we examined their efficiency in predation of fruit fly larvae *Anastrepha* Schiner, 1868 (Diptera: Tephritidae). Hence, we considered the differences among species of fruit trees, the degree of soil compaction, and the content of soil moisture as variables that would explain predation by ants because these variables affect burying time of larvae. We carried out the experiment in an orchard containing various fruit bearing trees, of which the guava (*Psidium guajava* Linn.), jaboticaba (*Myrciaria jaboticaba* (Vell.) Berg.), and mango trees (*Mangifera indica* Linn.) were chosen for observations of *Anastrepha*. We offered live *Anastrepha* larvae on soil beneath the tree crowns. We observed for 10 min whether ants removed the larvae or the larvae buried themselves. Eight ant species were responsible for removing 1/4 of the larvae offered. The *Pheidole* Westwood, 1839 ants were the most efficient genus, removing 93% of the larvae. In compacted and dry soils, the rate of predation by ants was greater. Therefore, this study showed that ants, along with specific soil characteristics, may be important regulators of fruit fly populations and contribute to natural pest control in orchards.

1. Introduction

The fruit fly *Anastrepha* spp., together with some rarer *Rhagoletis* Loew, 1862, and *Ceratitis capitata* (Wiedeman, 1824) (Tephritidae), cause damage to fruit crops in Brazil. Tephritids directly damage the fruit, because the orifice made to lay the eggs causes the fruit to rot and fall prematurely, and the larvae feeding destroy the fruit pulp [1]. Ants, a group of efficient insect predators that regulate populations of general insects [2–8], can be considered as agents of biological pest control in agroecosystems [9–11]. The predation by ants on fruit flies occurs when the larvae leave the fruit in order to bury themselves in the soil and transform into pupae. *Solenopsis geminata* (Fabricius, 1804) ants, for example, were responsible for predation of 95% of the *Anastrepha ludens* (Loew, 1873) larvae during the warm months in Mexico [7]. In Guatemala, these ants attacked 21.6% of the *C. capitata*

larvae in orange groves and 9.3% in coffee plantations [8].

Predation is strongly and indirectly influenced by the physical properties of the soil, because the larvae took longer in burying themselves in very dry soil, increasing the time in which they remained exposed and consequently the rate of ant predation [12]. In this study, we analyzed which factors were present and how they influenced the predation of fruit flies by ants, considering the different species of fruit trees, and the degree of soil compaction and moisture content.

2. Material and Methods

We conducted the experiment in a grove of the Universidade Federal da Grande Dourados (UFGD) (Mato Grosso do Sul state, Brazil, 22°13′16″S and 54°48′20″W), on the 8th, 10th, 11th, 14th, 18th and 21st of February 2007. The local

TABLE 1: Number and total percentage of larvae removed by ants beneath the crowns of 60 trees of three species of fruit, in a grove of the Universidade Federal da Grande Dourados, during an experiment offering groups of three larvae under each crown. Guava is *Psidium guajava*, jaboticaba *Myrciaria jaboticaba*, and mango *Mangifera indica*.

Subfamilies	Species or morphospecies	Fruit trees			Total (%)
		Guava (30 trees and 90 larvae)	Jaboticaba (11 and 33)	Mango (19 and 57)	
Myrmicinae	*Pheidole oxops* Forel, 1908	7	6	16	67.44
	Pheidole gertrude Forel, 1886	4	—	—	9.30
	Pheidole sp. 1	2	2	—	9.30
	Pheidole sp. 2	1	1	—	4.65
	Pheidole sp. 3	1	—	—	2.32
Dolichoderinae	*Dorymyrmex* sp. 1	—	—	1	2.32
Ponerinae	*Odontomachus chelifer* (Latreille, 1802)	1	—	—	2.32
Ectatomminae	*Ectatomma brunneum* Smith F., 1858	—	—	1	2.32
	Total	16	9	18	100

soil is red latosol eutrophic alic [13], and the climate is subtropical humid [14]. In the grove of 4 ha, there are various fruit trees, such as *Psidium guajava* Linn. (Myrtaceae) (popular name guava), *Myrciaria jaboticaba* (Vell.) Berg. (Myrtaceae) (popular name jaboticaba), *Mangifera indica* Linn. (Anacardiaceae) (popular name mango), which we used in this experiment, as well as all fruit trees of grove are arranged in blocks according to species, and only the guava trees had fruit at the time of the experiment. Sixty fruit trees were randomly chosen for the experiment: 30 guavas, 11 jaboticabas, and 19 mangoes. This number is referent to 50% of total individuals of these species of grove.

Beneath the trees' canopies, we delimited an area of 1 m² (quadrant) and we removed all vegetal biomass one day before the experimentation to facilitate observation and capture of ants. In each quadrant, we offered simultaneously three last instar larvae *Anastrepha* ssp., obtained from infested guava fruits in the same area of study. We released larvae individually from a height of ~30 cm above the ground, simulating the larva falling from a fruit. During 10 min, from the moment at which the larvae reached the ground, we recorded the time in which the larvae buried themselves, if the larva was attacked and removed by ants (larvae removed), and the time taken by ants to remove them (removal time). All experiments were done at the same period of the day (between 7:00 and 11:00 am), corresponding to the period of the highest incidence of larvae leaving the fruit. Our sampling unit consisted in each larva offered.

After the observations, we collected all ants active in removal of larvae and identified the species according to the dichotomous key of Bolton [15]. Then we stored the ant species in the Laboratório de Mirmecologia of UFGD.

To determine the degree of soil compaction, we used the measure of soil density. We collected 60 soil samples under the canopy of tree after each day of observation. The samples were oven-dried at 110°C for three days. Samples were collected using a metallic cylinder of 4.2 cm in diameter and 5 cm in height. We obtained soil density dividing the dry weight of soil (after 3 days) by the volume of the sample. To determine the soil moisture, we weighed the samples before and after three days in the dryer. The ratio between the initial and final weight multiplied by 100 corresponds to the percentage of moisture. During the days of field study, we recorded the weather conditions, such as daily temperature, relative moisture, and wind speed.

We performed the analysis of covariance (ANCOVA) to verify whether the removal time was dependent on the species of trees, the number of larvae removed, and the interaction between these variables. We used a multivariate analysis of variance (MANOVA) to test the difference in soil compaction and moisture in tree species. We also used models of multiple regression to evaluate if the average time to bury and rate of predation were related to soil moisture, soil compaction, or the interaction between these two variables.

The interaction between the soil characteristics and the species of fruit trees was considered as independent variables. We used multiple regression test to verify whether the time of larvae spent in burying themselves was related to soil moisture and compaction. For this test, we used 30 samples in which the larvae were not predated by ants.

3. Results

From 180 fly larvae used in the experiment, 43 (24%) were removed by ants, 88 (49%) buried themselves, and 49 (27%) did not bury themselves and were not removed by ants. Eight ant species in four genuses and four subfamilies were recorded removing larvae. *Pheidole* (Myrmicinae) accounted for 93% of predation upon larvae (40 records of removal), and individuals of *Pheidole oxops* Forel, 1908 were the most efficient, removing 67.44% of the larvae (Table 1).

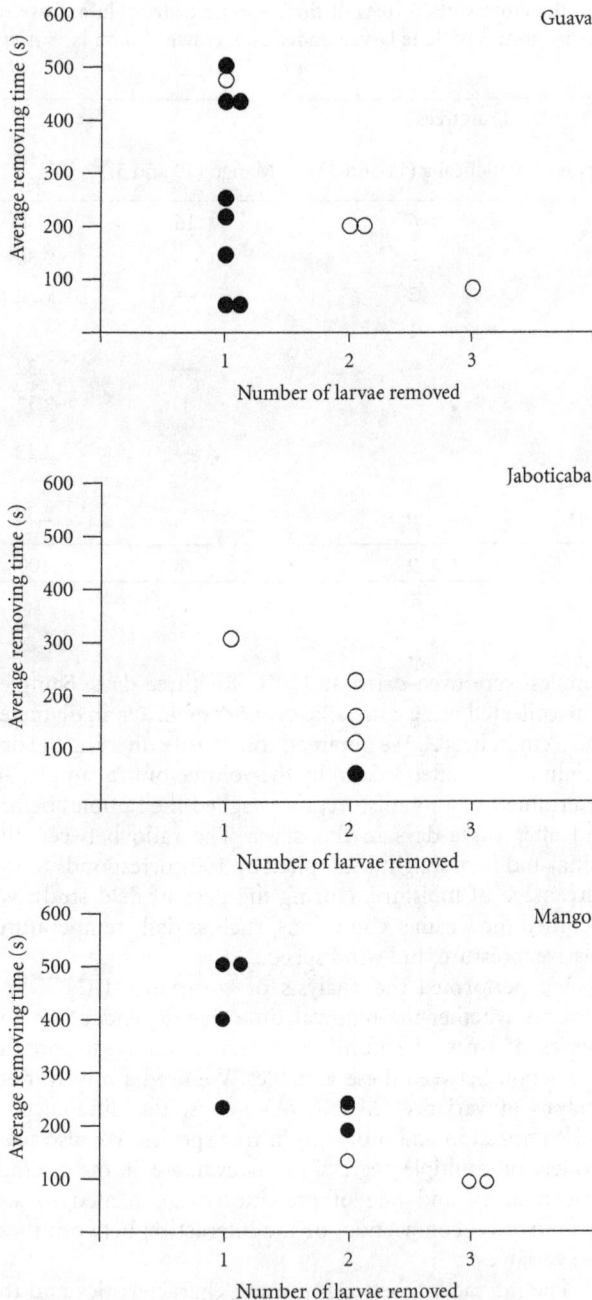

FIGURE 1: Relationship between amounts of fruit fly larvae (Tephritidae) removed by ants and average time for removal of each larva beneath the crowns of three species of fruit trees. Empty points are samples without larvae burying themselves. Guava is *Psidium guajava*, jaboticaba *Myrciaria jaboticaba*, and mango *Mangifera indica*.

The ants removed 16 of these larvae under the canopy of guava trees (all bearing fruit), nine under jaboticabas, and 18 under mangoes. The average time for the larvae to bury themselves was only obtained from 45 samples (26 guavas, six jaboticabas, and 13 mangos), because the larvae in 15 samples did not show this behavior. In 33 samples, there was no attack by ants and the mean of removal time was obtained

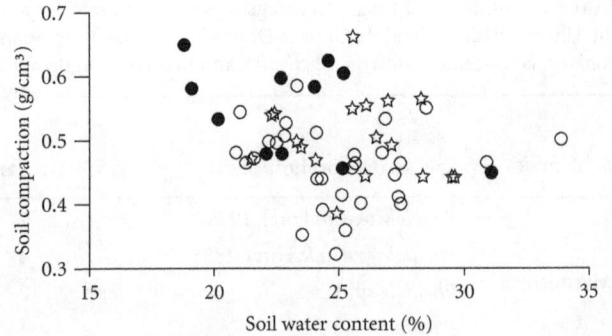

FIGURE 2: Soil moisture and compaction beneath of crowns the fruit trees. Open circles: guava (*Psidium guajava*); filled circles: jaboticaba (*Myrciaria jaboticaba*); stars: mango (*Mangifera indica*).

only from 27 samples (12 guavas, five jaboticabas, and 10 mangos).

The mean time required to remove a larva decreases as the number of larvae attacked and removed increased ($F = 7.356$; $P = 0.013$; gl = 1; Figure 1). This significant effect is more evident among samples in which the larvae did not bury themselves (open circles in Figure 1). Moreover, ANCOVA results showed that the removal time was independent of the tree species ($F = 0.894$; $P = 0.424$; gl = 2) and the interaction between number of larvae removed and tree species ($F = 0.449$; $P = 0.644$; gl = 2). The presence of fruit only in guava did not affect the removal time of the larvae.

Climatic data showed that weather conditions were constant throughout the study. The average daily temperature ranged between 23.7 and 25.5°C, relative moisture varied between 72.4 and 89.6%, and wind speed between 0.8 and 1.6 ms-1. It rained only on the nights of 7th (29.5 mm), 12th (0.3 mm), and 16th (14.2 mm).

The predation rate of larvae was affected by the different soil characteristics, as the larvae take longer to bury themselves in dry soil. Soil compaction and moisture were dependent on tree species (MANOVA: Pillai trace value = 1.195, $P < 0.001$, df = 6 and 112, Figure 2), being that the soil under the jaboticaba canopy had the highest compaction and lower moisture. The average time for burying itself under the different species of fruit trees was significantly related to the soil moisture ($F = 3.803$; $P = 0.037$; gl = 2; Figure 3), but not related to soil compaction ($F = 1.052$; $P = 0.366$; gl = 2), nor to the interaction between these two variables ($F = 0.553$; $P = 0.582$; gl = 2).

Soil characteristics affected the rate of larvae predation by ants (Figure 4). In soils with higher moisture, the predation was lower ($F = 4.753$, $P = 0.021$, df = 2), and in more compacted soil, the rate of predation was greater ($F = 5.989$, $P = 0.010$, df = 2). Interaction between these two independent variables also explained the predation rate ($F = 6.163$, $P = 0.009$, df = 2). In other words, ants were more efficient in preying on larvae on drier and more compact soil, despite compaction having no effect on the larvae burying time (Figure 5).

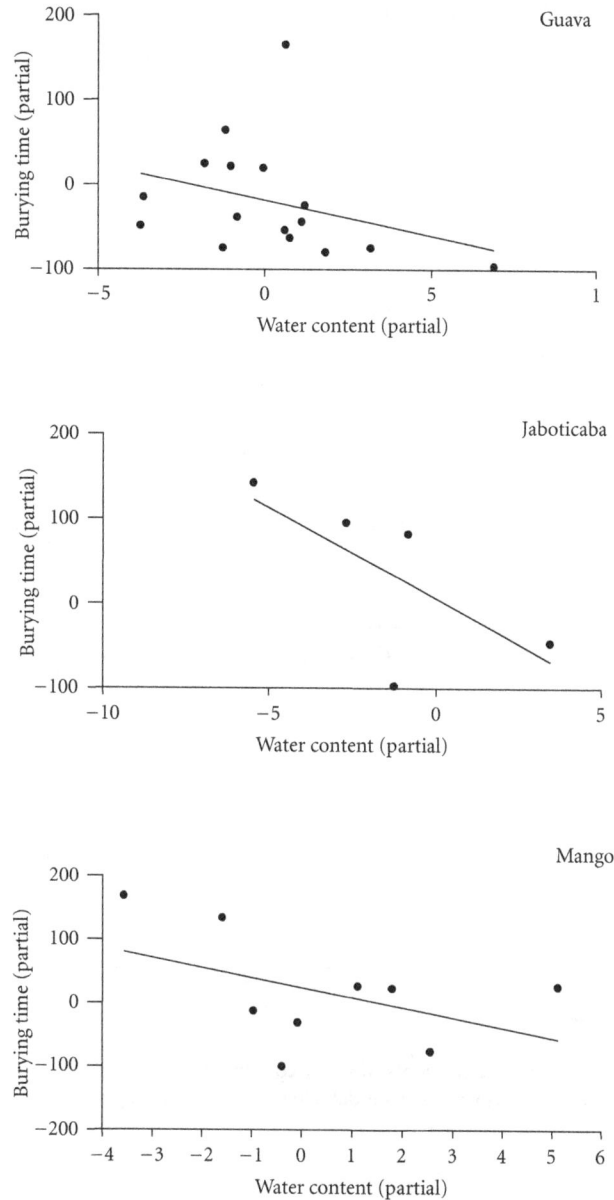

FIGURE 3: Average time until fruit fly larvae bury themselves through gradient of water content of soil beneath the crowns of three species of fruit trees. Only larvae that were not predated by ants were included. Partial residuals were obtained from a multiple-linear model that included compaction of soil (no significant effect). Guava is *Psidium guajava*, jaboticaba *Myrciaria jaboticaba*, and mango *Mangifera indica*.

4. Discussion

We observed that ants removed approximately 1/4 of the fruit fly larvae released on soil. This value is similar for biological control levels [3, 16, 17] and high for predation of the fruit flies by ants in most studies [8, 12]. Among the predatory ants genus, *Pheidole* individuals were more efficient, accounting for 93% of the larvae removed. The predominance of attacks by these ants evidenced their role as efficient predator, which is also due to their wide distribution, high species richness, and good adaptation to the physical conditions of the environment [18]. Its aggressive

behavior and efficient and massive recruitment increment this efficiency [19]. The potential performance of *Pheidole* as agents of biological pest control was also demonstrated in the fight against *Anthonomus grandis* Boheman, 1843 (Coleoptera: Curculionidae) in cotton fields in Brazil [10].

Strategies for predation and defense of organisms are among the most discussed topics in ecology and evolution [20, 21]. These relationships determine the survival or extinction of populations and the structure and maintenance of communities. Thus, if in the case of fruit flies, the rapid penetration into the soil is the best strategy to prevent their predation [7, 12], then the soil characteristics as well as the

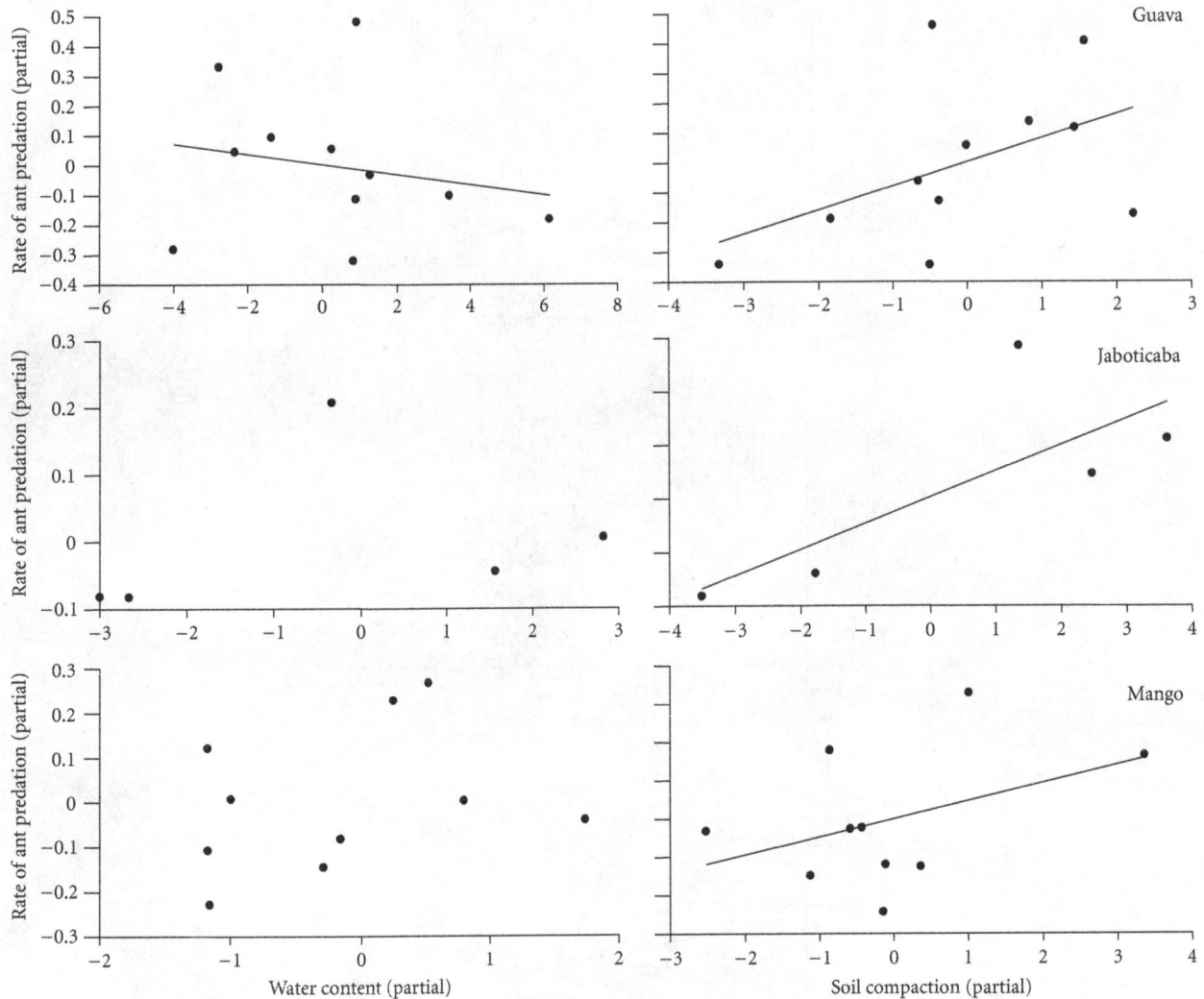

FIGURE 4: Ant predation on fruit fly larvae through gradients of water content and compaction of soil beneath the crowns of fruit trees of three species. Partial residuals obtained from multiple-linear model. Guava is *Psidium guajava*, jaboticaba *Myrciaria jaboticaba*, and mango *Mangifera indica*.

larvae ability of bury themselves are determinants for their survival. However, in this study, we found that larvae which were dropped on compacted and uncompacted soil took the same time to penetrate the soil. Although the time to drill the soil by larvae was not directly related to compaction, it was significantly associated with soil moisture, another determinant factor for the success of the burying behavior [22, 23] and for the development of pupae [24]. The lack of moisture in the soil can cause mortality of a large number of larvae, because the soils become more difficult to be bored [22]. Wet soils have greater tension between the particles resulting in larger particles and larger spaces among them [25]. Thus, wet soils are more easily bored by fruit fly larvae, as evidenced in this study.

Here we have evidence that both soil tilling and tree species influence the efficiency of ants in attacking the larvae. Several studies have shown that the abundance, not only of ants, but of other predators such as carabid beetles and spiders, increases with farming practices that reduce soil

turnover [26, 27]. This fact should be related to environment complexity and colony stability. Tillage systems in which the soil is not turned have a higher plant biomass on the soil surface [28], and this increases the availability of nutrients and shelter for many organisms. Thus, these communities have more local biodiversity [29, 30]. In addition, soil disturbance could have caused the death of various ant colonies, decreasing the number of individuals foraging for resources. Moreover, the tree species may also have influenced the soil characteristics through their complex canopy structure and root density.

Here we showed that rate of ant predation on fruit fly larvae was affected by soil, because larvae took longer to bury themselves in dry and compacted soil. Therefore, the moisture and compaction level of soil, resulting from the type of tillage and tree species, has a profound influence on the burying of larvae influencing the efficiency of ant predation (Figure 5). Nevertheless, the presence of fruit was not a determinant factor in the predation of larvae among the

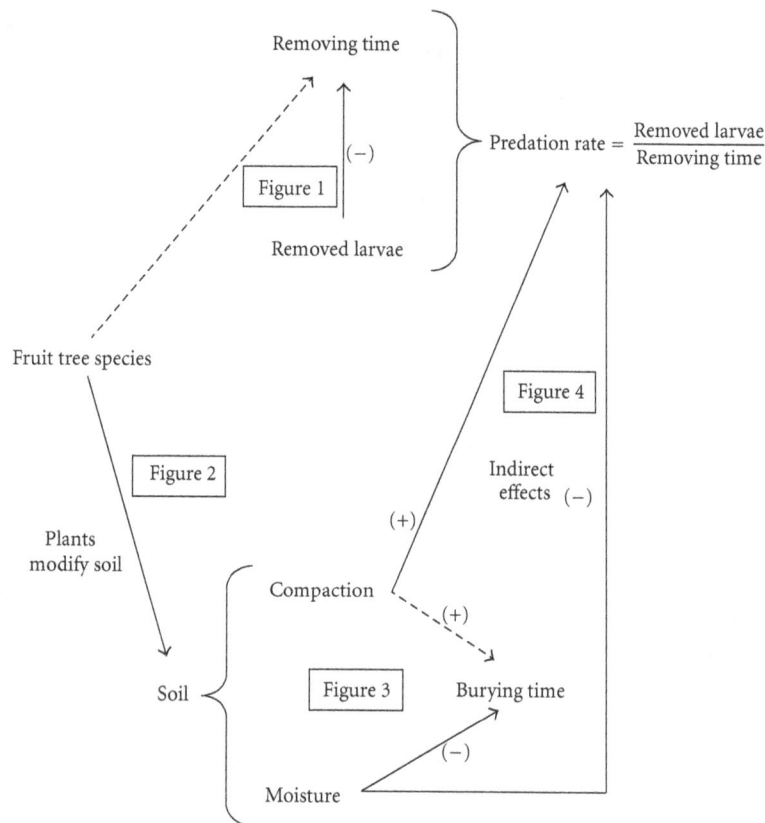

FIGURE 5: Effects' diagram for predation of fruit fly larvae by ants. Evidences for effects are in the indicated figures. Dashed lines: no statistical evidence; ($+$): positive effect; ($-$): negative effect.

fruit trees. This result was also evidenced by Aluja et al. [12]. Although we would expect that ants were more abundant in locations with higher density of fruit, for example, [31], due to the greater number of larvae, only guava trees were bearing fruits at the time of study, which could have masked the effect of fruit.

In this study, we showed that ants, mainly of *Pheidole* genus, are important predators of *Anastrepha* larvae, and can contribute to regulate this crop pest population. Furthermore, we also evidenced that the rate of ant predation depends on soil characteristics and fruit tree species. Thus, ants may have a beneficial impact on fruit growing and, together with other control methods, can reduce cost with insecticides and act as an important tool in integrated pest management.

Acknowledgments

The authors acknowledge S. A. Soares and R. Gutierrez for their collaboration in carrying out the field experiment, Professor J. L. Fornasieri for valuable guidance, B. C. Desidério for linguistic revision, and A. A. Vilela and E. Alves-Silva for valuable comments on the final version of the paper. This work was supported by Conselho Nacional de Desenvolvimento Científico e Tecnológico for financial support (Process no. 131999/2006-0 and AT/500868/2010-7).

References

[1] M. Aluja, "Bionomics and management of *Anastrepha*," *Annual Review of Entomology*, vol. 39, pp. 155–178, 1994.

[2] P. Radeghieri, "*Cameraria ohridella* (Lepidoptera Gracillariidae) predation by *Crematogaster scutellaris* (Hymenoptera Formicidae) in Northern Italy (Preliminary note)," *Bulletin of Insectology*, vol. 57, no. 1, pp. 63–64, 2004.

[3] V. M. Agarwal, N. Rastogi, and S. V. S. Raju, "Impact of predatory ants on two lepidopteran insect pests in Indian cauliflower agroecosystems," *Journal of Applied Entomology*, vol. 131, no. 7, pp. 493–500, 2007.

[4] V. Rico-Gray and P. S. Oliveira, *The Ecology and Evolution of Ant-Plant Interactions*, The University of Chicago Press, Chicago, Ill, USA, 2007.

[5] E. A. do Nascimento and K. Del-Claro, "Ant visitation to extrafloral nectaries decreases herbivory and increases fruit set in *Chamaecrista debilis* (Fabaceae) in a Neotropical savanna," *Flora*, vol. 205, no. 11, pp. 754–756, 2010.

[6] L. Nahas, M. O. Gonzaga, and K. Del-Claro, "Emergent impacts of ant and spider interactions: herbivory reduction in a tropical savanna tree," *Biotropica*, vol. 44, no. 4, pp. 498–505, 2012.

[7] D. B. Thomas, "Predation on the soil inhabiting stages of the Mexican fruit fly," *Southwestern Entomologist*, vol. 20, no. 1, pp. 61–71, 1995.

[8] F. M. Eskafi and M. M. Kolbe, "Predation on larval and pupal *Ceratitis capitata* (Diptera: Tephritidae) by the ant *Solenopsis*

geminata (Hymenoptera: Formicidae) and other predators in Guatemala," *Environmental Entomology*, vol. 19, no. 1, pp. 148–153, 1990.

[9] C. J. H. Booij and J. Noorlander, "Farming systems and insect predators," *Agriculture, Ecosystems and Environment*, vol. 40, no. 1–4, pp. 125–135, 1992.

[10] W. D. Fernandes, P. S. Oliveira, S. L. Carvalho, and M. E. M. Habib, "*Pheidole* ants as potential biological control agents of the boll weevil, *Anthonomus grandis* (Col, Curculionidae), in southeast Brazil," *Journal of Applied Entomology*, vol. 118, no. 4-5, pp. 437–441, 1994.

[11] W. D. Fernandes, L. A. G. Reis, and J. C. Parré, "Formigas como agentes de controle natural de pragas em plantações de milho com plantio direto e convencional," *Naturalia*, vol. 24, pp. 237–239, 1999.

[12] M. Aluja, J. Sivinski, J. Rull, and P. J. Hodgson, "Behavior and predation of fruit fly larvae (*Anastrepha* spp.) (Diptera: Tephritidae) after exiting fruit in four types of habitats in tropical Veracruz, Mexico," *Environmental Entomology*, vol. 34, no. 6, pp. 1507–1516, 2005.

[13] M. A. P. Pierangeli, L. R. G. Guilherme, N. Curi, M. L. N. Silva, L. R. Oliveira, and J. M. Lima, "Efeito do pH na adsorção-dessorção de chumbo em Latossolos brasileiros," *Revista Brasileira de Ciência do Solo*, vol. 25, pp. 269–277, 2001.

[14] M. C. Peel, B. L. Finlayson, and T. A. McMahon, "Updated world map of the Köppen-Geiger climate classification," *Hydrology and Earth System Sciences*, vol. 11, no. 5, pp. 1633–1644, 2007.

[15] B. Bolton, "Synopsis and classification of Formicidae," *Memoirs of the American Entomological Institute*, vol. 71, pp. 1–370, 2003.

[16] S. Mansfield, N. V. Elias, and J. A. Lytton-Hitchins, "Ants as egg predators of *Helicoverpa armigera* (Hübner) (Lepidoptera: Noctuidae) in Australian cotton crops," *Australian Journal of Entomology*, vol. 42, no. 4, pp. 349–351, 2003.

[17] T. Robyn, *Effects of ant predation on the efficacy of biological control agents: Hypena laceratalis Walker (Lepidoptera: Noctuirdae), Falconia intermedia Distant (Hemiptera: Miridae) and Teleonemia scrupulosa Stal (Hemiptera: Tingidae) on Lantana camara (Verbenaceae) in South Africa [M.S. thesis]*, Rhodes University, 2010, http://eprints.ru.ac.za/1892/1/RobynTourleMSCThesis.pdf.

[18] A. N. Andersen, "Parallels between ants and plants: implications for community ecology," in *Ant-Plant Interactions*, C. R. Husley and D. F. Cutler, Eds., pp. 539–558, Oxford University Press, Oxford, UK, 1991.

[19] B. Hölldobler and E. O. Wilson, *The Ants*, Harvard University Press, Cambridge, Mass, USA, 1990.

[20] J. N. Thompson, *The Coevolutionary Process*, University of Chicago Press, Chicago, Ill, USA, 1994.

[21] J. N. Thompson, "Conserving interaction biodiversity," in *The Ecological Basis of Conservation: Heterogeneity, Ecosystems, and Biodiversity*, S. T. A. Pickett, R. S. Ostfeld, M. Shachak, and G. E. Likens, Eds., pp. 285–293, Chapman & Hall, New York, NY, USA, 1997.

[22] M. A. Baker, W. E. Stone, C. C. Plummer, and M. McPhail, "A review of studies on the Mexican fruit fly and related Mexican species," USDA, Miscellaneous Publication 531, 1944.

[23] P. G. Mulder Jr. and R. A. Grantham, *Biology and Control of the Pecan Weevil in Oklahoma*, Division of Agricultural Sciences and Natural Resources, Oklahoma State University, 2012, http://pods.dasnr.okstate.edu/docushare/dsweb/Get/Document-4530/EPP-7079web.pdf.

[24] F. D. M. M. Bento, R. N. Marques, M. L. Z. Costa, J. M. M. Walder, A. P. Silva, and J. R. P. Parra, "Pupal development of ceratitis capitata (Diptera: Tephritidae) and diachasmimorpha longicaudata (Hymenoptera: Braconidae) at different moisture values in four soil types," *Environmental Entomology*, vol. 39, no. 4, pp. 1315–1322, 2010.

[25] J. D. Anderson and J. S. I. Ingam, *Tropical Soil Biology and Fertility: A Handbook of Methods*, CAB International, Wallingford, UK, 1996.

[26] M. S. Clark, S. H. Gage, and J. R. Spence, "Habitats and management associated with common ground beetles (Coleoptera: Carabidae) in a Michigan agricultural landscape," *Environmental Entomology*, vol. 26, no. 3, pp. 519–527, 1997.

[27] D. Lange, W. D. Fernandes, J. Raizer, and O. Faccenda, "Predacious activity of ants (Hymenoptera: Formicidae) in conventional and in no-till agriculture systems," *Brazilian Archives of Biology and Technology*, vol. 51, no. 6, pp. 1199–1207, 2008.

[28] D. N. Gassen, *Insetos Subterrâneos Prejudiciais às Culturas no sul do Brasil*, Passo Fundo, Rio Grande do Sul, Brazil, 1996.

[29] A. C. Castro and M. V. B. Queiroz, "Estrutura e organização de uma comunidade de formigas em agroecossistema neotropical," *Anais da Sociedade Entomológica do Brasil*, vol. 16, pp. 363–246, 1987.

[30] J. C. Perdue and D. A. Crossley, "Seasonal abundance of soil mites (Acari) in experimental agroecosystems: effects of drought in no-tillage and conventional tillage," *Soil and Tillage Research*, vol. 15, no. 1-2, pp. 117–124, 1989.

[31] T. T. Y. Wong, D. D. Mcinnis, J. L. Nishimoto, A. K. Ota, and V. C. S. Chang, "Predation of the Mediterranean fruit fly (Diptera: Tephritidae) by the Argentine ant (Hymenoptera: Formicidae) in Hawaii," *Journal of Economic Entomology*, vol. 77, pp. 1454–1438, 1984.

3

Management of Climatic Factors for Successful Silkworm (*Bombyx mori* L.) Crop and Higher Silk Production: A Review

V. K. Rahmathulla

P3 Basic Seed Farm, National Silkworm Seed Organization, Central Silk Board, Ring Road, Srirampura, Mysore, Karnataka 570 008, India

Correspondence should be addressed to V. K. Rahmathulla, rahmathullavk@yahoo.co.in

Academic Editor: Martin H. Villet

The seasonal differences in the environmental components considerably affect the genotypic expression in the form of phenotypic output of silkworm crop such as cocoon weight, shell weight, and cocoon shell ratio. The variations in the environmental conditions day to day and season to season emphasize the need of management of temperature and relative humidity for sustainable cocoon production. The present review paper discuss in details about the role of temperature and humidity on growth and development of silkworm including recent studies on heat shock protein. Study also discusses the influence of air and light on silkworm development. In addition to this study emphasis on the role of various environmental factors on embryonic development of silkworm egg, nutritional indices of silkworm larva and reproductive potential of silkworm moth. The study also highlights about the care to be required during silkworm spinning and influence of temperature and humidity on post cocoon parameters of silkworm. The study included future strategies to be taken for the management climatic condition for successful cocoon crop. The paper covers 140 references connected with the topic.

1. Introduction

Sericulture is the science that deals with the production of silk by rearing of silkworm. Silk is called the queen of textiles due to its glittering luster, softness, elegance, durability, and tensile properties and is discovered in China between 2600 and 2700 BC. Silk originating in the spittle of an insect is a natural fibrous substance and is obtained from pupal nests or cocoons spun by larvae known as silkworm. The silk is preferred over all other types of fibres due to its remarkable properties like water absorbency, heat resistance, dyeing efficiency, and luster. Factors mainly influence the physiology of insects are temperature and humidity. Despite wide fluctuations in their surroundings, insects show a remarkable range of adaptations to fluctuating environmental conditions and maintain their internal temperature and water content within tolerable limits. Adaptation is a complex and dynamic state that widely differs from species to species. Surviving under changing environment in insects depends on dispersal, habitat selection, habitat modification, relationship with water, resistance to cold, diapause and developmental rate,

sensitivity to environmental signals, and syntheses of variety of cryoprotectant molecules. The mulberry silkworm (*Bombyx mori* L.) is very delicate, highly sensitive to environmental fluctuations, and unable to survive extreme natural fluctuation in temperature and humidity because of their long years of domestication since 5000 years. Thus, the adaptability to environmental conditions in the silkworm is quite different from those of wild silkworm and other insects. Temperature, humidity, air circulation, gases, light, and so forth, show a significant interaction in their effect on the physiology of silkworm depending upon the combination of factors and developmental stages affecting growth, development, productivity, and quality of silk.

Silkworm is one of the most important domesticated insects, which produces luxuriant silk thread in the form of cocoon by consuming mulberry leaves during larval period. The growth and development of silkworm is greatly influenced by environmental conditions. The biological as well as cocoon-related characters are influenced by ambient temperature, rearing seasons, quality mulberry leaf, and genetic constitution of silkworm strains. Different seasons affect the performance of *Bombyx mori* L. The seasonal differences in the environmental components considerably affect the genotypic expression in the form of phenotypic output such as cocoon weight, shell weight, and cocoon shell ratio. The variations in the environmental conditions during the last decade emphasize the need of management of the temperature and relative humidity for sustainable cocoon production. India enjoys the comfortable second position for the production of silk in the world next only to China. Traditionally sericulture in India is practiced in tropical environmental regions such as Karnataka, Tamil Nadu, Andhra Pradesh, and West Bengal and to a limited extent in temperate region of Jammu and Kashmir. The existing tropical condition provides scope for exploiting the multivoltine × bivoltine hybrid at commercial venture as they are very hardy and have tremendous capacity to survive and reproduce under fluctuating environmental climatic conditions. Bulk share of the silk production (95%) is accounting from multivoltine hybrids. The researchers [1] evaluated the genetic potential of the multivoltine silkworm and identified suitable parents for breeding programmes. However, the silk quality produced from multivoltine strain is at low level when compared to the existing international standards. China account for over 80% of the world silk production while India, which is the second largest producer, accounts for about 15-16% of the total production. China silk is of superior quality as they are of the bivoltine strain. Considering these drawbacks, adoption of bivoltine sericulture became imperative and imminent considering its potentiality even under Indian tropical conditions. In this line, many productive and qualitatively superior bivoltine hybrids have been developed by utilizing Japanese commercial hybrids as a breeding resource material. However, the hot climatic conditions prevailing particularly in summer are not conducive to rear these high yielding bivoltine hybrids throughout the year. It is a well-established fact that under tropical condition, unlike polyvoltines, bivoltines are more vulnerable to various stresses like hot climatic conditions of tropics, poor leaf quality, and improper management of silkworm crop during summer that is not conducive for bivoltine rearing for technologically and economically poor farmers of India [2–4]. This paper discusses the role of different environmental factors affecting the growth, survivability, productivity, and disease incidence in silkworm. The paper also discusses the optimum conditions of environmental factors required for higher productivity in sericulture and, further, the paper also reviews the results and findings of various researchers who studied the effect of environmental factors on growth, development, feed conversion, reproductive potential, and postcocoon parameters of silkworm.

2. Role of Temperature on Growth of Silkworm

Temperature plays a vital role on the growth of the silkworms. As silkworms are cold-blooded animals, temperature will have a direct effect on various physiological activities. In general, the early instar larvae are resistant to high temperature which also helps in improving survival rate and cocoon characters. The temperature has a direct correlation with the growth of silkworms; wide fluctuation of temperature is harmful to the development of silkworm. Rise in temperature increases various physiological functions and with a fall in temperature, the physiological activities are decreases. Increased temperature during silkworm rearing particularly in late instars accelerates larval growth and shortens the larval period. On the other hand, at low temperature, the growth is slow and larval period is prolonged. The optimum temperature for normal growth of silkworms is between 20°C and 28°C and the desirable temperature for maximum productivity ranges from 23°C to 28°C. Temperature above 30°C directly affects the health of the worm. If the temperature is below 20°C all the physiological activities are retarded, especially in early instars; as a result, worms become too weak and susceptible to various diseases. The temperature requirements during the early instars (I, II, III) are high and the worms feed actively, grow very vigorously, and lead to high growth rate. Such vigorous worms can withstand better even at adverse conditions in later instars. The optimum temperature required for rearing silkworms of different early instars are as described in Table 1.

Generally, the room temperature is low during winter and rainy season, which should be regulated by heating the room with electric heaters or charcoal fires. Thermoregulator-fitted electrical heaters are best since they do not emit any injurious gases. When electricity becomes costly and not available in many rural areas of sericulture belt, properly dried charcoal can be used. However, the carbon dioxide and other gases emitted in this burning process are injurious to silkworms and they can be regulated by providing more ventilation particularly in daytime. Besides this, the doors and windows should be kept closed particularly during night. Late in the day, as the outside temperature goes up, doors and windows should be opened to allow warm air to the room. During summer season when day temperature is high, all the windows should be kept open. Simultaneously, windows and doors are covered with wet gunny cloth during hot days to reduce the temperature and increase humidity [5, 6]. Otherwise, suitable air coolers can be used for this purpose.

The success of the sericulture industry depends upon several variables, but environmental conditions such as biotic and abiotic factors are of particular importance. Among the abiotic factors, temperature plays a major role on growth and productivity of silkworms [7, 8]. There is ample literature stating that good quality cocoons are produced within a temperature range of 22–27°C and that cocoon quality is poorer above these levels [9, 10]. However, polyvoltine breeds reared in tropical countries are known to tolerate slightly higher temperature and adjust with tropical climatic conditions [11]. In order to use bivoltine races in a tropical country like India, it is necessary to have a stable cocoon crop in a high temperature environment. High temperature adversely affects nearly all biological processes including the rates of biochemical and physiological reactions [12], and can eventually affect the quality or quantity of cocoon crops in the silkworm and subsequently silk produced. Several studies [13–15] demonstrated that silkworms were more sensitive to high temperature during the fourth and fifth stages. It is well

TABLE 1: Optimum temperature and humidity requirements of silkworm during various stages.

Environmental factors	Incubation	I instar	II instar	III instar	IV instar	V instar	Spinning	Cocoon preservation
Temperature	25°C	28°C	27°C	26°C	25°C	24°C	25°C	25°C
Relative humidity	75–80%	85–90%	85%	80%	70–75%	65–70%	70%	80%

understood that the majority of the economically important genetic traits of silkworms are qualitative in nature and that phenotypic expression is greatly influenced by environmental factors such as temperature, relative humidity, light, and nutrition [16–23]. The Figures 1(a), 1(b), 1(c), 1(d), and 1(e) describe the contribution of various factors such as environmental, racial, and other factors on important cocoon traits. The figures illustrate that cocoon yield and reelability are the most affected traits due to the adverse environmental factors. The length of 1st instar period of silkworm larvae was increased by decreasing the temperature for 10 days [24]. Similarly, researchers [25] found that temperature and RH exert synergistic impact regarding silkworm larval period. These results are in conformity with those of earlier workers [26] and reported that change in temperature along with RH has pronounced effect on moulting period. Similarly, the reports of different researchers [27, 28] recorded that decrease in temperature enhances the moulting duration in silkworm. Table 2 shows the effect of different temperature during late instars on various cocoon traits of silkworm. In a serious of experiments, researchers [29] observed that resistance to high temperature is a heritable character and it may be possible to breed silkworm races tolerant to high temperature. Several studies of different workers [30–32] reported that the quantitative characters of silkworms in a known environment are of utmost importance in sericulture. The effect of low temperature during rearing on some characters of breeding line races were studied [33]. Similarly, studies of silkworm breeders [34, 35] found that low temperature is always better than higher temperature with reference to productivity of silkworm and larval duration for different instars. Another studies reported that the F1 bivoltine hybrids is more adjustable for high temperature and high humidity when compared with their parents [36]. In an important review, work scientists [37] also reported that temperature and relative humidity were among the various factors that influence growth, behavior, and instar larval periods in silkworm and discussed the adaptation of insects to varying environmental conditions. The acclimatization of silkworm races with environmental condition especially temperature was studied in detail [38]. Seasonal effect on silkworm under Kashmir and subtropical conditions were studied [39]. Different workers demonstrated the influence of humidity during rearing time and its role on disease incidence [40]. Suitable hardy bivoltine races evolved and were tested in the field for summer and, rainy seasons of Uttarpradesh, India [41]. Recently the influence of various seasons such as summer, rainy, and winter on cocoon and grainage traits of popular bivoltine races of India were studied and concluded that temperature and humidity affect all characters of these races [42]. High temperature affects nearly all biological processes including the rates of biochemical and

TABLE 2: Effect of temperature during late-age rearing on cocoon traits of silkworm.

Traits	Temperature °C				Control (24°C)
	36–22	36–22	32–22	32–26	
Pupation (%)	86.00	66.10	92.50	91.40	95.0
Cocoon weight [g]	1.58	1.62	1.96	1.84	2.10
Shell weight [g]	0.316	0.337	0.470	0.441	0.480

Source, Suresh Kumar et al. [45]

physiological reactions [43] ultimately affecting the quality and quantity of cocoon crops.

The silk cocoon production is determined by various factors including environment and genotype of the silkworm. Figure 2 describes the contribution of genetic and environmental factors on growth and development of silkworm and its important economic traits.

3. Studies on Heat Shock Protein

The tropical Indian multivoltine races of Bombyx mori (Pure Mysore, C. Nichi, and Nistari) are more tolerant to high temperature, as against the exotic bivoltine races of temperate origin. Unlike multivoltine, bivoltine races have better yield potential and produce superior quality of silk but do not survive under extreme climatic conditions prevailing in India. It is known that when subjected to high temperature, the cells display a heat shock response by synthesizing a novel set of protein, called heat shock protein (HSP) and it is interesting to note that thermotolerant always accompanied by the presence of HSPs [46]. An eminent scientist [47] attempted to understand the difference of thermotolerance in a multivoltine (C.Nichi) and bivoltine race (NB18). The heat shock response in mulberry silkworm races with different thermotolerances were also studied [48–50]. Many quantitative characters decline sharply at higher temperatures and, therefore, one of the key considerations in developing bivoltine hybrids for tropics could be the need for thermotolerant bivoltine strains. The recent advances in silkworm breeding and those in stress-induced protein synthesis have opened up a new avenues to evolve robust productive silkworm hybrids [45, 51–57]. On thermal treatment, all genetic traits of silkworms showed a decline with the increase of temperature above standard level. Workers [58] reported similar results and they found that biological molecules like DNA, RNA, lipids, and so forth were vulnerable to heat stress. Temperature stress causes a number of abnormalities at the cellular level as the normal pattern of protein synthesis halts. However, a brief exposure of cells to sublethal high temperature was found to render protection to the organism of subsequent and more severe temperature changes [59, 60].

Influence of various factors on cocoon weight   Influence of various factors on cocoon yield   Influence of various factors on cocoon shell weight

(a)   (b)   (c)

Influence of various factors on reelability   Influence of various factors on neatness of fibre

- Race
- Rearing environment
- Other factors

(d)   (e)

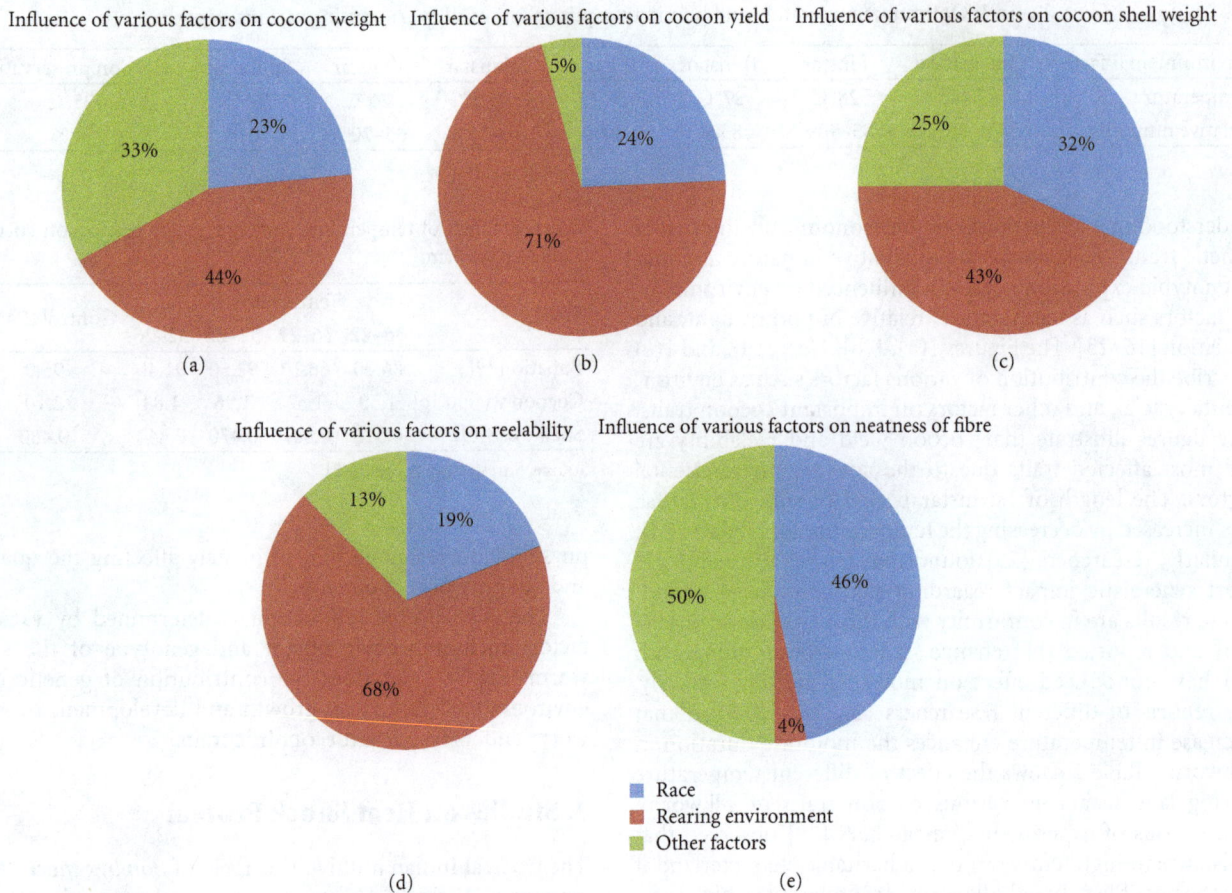

FIGURE 1: Influence of various factors on cocoon yield, weight, shell weight, reelability, and neatness source: [44].

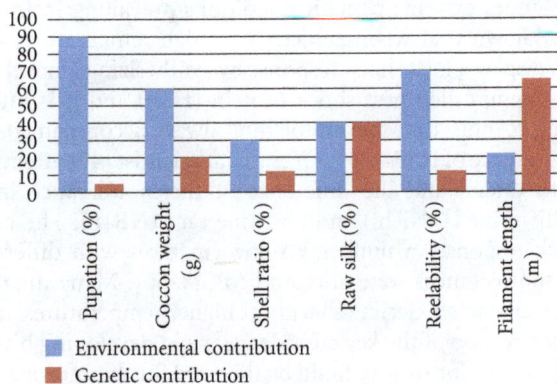

FIGURE 2: Environmental and genetic contribution to various economic traits of silkworm, *Bomby mori* L. (percentage).

Therefore, it is very much essential to gauge the degree of phenotypic difference of the economical traits to understand the genetic steadiness under the varied environmental conditions and the productivity of different breed under various environmental conditions. Intensive and careful domestication over centuries has apparently deprived the insect of opportunities to acquire thermotolerance. Among many factors responsible for poor performance of the

bivoltine strains under tropical conditions, the main culprit is temperature. Indeed, many quantitative characters decline sharply when temperature is higher than 28°C. Researchers, especially silkworm breeders in the sericulture field, always agree that it is a difficult task to breed such a bivoltine breeds that are suitable to the high temperature and fluctuating climatic conditions of India.

4. Role of Humidity on Growth of Silkworm

Humidity plays a vital role in silkworm rearing and its role is both direct and indirect. The combined effect of both temperature and humidity largely determines the satisfactory growth of the silkworms and production of good-quality cocoons. It directly influences the physiological functions of the silkworm. The young-age silkworms can withstand to high humidity conditions than later-age worms and under such condition, the growth of worm is vigorous. The optimum humidity conditions required for different early-age worms and late-age worms are as described in Table 1.

Humidity also indirectly influences the rate of withering of the leaves in the silkworms rearing beds. Under dry conditions especially winter and summer the leaves wither very fast and consumption by larvae will be less. This affects growth of the larvae and results in wastage of leaf in the rearing bed. Retarded growth of young larvae makes them weak and susceptible to diseases. At a humidity of 90 percent or higher, if temperature is maintained at 26°C–28°C, they can grow without being greatly affected. Like temperature, humidity also fluctuates widely not only from season to season but also within the day itself. Therefore, it is necessary for the silkworm rearers to regulate it for their successful crop. For this purpose, wax coated (paraffin) paper is used to cover the rearing beds during young-age rearing to raise humidity and to avoid leaf dryness. Otherwise, wet foam rubber pads or paper pads soaked in water can also be used to increase humidity in the rearing beds. Rich famers can use humidifier with humidistat to regulate humidity in the rearing room. However, it is important to lower humidity to 70 percent or below during the moulting time in each instar to facilitates uniform and good moulting. Water forms a large proportion of insect tissues and survival depends on the ability to maintain and to balance water in the body. There is no limiting range of humidity and most insects can develop at any humidity provided they are able to control their water balance [61]. The effect of high humidity on weight of larva of silkworm was studied [62]. The water content in insects ranges from less than 50% to more than 90% of the total body weight and there may be much variation within the same species even when reared at identical conditions [63–65]. The environmental factors, in particular temperature and humidity at the time of rearing and moisture content of mulberry leaf, affect growth of the silkworm [66, 67]. The role of water and humidity in sericulture were well studied by workers [68]. The seasonal effect and role of humidity on growth and nutritional efficiency of silkworm were also studied [69]. Researchers evolved a suitable race particularly to spring and autumn rearing [70]. The role of humidity on seed production parameters were well-studied by different workers [71]. Researchers [72] also listed out the phenotypic and genotypic characters during different season. The performances of polyvoltine races under dry climatic condition of Rayala seema areas of Andhra Pradesh, India [73], were also studied. Abiotic factor that has significant impact on the performance of insects in terrestrial environments is humidity. Humidity interacts with the availability of free water and with the water content of the food and it mostly shows indirect effect on growth and development. Demands in humidity vary depending on the biological circle. The effect of adverse climatic conditions for successful bivoltine cocoon crop was studied [74, 75]. In Italy, researchers studied the effect of various environmental factors on growth of silkworm larva [76]. The seasonal changes, atmospheric humidity, and soil moisture percentage have profound effect on the growth and quality of mulberry leaves, which in turn influence the silkworm health and cocoon crop production and, therefore, suggest the importance of leaf moisture both in palatability and assimilation of nutritive components of the leaf.

5. Role of Air and Light on Growth of Silkworm

Like other animals, silkworms also require fresh air. By respiration of silkworms, carbon dioxide gas is released in the rearing bed. The freshness of air can be determined by its CO_2 contents. Although atmospheric CO_2 content is generally 0.03-0.04% in the rearing room, carbon monoxide, ammonia, sulphur dioxide, and so forth are also released in the rearing room when farmers burn charcoal to raise temperature. These gases are injurious to silkworms; therefore, care should be taken to allow fresh air through proper ventilation to keep the toxic gases at a low level. If CO_2 exceeds above 2 percent concentration, the growth of silkworm is retarded. Insecticides and disinfectants are also avoided in the rearing room. Young silkworm larvae are more susceptible to the poisonous gases and hence artificial circulation of air is extremely useful in bringing down the contaminated air. The air current of 1.0 m/sec during 5th-age rearing reduces the larval mortality and improves ingestion, digestibility, larval weight, cocoon weight, and pupation rate compared to those recorded under zero ventilation condition [77].

Silkworms are photosensitive and they have a tendency to crawl towards dim light. They do not like either strong light or complete darkness. Rearing of silkworms in continuous light delays the growth. Further, it causes pentamoulters and reduces both larval and cocoon weights. Silkworms are fond of dim light of 15 to 20 lux and avoid strong light and darkness. Late-age worms survive better in 16-hour light and 8-hour dark periods. However, young-age worm prefers 16 hr darkness and 8 hr light period. Larvae of silkworm do not prefer either strong light or complete darkness but usually light phase, in contrast to the dark phase, activates the larvae. Silkworm is an insect of small lifespan with positive phototactism [78]; the silkworm larvae are fed in complete darkness during the life cycle, their larval duration is longer, and cocoon quality becomes poor [79]. Rearing in either complete darkness or in bright light leads to irregularity in growth and moulting. Light phase usually makes larval duration longer than the dark phase. The influence of light and temperature on growth of silkworm was studied in detail [80].

6. Role of Environmental Factors on Embryonic Development

The effect of temperature on the growth and development of silkworm has been studied extensively; however, much attention has not been paid on the effect of temperature on embryonic development. It has been reported that in exothermic organisms, when rate of development is plotted against temperature, a sigmoidal curve is obtained with an almost linear correlation in central temperature range. Temperature is a parameter in developmental cycle, which can be manipulated experimentally, but its effect is very complex for interpretation. The physiological explanation for embryonic death after exposure to lethal temperature is likely to be highly complex and probably species specific. Improper egg incubation results in various problems during the hatching and rearing period. If silkworm egg is subjected to incubation at high temperature and low humidity the hatching of larvae severely affected. It is well known that the environmental conditions during embryonic development not only affect the diapauses nature of eggs but also larval/pupal duration, cocoon weight, and egg production [81]. Among the development stage of silkworm, *Bombyx mori,* the egg stage has the lowest tolerance to high temperature. Temperature during incubation also affects voltinism character, as the embryonic stage is the most sensitive to temperature [82]. Bivoltine eggs incubated at a temperature above 25°C produce moths that lay hibernating eggs, while those incubated at lower temperature (below 25°C) produce moths that lay mixed and fully nonhibernating eggs. Development rate is directly influenced by temperature and is modified by humidity. At high temperature the embryo grows faster up to the setae formation stage and succumbs to death as the yolk cannot be utilized in pace with the high rate of development and comes in way of normal development [83]. Kittlans [84] stated that temperature above 33°C and abnormal cold treatment of embryos might also cause embryonic death or abnormal development. Cold treatment of silkworm *(Bombyx mori)* eggs leads to formation of tetraploid individuals, which lay large eggs [85].

During silkworm egg incubation, it is important that humidity should be maintained at 80% on an average for normal growth of embryo. If humidity falls below 70% during incubation, the hatching invariably is low. A number of factors complicate the effect of humidity on respiration; most important of which is the water content of the insect. The effects of humidity are very closely associated with temperature, that is, water loss by desiccation, spiracular diffusion, retention of ingested water, and production of metabolic water [63]. Humidity less than 60% results in loss of water from silkworm eggs and high humidity of 90% and above leads to retention of physiological waste water inside the egg resulting in poisoning of embryos. The effect of light on incubation and growth of embryo, requirement of black boxing at pinhead stage, and exposure to light on 10 or 11 day was studied in detail for bivoltine silkworm [78]. Humidity in combination with temperature has a very strong effect on physiology of the egg [86]. However, the ideal temperature is 25°C and coupled with 75–80% relative humidity for both bivoltine and multivoltine eggs (Table 1). The occurrence of unfertilized eggs in silkworm and its reason were studied in detail [87]. Workers also studied the effect of refrigeration of multivoltine eggs at blue stage and its effects on hatching and rearing parameters [88].

7. Influence of Environmental Factors on Nutritional Indices

Variation in the fluctuations of temperature prevents insects from attaining their physiological potential performance and they achieve it only if placed in an ideal and favorable environment. Because of natural selection imposed by less ideal environmental conditions, insects have evolved certain abilities to evaluate their environment and to make decisions involving physiological, behavioral, and genetic responses. These responses frequently involve changes in the consumption and utilization of food, rate, and time of feeding

TABLE 3: Combined effect of humidity and air current during spinning time on cocoon reelabilty of silkworm.

Temperature (°C)	Humidity (%)	Air current (cm/sec)	Reelability (%)
20°C	65	0	75
		50	90
	90	0	78
		50	92
25°C	65	0	92
		50	96
	90	0	55
		50	90
30°C	65	0	85
		50	93
	90	0	30
		50	80

behavior, metabolism, enzyme synthesis, nutrient storages, flight behavior, and other physiological and behavioral processes. Natural environments exhibit large amount of variation in abiotic components (temperature, humidity, etc.) which play an important role on the consumption and utilization of food. Variation in environmental factors away from the conditions that allow insects to achieve their ideal performance may reduce their performance unless compensated for changes in their physiology and behavior. Variable temperature regimes may influence performance differently compared to constant temperature; growth performance is often stimulated in fluctuating temperature regime [89]. Insects have also evolved various enzymatic and metabolic adaptations that allow them to survive and develop in a broad range of temperatures. Temperature acclimation and physiological and behavioral thermoregulation allow individual insects to compensate to various degrees of changes in ambient temperature [90]. The silkworm growth is manifested by the accumulation of organic matter resulting from the balance between anabolic and catabolic reactions fuelled by the nutritive substances absorbed after digestion of food. The silkworms from the same genetic stock responded variedly when fed on the leaves of different nutritional quality, which is an indicator of efficient utilization and conversion of food into silk substance. When a temperature exceeds 30°C, metabolic functions become erratic and result in poor health. At the same time when temperatures fall below 20°C, metabolic functions becomes inactive again leading to irregular growth and health becomes poor [91].

Variation of the ingesta and digesta values among the different breeds and same breed in different seasons has been reported [92, 93]. The rate of food consumption and leaf quality significant influences on larval growth, weight, and survival were studied [94]. Analysis of the nutritional indices like the rates of ingestion, digestion, assimilation, and conversion in the growing larvae would be useful in understanding the racial differences in the digestive and assimilation abilities of the silkworm. The evaluation of the strains must be made based on food utilization efficiency at different feeding amounts under favorable conditions for each sex [95]. Researchers [44] reported that with the increase in temperature (20–30°C) leaf-silk conversion rate decreases. The physiological activities, food intake, and economic parameters are influenced by body temperature of the silkworm. An increase in intake of mulberry leaves during late age with decrease in rearing temperature was reported [96]. Effects of various environmental factors on nutritional and water requirements of insects were studied [97–105]. The effect of temperature on leaf-silk conversion in silkworm reported that low temperature (26°C) throughout the rearing period favoured higher silk conversion with better survival in bivoltine silkworm [106]. A high efficiency of conversion in larvae reared at low temperature was reported [107]. Workers analyzed the nutritional indices and nutritional efficiency parameters of 5th instar larvae of popular India bivoltine races under different environmental conditions [108]. They demonstrated that the nutritional indices parameters like ingesta, digesta, approximate digestibility, and reference ratio were superior under optimum temperature (23–25°C) and humidity conditions (65–70%). Different authors [109–111] studied the effect of leaf moisture on nutritional parameters and development of silk gland of silkworm. However, the efficiency parameters like ECI cocoon were higher for higher-temperature-maintained silkworm batches and this might be because of less choice of feed that leads to some physiological adoptions to overcome nutritional stress (Table 4).

8. Influence of Temperature on Reproductive Potential of Silkworm Moth

The reproductive performance of silkworm varies with impertinent climatic factors in addition to physiological status of the parent. The commercial viability of silkworm is dependent on correlation between cocoons, moths, and reproductive potential of the strains. The cocoon weight and reproductive characters were greatly influenced by different temperature regimes [112]. The rate of egg production varies with temperature, accelerated up to a point of optimum temperature and humidity conditions. In the silkworm, *Bombyx mori*, maximum ovulation and fecundity with minimum retention were observed at temperature $25.36 \pm 0.17°C$ and any fluctuation from this level decreased ovulation, oviposition fecundity and increased retention of eggs [113]. Reports are available on temperature-induced sterility in silkworm. Male silkworm moths almost became sterile when kept at 32°C for 4 days after spinning even though pupae were preserved at moderate temperature of 23°C throughout the remaining period [114]. Workers [115] reported that male silkworm become sterile at high temperature (above 33°C) during spinning and demonstrated that 19 hr continuous exposure to high temperature induces sterility in male silkworm [116]. The Induction of sterility at high temperature in different Indian races was studied [117, 118]. The quality of seed cocoon and that of egg yield are directly related and the number of dead pupae varies from race to race in different seasons [119]. The effect of temperature on

reproductive potential of Bulgarian races was studied [120]. The effect of different temperature and humidity on preservation of seed cocoon and pupae of the new bivoltine hybrids (CSR races) was also studied [121]. The role of humidity on preservation of seed cocoon and potentiality of egg production was studied [122] and researchers studied [123] the refrigeration of cocoon and its effects on grainage parameters and egg production.

9. Environmental Care during Silkworm Spinning

If the temperature rises beyond 22°–25°C during spinning, the shell becomes very loose and folded with wrinkles and knots. It also changes the properties of sericin. This induces cohesion of silk filaments and causes difficulties in reeling. Low temperature slows down the secretion of silk thread and resulting in large-sized cocoons. Further, it takes very long-time for spinning. Relative humidity (60–70%) induces good health, good reelability, and good-quality cocoon. When it rises above optimum level the larvae and pupae cease to death. Low humidity causes double-layered cocoons and loose cocoons. Excessive moisture and harmful gases are released from the faeces and urine of silkworms during spinning. Air current speed should be less than one meter per second and fast or strong air current causes crowding of mature silkworms resulting in more number of double cocoons. Mounting room requires moderate, even illumination and strong light causes crowding of silkworms at one side and finally results in double cocoons or uneven-thickness cocoons. Complete darkness will slow down the spinning process resulting in low-quality cocoons. The effect of temperature and humidity on spinning behavior of silkworm was studied [124, 125]. Table 3 describes the combined effect of humidity and air current during spinning time of silkworm and the results show that low humidity (65%) and air current of 50 cm/sec are adequate for obtaining a good reelability of above 95%.

10. Influence of Temperature on Postcocoon Parameters

Cocoon quality parameters play an important role on the quality of the raw silk reeled. A large number of parameters, some of which are important for the parent cocoon race maintenance, define the cocoon properties and some are important for cocoon reeling. For a reeler, the technological parameters of the cocoon are significantly important, since they determine the quality, quantity, and efficiency of the reeling process. Significant variation in cocoon shape and cocoon size in hybrids results in variation in filament size as well as the quality of the reeled thread [126]. It was also mentioned that when reeling is carried out with irregular and nonuniform cocoons it results in thread breakage, hindrance due to slugs, poor reelability, poor cooking, decreased raw silk recovery, variation in raw silk denier, and poor neatness [127]. To obtain uniform filament size in auto and semiautomatic reeling units, cocoon size uniformity is very important

TABLE 4: Effect of different temperature and humidity conditions on nutritional indices of silkworm.

Environmental condition	Ingesta (g)	Digesta (g)	Approximate digestibility (%)	ECI cocoon (%)	ECI shell (%)
Temp 36°C and humidity 40%	3.04	0.850	27.99	18.27	8.94
Temp 20°C and humidity 90%	4.45	1.340	30.11	16.91	8.60
Temp 25°C and humidity 70%	4.41	1.430	32.44	16.86	9.17

Source [108].

[128]. Extensive studies have been carried out on cocoon shape variation in parental silkworm breeds and their hybrids [129–135]. Limited information is available on the combined effect of different temperature and humidity on various cocoon characters and reeling parameters at different stages during rearing and spinning of silkworm larvae which in turn will provide valuable information to the technology developers who are engaged in the improvement of quality and quantity of silk acceptable to the level of international standard [136]. The different temperature during spinning period and its effects on cocoon and reeling parameters of new bivoltine hybrids were studied [137]. The researchers evaluated the influence of various nutritional and environmental stress factors on silk fibre characters of bivoltine silkworm [138]. Researchers also [139] have shown that a relation exists between water content of cocoon layers during the spinning stage and reelability of cocoons and recommended that water content of the cocoon layer should be below 20% in order to obtain good-quality cocoons with better reelability. It is to be noted that when the humidity of ambient conditions is high during cocoon spinning, water present in the spinning solution, silkworm urine, and faeces is evaporated slowly influencing the structure of sericin. Various workers also described care to be taken after cocooning and before harvesting the cocoon [140].

11. Future Strategies

India enjoys the patronage of second position for the production of silk in the world next only to China. Sericulture in India is practiced predominantly in tropical environmental regions. Though, the introduction of robust and thermotolerant races in the field during summer months had considerable impact on the productivity level and returns in some selected areas, later planners realized this does not match to that of other productive bivoltine hybrids. Therefore, the acceptance level of this hybrid with the farmers was not up to the expected level because of the low productivity traits. This has necessitated the development of a more suitable temperature tolerant hybrid with better productivity traits than existing races. Mulberry and silkworm improvement programmes are continuous processes for evolving newer and high-yielding genotypes, which can sustain productivity under biotic and abiotic stresses. Genetic transformation techniques need to be further fine-tuned for developing transgenic silkworm with stable expression of cloned genes of commercial importance.

Some of the earlier studies addressed the selection of silkworm breeds in respect of thermotolerance by identifying thermotolerant silkworm breeds. However, a clear understanding of the genetic basis and variability in the expression of quantitative and qualitative genetic traits during exposure to high temperatures is an important step for the selection of potential thermotolerance parental resources for breeding programmes. To achieve greater success, there is a necessity of understanding the molecular mechanism of temperature tolerance in silkworm, identification of various groups of heat shocking proteins (HSPs), understanding of different expression pattern of various HSPs, in polyvoltine and bivoltine races to locate the genes responsible for the heat inducible HSPs, and subsequent steps to introduce the same into the silkworm genome.

References

[1] C. G. Rao, S. V. Seshagiri, C. Ramesh, K. Ibrahim Basha, H. Nagaraju, and Chandrashekaraiah, "Evaluation of genetic potential of the polyvoltine silkworm (Bombyx mori L.) germplasm and identification of parents for breeding programme," Journal of Zhejiang University B, vol. 7, no. 3, pp. 215–220, 2006.

[2] N. Suresh Kumar, T. Yamamoto, H. K. Basavaraja, and R. K. Datta, "Studies on the effect of high temperature on F1 hybrids between polyvoltine and bivoltine silkworm races of Bombyx mori L," International Journal of Industrial Entomology, vol. 2, no. 2, pp. 123–127, 2001.

[3] H. Lakshmi and M. Chandrashekaraiah, "Identification of breeding research material for the development of Thermotolerant breeds of silkworm Bombyx mori," Journal of Experimental Zoology India, vol. 10, no. 1, pp. 55–63, 2007.

[4] N. A. R. Begum, H. K. Basavaraja, P. G. Joge, and A. K. Palit, "Evaluation and identification of promising bivoltine Breeds in the silkworm, Bombyx mori L," International Journal of Industrial Entomology, vol. 16, no. 1, pp. 15–20, 2008.

[5] V. K. Rahmathulla, "Management of climatic factors during silkworm rearing," The Textile Industry and Trade Journal, pp. 25–26, 1999.

[6] B. M. Sekarappa and C. S. Gururaj, "Management of silkworm rearing during summer," Indian Silk, vol. 27, no. 12, p. 16.

[7] S. Ueda, R. Kimura, and K. Suzuki, "Studies on the growth of the silkworm Bombyx mori. IV mutual relationship between the growth in the fifth instar larvae and productivity of silk substance and eggs," Bulletin of the Sericultural Experiment Station, vol. 26, no. 3, pp. 233–247, 1975.

[8] K. V. Benchamin and M. S. Jolly, "Principles of silkworm rearing," in Proceedings of Seminar on Problems and Prospects of Sericulture, S. Mahalingam, Ed., pp. 63–106, Vellore, India, 1986.

[9] S. Krishanswami, M. N. Narasimhanna, S. K. Suryanarayana, and S. Kumararaj, Silkworm rearing Bulletin " 15/2 FAO

Agricultural Services, United Nations Organizations, Rome, Italy, 1973.

[10] R. K. Datta, *Guidelines for Bivoltine Rearing*, Central Silk Board, Bangalore, India, 1992.

[11] F. K. Hsieh, S. Yu, S. Y. Su, and S. J. Peng, "Studies on the thermo tolerance of the silkworm, *Bombyx mori L*," *Zsongriva*, 1995.

[12] C. W. Willmer, G. Stone, and I. Johnston, *Environmental Physiology of Animals*, Blackwell Science, Oxford, UK, 2004.

[13] S. Ueda and H. Lizuka, "Studies on the effects of rearing temperature affecting the health of silkworm larvae and upon the quality of cocoons-1 Effect of temperature in each instar," *Acta Sericologia in Japanese*, vol. 41, pp. 6–21, 1962.

[14] T. Shirota, "Selection of healthy silkworm strains through high temperature rearing of fifth instar larvae," *Reports of the Silk Science Research Institute*, vol. 40, pp. 33–40, 1992.

[15] Y. Tazima and A. Ohuma, "Preliminary experiments on the breeding procedure for synthesizing a high temperature resistant commercial strain of the silkworm, *Bombyx mori L*," *Japan Silk Science Research Institute*, vol. 43, pp. 1–16, 1995.

[16] Matsumara and Y. Ihizuka, "The effect of temperature on development of *Bombyx mori L*," *Representative Nagano Sericultural Experimental Station, Japan*, vol. 19, 1929.

[17] S. Krishnaswami, *New Technology of Silkworm Rearing. CSR&TI, Bulletin No. 2*, Central Silk Board, Bangalore, India, 1978.

[18] V. S. Pillai and S. Krishnaswami, "Effect of high temperature on the survival rate, cocoon quality and fecundity of *Bombyx mori L*," in *Sericulture Symposium and Seminar*, pp. 141–148, Tamil Nadu Agriculture University, Tamil Nadu, India, 1980.

[19] D. J. Wu and R. F. Hou, "The relationship between thermo tolerancy and heat stable esterase in the silkworm *Bombyx mori L.*, (Lepidoptera: Bombycidae)," *Applied Entomology and Zoology*, vol. 28, pp. 371–377, 1993.

[20] Y. Tazima, *Silkworm Moth, Evolution of Domesticated Animals*, Longman, New York, NY, USA, 1984.

[21] V. Thiagarajan, S. K. Bhargava, M. Ramesh babu, and B. Nagaraj, "Differences in seasonal performance of twenty-six strains of silkworm, *Bombyx mori* (Bombycidae)," *Journal of the Lepidopterists Society*, vol. 47, pp. 331–337, 1993.

[22] C. Ramesha, S. V. Seshagiri, and C. G. P. Rao, "Evaluation and identification of superior polyvoltine crossbreeds of mulberry silkworm, *Bombyx mori L*," *Journal of Entomology*, vol. 6, no. 4, pp. 179–188, 2009.

[23] S. N. Chatterjee, C. G. P. Rao, G. K. Chatterjee, S. K. Ashwath, and A. K. Patnaik, "Correlation between yield and biochemical parameters in the mulberry silkworm, *Bombyx mori L*," *Theoretical and Applied Genetics*, vol. 87, no. 3, pp. 385–391, 1993.

[24] K. V. Benchamin, M. S. Jolly, and D. A. I. Benjamin, "Study on the reciprocal crosses of multivoltine × bivoltine with special reference to the use of bivoltine hybrid as a parent," National Seminar on Silk Research and Development, Bangalore, India, 1983.

[25] P. L. Reddy, S. S. Naik, and N. S. Reddy, "Implications of temperature and humidity on the adult eclosion patterns in silkworm *Bombyx mori L*," *Journal of the Entomological Research Society*, vol. 26, pp. 223–228, 2002.

[26] Mishra and V. B. Upadhyay, "Influence of relative humidity on the silk producing potential of multi-voltine *Bombyx mori L*. race nistari," *Journal of Ecophysiology & Occupational Health*, vol. 2, no. 3-4, pp. 3275–4280, 2002.

[27] S. Morohoshi, "The control of growth and development in *Bombyx mori L*. Relationship between environmental

[28] S. Kamili and M. A. Masoodi, *Principls of Temperate Sericulture*, Kalyani, New Dehli, India, 2004.

[29] M. Kato, K. Nagayasu, O. Ninagi, W. Hara, and A. Watanabe, "Studies on resistance of the silkworm, *Bombyx mori L*. for high temperature," in *Proceedings of the 6th International Congress of SABRAO (II)*, pp. 953–956, 1989.

[30] H. Watanabe, "Temperature effects on the heterosis of starvation resistance in the silkworm, *Bombyx mori*," *Journal Sericult Science Japan*, vol. 29, pp. 59–62, 1960.

[31] H. Watanabe, "Studies on difference in the variability of larval body and cocoon weights between single cross and three-way cross or double cross hybrids in the silkworm, *Bombyx mori*," *Journal Sericult Science Japan*, vol. 30, pp. 463–467, 1961.

[32] K. Kogure, "The influence of light and temperature on certain characters of silkworm, *Bombyx mori*," *Journal Department Agriculture, Kyushu Imperial University*, vol. 4, pp. 1–93, 1932.

[33] K. Kremkyremky and E. Michalska, "Effect of temporary reduced air temperature during silkworm *Bombyx mori L*. rearing on some characters of the inbred lines," *Sericologia*, vol. 24, pp. 29–42, 1984.

[34] R. K. Datta, N. Suresh Kumar, H. K. Basavaraja, C. M. Kishor Kumar, and N. Mal Reddy, "'CSR18 × CSR19'-a robust bivoltine hybrid suitable for all season rearing in the tropics," *Indian Silk*, vol. 39, pp. 5–7, 2001.

[35] P. Pandey and S. P. Tripathi, "Effect of humidity in the survival and weight of *Bombyx mori L*. Larvae," *Malaysian Applied Biology*, vol. 37, pp. 37–39, 2008.

[36] S. Das, A. K. Saha, and M. Shamsuddin, "High temperature induced sterility in silkworm?" *Indian Silk*, vol. 35, pp. 26–28, 1996.

[37] S. Tribhuwan and T. Singh, "Behavioural aspects of oviposition in the silkworm *Bombyx mori L*—A review," *Indian Journal of Forestry*, vol. 37, pp. 101–108, 1998.

[38] F. Dingley and J. Maynard Smith, "Temperature acclimatization in the absence of *Bombyx mori* (Bombicidae)," *Journal of Lepidopteron Society*, vol. 47, pp. 321–337, 1968.

[39] M. A. Malik, *Studies on the performance and adaptation of bivoltine races of silkworm Bombyx mori L of Kashmir and evolution of Heterosis in their hybrids under temperature and sub tropical climates [Ph.D. thesis]*, University of Mysore, Mysore, India, 1992.

[40] K. M. Vijaya Kumari, M. Balavenkatasubbiah, R. K. Rajan, H. T. Himantharaj, B. Nataraj, and M. Rekha, "Influence of temperature and relative humidity on the rearing performance and disease incidence in CSR hybrid silkworms, *Bombyx mori L*," *International Journal of Industrial Entomology*, vol. 3, no. 2, pp. 113–116, 2001.

[41] A. Siddiqui, B. D. Singh, and T. P. S. Chauhan, "Evolution of hardy bivoltine silkworm breeds for summer and monsoon seasons," in *Advances in Tropical Sericulture, National Conference on Tropical Sericulture*, pp. 125–129, CSR&TI, Mysore, India, November 2005.

[42] V. K. Rahmathulla, C. M. Kishor Kumar, A. Manjula, and V. Sivaprasad, "Effect of different season on crop performance of parental stock races of bivoltine silkworm (*Bombyx mori L.*)," *Munis Entomology & Zoology*, vol. 6, no. 2, pp. 886–892, 2011.

[43] J. R. Hazel, "Thermal adaptation in biological membranes: is homeoviscous adaptation the explanation?" *Annual Review of Physiology*, vol. 57, pp. 19–42, 1995.

[44] S. Ueda and K. Suzuki, "Studies on the growth of the silkworm *Bombyx mori* L. 1. Chronological changes on the amount of food ingested and digested, body weight and water content of the body and their mutual relationship," *Bulletin of the Sericultural Experiment Station of Japan*, vol. 22, pp. 33–74, 1967.

[45] H. K. Basavaraja, S. K. Aswath, N. Suresh Kumar, N. Mala Reddy, and G. V. Kalpana, *A Text Book on Silkworm Breeding and Genetics*, Central Silk Board, Bangalore, India, 2005.

[46] E. A. Craig, "The heat shock response," *CRC Critical Reviews in Biochemistry*, vol. 18, no. 3, pp. 239–280, 1985.

[47] J. Nagaraju, "Application of genetic principles for improving silk production," *Current Science*, vol. 83, no. 4, pp. 409–414, 2002.

[48] O. Joy and K. P. Gopinathan, "Heat shock response in mulberry silkworm races with different thermotolerances," *Journal of Biosciences*, vol. 20, no. 4, pp. 499–513, 1995.

[49] S. Chitra and N. Sureshkumar, "Heat shock response in bivoltine races of silkworm, *Bombyx mori* L," in *Advances in Tropical Sericulture, National Conference on Tropical Sericulture*, pp. 118–121, CSR&TI, Mysore, India, November 2005.

[50] M. H. Manjunatha, "Impact of heat shock on heat shock proteins expression, biological and commercial traits of *Bombyx mori*," *Insect Science*, vol. 13, p. 243, 2006.

[51] M. B. Evgen'ev, T. Y. Braude-Zolotareva, E. A. Titarenko et al., "Heat shock response in *Bombyx mori* cells infected by nuclear polyhedrosis virus (NPV)," *MGG Molecular & General Genetics*, vol. 215, no. 2, pp. 322–325, 1989.

[52] M. Coulon-Bublex and J. Mathelin, "Variations in the rate of synthesis of heat shock proteins HSP70, between laying and neurula, the diapausing embryo of the silkworm, *Bombyx mori*," *Sericologia*, vol. 3, pp. 295–300, 1991.

[53] D. J. Wu and R. F. Hou, "The relation ship between thermo tolerancy and heat stable esterase in the silkworm *Bombyx mori* L., (Lepidoptera: Bomycidae)," *Applied Entomology and Zoology*, vol. 28, pp. 371–377, 1993.

[54] P. L. Reddy, S. S. Naik, and N. S. Reddy, "Implications of temperature and humidity on pupation patterns in the silkworm, *Bombyx mori* L," *International Journal of Industrial Entomology*, vol. 5, pp. 67–71, 2002.

[55] B. Vasudha Chavadi, H. Aparna, S. Gowda, and M. H. Manjunatha, "Impact of heat shock on heat shock proteins expression, biological and commercial traits of Bombyx mori," *Insect Science*, vol. 13, p. 243, 2006.

[56] P. P. Srivastava, P. K. Kar, A. K. Awasthi, and S. Raje Urs, "Identification and association of ISSR markers for thermal stress in polyvoltine silkworm *Bombyx mori*," *Russian Journal of Genetics*, vol. 43, no. 8, pp. 858–864, 2007.

[57] S. H. H. Moghaddam, X. Du, J. Li, J. Cao, B. Zhong, and Y. Y. Chen, "Proteome analysis on differentially expressed proteins of the fat body of two silkworm breeds, *Bombyx mori*, exposed to heat shock exposure," *Biotechnology and Bioprocess Engineering*, vol. 13, no. 5, pp. 624–631, 2008.

[58] N. S. Kumar, H. K. Basavaraja, C. M. K. Kumar, N. M. Reddy, and R. K. Datta, "On the breeding of "CSR18 × CSR19"-a robust bivoltine hybrid of silkworm, *Bombyx mori* L. for the tropics," *International Journal of Industrial Entomology*, vol. 5, pp. 155–162, 2002.

[59] D. L. Denlinger and G. D. Yoccum, "Physiology of heat sensitivity in insects," in *Lethal Temperatures in Integrated Pest Management*, G. J. Hallman and D. L. Denlinger, Eds., p. 7, Westview Press, 1998.

[60] G. W. Gilchrist and R. B. Huey, "The direct response of *Drosophila melanogaster* to selection on knockdown temperature," *Heredity*, vol. 83, no. 1, pp. 15–29, 1999.

[61] V. K. Rahmathulla, M. T. Himantharaj, G. Srinivasa, and R. K. Rajan, "Association of moisture content in mulberry leaf with nutritional parameters of bivoltine silkworm (*Bombyx mori* L.)," *Acta Entomologica Sinica*, vol. 47, pp. 701–704, 2004.

[62] P. Pandey, S. P. Tripathi, and V. M. S. Shrivastav, "Effect of ecological factors on larval duration of silkworm (*Bombyx mori* L.)," *Journal of Ecophysiology and Occupational Health*, vol. 6, pp. 3–5, 2006.

[63] S. K. Mathur and S. B. Lal, "Effects of temperature and humidity on the adaptability of insects?" *The Indian Textile Journal*, vol. 136, pp. 34–47, 1994.

[64] D. C. Deb, D. C. Paul, T. P. Kumar, and B. P. Nair, "Role of foliar moisture on consumption and conversion efficiency of dry matter of food into cocoon and shell by the 5th instar larvae of *Bombyx mori* L," in *Proceedings of the Zoological Society*, vol. 53, pp. 31–40, Calcutta, India, 2000.

[65] V. K. Rahmathulla, Tilakraj, and R. K. Rajan, "Influence of moisture content in mulberry leaf on growth and silk production in *Bombyx mori*," *Caspian Journal of Environmental Sciences*, vol. 4, no. 1, pp. 25–30, 2006.

[66] H. R. Rapusa and B. P. T. Gabrieal, "Suitable temperature and humidity and larval density in the rearing of *Bombyx mori* L," *Philippine Department of Agriculture*, vol. 60, pp. 130–138, 1975.

[67] V. K. Rahmathulla, T. Raj, M. T. Himanthraj, G. S. Vindya, and R. G. G. Devi, "Effect of feeding different maturity leaves and intermixing of leaves on commercial characters of bivoltine hybrid silkworm (*Bombyx mori*)?" *International Journal of Industrial Entomology*, vol. 6, no. 1, pp. 15–19, 2003.

[68] P. M. Rajendran, M. T. Himantharaj, A. Meenal, R. K. Rajan, C. K. Camble, and R. K. Datta, "Importance of water in sericulture," *Indian Silk*, vol. 31, no. 4, pp. 46–47, 1993.

[69] K. V. Anantharaman, V. R. Mala, S. B. Magadum, and R. K. Datta, "Effect of season and mulberry varieties on the feed conversion efficiencies of different silkworm hybrids of *Bombyx mori* L?" *Uttar Pradesh Journal of Zoology*, vol. 15, pp. 157–161, 1995.

[70] Y. Zhao, K. Chen, and S. He, "Key principles for breeding spring and autumn using silkworm varieties: from our experience of breeding 873 × 874," *Caspian Journal of Environmental Science*, vol. 5, pp. 57–61, 2007.

[71] M. Hussain, S. A. Khan, M. Naeem, and A. U. Mohsin, "Effect of relative humidity on factors of seed cocoon production in some inbred silk worm (*Bombyx mori*) lines," *International Journal of Agriculture and Biology*, vol. 13, no. 1, pp. 57–60, 2011.

[72] J. Nacheva and Junka, "Phenotypic and genotypic characterization of silkworm character during the different seasons of silkworm feeding," *Genetics Selection Evolution*, vol. 22, no. 3, pp. 242–247, 1989.

[73] K. Nagalakshmamma and P. Naga Jyothi, "Studies on commercial exploitation of selected multivoltine races of the silkworm *Bombyx mori* L. in different seasons of Rayalaseema region (A. P.) in India," *The Bioscan*, vol. 5, no. 1, pp. 31–34, 2010.

[74] A. K. Saha, T. Datta, S. K. Das, and S. M. Moorthy, "Bivoltine rearing during adverse season in West Bengal," *Indian Silk*, vol. 47, no. 1, pp. 5–7, 2008.

[75] T. Singh, M. M. Bhat, and M. K. Ashraf, "Insect adaptations to changing environments—temperature and humidity," *International Journal of Industrial Entomology*, vol. 19, no. 1, pp. 155–164, 2009.

[76] N. Venturia, "The effect of environmental conditions on the growth of larvae of silkworm Lucrai., Stiinifice Medicinia Veterinara," *Uiversttaea de Stiinte Agricole si Mdeicinia*, vol. 45, no. 2, pp. 544–546, 2002.

[77] R. K. Rajan and M. T. Himantharaj, *A Text Book on Silkworm Rearing Technology*, Central Silk Board, Bangalore, India, 2005.

[78] M. V. B. Mathur and R. K. Rajan, "Effect of light on incubation," *Indian Silk*, vol. 33, no. 8, pp. 45–46, 1991.

[79] C. M. Patil and B. L. Vishweshwara Gowda, "Environmental adjustment in sericulture," *Indian Silk*, vol. 25, no. 8, pp. 11–14, 1986.

[80] M. Kogure, "Influence of light and temperature on certain characters of silkworm (*Bombyx mori* L.)?" *Journal of the Faculty of Agriculture, Kyushu University*, vol. 4, pp. 1–93.

[81] H. Kai and K. Hasegawa, "Studies on the mode of action of the diapauses hormone with special reference to the protein metabolism in the silkworm, *Bombyx mori* L.The diapuse hormone and the protein suble in ethanol containing trichloro acetic acid in mature eggs of adult ovaries," *Journal of Sericultural Science of Japan*, vol. 40, pp. 199–208, 1971.

[82] J. Kobayashi, H. E. Edinuma, and N. Kobayashi, "The effect of diapause egg production in the tropical race of the silkworm, *Bombyx mori* L," *Journal of Sericultural Science of Japan*, vol. 55, pp. 345–348, 1986.

[83] G. Vemananda Reddy, V. Rao, and C. K. Kamble, *Fundamentals of Silkworm Egg Bomby mori, L.*, Edited by G. K. Kamble, Silkworm Seed technology Laboratory, Bangalore, India, 2003.

[84] E. Kittlans Die, "Ebmryohalant wicklung von Leptinotarsa decemlineata, Epilachna sparsa and Epilachna vigintiocto maculata in abhangigkeit von der temperature," *Deutsche Entertainment*, vol. 8, pp. 41–52, 1961.

[85] O. Yamasita and K. Hasegawa, "Embryyonic diapauses," in *Comprehensive Insect Physiology Biochemistry and Pharmacology*, G. A. Kerkut and G. A. Gilbert, Eds., vol. 1, pp. 407–430, Pergaman Press, Oxford, UK.

[86] S. N. M. Biram and P. Gowda, "Silkworm seed technology," in *Appropriate Sericulture Techniques*, M. S. Jolly, Ed., pp. 35–62, Central Silk Board, Bangalore, India, 1987.

[87] S. N. M. Biram, S. Tribhuwan, and S. Beera, "Occurrence of unfertilized eggs in the mulberry silkworm, *Bombyx mori* L., (Lepidoptera: Bombycidae)," *International Journal of Industria*, vol. 18, pp. 1–7, 2009.

[88] R. Govindan and T. K. Narayanaswamy, "Influence of refrigeration of eggs of multivoltine silkworm, *Bombyx mori* L. at eye spot stage on rearing performance," *Sericologia*, vol. 26, no. 2, pp. 151–155, 1986.

[89] J. M. Scriber and F. Slarisky Jr., "The nutritional ecology of immature insects," *Annual Review of Entomology*, vol. 26, pp. 181–211, 1981.

[90] B. Heinrich, "A brief historical survey," in *Insect Thermo Regulation*, B. Heinrich, Ed., pp. 7–17, John Wiley, New York, NY, USA, 1981.

[91] K. Tanaka, A. Lino, C. Naguro, and H. Fukudome, Collection of papers presented at the 31st congress at Chubu, Japan, 1973.

[92] T. Yamamoto and T. Fujimaki, "Inters train differences in food efficiency of the silkworm *Bombyx mori* L. reared on artificial diet," *Journal of Sericultural Science of Japan*, vol. 54, pp. 312–315, 1982.

[93] V. K. Rahmathulla and H. M. Suresh, "Feed consumption and conversion efficiency in male and female bivoltine silkworms (*Bombyx mori* L.)—a comparative study," *Journal of the Entomological Research Society*, vol. 10, no. 1, pp. 59–65, 2008.

[94] K. Murugan and A. George, "Feeding and nutritional influence on growth and reproduction of *Daphnis nerii* (Linn.) (Lepidoptera: Sphingidae)," *Journal of Insect Physiology*, vol. 38, no. 12, pp. 961–967, 1992.

[95] A. Kafian and Ocenka, "Produktivnosti Samiovisamok tutovogo Shelkopriada Vzavisimosti Ot norm Kormlenia," *Nauch. Trudi. Gruz. Sel.Hoz.in-ta*, vol. 1–5, pp. 34–43, 1982.

[96] H. H. Sigematsu and Takeshita, "On the growth of silk glands and silk proteins production by silkworm reared at various temperature during fifth instar," *Acta Sericologia of Japan*, vol. 65, pp. 125–128, 1967.

[97] E. Hiratsuka, "Researches on the nutrition of silkworm," *Bulletin of the Sericultural Experiment Station*, vol. 1, pp. 257–315, 1920.

[98] Matsumara and Y. Ihizuka, "The effect of temperature on development of *Bombyx mori* L," *Rep. Nagano Seri. Experimental Station, Jpn*, vol. 19, 1929.

[99] J. M. Legay, "Recent advances in silkworm nutrition," *Annual Review of Entomology*, vol. 3, pp. 75–86, 1958.

[100] T. ITO, "Silkworm nutrition," in *The Silkworm an Important Laboratory Tool*, Y. Tazima, Ed., pp. 121–157, Kodansha Ltd, Tokyo, Japan, 1978.

[101] R. P. Kapil, "Quantitative feeding of larvae of Philosamia ricini," *Indian Journal of Entomology*, vol. 25, pp. 233–241, 1963.

[102] R. P. Kapil, "Effect of feeding different host plants on the growth of larvae and weight of cocoons of Philosamia ricini," *Indian Journal of Entomology*, vol. 29, pp. 295–296, 1967.

[103] A. N. Verma and A. S. Atawal, "Effect of constant and variable temperature on the development and silk production of *Bombyx mori* L," *Journal of Research Punjab Agricultural University*, vol. 4, pp. 233–239, 1967.

[104] A. B. Mishra and V. B. Upadhaya, "Effect of temperature on the nutritional efficiency of food in mulberry silkworm (*Bombyx mori*) larvae," *Justice Standards, Evaluation & Research Initiative*, vol. 3, pp. 50–58, 1995.

[105] V. K. Rahmathulla and R. G. Geetha Devi, "Nutritional efficiency of bivoltine silkworm (*Bombyx mori* L.) under different temperature and humidity conditions," *Insect Environment*, vol. 6, pp. 171–172, 2001.

[106] E. Muniraju, B. M. Sekharappa, and R. Raghuraman, "Effect of temperature on leaf silk conversion in silkworm *Bombyx mori* L," *Sericologia*, vol. 39, pp. 225–231, 1999.

[107] W. D. Shen, "Effect of different rearing temperature on fifth instar larvae of silkworm on nutritional metabolism and dietary efficiency. 2. Digestion and utilization of dietary protein," *Journal of Sericulture Science of Japan*, vol. 12, pp. 72–76, 1986.

[108] V. K. Rahmathulla, H. M. Suresh, V. B. Mathur, and R. G. Geethadevi, "Feed conversion efficiency of elite bivoltine CSR hybrids silkworm *Bombyx mori* L. reared under different environmental conditions," *Sericologia*, vol. 42, pp. 197–203, 2002.

[109] R. Narayana Prakash, K. Periaswamy, and S. Radhakrishnan, "Effect of dietary water content on food utilization and silk production in *Bombyx mori* L., (Lepidoptera: Bombycidae)," *Indian Journal of Sericulture*, vol. 24, pp. 12–17, 1985.

[110] D. C. Paul, G. Subba Rao, and D. C. Deb, "Impact of dietary moisture on nutritional indices and growth of Bombyx mori and concommitant larval duration," *Journal of Insect Physiology*, vol. 38, no. 3, pp. 229–235, 1992.

[111] V. K. Rahmathulla, "Growth and development of silk gland in mulberry silkworm (*Bombyx mori* L.) fed with different maturity leaves," *Insect Environment*, vol. 9, no. 2, pp. 92–93, 2003.

[112] T. Singh and M. V. Samson, "Embryonic diapause and metabolicchanges during embryogenesis in mulberry silkworm, *Bombyx mori* L," *Journal of Sericulture*, vol. 7, pp. 1–11, 1999.

[113] S. K. Mathur, D. R. Pramanik, S. K. Sen, and G. S. Rao, "Effect of seasonal temperature and humidity on ovulation, fecundity and retention of eggs in silk moth, *Bombyx mori* L.(Lepidoptera:Bombycidae)," in *Proceedings of the National Seminar on Advances in Economic Zoology*, p. 48, Jodhpur, India, 1988.

[114] E. Sugai and Ashoush, "Sterilization effect of high temperature on the male silkworm, *Bombyx mori* L.(Lepidoptera: Bombycidae)," *Applied Entomology and Zoology*, vol. 3, pp. 99–102, 1968.

[115] E. Sugai and T. Takahanshi, "High temperature environment at the spinning stage and sterilization in the males of silkworm, *Bombyx mori* L," *The Journal of Sericultural Science of Japan*, vol. 50, pp. 65–69, 1981.

[116] E. Sugai and A. Hanoaka, "Sterilization of the male silkworm, *Bombyx mori*.L by the high temperature environment," *The Journal of Sericultural Science of Japan*, vol. 41, no. 1, pp. 51–56, 1972.

[117] G. Vermana Reddy, M. Venkatachalapathy, A. Manjula, and T. M. Veeriah, "Influence of temperature during spinning and its impact on the reproductive performance of silkworm, *Bombyx mori* L. In summer months," in *Proceedings of the National Seminar on Mulberry Sericulture Research in India*, abstract 183, KSSRDI, Thalagattapura, Bangalore, November, 2001.

[118] G. Vermana Reddy, M. Venkatachalapathy, and C. K. Kamble, "Temperature induced sterility in silkworm, *Bombyx mori* L," in *Proceedings of the National Seminar on Silkworm Seed Production*, Kodathi, Bangalore, 2003.

[119] P. Ram Mohana Rao, M. K. Noamani, and H. K. Basavaraja, "Some observations on melting in bivoltine breeds of the silkworm *Bombyx mori* L," *Sericolgia*, vol. 30, pp. 876–879, 1989.

[120] H. Greiss and N. Patkov, "Effect of temperature on silkworm moth development and productivity," *Bulgarian Journal of Agricultural Science*, vol. 7, no. 4-5, pp. 471–474, 2001.

[121] A. Manjula, P. Jayarama Raju, S. Vijaya Kumar, S. T. Christiana, and C. K. Kamble, "Effect of preservation of seed cocoon /pupae at different environmental conditions on the reproductive efficiency of new bivoltine silkworms, *Bombyx mori* L," in *Proceedings of the Advances in Tropical Sericulture, National Conference on Tropical Sericulture*, pp. 257–261, CSR&TI, November 2005.

[122] M. Hussain, S. A. Khan, M. Naeem, and A. U. Mohsin, "Effect of relative humidity on factors of seed cocoon production in some inbred silk worm (*Bombyx mori*) lines," *International Journal of Agriculture and Biology*, vol. 13, no. 1, pp. 57–60, 2011.

[123] V. B. Upadhaya, R. Singh, and S. Prasad, "Effect of the refrigeration of cocoon on the fecundity of moth and hatchability of eggs of multivoltine mulberry silkworm (*Bombyx mori* L)," *Malaysian Applied Biology*, vol. 35, pp. 13–19, 2006.

[124] Y. L. Ramachandra, G. Bali, and S. Padmalatha Rai, "Effect of temperature and relative humidity on spinning behaviour of silkworm (*Bombyx mori*.L)," *Indian Journal of Experimental Biology*, vol. 39, no. 1, pp. 87–89, 2001.

[125] G. Manisankar, M. Ujjal, and M. Aniruddha, "Effect of environmental factors (temperature and humidity) on spinning worms of silkworm (*Bombyx mori* L)," *Research Journal of Chemistry and Environment*, vol. 12, no. 4, pp. 12–18, 2008.

[126] T. Nakada, "Genetic differentiation of cocoon shape in silkworm, *Bombyx mori*. L," *International Congress of Genetics*, p. 224, 1993.

[127] C. Takabayashi, *Manual on Bivoltine Silk Reeling*, Central Silk board, Bangalore, India, 1997.

[128] Y. Mano, *Comprehensive Report on Silkworm Breeding*, Central Silk Board, Bangalore, India, 1994.

[129] T. Hirashi, "On the cocoon shape of hybrids in the silkworm," *Dainihon-Sanshikaihou*, vol. 21, pp. 22–28, 1912.

[130] K. Katsuki and S. Nagasawa, "Cocoon shape of the hybrids?" *Dainihon-Sanshikaihou*, vol. 26, pp. 8–15, 1917.

[131] T. Gamo, S. Saito, Y. Otsuka, T. Hirobe, and Y. Tazima, "Estimation of combining ability and genetic analysis by diallel crosses between regional races of the silkworm (2) Shape and size of cocoons," *Technical Bulletin of Sericultural Experiment Station*, vol. 129, pp. 121–135, 1985.

[132] T. Gamo and S. Ichiba, "Selection experiments on the fibroin hydrolyzing ratio in silkworm cocoons and its effects upon the economical characters," *Japanese Journal of Breeding*, vol. 21, no. 2, pp. 87–92, 1971.

[133] T. Nakada, "On the measurement of cocoon shape by use of image processing method with an application to the sex determination of silkworm, *Bombyx mori*. L," in *Proceedings of the International Congress, SABRO*, pp. 957–960, 1989.

[134] T. Nakada, "On the cocoon shape measurement and its statistical analysis in the silkworm, *Bombyx mori* L," *Indian Journal of Sericulture*, vol. 33, no. 1, pp. 100–102, 1994.

[135] R. Singh, G. V. Kalpana, P. Sudhakara Rao, and M. M. Ahsan, "Studies on cocoon shapes in different crosses of the mulberry silkworm, *Bombyx mori* L," *Indian Journal of Sericulture*, vol. 37, no. 1, pp. 85–88, 1998.

[136] V. B. Mathur, A. Rahman, R. G. Geetha Devi, and V. K. Rahmathulla, "Influence of environmental factors on spinning larvae and its impact on cocoon and reeling characters?. *Advances in sericulture research*," in *Proceedings of National Conference on Strategies for Sericulture Research and Development*, p. 2, Central Sericultural Research and Training Institute, November 2000.

[137] B. N. Gowda and N. M. Reddy, "Influence of different environmental conditions on cocoon parameters and their effects on reeling performance of bivoltine hybrids of silkworm, *Bombyx mori*," *International Journal of Industrial Entomology*, vol. 14, no. 1, pp. 15–21, 2007.

[138] V. K. Rahmathulla, G. Srinivasa, M. T. Himantharaj, and R. K. Rajan, "Influence of various environmental and nutritional factors during fifth instar silkworm rearing on silk fibre characters," *Man-Made Textiles in India*, vol. 47, no. 7, pp. 240–243, 2004.

[139] T. Akahane and K. Subouchi, "Reelability and water content of cocoon layer during the spinning stage," *Journal of Sericultural Science of Japan*, vol. 63, pp. 229–234, 1994.

[140] S. T. Wu, "Management after cocooning process I," *Journal of Sericultural Science of Japan*, vol. 15, pp. 62–65, 1976.

4

Observations of the Biology and Ecology of the Black-Winged Termite, *Odontotermes formosanus* Shiraki (Termitidae: Isoptera), in Camphor, *Cinnamomum camphora* (L.) (Lauraceae)

Arthur G. Appel,[1] Xing Ping Hu,[1] Jinxiang Zhou,[2] Zhongqi Qin,[2] Hongyan Zhu,[2] Xiangqian Chang,[3] Zhijing Wang,[2] Xianqin Liu,[2] and Mingyan Liu[2]

[1] *Department of Entomology and Plant Pathology, Auburn University, 301 Funchess Hall, Auburn, AL 36849-5413, USA*
[2] *Fruit and Tea Institute, Hubei Academy of Agricultural Sciences, Wuhan 430209, China*
[3] *Plant Protection and Fertilizer Institute, Hubei Academy of Agricultural Sciences, Wuhan 430070, China*

Correspondence should be addressed to Arthur G. Appel, appelag@auburn.edu

Academic Editor: Deborah Waller

Aspects of the biology and ecology of the black-winged termite, *Odontotermes formosanus* Shiraki, were examined in a grove of camphor trees, *Cinnamomum camphora* (L.), located at the Fruit and Tea Institute, Wuhan, China. Of the 90 trees examined, 91.1% had evidence of termite activity in the form of exposed mud tubes on the bark. There was no relationship between tree diameter and mud tube length. Mud tubes faced all cardinal directions; most (60%) trees had multiple tubes at all directions. However, if a tree only had one tube, 22.2% of those tubes faced the south. The majority (>99%) of mud tubes were found on the trunk of the tree. Approximately 35% of all mud tubes had termite activity. Spatial distribution of termite activity was estimated using camphor and fir stakes installed throughout the grove. Camphor stakes were preferred. Kriging revealed a clumped distribution of termite activity.

1. Introduction

The black-winged termite, *Odontotermes formosanus* Shiraki, is distributed throughout Southeast Asia including Burma, China, India, Japan, Thailand, and Vietnam where it is an economically important pest of crops, forests, and various wooden structures [1]. Although this species consumes wood and other cellulosic material, it does not directly use these for food. Rather, masticated cellulosic material is used to grow fungus gardens which are the termite food. Termite infestation may result in weakened trees and reduction of yield in fruit trees, or even death of tress, without proper prevention and management [1]. Foraging areas of this species range from 4.2 to 35 m; foraging territories are 13–367.9 m^2 [2]. Damage to camphor trees appears as areas of removed bark that may extend from the soil line and roots to the tree crown. These termites move up the tree by building mud tubes along the trunk and removing the bark beneath. In severe infestations, these termites can infest and hollow out branches resulting in severe limb drop especially during windy or icy weather. The mud tubes are thought to provide protection from predators such as ants and from the environment by allowing the creation of a dark and humid microclimate.

Camphor, *Cinnamomum camphora* (L.), is an aromatic tree in the laurel family (Lauraceae) that is native to Southeast Asia including southern China. It is an economically important species because it is used for construction and furniture making, as a spice in cooking, as incense, as a medicine, as an ornamental plant (pers. comm.), and as a repellent for several insect pests [3]. Extracts of camphor trees include several essential oils including camphor, linalool, and 1–8 cineole [4]. The latter two essential oils have toxic and repellent properties to a number of insects

FIGURE 1: Satellite image of Fruit and Tea Institute, Wuhan, Hubei, China. Top of the image is the north (Google Earth, version 6.0.3.2197, 2011).

including cockroaches [5, 6]. Interestingly, even with its toxic and repellent characteristics, camphor is a preferred indirect food plant of the black-winged termite [1].

The objectives of this study were to examine the distribution of black-winged termites in camphor groves and on camphor trees, determine if there is a relationship between tree size and the length of termite tubes, measure feeding preferences, and observe various aspects of termite tubing behavior.

2. Materials and Methods

This study was conducted at the Fruit and Tea Institute of the Hubei Academy of Agricultural Sciences, at Jin Shui Zha, Jiang Xia District, Wuhan, China, the Institute is situated on approximately 2.3 km^2 (230 hectares) at 30°17′53.74″N 114°08′29.76″E and an elevation of 55 m above sea level. It is located in a relatively rural area about 38 km from the urban areas of Wuhan. There are a number of brick and cement buildings, fields of tea, and orchards of kiwis, oranges, and pears. There are also several groves of camphor trees, *Cinnamomum camphora* (L.), situated between buildings (Figure 1). The camphor trees were planted in a regular grid pattern with almost equal distance between individual trees. A cursory inspection of the groves and nearby (ca. 100 m) wooden structures for the presence of tunnels and live termites and pieces of wood was conducted. Observations were made on 25-26 October 2009 with subsequent visits in March 2010 and July 2011.

Approximately one month before our observations (24–29 August 2009), pairs of fir, *Cunninghamia lanceolata* (Lambert) Hooker, and camphor wood stakes (5 by 3 by 40 cm) were installed 20 cm deep into the soil in the camphor groves (Figure 2). Fir was selected because it is a common wood species in the test area and stakes were readily available. Stakes were installed at 5 m intervals forming a uniform sampling grid. There were a total of 85 pairs of camphor-fir stakes installed in two groves (east and west). The percentage of stake locations with termite feeding, preferences in feeding between the types of wood, and the distribution of termite activity were recorded.

Measurements of termite infestations included visual inspection of 90 camphor trees located in the two adjacent

groves of 45 trees each. Each tree was inspected for termite tubes and the location (trunk, roots, leaves, etc.), length (from the ground to the highest point on the trunk or branch), direction (cardinal direction), and presence of termites in each tube were recorded. The size of each tree was also estimated by measuring the circumference of the trunk 1.5 m above the ground with a tape measure. Presence of termites was determined by manually removing 5–10 cm section of each tube at several heights above the ground and noting the presence of termites.

The bark on a number of trees along the northernmost edge of each grove was covered with moss. When the moss-covered bark was removed it was evident that termites had formed a ca. 0.5 cm foraging space under the bark. There were no mud tubes; the moss-covered bark likely served the same protective functions as tubes. To determine the behaviors of these termites, sections of moss-covered bark (and active mud tubes) were removed and the areas were observed over a 2–4 h period.

Data are expressed as means ± standard errors, and differences were considered at $P \leq 0.05$. Regression and correlation analysis was used to determine the relationship between tree trunk diameters and termite tunnel length, analysis of variance (ANOVA) was used to determine differences in the orientation of termite tunnels on tree trunks, t-tests were used to compare wood preference, the length and areas of mud and moss-covered sections and repair times, and Kriging was used to estimate the distribution of termite activity in camphor groves. SigmaPlot 12.1 [7] was used for ANOVA, correlation, regression, and t-test analyses, and Surfer 10.0 [8] was used for Kriging.

3. Results

Of the 85 pairs of monitoring stakes, only 10 (11.8%) were infested by termites. There was no significant preference between the camphor (13.3%) and fir (8.9%) stakes (t-test; $P > 0.05$). The distribution of termite activity in each of the two camphor groves is illustrated in Figure 3. There are several concentrated areas of activity, but most stakes were not attacked. Termite activity was concentrated at both the northern and southern portions of both the eastern and western groves with relatively little activity in the center of the western grove and the southeaster and northwestern portions of the eastern grove. In contrast, virtually every tree (93.3%) had termite tubes indicating that the entire area of both groves was foraging territory of one or more termite colonies.

Camphor trunk circumference ranged from 42 to 183 cm with a mean of 109.73 ± 3.05 cm. Termite tube length ranged from 0 to 6 m with a mean of 3.31 ± 0.20 m. There was no significant relationship between camphor tree trunk circumference and the length (height) of termite tubes (regression; $P > 0.05$) (Figure 4).

Over 90% (91.1%) of the 90 camphor trees examined showed signs of current or prior infestation as determined by the presence of mud tubes. There was no obvious directional preference by these termites for the location of their tubes on tree trunks. Mud tubes were found facing all four cardinal

Observations of the Biology and Ecology of the Black-Winged Termite, Odontotermes formosanus Shiraki
(Termitidae: Isoptera), in Camphor, Cinnamomum camphora (L.) (Lauraceae)

29

(a)

(b)

FIGURE 2: Grove of camphor trees with pairs of sampling stakes: (left) pair of sampling stakes.

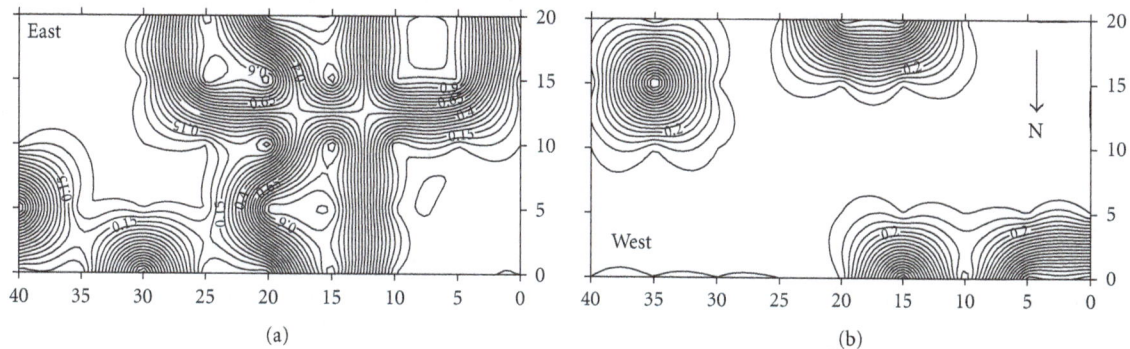

(a)

(b)

FIGURE 3: Distribution of termite activity as measured by attacked stakes. (a) The eastern grove, (b) western grove. Increasing contour values indicate increasing probability of termite activity.

directions on 60% of the infested trees. Of those trees having only one tube, 11.1% of the tubes were formed on the north facing side of a tree compared with 22.2% formed on the south facing side (Figure 5), these proportions were not significantly different (ANOVA; $P > 0.05$). All mud tubes were broken open to determine termite activity. Only 34.4% of tubes were active, and there was no significant directional preference of the active tubes (ANOVA; $P > 0.05$).

A total of four mud tubes and four moss-covered areas were selected and the mud or moss removed to reveal active termites. There was no difference in the mean length of the areas removed (ca. 5.6 cm); however, the moss-covered areas (2.25 ± 0.26 cm) were significantly wider (ca. 2.2 times) than mud tubes (1.18 ± 0.10 cm) (t-test; $t = 3.3806$, df $= 6$; $P = 0.0089$). All exposed areas were repaired by the termites within 1 h. Mud tubes (18.23 ± 3.40 min) were repaired significantly faster than moss-covered areas (41.30 ± 5.51 min) (t-test; $t = 3.5656$, df $= 6$; $P = 0.0118$).

4. Discussion

Black-winged termites were distributed throughout both adjacent groves of camphor trees. Inspection of nearby (within 100 m) trees, small wooden structures, and even relatively small (<2 cm diameter) sticks revealed the presence of these termites. Nearly all camphor trees (93.3%) in both groves had been attacked by termites as evidenced by the

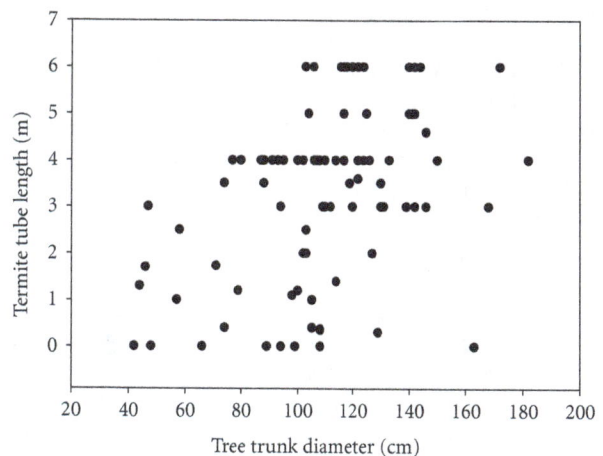

FIGURE 4: Relationship between tree trunk diameter measured 1.5 m above the ground and maximum termite tube length on the tree trunk.

presence of mud tubes on the bark. The size and therefore age [3] of camphor trees were not related to termite attack. A variety of studies have shown that the concentration and composition of protective compounds, such as essential oils, in plants change with age. Also, as trees age and grow, they increase in height and circumference. There was no relationship between tree size and length of termite tubes

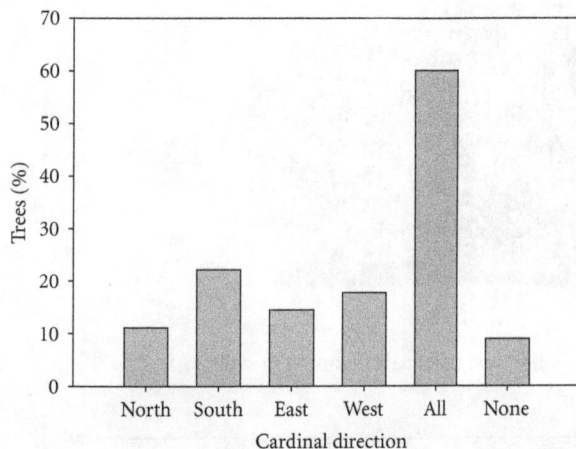

FIGURE 5: Percentage of trees with termite tubes oriented in each direction. All indicate that termite tubes were present in all four directions.

(Figure 4) indicating a probable lack of effective change in protective compounds with camphor tree age, or black-winged termites are not affected by these compounds.

Even though there was evidence of almost complete infestation of trees, only 11.8% of the pairs of monitoring stakes were infested. It is possible that the approximately one-month period between installation of the stakes and inspection was not sufficient for the termites to locate the stakes. It is more likely, however, that the termites had sufficient wood on the camphor trees and only attacked the stakes if a subterranean foraging tunnel directly contacted them. The condition of the wood stakes may also affect feeding activity and it is possible that the stakes had to age and decay to become as attractive as bark. Decomposing wood has greater concentrations of sucrose and more associated yeasts than sound wood [9]. In addition, fungi that are associated with wood decay are often consumed by termites [10], and these fungi are rich in urea [11]. Laboratory feeding preference studies that compare black-wing termite feeding on camphor and fir are clearly indicated. Many factors including wood density, presence of protective compounds, and concentration of glucose affect black-winged termite feeding preferences [12]. In a recent study with the black-winged termite, Kasseney et al. [12] found that solid wood consumption was inversely correlated with wood density and positively correlated with glucose concentration. Although camphor had moderately dense wood, it also had the greatest concentration of glucose among the wood species tested [12]. Perhaps feeding on bark allows these termites to avoid the dense solid wood while feeding on more glucose-rich cellulose. Unfortunately, the Kasseney et al. [12] study did not use fir as one of the wood choices. Interestingly, there was no preference between camphor and fir wood stakes in this study. However, in a similar field study about 0.5 km from the camphor groves, black-winged termites showed a decided (2:1) preference for camphor over fir wood stakes when the stakes were installed in a kiwi field (unpublished). Camphor is probably a preferred wood, and there is sufficient

camphor in the camphor groves to interfere with relatively small camphor stakes.

The spatial analysis of the termite activity (Figure 3) indicated several areas of greater activity in both the east and west camphor groves. These areas of greater activity tended to be nearer to the northern and southern borders of the groves rather than in the center of the groves. Since many termites, and other insects, are known to follow structural guidelines in their foraging patterns [13, 14], it is possible that the greater activity along the northern and southern borders is due to the presence of cement sidewalks and brick walls that enclose the camphor groves. It is also possible that the greater activity in certain areas is a result of greater termite density in those areas or closer proximity to a primary nest. There was no correlation between termite activity in stakes and activity on trees.

The majority (60%) of trees had termite tubes facing all cardinal directions (Figure 5). If trees had only one termite tube, there was no preference for direction. Cardinal direction and therefore light and temperature exposure could affect the distribution of termite mud tubes on trees. Exposure to increased temperature could cause the conditions in tubes on one side of a tree, such as the side facing south, to become too hot for foraging workers. Increasing heat could cause an increase in desiccation or reach the critical thermal maximum. Water and temperature relations have not been studied in this species, but information on these aspects of termite physiology could help explain their micro- and macrodistribution patterns.

Black-winged termites usually construct mud or soil tubes when they are foraging above ground and exposed to the environment. Hundreds of meters of mud and soil tubes were observed in this study, most on camphor tree trunks and branches. These termites will also forage in other protected structures such as moss- and bark-covered voids. These voids ranged in size from 1 cm in width to >10 cm and could extend >1 m in length and were about 0.5 cm in height. The surfaces of these voids were very smooth and did not contain mud or soil. When a section of moss-covered bark was removed to expose the void, termite workers immediately began to seal the exposed areas by bringing soil and mud and depositing it along the exposed area. Rather than resealing the entire exposed area, the termites rapidly constructed a mud tube that provided an enclosed foraging corridor. Termites required about twice the time to repair exposed moss-covered bark foraging areas as they did to repair similar sized damage to mud tubes. It is likely that mud tubes could be repaired more quickly because there were mud tubes close to the broken area and termites removed some of this mud to repair the break. Termites foraging under moss-covered bark probably did not have ready access to mud or soil and had to transport it from the ground up to the broken area. The rapid repair of all broken foraging tubes and areas indicates the importance of these structures to the biology of the black-winged termite and the size of their colonies.

In conclusion, the black-winged termite is an important pest of camphor trees particularly in dense groves. Most trees in an area will be attacked, but sampling studies that rely

Observations of the Biology and Ecology of the Black-Winged Termite, Odontotermes formosanus Shiraki (Termitidae: Isoptera), in Camphor, Cinnamomum camphora (L.) (Lauraceae)

31

on wooden stakes may require extended periods to yield results. This termite builds mud tubes on all age and size trees and shows no preference for the direction of these tubes. Further studies on the physiological ecology of this species will provide insight into foraging and above-ground tubing activities. Additional studies on the distribution patterns of this species will aid control strategies by accurately providing locations for insecticidal bait placements.

References

[1] M. Cheng, J. Mo, T. Deng, W. Mao, and D. Li, "Biology and ecology of *Odontotermes formosanus* in China (Isoptera: Termitidae)," *Sociobiology*, vol. 50, no. 1, pp. 45–61, 2007.

[2] J. Hu, J.-H. Zhong, and M.-F. Guo, "Foraging territories of the black-winged subterranean termite *Odontotermes formosanus* (Isoptera: Termitidae) in southern China," *Sociobiology*, vol. 50, pp. 1–12, 2007.

[3] P. H. Raven, R. F. Evert, and S. E. Eichhorn, *Biology of Plants*, Worth Publishers, New York, NY, USA, 5th edition, 1992.

[4] V. Rozman, I. Kalinovic, and Z. Korunic, "Toxicity of naturally occurring compounds of Lamiaceae and Lauraceae to three stored-product insects," *Journal of Stored Products Research*, vol. 43, no. 4, pp. 349–355, 2007.

[5] A. K. Phillips, A. G. Appel, and S. R. Sims, "Topical toxicity of essential oils to the German cockroach (Dictyoptera: Blattellidae)," *Journal of Economic Entomology*, vol. 103, no. 2, pp. 448–459, 2010.

[6] A. K. Phillips and A. G. Appel, "Fumigant toxicity of essential oils to the German cockroach (Dictyoptera: Blattellidae)," *Journal of Economic Entomology*, vol. 103, no. 3, pp. 781–790, 2010.

[7] *SigmaPlot Version 12.0*, Systat Software, San Jose, Calif, USA, 2011.

[8] *Surfer Surface Mapping System, Version 10.3*, Golden Software, Golden, Colo, USA, 2011.

[9] D. A. Waller, S. E. Morlino, and N. Matkins, "Factors affecting termite recruitment to baits in laboratory and field studies," in *Proceedings of the 3rd International Conference on Urban Pests*, W. H. Robinson, W. F. Rettich, and G. W. Rambo, Eds., pp. 597–600, 1999.

[10] D. A. Waller, J. P. LaFage, R. L. Gilbertson, and M. Blackwell, "Wood decay fungi associated with subterranean termites in southern Louisiana," in *Proceedings of the Entomological Society of Washington*, vol. 89, pp. 417–424, 1987.

[11] M. M. Martin, "Biochemical implications of insect mycophagy," *Biological Reviews*, vol. 54, pp. 1–21, 1979.

[12] B. D. Kasseney, T. Deng, and J. Mo, "Effect of wood hardness and secondary compounds on feeding preference of *Odontotermes formosanus* (Isoptera: Termitidae)," *Journal of Economic Entomology*, vol. 104, no. 3, pp. 862–867, 2011.

[13] T. L. Pitts-Singer and B. T. Forschler, "Influence of guidelines and passageways on tunneling behavior of *Reticulitermes flavipes* (Kollar) and *R. virginicus* (Banks) (Isoptera: Rhinotermitidae)," *Journal of Insect Behavior*, vol. 13, no. 2, pp. 273–290, 2000.

[14] J. H. Klotz and B. L. Reid, "The use of spatial cues for structural guideline orientation in *Tapinoma sessile* and *Camponotus pennsylvanicus* (Hymenoptera: Formicidae)," *Journal of Insect Behavior*, vol. 5, no. 1, pp. 71–82, 1991.

5

Temperature-Driven Models for Insect Development and Vital Thermal Requirements

Petros Damos and Matilda Savopoulou-Soultani

Laboratory of Applied Zoology and Parasitology, Department of Plant Protection, Faculty of Agriculture, Aristotle University of Thessaloniki, 54124 Thessaloniki, Greece

Correspondence should be addressed to Petros Damos, damos@agro.auth.gr

Academic Editor: Nikos T. Papadopoulos

Since 1730 when Reaumut introduced the concept of heat units, many methods of calculating thermal physiological time heat have been used to simulate the phenology of poikilothermic organisms in biological and agricultural sciences. Most of these models are grounded on the concept of the *"law of total effective temperatures"*, which abstracts the temperature responses of a particular species, in which a specific amount of thermal units should be accumulated above a temperature threshold, to complete a certain developmental event. However, the above temperature summation rule is valid within the species-specific temperature range of development and therefore several empirical linear and nonlinear regression models, including the derivation of the biophysical models as well, have been proposed to define these critical temperatures for development. Additionally, several statistical measures based on ordinary least squares instead of likelihoods, have been also proposed for parameter estimation and model comparison. Given the importance of predicting distribution of insects, for insect ecology and pest management, this article reviews representative temperature-driven models, heat accumulation systems and statistical model evaluation criteria, in an attempt to describe continuous and progressive improvement of the physiological time concept in current entomological science and to infer the ecological consequences for insect spatiotemporal arrangements.

1. Introduction

Climate has a profound effect on the distribution and abundance of invertebrates such as insects, and the mathematical description of the climatic influence on insect development has been of considerable interest among entomologists. Additionally, as temperature exerts great influence among the climate variables, by directly affecting insect phenology and distribution, most of the models that describe insect development are temperature driven [1–5].

This first effort for a formal description of the relation between temperature and developmental rate was taken by botanists, to model the effect of temperature on plant growth and development [6–10]. However, similar modeling procedures extended to most of the poikilothermic organisms, including insects as well [1–3]. To date, the earliest experiment that related the velocity of insect development and heat, was made by Bonnet (1779) [11] on the study of

the reproduction rate of *Aphis evonymi*, F. [12], while the major assumption and principles that have been brought out by these earlier works, constituted the basis for all future research. Nevertheless, since then, several theoretical and experimental works have been carried out and current progress in entomology, mathematics and computation offers new means in describing the relation of temperature to insect development [13–20].

Thus, although simple predictive models have been developed during the last century, the development and broader availability of personal computers in the 70s and 80s resulted in the rapid development of computer-based models to predict responses of insects in relation to climate [21, 22].

Insects are adapted to particular temperature ranges and temperature is often the most detrimental environmental factor influencing their populations and distribution. In general, within optimum ranges of development and as environmental temperature decreases, their rates of development slow and cease at the lowest (base) temperature, while as temperature rises, developmental rates increase up to an optimum temperature, above which they again decrease and eventually cease at their temperature maximum [4, 5, 15, 23, 24].

It is proposed that this effect of temperature on poikilothermic organism functioning is related to the effect on enzymatic activities. For instance, the conformation of enzymes is the essential step in the enzymatic reaction and this conformation depends on temperature [22, 25, 26].

One common approach to model temperature effects on insect development is to convert the duration of development to their reciprocals. This simple transformation is used to reveal the relationship type, as it will be shown later close to linear, between temperature and rate of development and permits the determination of two vital parameters of development namely, the thermal constant (K) and the base or lower temperature of development (T_{min}). The thermal constant is expressed as the number of degree-days (in °C) and provides an alternative measure of the physiological time required for the completion of a process or a particular developmental event [4, 5, 21, 27].

Attempts to quantify the influence of temperature on insect development rates, growth, and fecundity have been carried out by several studies for species of economic significance [16, 27–32]. Entomologists have strong interest on this kind of relationships, since they are prerequisite to predicting timing and phenology of insect life cycle events and to initiating management actions [33–35], while application of temperature driven models are also essential in epidemiology modeling, development of effective vector control programmes [36] and prediction of biological invasions [37, 38]. From an agronomical standpoint, empirical models are often used to predict specific population events and provide means for precisely applied control methods, reducing costs as well as insecticide use [39, 40]. Furthermore, the determination of insect-specific vital thermal requirements provides evidence to infer on observed geographical distributions and predict future dynamics [8, 41].

The current review highlights the importance of the relationship between insect development and its vital thermal requirements and outlines important constraint and challenges regarded to model selection and applicability in pest management and insect ecology. Within our aims, building on preview reviews, was to provide a simple account for applied entomologists and field ecologists by avoiding complex and technical details. Furthermore, efforts are also made to present a short example of the linear model and to propose a simple three parameter non-linear equation for modeling temperature effects on insect developmental rates.

The rest of this article is structured as follows. The first section describes and explains the concept of the *law of total effective temperatures* and how it is related to the linear models of insect development. A paradigm of the *x-intercept* method is presented in defining lower developmental threshold for *Grapholitha molesta* (Lepidoptera: Gelechiidae). This threshold is vital in applying phenology models in field, and to our knowledge estimated for first time in a laboratory

trial. The next section summarises the most common non-linear regression models, including the derivation of the biophysical ones, which have been proposed by researchers in order to estimate cardinal temperatures of insect development. Additionally, among the given functions, a new 3-parameter equation is proposed and its general shape is also presented. Section 3 lists principal statistics that are used for parameter estimation in regression analysis and criteria for model selection among candidate equations. Section 4 briefly outlines the major heat accumulation systems for estimating species-specific heat energy in field during the growth season. Finally, there is extensive discussion regarding constraints and challenges of the models for pest management while efforts have been made to discuss how the estimated insects vital thermal requirement are related to the species environmental adaptation and field distribution.

2. Mathematical Models and Insect Development

Mathematical models represent a language for formalizing the knowledge on live systems obtained after experimental observation and hypothesis testing. An empirical model, if successful, determines result and cause and can be further used to describe the behavior of the system under different conditions [39, 40, 42].

Since temperature is considered as the most critical factor affecting insect development, numerous efforts have been made by researchers to propose models to describe such relations either in laboratory or field [6, 16, 22, 28, 29, 39, 40, 43–45]. Moreover, several of these models have been constructed in the view to be applicable for pest management [1, 21, 23, 27, 39, 42, 43, 46–48].

The term model emphasises some qualitative and quantitative characteristics of the process, which are actually abstracted, idealized, and described mathematically rather than the system itself.

Most of these approaches are based on the empirical detection of relationships and the construction of relative models that in brief capture all information about the response variable in relation to temperature. It should be noted that the presented temperature relationships can be judged as deterministic or empirical, by the sense that they consist of descriptions in which processes are not known, but where relations are established. However, all regression procedures that are followed, for parameter estimation, are purely probabilistic.

In applied entomology, empirical approaches are often used in the construction of developmental models. In general, the procedures include the delimitation of all the factors that affect development to the most limiting one, which is further chosen (i.e., temperature), in order to reveal empirical dependence of the developmental variable upon the limiting factor. A function which describes the data with higher accuracy is plugged to this relation, and its prediction power is further evaluated by using new datasets.

2.1. The Law of Total Effective Temperatures and the Linear Model. All poikilothermic organisms are related to a species-specific thermal constant that corresponds to time units that must be accumulated to complete a particular developmental event. The above principle forms the basis for all modeling approaches that have been developed since the first introduction of the heat units concept by Reaumut on 1730 and the following initiation of the temperature summation rule [20, 49]. This rule, which was first proposed by Candolle [6] and characterized the development of all poikilothermic species, is referred to as the *law of total effective temperatures* and consists of the first effort in modeling temperature-dependent developmental rates instead of developmental times [7, 31, 50].

The model is characterized by universality, since development of all species is addressed by a thermal constant which corresponds to the accumulated degree-days that are needed to complete a particular developmental stage. This principle is further related to most other cumulative degree-day approaches.

According to the *law of total effective temperatures*, it is possible to estimate the emergence and number of generations for a given duration, of the organism of interest, according to the following fundamental equation:

$$K = D(T - T_0), \tag{1}$$

where K is the species (or stage-specific) thermal constant of the poikilothermic organism, T temperature, and T_0 developmental zero temperature. This thermal constant provides a measure of the physiological time required for the completion of a developmental process and is measured in degree-days (DD).

One popular method of estimating the above parameters is to use a linearizing transformation of the above function by calculating the rate of development $y = 1/D$ for the day variable resulting to the following equation [44]:

$$\frac{1}{D} = -\frac{T_0}{K} + \frac{1}{K}T. \tag{2}$$

Equation (2) is often referred to as the linear degree-day model or as the *x-intercept* method [24, 51], which is simply derived after growth rate fitting to a simple linear equation and then extrapolated to zero:

$$y = a + bT. \tag{3}$$

The lower theoretical temperature threshold (i.e., base temperature) is derived from the linear function as $T_b = -a/b$ whereas 1/slope is again the average duration in degree days or thermal constant K.

Equation (3) simply means that the thermal constant is a product of time and the degrees of temperature above the threshold temperature.

2.2. Lower Developmental Threshold for Grapholitha Molesta (Lepidoptera: Gelechiidae).
Figure 1, for instance, describes a typical temperature effect on the developmental time of the pupae of *G. molesta* as well as the respective linear relationship between temperature and developmental rate according to (3). To reveal the above relations, larvae were reared in the laboratory at the Aristotle University of Thessaloniki

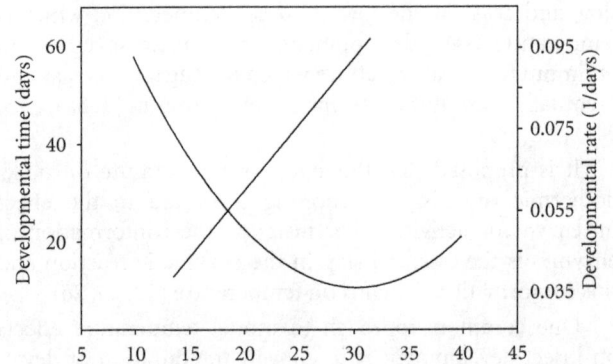

FIGURE 1: Typical response and temperature effect on the developmental time ($y = 115.5 - 6.9x + 0.1134x^2$) of an insect (i.e., pupal stage of *G. molesta*) and respective linear relationship between temperature and developmental rate according to the linear model ($y = 0.041x - 0.0412$, $T_{min} = 10°C$).

and respective pupae were incubated at different constant temperatures at constant laboratory conditions (15, 20, 25, and 30°C, and 65 ± 5% R.H., 16:8 h L : D).

The need for inverse regression, as also displayed in the above paradigm, arises most often when the observed variable (developmental time) is the result of the major factorial cause variable (temperature) which is not subjected to error. Thus, in order to measure the predicted variable with negligible error and avoid bias, such kind of "physical problems" should be treated as inverse even if causality is not known or not considered [21, 27, 39, 52, 53].

However, if the dependent variable is measured with negligible error (relative to error in the measurement of the factorial variable), or is much smaller than that of the response variable, the direct prediction will involve bias, unless the two variables are perfectly correlated [53]. Therefore, regressions in which both variables are subjected to error have been also proposed [12] and are applied to insect temperature-dependent development to improve prediction precision [21, 27]:

$$DT = K + T_b D, \tag{4}$$

where D is development time (days) and T is temperature.

One of the major advantages of this equation, as in the case of the *x-intercept* method, is simplicity and the existence of biological interpretation over the estimated parameters: thermal constant and lower temperature threshold. Its added value, however, is increased precision in parameter estimation and the detection of outliers that reside on the non linear response curve and should be eliminated by the regression [44].

2.3. Nonlinear Regression Models.
Although in practice the linear models are quite adequate over a range of favourable temperatures, they proved unsecure in predicting development in extreme conditions and temperatures in which the relationship becomes non linear [21, 27, 48, 55, 57]. Hence, ideally one should know the response of the organism

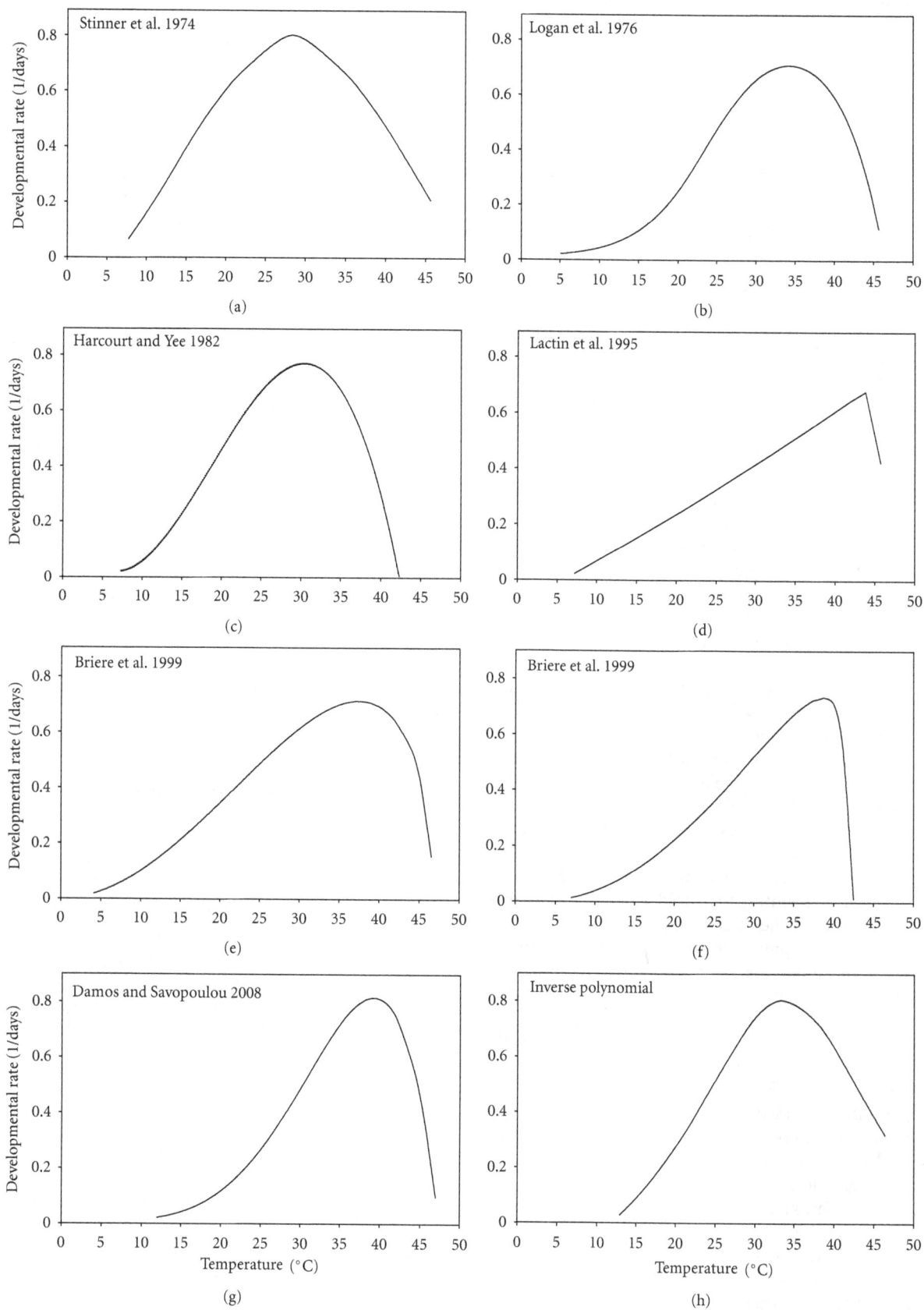

Figure 2: Typical relationships between temperature and insect developmental rates according to several representative non-linear models.

TABLE 1: Some representative regression models that have been created for the description of temperature-dependent development of insects and related arthropods.

Non-linear model	Equation	Description	Reference
$1/D = c/(1 + e^{(a+b \cdot T)})$, if $T \leq T_{opt}$ $1/D = c/(1 + e^{[(a+b \cdot (2 \cdot T_{opt} - T))]})$, if $T > T_{opt}$	(1)	"Stinner" (non-linear)	Stinner et al. 1974 [54]
$1/D = \psi \cdot [1/(1 + k \cdot e^{-\rho \cdot T}) \cdot e^{-((T_{max} - T)/\Delta)}]$	(2)	"Logan 10"	Logan et al. 1976 [55]
$1/D = a \cdot T^3 + b \cdot T^2 + c \cdot T + d$	(3)	"3rd-order polynomial" (non-linear)	Harcourt and Yee 1982 [56]
$1/D = e^{\rho \cdot T} - e^{(\rho \cdot T_{max} - (T_{max} - T/\Delta))} + \lambda$	(4)	"Lactin" (non-linear)	Lactin et al. 1995 [57]
$1/D = a \cdot T \cdot (T - T_{min}) \cdot (\sqrt{T_{max} - T})$	(5)	"Briere 1" (non-linear)	Briere et al. 1999 [29]
$1/D = a \cdot T \cdot (T - T_{min}) \cdot (\sqrt{T_{max} - T})^{(1/m)}$	(6)	"Briere 2" (non-linear)	Briere et al. 1999 [29]
$1/D = \rho \cdot (a - T/10) \cdot (T/10)^{\beta}$	(7)	"Simplified beta type" (non-linear)	Damos and Savopoulou-Soultani 2008 [27]
$1/D = a/(1 + bT + cT^2)$	(8)	"Inverse second-order polynomial 1"	This study

over the entire range of temperatures to compute accurately developmental rates over all temperature range.

Several non linear models have been proposed to describe developmental rate response curves over the full range of temperatures, aimed either to build general insect phenology models, or to be used as forecasting tools for pest management [4, 5, 20, 21, 27, 29, 31, 34, 45, 50, 57–60]. Although the procedure can be easily generated using several different softwares, one important limitation is that the optimization procedure is performed only for the dependent variable and assumes that the residual errors of the independent variable are negligible.

Table 1 presents some of the most common non-linear models that have been developed to describe insect development rates over the whole range of temperature. Figure 2 depicts typical temperature response curves according to some common non-linear equations that are presented in Table 1. The models have been abstracted by the respective references and are additionally generated for representative selected empirical data.

Typically, and according to all models, there is no growth below the lower temperature threshold, while developmental rate increases and reaches a maximum at optimal temperature and declines rapidly approaching zero at the higher temperature threshold that is often considered as lethal temperature.

2.4. Biophysical Models. Biophysical models predict the behavior of insect developmental rate in physical terms. Since "*temperature rate biophysical models*" are representations of temperature-depended development and based on the primitive rules of temperature dependence of reaction rates narrowed by biophysics, they are differentiated to all the other non-linear models.

The conformation of enzymes is the essential step in the enzymatic reaction and this conformation depends on temperature. Because poikilothermic development can be considered as a macroscopic revelation of enzyme reactions, in which temperature exerts a catalytic effect at a molecular level, these equations have been applied in modeling microorganism growth and in describing temperature-dependent development of arthropods.

Traditionally, such kinds of relations are based on the empirical equations of *Van't Hoff's law* [7], Arrhenius [46], and Eyring [50, 60–62]; and these relationships provided the principal foundation of later works.

Van't Hoff, based upon the experimental results of the botanist and pharmacist Pfeffer (who first measured osmotic pressure in 1877), concluded that the osmotic pressure π of a sugar solution in relation to its volume is constant and directly related to the absolute temperature T:

$$\pi = kT, \tag{5}$$

where k is a constant of analogy. Furthermore, by applying the ideal gas state equation to describe the osmotic pressure, as in the case of ideal gas, results in

$$\pi = RT\Sigma c_i, \tag{6}$$

where R is the universal gas constant, T is the absolute temperature, and c_i is the molar concentration of solute i. Interpretation of (5) and (6) simple states that the rate of chemical reactions increases between two- and threefold for each $10°C$ rise in temperature. This conclusion, according to *Van't Hoff's law*, that an increase in temperature will cause an increase in the rate of an endothermic reaction had a huge impact in chemistry, biochemistry, and physiology.

The Arrhenius equation relates the chemical reaction rate constant to temperature T (in Kelvins or degrees Rankin) and the activation energy of the reaction E_α as follows:

$$k = k_0 e^{-E_a/RT}, \tag{7}$$

where K_0 is the rate coefficient, E_a the activation energy, R the universal gas constant, and T absolute temperature. According to the Eyring function [61] any biochemical reaction rate (without prior enzyme activation) increases exponentially while in the equation parameterized by Schoolfield et al. [60] the reaction rate $r(T)$ is given as a modification of a reference reaction rate to a respective reference temperature:

$$r(T) = \rho \frac{T}{T_{ref}} e^{[H_\alpha/R \, (1/T_{ref} - 1/T)]}. \tag{8}$$

In (8), ρ is considered as *1/time* (reference rate) and H_α corresponds to the temperature sensitivity coefficient (or activation enthalpy in J/mol) and R is the universal gas constant ($8.314\,\mathrm{J\,K^{-1}\,mol^{-1}}$). The above equation can be applied to any intended temperature sensitive rates including developmental rates as well.

However, when dealing with biological rates, exponential increase is observable on a limited range and not throughout all temperature regimes. Sharp and DeMichele [63] considered activation process of the two extreme temperatures as independent and proposed a modification of the Arrhenius equation. This result to an equation having two components in the denominator, each for the description of the reversible inactivation of the rate-controlling enzyme considering both low and high temperatures and including "linearity" at middle temperatures:

$$r(T) = \left[\frac{T \cdot \exp\left[\left(\Phi - \Delta H_A^{\neq}/T \right)/R \right]}{1 + \exp[(\Delta S_L - \Delta H_L/T)/R] + \exp[(\Delta S_H - \Delta H_H/T)/R]} \right], \tag{9}$$

where $r(T)$ is the mean developmental rate at temperature T (1/time), T is the temperature in K, R is the universal gas constant ($1.987\,\mathrm{cal\,deg^{-1}\,mol^{-1}}$), while the other parameters are associated with the rate-controlling enzyme reaction: ΔH_A is the activation enthalpy of the enzyme reaction while ΔH_H is the change in enthalpy associated with high-temperature inactivation of the enzyme ($\mathrm{cal\,mol^{-1}}$), ΔS_L is the change in entropy associated with low-temperature inactivation ($\mathrm{cal\,deg^{-1}\,mol^{-1}}$), and Φ is a conversion factor having no thermodynamic meaning.

Figure 3 gives the biophysical model (9) for representative datasets as well as the respective Arrhenius plot. The biological interpretation of the above function has analogies to those of the Arrhenius function in which the dominator represents the fraction of rate-controlling enzyme that is in the active state. Derivation of the above mathematical function as well as the basic assumptions and modifications of the original formula are covered in details in [60, 63].

3. Statistics for Parameter Estimation and Model Comparison

3.1. Parameter Estimation.
Numerous procedures have been developed for parameter estimation and inference in regression analysis.

Campbell et al., 1974 [43, 64], provide statistics for the Standard error (SE) of the lower developmental threshold ($T_{\min}$) and the thermal constant K for the linear model based on "principal-manually" derived statistics:

$$\mathrm{SE}_{T_{\min}} = \frac{\bar{r}}{b} \sqrt{\frac{s^2}{N \cdot \bar{r}^2} + \left[\frac{\mathrm{SE}_b}{b} \right]^2}, \tag{10}$$

where s^2 is the residual mean square of r, $\bar{r}$ is the sample mean, and N is the sample. Additionally, the size of the SE_K for the thermal constant K for the linear model having slope b is, respectively [64],

$$\mathrm{SE}_K = \frac{\mathrm{SE}_b}{b^2}. \tag{11}$$

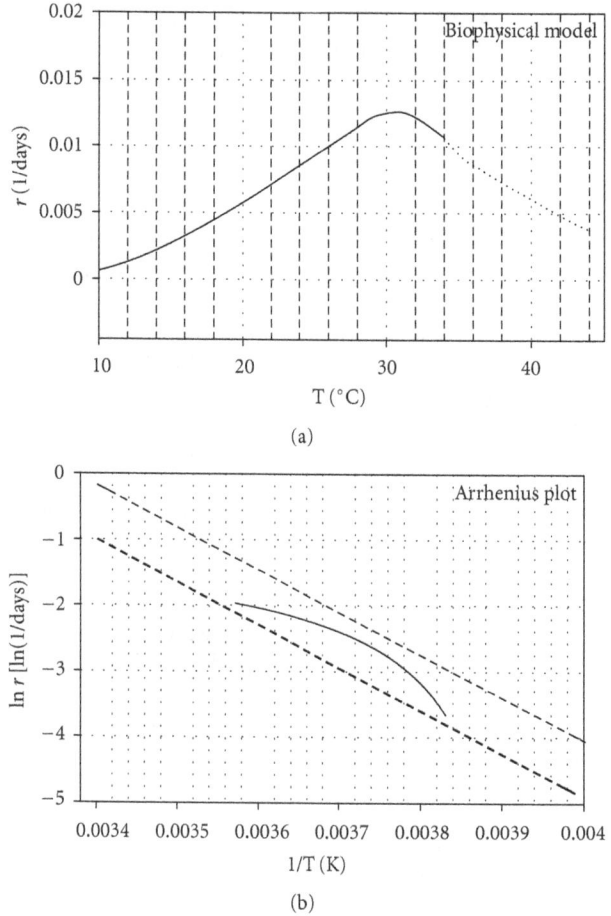

FIGURE 3: Curve shape of the biophysical model of sharp and DeMichelle [63] as modified by Schoolfield et al. [20] (a) and the respective Arrhenius plot (b).

However, several other procedures are also proposed for parameter estimation and relative statistics. The most common are the maximum likelihood (ML) and the ordinary least square (OLS) estimation, and they are used for both linear and non linear models [65].

Point and interval estimation using ML relies on distributional assumptions (here a specific probability function for error dispersion must be specified), in contrast to OLS point estimates, which generally do not require hidebound distributional assumptions, are unbiased, and have minimum variance.

The OLS minimise the sum of square residuals of the regression function of interest. Additionally, most statistical packages of parameter estimation are based on the Levenberg-Marquardt algorithm (LMA) which provides a numerical-iterative solution of curve fitting over a space of parameters of the function.

The Marquardt algorithm [66] is a least squares method based on successive iterations for parameter optimization. Thus, if (x_i, y_i) is the given set of n empirical observation pairs of the independent (temperature) and dependent (developmental times) variables, the algorithm optimizes the parameters p of the model curve $f(x, p)$ so that the sum of the squares of the deviations is minimum:

$$g(p) = \sum_{i=1}^{n} [y_i - f(x_i, p)]^2 \qquad (12)$$

The method is that the analyst has to provide an initial starting guess for final parameter estimation. This is an important constrain of the method and especially in curves with multiple minima the initial guess must already to be closed to the final solution. Furthermore, problems can arise in the case of observational data (i.e., time series) in which covariates can exist between observed and response variables.

The methods described above for calculating standard error and confidence intervals for a parameter relay on the assumption that the statistic of interest is assumed to be normal distributed. Thus, there is no need whatsoever for bootstrapping in regression analysis if the OLS assumptions are met. However, in the case of estimating population values in the absence of any information (i.e., variables in which sampling distributions and variances are unknown due to limited data), or in the case in which the variable is the final result of several observations (as in the case of life table statistics), parameter estimation and standard errors can be based on resampling methods such as the Bootstrap and/or the Jackknife method, or even based on Bayesian inference estimation.

For more details on resampling the reader should consider the references cited [67, 68].

3.2. Model Comparison. Since several regression models are available it is convenient to provide criteria or goodness of fit tests for model comparison. For instance, a common question that applied entomologists are facing is how to compare two different models for a given species and/or how to compare two different species with a given model.

Generally, several criteria have been proposed to evaluate model performance including the root mean square error (RMSE), the Pearson x^2, the deviance (G^2) statistics, regular and adjusted to the parameter numbers regression coefficients, and information criteria such as the *Akaikes* and *Bayes-Schwarz* information criteria [21, 27, 39, 69].

The idea behind most of these criteria is to measure the "range" of which the predicted values of a given model match the observed and can be applied in evaluating prediction capability for a particular dataset (i.e., one species-several models). Some of them are described in brief.

The Pearson x^2 statistic is based on observed (O) and expected fitted or predicted (e) observations and has similarities to the Root Mean Square Error [27, 65]:

$$x^2 = \sum_{i=1}^{n} \frac{(o - e)^2}{e} = \sum_{i=1}^{n} \frac{(y_i - n\hat{\pi}_i)^2}{n\hat{\pi}_i(1 - \hat{\pi}_i)} \qquad (13)$$

Where y_i is the observed value of Y, $\hat{\pi}_i$ is the predicted or fitted value of x_i and n is the number of observations. Additionally, based on the same concept a "prediction capability" index d can be addressed to be used to compare candidate models and rank them according to the degree to which the predictions are error-free:

$$d = 1 - \frac{\left[\sum (P_i - O_i)^2 \right]}{\sum \left[\left(\left| P_i - \overline{O_i} \right| \right) + \left(\left| O_i - \overline{O_i} \right| \right) \right]^2}, \qquad (14)$$

where $\overline{O_i}$ is the average of the observed values [27, 70].

For a comparison of only two models, an efficacy ratio can be calculated as follows [27, 70]:

$$E_{1,2} = \frac{\text{MSE}_1}{\text{MSE}_2}. \qquad (15)$$

Where the respective to the models efficacy ratio E is based on the mean square errors (MSE) and can be used as evaluation index [70]. Values close to 1 indicating very low differences between the selected models in predicting a particular dataset [21, 27].

Considering that there are cases in which different datasets (i.e., two different species) are described with a particular model and cases in which there is model selection among equations that differ on the number of parameters, model performance comparisons can be made according to the adjusted coefficient of determination (Adj $\cdot r^2$) and on the Akaike's information criteria [71].

The Adj $\cdot r^2$ is a modification of r^2 that adjust for the number of explanatory terms in a model. Unlike r^2, Adj $\cdot r^2$ increases only if an additional new term improves the model more than would be expected by change [21, 39]. The Adj $\cdot r^2$ is defined as

$$\text{Adj} \cdot r^2 = 1 - \frac{\text{RSS}/(n - (\vartheta + 1))}{\text{SS}/(n - 1)}. \qquad (16)$$

Akaike's information criterion (AIC) developed and proposed by Akaike in 1974 [39] is

$$\text{AIC} = n \cdot [\ln(\text{RSS})] - [n - 2 \cdot (\theta + 1)] - n \cdot \ln(n) \qquad (17)$$

and the Bayesian-Schwartz information criterion (BIC or SIC) was proposed on 1978 and is [39]

$$\text{BIC} = n \cdot [\ln(\text{RSS})] + (\theta + 1) \cdot \ln(n) - n \cdot \ln(n), \qquad (18)$$

where RSS is the residual sum of squares and SS total sum of squares, θ number of parameters and n observation number. These criteria permit to infer on how the different number of parameters add to the explanatory power of the candidate model.

4. Physiological Time and Heat Unit's Accumulation Systems

Considering the above models in defining cardinal temperatures of development in the laboratory, as well as the respective for each stage and species thermal constants, the interest is to apply this knowledge in order to make field predictions of temperature effects on insect phenology in time and space, according to the physiological time and related heat accumulation systems [50, 72–75].

Often referred to also as thermal time, the progress of the development of an organism is viewed as a biological clock that measures time units. Thus, although physiological time accelerates or slows according to prevailing temperatures, the time units to complete a particular developmental event in field should be the same as defined in the laboratory and equals the species specific thermal constant.

Thus, since the *law of effective temperatures* suggest that the completion of a given stage in development requires an accumulation of a definite amount of heat energy, similar approaches can be followed in which effective accumulated temperatures are estimated by the respective heat energy in field during the growth season.

According to this approach the amount of age or development accumulated from time 0 to t, and for discrete time intervals is

$$\Delta \alpha = \sum f\,[T(t)]\Delta t, \quad [T(t)] > 0. \quad (19)$$

According to this function the species integrate temperature effects according to some function, f, peculiar to their species. This function, $f[T(t)]$, can be either linear or non-linear. If $f[T(t)]$ is assumed to be linear, then the developmental rate is proportional to temperatures above threshold (as defined according to the *x-intercept* method and apart from the linearity check of the rate-temperature curve), on the other hand, several non linear relations exist such as the logistic curve. However, in order to be effective, heat summation takes into account only the active temperatures within the species-specific range of development [24, 51].

Several methods have been proposed in calculating degree days accumulated in field, as well as related software. However, for the sake of brevity, in this review, the following three widely applied methods the average method, the modified average method, and the modified sine wave method, are briefly discussed.

4.1. Average Method. According to the average method developed by Baskerville and Emin [14], which is the simplest one, the number of daily degree-days is calculated by subtracting the base temperature from the average daily temperature as follows:

$$DD = \left[\frac{\min T + \max T}{2}\right] - T_{\min}. \quad (20)$$

Among the disadvantages of the above approach is that it does not take into account those daily minimum temperatures that can fall below the species lower temperature thresholds. This situation is very common in spring and results in bias and underestimation of degree-days accumulated by the insect since not all hourly temperatures during a day are above the threshold level. Thus, during this short period, development proceeds but is not taken into account by the proposed heat accumulation system.

4.2. Modified Average Method. In order to avoid the above-mentioned disadvantage it is convenient to modify the first component of (20) by substituting minimum temperature with lower temperature threshold, thereby approximating closer reality by calculating the daily temperature accumulation that corresponds to the interval between maximum temperature and that which is higher than the lower threshold of the species, or

$$DD = \left[\frac{T_{\min} + \max T}{2}\right] - T_{\min}. \quad (21)$$

This approach will result in a higher number of degree-days by taking into account development during the short periods in which temperature is slightly above the lower developmental threshold.

4.3. Modified Sine Wave Method. In principle mathematical relationships for this technique were given by Baskerville and Emin [14], Allen, and Watanabe [2]. Arnold [24, 51, 76] showed that the area under the temperature curve, the amplitude of which has been adjusted to the daily maximum and minimum temperatures for a given day, can be approximated according to sine curve.

Thus, according to the modified sine wave method, proposed by Allen [51], a trigonometric sine function is being used to describe this kind of daily temperature fluctuations. Based on the same principle as previously stated, heat accumulations during a day correspond to the area above the species lower temperature threshold. It is also noteworthy to state that this method leads to similar results as the modified average method in the case where minimum temperature is higher than the base temperature.

All these methods that are briefly described are based on the principle that the specimen is accumulating climate temperatures that are limited within its thresholds. Heat units are expressed as accumulated degree-days that correspond to a 24-hours daily interval that is limited between minimum and maximum temperature range and the predetermined species-specific thresholds.

5. Discussion

Among the scopes of this article was the description of representative models that have been proposed to model insect temperature dependent development either in the laboratory or field. However, a tremendous amount of prior work has been done in the field of insect temperature modelling since the first defined principles and the reader should consider the work of Ludwig [18], Uvarov [49], Powsner [19], Wigglesworth [26], Laudien [25] and Wagner et al. 1984 [20] for additional information.

Nevertheless, among the purposes of this review was to popularise prior studies. Several statistical criteria for model comparison are also gathered in order to integrate and familiarise most current approaches and tools for modelling the effect of temperature on insect development. This is an essential step to be made in order to draw inference upon the species ecology, spatiotemporal arrangement, and abundance.

According to selected linear and non linear models, that are presented in brief, developmental responses can be summarized in terms of the three critical, or cardinal, temperatures of development. In addition, since calculation of physiological time by temperature-driven field models is related to the area summated by the chosen heat-accumulation system, the definition of these temperatures is a prerequisite for accurate phenology prediction. Thus, apart from the ecological concerns, the importance of finding a mathematical/statistical model which describes and then simulates the phenology of individuals under field conditions is a prior constraint for further successful timing of pest management practices in field.

Depending on their parameters, the presented models can be judged more or less complex and several algorithms for least squares estimation have been proposed for nonlinear parameters [66, 77]. By incorporating several more factors-parameters on the equations, the authors search to gain higher accuracy on data description. However, complexity does not assure more accuracy in all cases. Prior comparative approaches should be followed to choose among most appropriate models that are available. To put forward, since most model shapes are quite similar, comparative differences of model performances can be only indicated by detailed statistical measures [39].

Hence, not all models display the same fit behaviour when carefully observed while very few provide a detailed biological interpretation of the estimated parameters. For instance, the advantage of the models proposed by Logan and Lactin over the other equations is due to the fact that they incorporate parameters that have direct biological interpretation and this is a major asset. In addition, the models proposed by Sharpe and DeMichele [63] and Schoolfield et al. [60], based on enzyme kinetic reactions, display a radical departure from those based on empirical fits to data. Nevertheless, it is common that temperature affects not only the rate of chemical reactions, but also induces conformational changes in biological systems [49].

Moreover, one disadvantage of complexity in models is that it strongly influences parameters estimation [39]. For example, although most of the polynomial models do not have any biological interpretation, probably the most important advantage they have is that parameter estimation can be easily done [56].

One other characteristic, among the presented models, is that not all of them are able to make predictions that are matched over, the experimentally derived, observed values. Unfortunately, there are instances in which optimum and upper threshold temperature predictions are quite overestimated when compared to real data [21, 27]. For instance the lower temperature threshold for *G. molesta*, as estimated in the current laboratory trial, slightly deviates from that estimated by prior field studies [47]. Nevertheless, differences in respect to insect stage can also exist so it is important to model all development of *G. molesta* for safer interpretations. Thus, a good fit for a respective model has no utility if it predicts temperature thresholds that have no biological meaning. Such false predictions can result in bias on the estimation of cardinal temperatures. In most cases overestimation of optimum and maximum temperature thresholds is the result of skewed curve, although coefficients of determination are quite high but can be the result of a good data fit on the intermediate temperature range. In other words, a good fit is not always a guarantee for biologically significant model performance and a reliable and accurate data description over all temperature range [21, 27, 44].

On the other hand, not all models can predict lower temperature thresholds, since there is no intersection with the temperature axis, when rate of development is zero [27], while in some cases cardinal temperatures are derived graphically and not numerically. In addition, the assumption of a base temperature close to 0°C, in the cases in which the curve approximates origin may seem unreasonable, considering that it is well accepted that lower temperature thresholds for most arthropods are well above 0°C, usually around 6–10°C, or higher. This is also displayed for the dataset used to model *G. molesta* in the current study. Thus, the most currently used non-linear temperature models describe only part of the whole picture of insect temperature-dependent development. The equation of Logan et al. [55], as modified by Lactin et al. [57], due to the constant factor that intersects with the temperature axis, as well as the equation proposed by Hilbert and Logan [16], proposes a lower threshold as well, although proved rigid in describing particular datasets [27].

The above reasons, as well as the species and stage-specific plasticity on temperature responses, give important reasons that should be taken into account to choose among several available formulas. These trends have been pointed out by several researchers and are probably the major cause that resulted to the development of plethora of non-linear models in the literature [27, 31, 55–57, 78, 79].

Another important constraint is that most of these models are directly related to temperature and do not take into account other climatic variables. For insects in particular, temperature is probably the most critical abiotic factor that influences their developmental rates and their life cycles, although other factors such as photoperiod, humidity, and nutrition should not be excluded, as well as crowding or density and competition [13, 40, 44].

Furthermore, in most cases it is virtually impossible to measure the temperature that an insect experiences in its original microenvironment. For example, most plant-feeding insects display a species-specific behavior in relation to their host (i.e., crawling inside of shoots or barks at the larval stage) while others exert some control over their body temperature through their behavior (i.e., they rest at shadowed and cool places when temperature is high) [21, 40].

Considering that the existence of alternating temperatures is more probable in reality [80], there are cases in which models displayed considerable inaccuracy in predicting insect development and phenology under field conditions [21, 39, 42, 44, 58, 81, 82, 82, 83].

Hence there is no perfect model, but we rely on the available ones that best describe our datasets, under certain conditions, and even though most models are oversimplifications, they are acceptable for empirical predictions in some defined ranges and instances.

Thus, if the model is proved reliable after seedily experimental evaluation, heat accumulations of a phenological event that occurs in field should reflect that which have been estimated by the model and thereby provide means of accurate timing of pesticides and initiation of pest management tactics. Therefore, it is not risky to claim that temperature has a prominent role in insect biology and by understanding the temperature effects on insect development we are able to describe and predict the distribution and abundance of insect species in any locality [83–85].

From an ecological standpoint, insect vital thermal requirements, as described in this article (i.e., thermal constant and temperature thresholds) provide ecologically and practically useful information [34, 66, 86]. For instance, as the thermal constant differs among genera, species or even stages, their study reveals various aspects of temperature adaptation and in particular the adaptation of each to its environment. On the other hand, species specific thermal requirements can also be used as indicators of the distribution and abundance of insect populations [32].

The effect of a climatic factor, such as temperature for instance, sets the tolerance limits for a species, and this has been acknowledged by earlier studies (i.e., Shelfold, 1913: The Law of Tolerance). Later studies [13, 87, 87, 88, 88] discuss how the species-specific "environmental boundaries" are determined by the ultimate tolerance factor (i.e., temperature) which may further restrict geographic distribution [8, 37, 41, 89].

Moreover, is it though for species whose geographical distributions ultimately are determined by temperature, global warming should result in spatial range shift [33]. Thus, the speculations on the effects of climatic change on the spatial dynamics of insect species have been quite general and populations are expected to extend their ranges to higher latitudes and elevations [37, 38, 90–94].

However, contrasting results concerning future projecting of species distribution have been also reported [90, 95], and one cannot exclude a progressive temperature selection of individuals that are adapted to the new temperature environment and especially for species with high reproductive potential [96–98] and host alternatives. Furthermore, the rate of temperature change affects species acclimation potential which further results in different conclusions regarding the responses of the species to acclimation [38, 99] and that thermal tolerances of many organisms to be proportional to the magnitude of temperature variation they experience.

Since genetic variation and potential response to selection should be positively correlated with population size, species with restricted ranges, or smaller populations, are predicted to have reduced capacity to adapt to environmental change [96, 97, 100]. On the other hand, it is more likely that temperature alteration can affect the reproductive potential of a species (i.e., abundance) and its life cycle, since additional generations or/and outbreaks are possible during the growth season [101] when not limited by photoperiod [48].

For a particular species, there is an inverse relation between the thermal constant and the lower developmental threshold and it is suggested that this trade modifies the fitness of the species and finally influences the outcome of competition between related species and their distributions [85, 88, 102–104]. Moreover, tropical species and warm-adapted species tend to have higher values on their lower temperature thresholds when compared to cold-adapted species that had greater DD requirements and much lower temperature ranges [85, 88, 102, 104].

Based on such linear relationships, between thermal constants and lower temperature thresholds, for several cold-blood species, it is suggested that there is an inverse relationship between lower temperature thresholds and the thermal constant associated with latitude and/or habitat that adapts each species to its thermal environment [85, 103]. Thermal constant and respective DD requirements are also based on the particular morphology and size of the species. For example, size at maturity is a function of the rate and duration of growth, and large size at maturity implies a long generation time and a correspondingly large DD requirements [17, 102, 105].

Hence, insect thermal requirements have a strong physiological and ecological interpretation since they modify species-specific ecological strategy which is adapted to a particular thermal environment [26, 49, 74, 84, 104, 106].

Thus, any model which provides biologically important parameters is useful in modeling population dynamics under several temperature regime alterations. In addition, by incorporating more factors in the equations, climate-driven models have the potential to describe the general ecological behaviour, abundance, distribution, and outbreaks of insects on a regional or even global scale, with important practical applications.

Finally, future research must be carried out in the direction of insect thermal adaptation in order to assess the species reproduction potential and related evolutionary properties as they respond to short- and long-term temperature alterations. The development of more sophisticated models, such as demographic system models and ecological niche models, that incorporate species-specific vital thermal requirements as well, is also an urgent necessity to improve and complete all current models. Thus models that are based on weather and other factors can more realistically estimate the spatiotemporal population evolution and invasive potential of native and nonindigenous species in new areas.

Acknowledgment

Part of this work was supported by IKY (Greek Scholarship Foundation) through a postdoctorate scholarship awarded to P. Damos.

References

[1] R. C. Akers and D. G. Nielsen, "Predicting *Agrilus anxius* adult emergence by heat unit accumulation," *Journal of Economic Entomology*, vol. 77, pp. 1459–1463, 1984.

[2] J. C. Allen, "A modified sine wave method for calculating degree-days," *Environmental Entomology*, vol. 5, pp. 388–396, 1976.

[3] P. G. Allsopp and D. G. Butler, "Estimating day-degrees from daily maximum-minimum temperatures: a comparison of techniques for a soil-dwelling insect," *Agricultural and Forest Meteorology*, vol. 41, no. 1-2, pp. 165–172, 1987.

[4] S. Analytis, "Über die relation zwischen biologischer entwicklung und temperatur bei phytopathogenen Pilzen," *Phytopathologische Zeitschrift*, vol. 90, pp. 64–76, 1977.

[5] S. Analytis, "Study on relationships between temperature and development times in phytopathogenic fungus: a mathematical model," *Agricultural Research*, vol. 3, pp. 5–30, 1979.

[6] A. P. Candolle, *Geographique botanique*, Raisonee, Paris, France, 1855.

[7] J. H. Van't Hoff, "Osmotic pressure and chemical equilibrium Nobel Lecture," December 1901.

[8] M. T. Vera, R. Rodriguez, D. F. Segura, J. L. Cladera, and R. W. Sutherst, "Potential geographical distribution of the Mediterranean fruit fly, Ceratitis capitata (Diptera: Tephritidae), with emphasis on Argentina and Australia," *Environmental Entomology*, vol. 31, no. 6, pp. 1009–1022, 2002.

[9] Y. Weikai and L. A. Hunt, "An equation for modelling the temperature response of plants using only the cardinal temperatures," *Annals of Botany*, vol. 84, no. 5, pp. 607–614, 1999.

[10] S. Yang, J. Logan, and D. L. Coffey, "Mathematical formulae for calculating the base temperature for growing degree days," *Agricultural and Forest Meteorology*, vol. 74, no. 1-2, pp. 61–74, 1995.

[11] A. Bonnet, "Euvres d'histoire et de philosophie. i. Traité d'insectologie. Neuchâtel," 1779.

[12] T. Ikemoto, "Intrinsic optimum temperature for development of insects and mites," *Environmental Entomology*, vol. 34, no. 6, pp. 1377–1387, 2005.

[13] J. H. Brown, "On the relationship beween abundance and distribution of species," *American Naturalist*, vol. 124, no. 2, pp. 255–279, 1984.

[14] G. L. Baskerville and P. Emin, "Rapid estimation of heat accumulation from maximum and minimum temperatures," *Ecology*, vol. 50, pp. 514–517, 1969.

[15] J. Davidson, "On the relationship between temperature and the rate of development of insects at constant temperatures," *Journal of Animal Ecology*, vol. 13, pp. 26–38, 1944.

[16] D. W. Hilbertand and J. A. Logan, "Empirical model of nymphal development for the migratory grasshopper, Melanoplus saguinipes (Orthoptera: Acrididae)," *Environmental Entomology*, vol. 12, pp. 1–5, 1983.

[17] E. Janisch, "The influence of temperature on the life history of insects," *Transactions of the Entomological Society of London*, vol. 80, pp. 137–168, 1932.

[18] D. Ludwig, "The effects of temperature on the development of an insect (Popilia japonica New-man)," *Physiological Zoology*, vol. 1, pp. 358–389, 1928.

[19] L. Powsner, "The effects of temperature on the durations of the developmental stages of *Drosophila melanogaster*," *Physiological Zoology*, vol. 8, pp. 474–450, 1935.

[20] T. L. Wagner, H. I. Wu, P. J. H. Sharpe, R. M. Schoolfield, and R. N. Coulson, "Modeling insect development rates: a literature review and application of a biophysical model," *Annals of the Entomological Society of America*, vol. 77, pp. 208–225, 1984.

[21] P. Damos, *Bioecology of microlepidopterous pests of peach trees and their management according to the principles of integrated fruit production*, Ph.D. thesis, Aristotle University of Thessaloniki, Greece, 2009.

[22] L. G. Higley, L. P. Pedigo, and K. R. Ostlie, "DEGDAY: a program for calculating degree-days, and assumptions behind the degree-day approach," *Environmental Entomology*, vol. 15, pp. 999–1016, 1986.

[23] A. Arbab, D. C. Kontodimas, and M. R. Mcneill, "Modeling embryo development of Sitona discoideus Gyllenhal (Coleoptera: Curculionidae) under constant temperature," *Environmental Entomology*, vol. 37, no. 6, pp. 1381–1388, 2008.

[24] C. Y. Arnold, "The determination and significance of the base temperature in a linear heat unit system," *Proceedings of the American Society of Horticultural Science*, vol. 74, pp. 430–445, 1959.

[25] H. Laudien, "Changing reaction systems," in *Temperature and Life*, J. C. Precht, H. Hensel, and W. Larcher, Eds., pp. 355–399, Springer, New York, NY, USA, 1973.

[26] V. B. Wigglesworth, *The principles of insect physiology*, Chapman and Hall, London, UK, 1972.

[27] P. T. Damos and M. Savopoulou-Soultani, "Temperature-dependent bionomics and modeling of Anarsia lineatella (Lepidoptera: Gelechiidae) in the laboratory," *Journal of Economic Entomology*, vol. 101, no. 5, pp. 1557–1567, 2008.

[28] M. Bieri, J. Baumgärtner, G. Bianchi, V. Delucchi, and R. von Arx, "Development and fecundity of pea aphid (Acyrtosiphon pisum Harris) as affected by constant temperatures and pea varietes," *Mitteilungen der Schweizerischen Entomologischen Gesellschaft*, vol. 56, pp. 163–171, 1983.

[29] J. F. Briere, P. Pracros, A. Y. Le Roux, and J. S. Pierre, "A novel rate model of temperature-dependent development for arthropods," *Environmental Entomology*, vol. 28, no. 1, pp. 22–29, 1999.

[30] M. E. Cammell and J. D. Knight, "Effects of climatic change on the population dynamics of crop pests," *Advances in Ecological Research*, vol. 22, pp. 117–162, 1992.

[31] A. Campbell, B. D. Frazer, N. Gilbert, A. P. Gutierrez, and M. Mackauer, "Temperature requirements of someaphids and their parasites," *Journal of Applied Ecology*, vol. 11, 1974.

[32] P. S. Messenger and N. E. Flitters, "Effect of constant temperature environments on the egg stage of three species of Hawaiian fruit flíes," *Annals of the Entomological Society of America*, vol. 51, pp. 109–119, 1958.

[33] S. A. Estay, M. Lima, and F. A. Labra, "Predicting insect pest status under climate change scenarios: combining experimental data and population dynamics modelling," *Journal of Applied Entomology*, vol. 133, no. 7, pp. 491–499, 2009.

[34] H. Feng, F. Goult, Y. Huang, Y. Jiang, and K. Wu, "Modeling the population dynamics of cotton bollworm Helicoverpa armigera (Hübner) (Lepidoptera: Noctuidae) over a wide area in northern China," *Ecological Modeling*, vol. 221, pp. 1819–1830, 2010.

[35] K. P. Pruess, "Day-degree methods for pest management," *Environmental Entomology*, vol. 12, pp. 613–619, 1983.

[36] M. A. Oshaghi, N. M. Ravasan, E. Javadian et al., "Application of predictive degree day model for field development of sandfly vectors of visceral leishmaniasis in northwest of Iran," *Journal of Vector Borne Diseases*, vol. 46, no. 4, pp. 247–254, 2009.

[37] A. P. Gutierrez and L. Ponti, "Assessing the invasive potential of the Mediterraneanfruit fly in California and Italy," *Biological Invasions*. In press.

[38] J. H. Porter, M. L. Parry, and T. R. Carter, "The potential effects of climatic change on agricultural insect pests," *Agricultural and Forest Meteorology*, vol. 57, no. 1–3, pp. 221–240, 1991.

[39] P. T. Damos and M. Savopoulou-Soultani, "Development and statistical evaluation of models in forecasting moth phenology of major lepidopterous peach pest complex for

Integrated Pest Management programs," *Crop Protection*, vol. 29, no. 10, pp. 1190–1199, 2010.

[40] P. Damos and M. Savopoulou-Soultani, "Microlepidoptera of economic significance in fruit production: challenges, constrains and future perspectives for inegrated pest management," in *Moths: Types, Ecological Significance and Control Methods*, Nova Science, 2011.

[41] M. B. Davis and C. Zabinski, "Changes in geographical range resulting from greenhouse warming: effects on biodiversity in forests," in *Global Warming and Biological Diversity*, R. L. Peters and T. E. Lovejoy, Eds., pp. 297–308, Yale University Press, New Haven, Conn, USA, 1992.

[42] P. Damos and M. Savopoulou-Soultani, "Population dynamics of *Anarsia lineatella* in relation to crop damage and the development of economic injury levels," *Journal of Applied Entomology*, vol. 134, no. 2, pp. 105–115, 2010.

[43] A. Campbell and M. Mackauer, "Thermal constants for development of the pea aphid (Homoptera: Aphidiidae) and some of ts parasites," *Canadian Entomologist*, vol. 107, pp. 419–423, 1975.

[44] P. Damos, A. Rigas, and M. Savopoulou-Soultani, "Application of Markov chains and Brownian motion models in insect ecology," in *Brownian Motion: Theory, Modelling and Applications*, R. C. Earnshaw and E. M. Riley, Eds., Nova Science Publishers, 2011.

[45] S. Hartley and P. J. Lester, "Temperature-dependent development of the Argentine ant, Linepithema humile (Mayr) (Hymenoptera: Formicidae): a degree-day model with implications for range limits in New Zealand," *New Zeeland Entomologist*, vol. 26, pp. 91–100, 2003.

[46] S. Arrhenius, "Über die Reactionsgeschwindigkeit bei der Inversion von Rohrzucker durch Sauren," *Zeitschrift für Physikalische Chemie*, vol. 4, pp. 226–692, 1889.

[47] B. A. Croft, M. F. Michels, and R. E. Rice, "Validation of a PETE timing model for the oriental fruit moth in Michigan and central California (Lepidoptera: Olethreutidae)," *Great Lakes Entomologist*, vol. 13, pp. 211–217, 1980.

[48] P. T. Damos and M. Savopoulou-Soultani, "Synchronized diapause termination of the peach twig borer *Anarsia lineatella* (Lepidoptera: Gelechiidae): Brownian motion with drift?" *Physiological Entomology*, vol. 35, no. 1, pp. 64–75, 2010.

[49] B. P. Uvarov, "Insect and climate," *Transactions of the Royal Entomological Society, London*, vol. 79, pp. 1–247, 1933.

[50] T. L. Wagner, R. L. Olson, and J. L. Willers, "Modeling arthropod development time," *Journal of Agricultural Entomology*, vol. 8, pp. 251–270, 1991.

[51] C. Y. Arnold, "Maximum-minimum temperatures as a basic for computing heat units," *Proceedings of the American Society for Horticultural Science*, vol. 76, pp. 682–692, 1960.

[52] P. Dagnélie, *Théorie et méthodes statistiques: applications agronomiques*, Les Presses Agronomiques de Gembloux, Belgique, 1973.

[53] J. L. Gill, "Biases in regression when prediction is inverse to causation," *Journal of Animal Science*, vol. 67, pp. 594–600, 1987.

[54] R. E. Stinner, A. P. Gutierez, and G. D. Butler Jr., "An algorithm for temperature-dependent growth rate simulation," *Canadian Entomologist*, vol. 106, pp. 519–524, 1974.

[55] J. A. Logan, D. J. Wollkind, S. C. Hoyt, and L. K. Tanigoshi, "An analytical model for description of temperature dependent rate phenomena in arthropods," *Environmental Entomology*, vol. 5, pp. 1133–1140, 1976.

[56] D. C. Harcourt and J. M. Yee, "Polynomial algorithm for predicting the duration of insect life stages," *Environmental Entomology*, vol. 11, pp. 581–584, 1982.

[57] D. J. Lactin, N. J. Holliday, D. L. Johnson, and R. Craigen, "Improved rate model of temperature-dependent development by arthropods," *Environmental Entomology*, vol. 24, no. 1, pp. 68–75, 1995.

[58] W. C. Cool, "Some effects of alternating temperatures on the growth and metabolism of cutworm larvae," *Journal of Economic Entomology*, vol. 20, pp. 769–782, 1927.

[59] D. J. Lactin and D. L. Johnson, "Temperature-dependent feeding rates of *Melanoplus sanguinipes* nymphs (Orthoptera: Acrididae) in laboratory trials," *Environmental Entomology*, vol. 24, no. 5, pp. 1291–1296, 1995.

[60] R. M. Schoolfield, P. J. H. Sharpe, and C. E. Magnuson, "Nonlinear regression of biological temperature-dependent rate models based on absolute reaction-rate theory," *Journal of Theoretical Biology*, vol. 88, no. 4, pp. 719–731, 1981.

[61] T. M. Van der Have, *Slaves to the Eyring equation?: Temperature dependence of life-history characters in developing ectotherms*, Ph.D. thesis, Department of Environmental Sciences, Resource Ecology Group, Wageningen University, The Netherlands, 2008.

[62] X. Yin, M. J. Kropff, G. McLaren, and R. M. Visperas, "A nonlinear model for crop development as a function of temperature," *Agricultural and Forest Meteorology*, vol. 77, no. 1-2, pp. 1–16, 1995.

[63] P. J. H. Sharpe and D. W. DeMichele, "Reaction kinetics of poikilotherm development," *Journal of Theoretical Biology*, vol. 64, no. 4, pp. 649–670, 1977.

[64] A. Campbell, B. D. Frazer, N. Gilbert, A. P. Guutierrez, and M. Mackauer, "Temperature requirements of some aphids and their parasites," *Journal of Applied Ecology*, vol. 11, pp. 431–438, 1974.

[65] V. Barnett, *Comperative Statistical Inference*, Willey, New-York, NY, USA, 3rd edition, 1999.

[66] D. V. Marquardt, "An algorithm for least squares estimation of nonlinear parameters," *Journal of the Society for Industrial and Applied Mathematics*, vol. 11, pp. 431–441, 1963.

[67] P. A. Eliopoulos, "Life tables of Venturia canescens (Gravenhorst) (Hymenoptera: Ichneumonidae) parasitizing Ephestia kuehniella Zeller (Lepidoptera : Pyralidae)," *Journal of Economic Entomology*, vol. 99, pp. 237–243, 2006.

[68] B. Effron, *The Jack-Knife, The Bootsrap and Other Resampling Methods*, CBMS-NSF Monograph 38, Society of Industrial and Applied Mathematics, Philadelphia, Pa, USA, 1982.

[69] H. Ranjbar Aghdam, Y. Fathipour, and D. C. Kontodimas, "Evaluation of non-linear models to describe development and fertility of codling moth (Lepidoptera: Tortricidae) at constant temperatures," *Entomologia Hellenica*, vol. 20, pp. 3–16, 2011.

[70] C. J. Willmot, "Some coments on the evaluation of model performance," *Bulletin of American Meterorological Society*, vol. 63, pp. 1309–1313, 1982.

[71] B. Zahiri, Y. Fathipour, M. Khanjani, S. Moharramipour, and M. P. Zalucki, "Preimaginal development response to constant temperatures in hypera postica (Coleoptera: Curculionidae) : picking the best model," *Environmental Entomology*, vol. 39, no. 1, pp. 177–189, 2010.

[72] V. I. Pajunen, "The use of physiological time in the analysis of insect stage-frequency data," *Oikos*, vol. 40, no. 2, pp. 161–165, 1983.

[73] P. J. H. Sharpe, G. L. Curry, D. W. DeMichele, and C. L. Cole, "Distribution model of organism development times," *Journal of Theoretical Biology*, vol. 66, no. 1, pp. 21–38, 1977.

[74] F. Taylor, "Ecology and evolution of physiological time in insects," *The American Naturalist*, vol. 117, pp. 1–23, 1979.

[75] F. Taylor, "Sensitivity of physiological time in arthropods to variation of its parameters," *Environmental Entomology*, vol. 11, pp. 573–577, 1982.

[76] C. Y. Arnold, "Maximum-minimum temperatures as a basis for computing heat units," *Proceedings of the American Society of Horticultural Science*, vol. 74, pp. 682–692, 1960.

[77] J. Aldrich, "Doing least squares: perspectives from Gauss and Yule," *International Statistical Review*, vol. 66, no. 1, pp. 61–81, 1998.

[78] D. C. Kontodimas, P. A. Eliopoulos, G. J. Stathas, and L. P. Economou, "Comparative temperature-dependent development of *Nephus includens* (Kirsch) and *Nephus bisignatus* (Boheman) (Coleoptera: Coccinellidae) preying on *Planococcus citri* (Risso) (Homoptera: Pseudococcidae): evaluation of a linear and various nonlinear models using specific criteria," *Environmental Entomology*, vol. 33, no. 1, pp. 1–11, 2004.

[79] J. Y. Wang, "A critique of the heat unit approach to plant response studies," *Ecology*, vol. 41, no. 4, pp. 785–790, 1960.

[80] S. P. Worner, "Performance of phenological models under variable temperature regimes: consequences of the Kaufmann or rate summation effect," *Environmental Entomology*, vol. 21, no. 4, pp. 689–699, 1992.

[81] J. Bernal and D. Gonzalez, "Experimental assessment of a degree-day model for predicting the development of parasites in the field," *Journal of Applied Entomology*, vol. 116, no. 5, pp. 459–466, 1993.

[82] B. S. Nietschke, R. D. Magarey, D. M. Borchert, D. D. Calvin, and E. Jones, "A developmental database to support insect phenology models," *Crop Protection*, vol. 26, no. 9, pp. 1444–1448, 2007.

[83] L. T. Wilson and W. W. Barnett, "Degree-days: an aid in crop and pest management," *California Agriculture*, vol. 37, pp. 4–7, 1983.

[84] W. G. Wellington, "The synoptic approach to studies of insects and climates," *Annual Review of Entomology*, vol. 2, pp. 143–162, 1957.

[85] D. L. Trudgill, A. Honek, D. Li, and N. M. Van Straalen, "Thermal time—concepts and utility," *Annals of Applied Biology*, vol. 146, no. 1, pp. 1–14, 2005.

[86] P. H. Crowley, "Resampling methods for computation-intensive data analysis in ecology and evolution," *Annual Review of Ecology and Systematics*, vol. 23, no. 1, pp. 405–447, 1992.

[87] G. Caughley, D. Grice, R. Barker, and B. Brown, "The edge of the range," *Journal of Animal Ecology*, vol. 57, no. 3, pp. 771–785, 1988.

[88] A. A. Hoffmann and M. W. Blows, "Species borders: ecological and evolutionary perspectives," *Trends in Ecology and Evolution*, vol. 9, no. 6, pp. 223–227, 1994.

[89] P. S. Messenger, "Bioclimatic studies with insects," *Annual Review of Entomology*, vol. 4, pp. 183–206, 1959.

[90] J. Régnière and V. Nealis, "Modelling seasonality of gypsy moth, *Lymantria dispar* (Lepidoptera: Lymantriidae), to evaluate probability of its persistence in novel environments," *Canadian Entomologist*, vol. 134, no. 6, pp. 805–824, 2002.

[91] S. H. Schneider, "Scenarios of global warming," in *Biotic Interactions and Global Change*, P. M. Kareiva, J. G. Kingslver, and R. B. Huey, Eds., pp. 9–23, Sinauer Associates, Sunderland, Mass, USA, 1993.

[92] V. E. Shelford, *Animal Communities in Temperate America*, University Chicago Press, Chicago, Ill, USA, 1913.

[93] R. W. Sutherst, "Pest risk analysis and the greenhouse effect," *Review of Agricultural Entomology*, vol. 79, pp. 1177–1187, 1991.

[94] S. P. Worner, "Ecoclimatic assessment of potential establishment of exotic pests," *Journal of Economic Entomology*, vol. 81, pp. 973–983, 1988.

[95] D. R. Mille, T. K. Mo, and W. E. Wallner, "Influence of climate on gypsy moth defoliation in southern New England," *Environmental Entomology*, vol. 18, pp. 646–650, 1989.

[96] S. L. Chown and J. S. Terblanche, "Physiological diversity in insects: ecological and evolutionary contexts," *Advances in Insect Physiology*, vol. 33, pp. 50–152, 2006.

[97] S. L. Chown, K. R. Jumbam, J. G. Sørensen, and J. S. Terblanche, "Phenotypic variance, plasticity and heritability estimates of critical thermal limits depend on methodological context," *Functional Ecology*, vol. 23, no. 1, pp. 133–140, 2009.

[98] D. W. Whitman, "Acclimation," in *Phenotypic Plasticity of Insects. Mechanisms and Consequences*, D. W. Whitman and T. N. Ananthakrishnan, Eds., pp. 675–739, Science Publishers, Enfield, NH, USA, 2009.

[99] C. Nyamukondiwa and J. S. Terblanche, "Within-generation variation of critical thermal limits in adult Mediterranean and Natal fruit flies Ceratitis capitata and Ceratitis rosa: thermal history affects short-term responses to temperature," *Physiological Entomology*, vol. 35, no. 3, pp. 255–264, 2010.

[100] N. Gilbert, A. P. Gutierrez, B. D. Frazer, and R. E. Jones, *Ecological Relationships*, W. H. Freeman and Co., Reading, UK, 1976.

[101] W. D. Williams and A. M. Liebhold, "Herbivorous insects and global change: potential changes in the spatial distribution of forest defoliator outbreaks," *Journal of Biogeography*, vol. 22, no. 4-5, pp. 665–671, 2009.

[102] A. Honěk, "Geographical variation in thermal requirements for insect development," *European Journal of Entomology*, vol. 93, no. 3, pp. 303–312, 1996.

[103] T. Ikemoto, "Possible existence of a common temperature and a common duration of development among members of a taxonomic group of arthropods that underwent speciational adaptation to temperature," *Applied Entomology and Zoology*, vol. 38, no. 4, pp. 487–492, 2003.

[104] D. L. Trudgill, "Why do tropical poikilothermic organisms tend to have higher threshold temperatures for development than temperate ones?" *Functional Ecology*, vol. 9, no. 1, pp. 136–137, 1995.

[105] A. Honěk and F. Kocourek, "Thermal requirements for development of aphidophagous *Coccinellidae* (Coleoptera), *Chrysopidae*, Hemerobiidae (Neuroptera), and *Syrphidae* (Diptera): some general trends," *Oecologia*, vol. 76, no. 3, pp. 455–460, 1988.

[106] T. M. Van Der Have and G. De Jong, "Adult size in ectotherms: temperature effects on growth and differentiation," *Journal of Theoretical Biology*, vol. 183, no. 3, pp. 329–340, 1996.

6

Investigations on the Effects of Five Different Plant Extracts on the Two-Spotted Mite *Tetranychus urticae* Koch (Arachnida: Tetranychidae)

Pervin Erdogan,[1] **Aysegul Yildirim,**[1] **and Betul Sever**[2]

[1] *Central Plant Protection Research Institute, Yenimahalle, 49.06172 Ankara, Turkey*
[2] *Faculty of Pharmacy, University of Ankara, Tandogan, 06100 Ankara, Turkey*

Correspondence should be addressed to Pervin Erdogan, pervin_erdogan@hotmail.com

Academic Editor: Kabkaew Sukontason

Two-spotted mite, *Tetranychus urticae* Koch (Arac.: Tetranychidae), is an economic pest worldwide including Turkey, causing serious damage to vegetables, flowers, and fruit crops. In recent years, broad-spectrum insecticides/miticides have been used to control this pest in Turkey. Control is difficult mainly due to resistance to conventional pesticides. This study was conducted to determine efficacy of pesticides extracted from five different plants [i.e., *Allium sativum* L. (Amaryllidaceae), *Rhododendron luteum* S. (Ericaceae), *Helichrysum arenarium* L. (Asteraceae), *Veratrum album* L. (Liliaceae), and *Tanacetum parthenium* L. (Asteraceae)] against this mite. Bioassays were tested by two different methods to determine the effects of varying concentrations. Experiments were performed using 3 cm diameter leaf disk from unsprayed bean plants (*Phaseolus vulgaris* L.). In addition, the effects of the extracts on reproduction and oviposition were investigated. The extract yielded high mortality. In the lowest-concentration bioassays, the adult mites laid lower numbers of eggs compared to the untreated control. No ovicidal effect was observed.

1. Introduction

Diseases and insect pests are the major limiting factors in the production of high quality agricultural products. Although conventional pesticides have become an indispensable tool in controlling some pests economically, rapidly, and effectively, extensive use of insecticides may lead to a number of undesirable side effects including the development of insect resistance and resurgence of primary and secondary pests outbreaks. Also they can have adverse effects on nontarget organisms and general environmental contamination [1–4]. The other problems with synthetic insecticides are environmental pollution and insect resistance. According to Nas [5] interest in the application of botanical pesticides for crop protection is on the rise. Many researchers are experimenting and developing alternative plant extracts as pesticides to be used against pest insects.

Plants have the richest source of renewable natural pesticides. Specifically, plant extracts provide a safe and viable alternative to synthetic pesticides and are compatible with the use of beneficial organisms, pest-resistant plants, and to preserving a healthy environment in an effort to decrease reliance on synthetic pesticides. There are many benefits of using botanical pesticides such as reduced environmental degradation, increased safety for farm workers, increased food safety, reduction in pesticide resistance, and improved profitability of production.

As a result, many plant compounds, the majority of which are alkaloids and terpenoids, have now been known to affect insects' behaviour, growth and development, reproduction, and survival [6–9]. Many investigations have recently been performed in relation to effects of plants such as *Chrysanthemum roseum* Web. and Mohr. (Compositae), *Nicotiana tabaccum* L. (Solanaceae), *Derris elliptica* Benth (Fabaceae), neem tree, *Azadirachta indica* A. Juss (Meliaceae), *Melia azaderach* L. (Meliaceae), and *Xanthium strumarium* L. (Solanaceae) on insects [10–13]. The seed kernel extract of neem, known as azadirachtin, has been most thoroughly tested, and it has been extracted in larger quantities than the other components of neem [14, 15]. High rates of mortality have been found on the two spotted mites fed on the leaves treated with *A. indica* extract. In addition, the same extract significantly reduced the reproductive capacity of mites and the survival of the progeny of treated females greatly diminished in comparison to the control [16].

T. urticae is a very important pest worldwide, causing serious damage to vegetables, flowers, and fruit crops. Many crops must be protected with synthetic acaricides during hot and dry seasons that favor severe outbreaks of *T. urticae*. It is able to transmit many of plants viruses [17].

R. luteum and *V. album* are poisonous plants. It is recorded that the extract of *V. album* has been used as insecticide or rodenticide since the Roman times. Also, today, plants containing toxic alkaloid are used successfully as insecticides and fungicides [18]. In one of the studies evaluating the effectiveness of plant extracts against house flies as indicated, *V. album* inhibited the development of the larvae and the high toxicity [19].

H. arenarium, *T. parthenium*, and *A. sativum* are important medicinal plants. *H. arenarium*, an infusion of the bright yellow flowers, is used in the treatment of gallbladder disorders and as a diuretic in treating rheumatism and cystitis. It is a component in *zahraa*, an herbal tea used for medicinal purposes in some countries [20]. *A. sativum* and *T. parthenium* have a broad spectrum of biological activity. They have been used for anti-inflammatory, antibacterial, and antifungal activities [21]. It is determined that the extract of *T. vulgare* inhibited the development of *Dermanyssus gallinae* (Mesostigmata: Dermanyssidae). In addition, the same plant extract cultivated showed that it is effective on *T. urticae* [22]. The extract of garlic leaves caused high mortality and reduced reproductive capacity on *T. urticae* [23]. According to the literature, no Works have been published on the acaricidal activity of *H. arenarium*. This study was undertaken in the laboratory at the Central Plant Protection Research Institute in 2009, and the miticidal effect of five plant extracts on *T. urticae* was tested.

2. Material and Methods

2.1. Plants and Preparation of Extracts. This study covered five plant species; *R. luteum*, *H. arenarium*, *A. sativum*, *V. Album*, and *T. parthenium* were tested as an alternative miticidal. Their leaves and stems were collected when plants were at the flowering stage during the years 2008 and 2009. Only the fruit garlic plant was used for this purpose. Ethanol was used as a solvent to extract the required material from five plants for use as an acaricide. The method of Brauer and Devkota [24] was used in preparation of five plants' ethanolic extract.

The materials were stored in the laboratory to dry up. The dried materials were grounded using a blender, and ethanol was added to the dried powder for 72 hours. This mixture was extracted in 5-6 hours using a Soxhlet machine. The ethanol was removed from the extract in a rotary evaporator (50–60°C). For each plant sample 200 g of dried materials were used to prepare the extract.

2.2. Mites. As a test organism, *T. urticae* was reared on green bean plants, *Phaseolus vulgaris*. The bean plants used in the experiment were grown in a greenhouse.

2.3. Effects of the Extracts of Five Plants on Tetranychus urticae. In all the experiments, first instar larvae and 3-day-old adults were used. Four concentrations and an untreated control were used for all bioassays. Test samples for bioassay were resuspended in distilated water with TritonX.100 at a rate of 0.1 mL/L. Vaseline was used so as to prevent the mites from escaping. Experiments were carried out using (3 cm diameter) leaf discs of green bean leaves. The leaf disks were placed on a moistened filter paper disk and each disk was infested with 10 individuals. Each treatment was replicated 10 times. The concentrations used for mites were 1%, 3%, 6%, and 12% [16].

2.4. Effect on Eggs. Green bean leaf discs were placed into petri dishes on moistened filter paper and females of the same age were put on leaf discs. The eggs were counted after two days. Ten eggs were placed in every petri dish and the other eggs removed. Then the eggs were sprayed with different concentrations of extract ($17-20\,\mu L/cm^2$) using a small hand-held sprayer. The numbers of hatched larvae were recorded.

2.5. Effect of the Extracts on Larvae and Adults

2.5.1. Leaf-Dipping Method. Green bean leaf discs were treated by dipping them into extract solutions of known concentrations, then left to dry for 30 minutes. The treated leaf discs and individual mites were placed in the petri dishes (9 cm in diameter) that were lined with moistened filter paper. The results were assayed after 1, 3, and 6 days by counting the number of living adults and larvae.

2.5.2. Leaf-Spraying Method. Green bean leaf discs were placed into Petri dishes on moisturized filter paper. Ten adults were placed in every Petri dish. Then eggs were sprayed with different concentrations of extract ($17-20\,\mu L/cm^2$) using a small hand-held sprayer. The results were assayed after 1, 3, and 6 days by counting the number of living adults.

2.6. Effect on Egg-Laying Capacity. Green bean leaf discs were dipped for 3–5 seconds in prepared concentrations (1, 3, 6, and 12%), then they were dried for 30 minutes and placed in petri dishes with ten adults. After 48 hours of feeding on treated green bean leaves, mites were given untreated green bean leaves. The experiment was repeated 10 times. Daily monitoring was done for fourteen days and the total number of eggs was recorded [25].

The experiments were conducted in a climate chamber at 25-26°C and under long daylight (18 h : 6 h, light : dark). The effect was calculated according to Abbott [26]. The obtained reasults were submitted to a variance analysis and the mean values were compared by Duncan's test ($P = 0.05$) calculated by the program SPSS 13.6). Mortality rate was calculated as; mortality = after treatment the number of died mites/before treatment the number of mites · 100).

Investigations on the Effects of Five Different Plant Extracts on the Two-Spotted Mite Tetranychus urticae Koch (Arachnida: Tetranychidae)

47

Table 1: Effect (mean ± SE) and mortality (%) of extracts obtained from different five plants on *T. urticae*.

Treatment	Concentration (%)	Leaf-dipping method				Leaf-spraying method	
		Larvae		Adult		Adult	
		Mortality (%)	Effect (%)	Mortality (%)	Effect (%)	Mortality (%)	Effect (%)
H. arenarium	1	46	31.59 ± 4.00^c	37	25.32 ± 4.10^c	52	39.76 ± 5.18^c
	3	53	41.59 ± 5.47^{bc}	47	37.22 ± 6.77^{bc}	66	59.88 ± 4.65^b
	6	58	46.09 ± 2.53^b	51	42.36 ± 5.61^b	76	71.82 ± 1.76^{ab}
	12	71	62.72 ± 2.28^a	64	56.85 ± 5.63^a	85	82.38 ± 1.92^a
A. sativum	1	46	31.30 ± 5.01^b	29	16.43 ± 2.43^b	66	59.76 ± 4.45^b
	3	50	37.80 ± 5.96^b	34	27.59 ± 5.17^{ab}	69	65.45 ± 5.16^{ab}
	6	56	43.37 ± 5.95^b	45	34.35 ± 6.76^a	77	72.79 ± 4.38^a
	12	68	58.35 ± 6.31^a	49	39.49 ± 5.07^a	78	73.92 ± 3.16^a
V. album	1	50	35.83 ± 4.33^c	51	29.07 ± 4.71^c	47	33.57 ± 4.12^c
	3	65	54.93 ± 5.22^b	61	42.41 ± 6.33^b	59	49.72 ± 3.39^b
	6	75	65.37 ± 3.15^{ab}	78	51.57 ± 5.37^a	70	62.58 ± 2.98^b
	12	77	70.55 ± 2.44^a	79	79.02 ± 3.76^a	81	75.77 ± 3.81^a
T. parthenium	1	49	38.22 ± 5.83^c	64	54.49 ± 4.34^c	47	33.61 ± 4.14^c
	3	64	54.06 ± 3.14^b	77	69.58 ± 1.52^b	60	49.75 ± 3.41^b
	6	75	67.89 ± 2.56^a	85	82.41 ± 1.94^a	71	62.62 ± 2.96^b
	12	82	76.54 ± 3.51^a	88	83.47 ± 1.95^a	81	75.68 ± 3.77^a
R. luteum	1	44	31.87 ± 3.31^b	27	31.66 ± 4.50^b	37	25.81 ± 2.94^c
	3	48	35.38 ± 4.05^b	44	34.02 ± 3.62^b	42	43.23 ± 3.40^b
	6	74	66.14 ± 4.50^a	53	44.35 ± 4.43^b	58	50.31 ± 3.28^{ab}
	12	81	75.62 ± 3.03^a	67	63.66 ± 2.44^a	68	61.97 ± 3.75^a
Control		22	0	15	0	15	0

Within columns, means ± SE followed by the same letter are not significantly different (DUNCAN's multiple *F*-test $P < 0.05$).

3. Results and Discussion

3.1. Effect on Eggs. All of the eggs treated were found to have hatched. It is determined that the ethanolic extracts of *R. luteum, H. arenarium, A. sativum, V. album,* and *T. parthenium* did not have an ovicidal effect. The hatched larvae continued to develop as it was in the control.

3.2. Effect of the Extracts on Larvae

3.2.1. Leaf-Dipping Methods. From Table 1, it can be observed that ethanol extracts of five plants had a significant mortality and the highest effect on *T. urticae* larvae. In all of the plant extracts, the highest effect occurred at a concentration of 12% while the smallest effect was at 1%. The increased concentration led to increased larval mortality. Statistical analysis showed $P < 0.05$ importance between the treatments. The extract of *T. parthenium* showed the highest effect on the *T. urticae* larvae. The smallest effect was at the extract of *A. Sativum*.

3.3. Effect of the Extracts on Adult

3.3.1. Leaf-Dipping Methods. As shown in Table 1, for the adults placed on leaf discs treated with different plant of extracts, the highest effect was determined at a concentration of 12% the extract of *T. parthenium*. Among the plant extracts, the extract of *T. parthenium* indicated the highest mortality. On the other hand, the smallest mortality was found at the extract of *A. sativum*. The increased concentration led to increased adult mortality.

3.3.2. Leaf Spraying Method. For the larvae placed on leaf discs treated with different plant of extracts at concentration of %12, mortality at the extract of *H. arenarium, A. sativum, V. album, T. parthenium,* and *R. luteum* was 85, 78, 81, 81, and 68%, respectively. In all of the extracts the highest effect was determined at a concentration of 12% while the smallest effect was at 1% (Table 1).

In both methods, similar results were obtained and there was not a significant difference on the mortality when leaf-dipping method was compared with direct spraying on the plant.

3.4. Effect on Egg-Laying Capacity. The numbers of eggs laid by mites feeding on extract-treated bean leaves were found to be statistically significant ($P < 0.05$) for all extracts with the maximum number of eggs obtained from the control. The lowest number of eggs was found at the 12% concentration of the extract of *R. luteum,* and the number of eggs laid was reduced significantly by increasing concentration (Table 2).

Ethanolic extracts were made from different plants and their effects were tested on two-spotted mite for the first time

TABLE 2: Effect of extracts from obteined different five plants on egg laying capacity of *T. urticae*.

Concentrations (%)	Treatment				
	H. arenarium	*A. sativum*	*V. album*	*T. parthenium*	*R. luteum*
			Number of eggs (mean ± SE)		
Control	162.5 ± 11.80[c]	162.5 ± 11.80[c]	162.5 ± 11.80[c]	162.5 ± 11.80	162.5 ± 11.80[c]
1	145.5 ± 5.91[c]	184.0 ± 12.10[b]	152.6 ± 10.50[c]	158.0 ± 12.1[b]	152.6 ± 10.50[c]
3	94.5 ± 6.0[b]	154.3 ± 10.3[b]	137.6 ± 13.43[c]	153.3 ± 10.3[b]	137.6 ± 13.40[c]
6	81.8 ± 6.40[b]	115.2 ± 9.13[a]	108.9 ± 19.9[b]	136.2 ± 9.12[b]	88.9 ± 19.92[b]
12	62.5 ± 6.33[ab]	98.2 ± 8.60[a]	96.4 ± 2.52[b]	87.2 ± 8.60[a]	18.4 ± 2.50[a]

Within columns, means ± SE followed by the same letter are not significantly different (DUNCAN's multiple F-test $P < 0.05$).

in the world. It was observed that some extracts showed a high rate of mortality and reduced fecundity on *T. urticae*.

There were no references in the literature of other studies using four plant extracts ethanolic extract on *T. urticae* except that *A. sativum*. However, other plant extracts have been investigated and the findings for *T. urticae* are similar to those of our study. Neem seed kernel extracts and its formulation are reported to influence mortality, repellency, and fecundity of mites [27–29]. It was found out that the two commercial preparations of neem seed extracts (Margosan-0 and Neem Azal S, Neem Azal T/S) were effective on *T. urticae* [16, 30]. Several herbal extracts of *Achillea millefolium* L. (Asteraceae), *Taraxacum officinales* F. H. (Asteraceae), *Matricaria chamomilla* L. (Asteraceae), and *Salvia officinalis* L. (Lamiaceae) demonstrated strong inhibition of the feeding activity of mites [31, 32]. It was determined that the extracts of yew showed a high mortality, decrease in female fecundity and shortened longevity [33, 34]. Shi et al. [35] revealed that the extract of *Bassia scoparia* (L.) A. J. Scott. (Chenopadiaceae) showed contact and systemic effects, and it caused high rates of mortality in all the three species (*T. urticae*, *T. cinnabarinus*, and *T. viennensis*). Pure azadirachtin reduced the reproductive capacity and feeding of *T. urticae* [36]. Crude foliar extracts of 67 species from six subfamilies of Australian Lamiaceae showed both contact and systemic toxicity to these mites [37]. The extracts of wild tomato leaf showed strong repellency effect on *T. urticae* [38]. The acaricidal activities of plant extracts on *T. urticae* were tested. The mortalities were high in extracts *Albizia coreana* Twig., *Pyracantha angustifolia* F. (Rosaceae), and *Ligustrum japonicum* Thunb. (Oleaceae) within 48 h treatment [39]. Attia et al. [23] revealed that the extract of garlic led to a rise in female mortality and a reduction in fecundity with the increasing of concentration. Essential oils of *Artemisia absinthium* L. (Asteraceae) and *Tanacetum vulgare* L. (Asteraceae) were extracted by three methods, a microwave-assisted process (MAP), distillation in water (DW), and direct steam distillation (DSD), and tested for their toxicity as contact acaricides to *T. urticae*. DSD and DW extracts of *T. vulgare* were more toxic (75.6 and 60.4% mite mortality, resp., at 4% concentration) to *T. urtica* than to the MAP extract (16.7% mite mortality at 4% concentration) [22]. The ethanol extracts of *Croton rhamnifolius* H.B.K. (Euphorbiaceae) *C. sellowi*, *C. jacobinensis*, and *C. micans* had a high mortality on *T. urticae*, whereas *C. sellowi* extract showed the

highest effect [40]. Garlic extract showed a mortality at 48–57% on *T. urticae* [41]. Wang et al. [42] revealed that the crude extract of walnut leaf had some contact and systemic effect on *T. cinnabarinus* and *T. viennensis*.

It was found out that the extract of *V. album* and *T. parthenium* had a high rate mortality and reduced fecundity for *T. urticae*. Ethanolic extracts of *V. album* and *T. parthenium* can be useful to control *T. urticae* populations on vegetable plants grown through Integrated Pest Management (IPM) and organic systems of agriculture.

References

[1] G. P. Georghiou, *Insecticides and Pest Résistance: The Conséquences of Abuse*, Faculty Research Lecture, Academie Senate, University of California, Riverside, Calif, USA, 1987.

[2] J. L. Metcalfe, "Biological water quality assessment of running waters based on macroinvertebrate communities: history and present status in Europe," *Environmental Pollution*, vol. 60, no. 1-2, pp. 101–139, 1989.

[3] F. N. Dempster, "Spacing effects in text recall: an extrapolation from the laboratory to the classroom," submitted.

[4] V. Ditrich, "A comperative study of toxicologial test methods on a population of the two spotted spider mite (*Tetranychus urticae*)," *Journal of Economic Entomology*, vol. 55, pp. 644–648, 1962.

[5] M. N. Nas, "In vitro studies on some natural beverages as botanical pesticides against *Erwinia amylovora* and *Curobacterium flaccumfaciens* subsp. Poinsettiae," *Turkish Journal of Agriculture and Forestry*, vol. 28, no. 1, pp. 57–61, 2004.

[6] J. T. Arnason, B. J. R. Philogene, and P. Morand, *Insecticides of Plants Origin*, vol. 387 of *American Chemical Society Symposium*, Washington, DC, USA, 1989.

[7] M. Jacobson, "Plants, insects, and man-their interrelationships," *Economic Botany*, vol. 36, no. 3, pp. 346–354, 1982.

[8] K. Nakanishi, "Structure of the insect antifeeedant Azadirachtin," *Recent Advances in Phytochemistry*, vol. 9, pp. 277–280, 1975.

[9] J. D. Warthen, E. D. Morgan, and N. B. Mandava, "Insect feeding deterrents," in *CRC Handbook of Natural Pesticides*, vol. 6 of *Insect Attractants and Repellents*, pp. 23–134, CRC Press, Boca Raton, Fla, USA, 1990.

[10] H. Martin and D. Woodcock, "The hydrocarbon oils," in *The Scientific Principles of Crop Protection*, pp. 212–220, Edward Arnold, London, UK, 7th edition, 1983, Matteoni, J., Chemical Effects on Greenhouse, 1993.

[11] C. L. Metcalf and W. P. Flint, *Destructive and Useful Ýnsects their Habits and Control M*, McGraw-Hill, 1951.

Investigations on the Effects of Five Different Plant Extracts on the Two-Spotted Mite Tetranychus urticae Koch
(Arachnida: Tetranychidae)

49

[12] H. Schmutterer, "Properties and potential of natural pesticides from the Neem Tree, *Azadirachta indica*," *Annual Review of Entomology*, vol. 35, no. 1, pp. 271–297, 1990.

[13] P. Erdogan and S. Toros, "Investigations on the effects of *Xanthium strumarium* L. extracts on Colorado potato betle, *Leptinotarsa decemlineata* Say (Col.: Chrysomelidae)," *Munis Entomology & Zoology*, vol. 2, no. 2, pp. 423–432, 2007.

[14] H. Schmutterer, K. R. S. Ascher, and H. Rembold, "Natural pesticides from Neem Tree (*Azadirachta indica* A. Juss).and other Tropical Plants," in *Proceedings of the 1st International Neem Conference*, p. 297, Rottach-Egern, Germany, 1981.

[15] H. Schmutterer and C. P. W. Zebitz, "Effect of in ethanolic extracts from seeds of single Neem Trees of African and Asian origin, on *Epilachna variivestis* and *Aedes aegypti*. In: n: natural pesticides from Neem Tree and other Tropical Plants," in *Proceedings of the 2nd International Neem Conference*, pp. 83–90, Rauischholzhausen, Germany, 1984.

[16] M. K. Miranova and E. G. Khorkhordin, "Effect of Neem Azal T/S on *Tetranychus urticae* Koch," in *Proceedings at the 5th Workshop*, pp. 22–25, Wetzlar, Germany, January 1996.

[17] C. E. Thomas, "Transmission of tobacco ringspot virus by *Tetranychus* sp.," *Phytopathology*, vol. 59, pp. 633–636, 1969.

[18] V. Gomilevsky, "Poisonous plants from which insecticides for orchard-pests may be prepared," in *Progressive Fruit-Growing and Market-Gardening 1915*, Orchard Library, p. 32, 2011.

[19] E. D. Bergmann, Z. H. Levinson, and R. Mechoulam, "The toxicity of *Veratrum* and *Solanum* alkaloids to housefly larvae," *Journal of Insect Physiology*, vol. 2, no. 3, pp. 162–177, 1958.

[20] H. E. Eroglu, E. Hamzaoglu, A. Aksoy, U. Budak, and S. Albayrak, "Cytogenetic effects of *Helichrysum arenarium* in human lymphocytes cultures," *Turkish Journal of Biology*, vol. 34, no. 3, pp. 253–259, 2010.

[21] K. Dancewicz and B. Gabrys, "Effect of extracts of garlic (*Allium sativum* L.), wormwood (*Artemisia absinthium* L.) and (*Tanacetum vulgare* L.) onthe behaviour of the peach potato aphid Myzus persicae (Sulzer) during the settling on plants," *Pesticides*, vol. 3-4, pp. 93–99, 2008.

[22] H. Chiasson, A. Bélanger, N. Bostanian, C. Vincent, and A. Poliquin, "Acaricidal properties of *Artemisia absinthium* and *Tanacetum vulgare* (Asteraceae) essential oils obtained by three methods of extraction," *Journal of Economic Entomology*, vol. 94, no. 1, pp. 167–171, 2001.

[23] S. Attia, K. L. Grissa, A. C. Mailleux, G. Lognay, S. Heuskin, and S. Mayoufi, "Effective concentrations of garlic distillate (*Allium sativum*) for the control of *Tetranychus urticae* Koch. (Tetranychidae)," *Journal of Applied Entomology*, vol. 136, no. 4, pp. 302–312, 2011.

[24] M. Brauer and, "Control of *Thaumatopoea piyocampa* (Den.&Schiff) by extrakts of *Melia azedarach* L., (Meliaceae)," *Journal of Applied Entomology*, vol. 110, no. 1–5, pp. 128–135, 1990.

[25] H. Schmutterer, "Fecundity reducing and sterilizing effects of Neem Seed kernel extracts in the Colorado potato beetle, *Leptinotarsa decemlineata*," in *Proceedings of the 3rd International Neem Conference*, pp. 351–360, Nairobi, Kenya, 1986.

[26] W. S. Abott, "A method of computing the effectiveness of an insecticide," *Journal of Economic Entomology*, vol. 18, pp. 265–267, 1923.

[27] F. A. Monsuer and K. R. S. Ascher, "Effects of Neem (*Azadirachta indica*) seed kernel extracts from different solvents on the carmine spider mite, *Tetranychus cinnabarinus*," *Phytoparasitica*, vol. 11, no. 3-4, pp. 3177–4185, 1983.

[28] F. A. Monsuer, K. R. S. Ascher, and F. Abo- Moch, "Effects of margosan-o, azatin and RD9-repelin on spiders, and on predacious and phytophagous mites," *Phytoparasitica*, vol. 21, no. 3, pp. 205–211, 1993.

[29] N. Z. Dimetry, S. A. A. Amer, and A. S. Reda, "Biological activity of 2 Neem Seed kernel extracts against the 2-spotted spider mite *Tetranychus urticae* Koch," *Journal of Applied Entomology*, vol. 116, no. 3, pp. 308–312, 1993.

[30] N. Z. Dimetry, S. A. A. Amer, and A. S. Reda, "Biological activity of 2 Neem Seed kernel extracts against the 2-spotted spider mite *Tetranychus urticae* Koch," *Journal of Applied Entomology*, vol. 116, no. 3, pp. 308–312, 1993.

[31] A. Tomczy and M. Szymanska, "Possibility of reduction of spider mite population by spraying with selected herb extracts," in *Proceedings of the 35th Scientific Session IOR, Part II*, pp. 125–128, 1995.

[32] B. Kawka and A. Tomczyk, "Influence of extract from sage (*Salvia officinalis* L.) on some biological parameters of *Tetranychus urticae* Koch. feeding on Algerian Ivy (*Hedera helix variegata* L.)," *IOBC/Bulletin OILB*, vol. 22, pp. 96–100, 2002.

[33] M. Furmanowa, D. Kroppczynska, A. Zobel et al., "Effect of water extreacts from needle surface of of *Taxus b baccata var. Ellegantisima* on life-history parameters of two spider mite (*Tetranychus urticae* Koch)," *Herba Polonica*, vol. 47, pp. 5–10, 2001.

[34] M. Furmanowa, D. Kropczyska, A. Zobel et al., "Influence of water extracts from the surface of two yew (*Taxus*) species on mites (*Tetranychus urticae*)," *Journal of Applied Toxicology*, vol. 22, no. 2, pp. 107–109, 2002.

[35] G. L. Shi, L. L. Zhao, S. Q. Liu, H. Cao, S. R. Clarke, and J. H. Sun, "Acaricidal activities of extracts of Kochia scoparia against *Tetranychus urticae, Tetranychus cinnabarinus*, and *Tetranychus viennensis* (Acari: Tetranychidae)," *Journal of Economic Entomology*, vol. 99, no. 3, pp. 858–863, 2006.

[36] K. M. S. Sundaram and L. Sloane, "Effects of pure and formulated azadirachtin, a Neem-based biopesticide, on the phytophagous spider mite, *Tetranychus urticae* koch," *Journal of Environmental Science and Health B*, vol. 30, no. 6, pp. 801–814, 1995.

[37] H. L. Rasikari, D. N. Leach, P. G. Waterman et al., "Acaricidal and cytotoxic activities of extracts from selected genera of Australian lamiaceae," *Journal of Economic Entomology*, vol. 98, no. 4, pp. 1259–1266, 2005.

[38] G. F. Antonious and J. C. Snyder, "Natural products: repellency and toxicity of wild tomato leaf extracts to the two-spotted spider Mite, *Tetranychus urticae* Koch," *Journal of Environmental Science and Health B*, vol. 41, no. 1, pp. 43–55, 2006.

[39] D. I. Kim, J. D. Park, S. G. Kim, H. Kuk, M. S. Jang, and S. S. Kim, "Screening of some crude plant extracts for their acaricidal and insecticidal efficacies," *Journal of Asia-Pacific Entomology*, vol. 8, no. 1, pp. 93–100, 2005.

[40] J. Pontes, J. Oliveira, C. Camara, C. Assis, M. Juniour, and R. Barros, "Effects of the ethanol extracts of leaves and branches from four species of the croton on *Tetranychus urticae* Koch (Acari: Tetranychidae)," *BioAssay*, vol. 6, pp. 3–14, 2011.

[41] Z. T. Dabrowsky and U. Seredynska, "Charactrisation of the two-spotted spider mite (*Tetranychus urticae* Koch.:Tetranychidae) responses to aqueous extracts from selected plant species," *Journal of Plant Protection Research*, vol. 47, no. 2, pp. 114–123, 2007.

[42] Y. N. Wang, G. L. Shi, L. L. Zhao et al., "Acaricidal activity of *juglans regia* leaf extracts on *Tetranychus viennensis* and *Tetranychus cinnabarinus* (acaris tetranychidae)," *Journal of Economic Entomology*, vol. 100, no. 4, pp. 1298–1303, 2007.

7

Life History of *Aricoris propitia* (Lepidoptera: Riodinidae)—A Myrmecophilous Butterfly Obligately Associated with Fire Ants

Lucas A. Kaminski[1] and Fernando S. Carvalho-Filho[2]

[1] *Departamento de Biologia Animal, Instituto de Biologia, Universidade Estadual de Campinas, CP 6109, Campinas 13083-970, SP, Brazil*
[2] *Laboratório de Ecologia de Invertebrados, Instituto de Biologia, Universidade Federal do Pará, Rua Augusto Corrêa 01, Guamá, Belém 66075-110, PA, Brazil*

Correspondence should be addressed to Lucas A. Kaminski, lucaskaminski@yahoo.com.br

Academic Editor: Jean Paul Lachaud

The immature stages of *Aricoris propitia* (Stichel) are described and illustrated for the first time, using both light and scanning electron microscopy. Females oviposit in at least seven host-plant families, always in the presence of fire ants (*Solenopsis saevissima* (Smith) complex), without being attacked by them. Larvae are tended by ants during all larval and pupal stages. From the fourth instar on, larvae feed at night and rest during the day inside underground shelters constructed by ants on the host plant roots, and where pupation occurs. Several observed features, including ant-mediated oviposition, persistent ant attendance throughout all instars, and high spatiotemporal fidelity indicate that *A. propitia* is a myrmecophile obligately associated with fire ants. We propose *A. propitia* as an extraordinary model for studies on ant-butterfly evolutionary history in the Neotropics.

1. Introduction

Symbiotic associations between butterfly larvae and ants have attracted the attention of early naturalists, both in Europe and North America, since the second half of the 18th century (see references in [1]). Nonetheless, these interactions are historically poorly studied in the Neotropical region despite their richness and abundance [2, 3]. An exception in this scenario is the classic paper by Bruch [4], which describes some aspects of the life history of an Argentinean species of *Aricoris* Westwood. In addition to being the first detailed description of a myrmecophilous larva from the Riodinidae family, the aforementioned study presents the first evidence of a butterfly larva living inside ant nests in the Neotropics. This behavior has been reported for a small number of Lycaenidae clades, such as the charismatic large blue *Maculinea* Van Eecke (*Phengaris* Doherty spp.), which parasitizes ant societies in Eurasia (see [5–7]). But unlike large blue butterflies, which today are model organisms in mutualism and parasitism studies, little progress has been

achieved on the biology of *Aricoris* since the initial work by Bruch [4] (but see [8–12]).

The riodinid genus *Aricoris* contains 24 described species [13, 14] typically found in open dry areas of South America [3]. *Aricoris propitia* (Stichel) is widespread in Central and Northern Brazil ([15], C. Callaghan, pers. comm.). Since its original description in 1910, no additional information was published for this species. The purpose of this paper is to fill that gap by presenting the natural history and morphological description of immature stages of *A. propitia*, with emphasis on their obligatory association with fire ants of the *Solenopsis saevissima* (Smith) complex (Formicidae: Myrmicinae).

2. Material and Methods

2.1. Study Sites. Four sites were sampled in central and northern Brazil (Figure 1): (1) cerrado *sensu stricto* and gallery forest areas in Alto Paraíso, Goiás (13°48′S, 47°54′W) (July 2009); (2) suburban areas of the city of Assis Brasil, Acre

Life History of Aricoris propitia (Lepidoptera: Riodinidae)—A Myrmecophilous Butterfly Obligately Associated
with Fire Ants

51

(a)

(b)

(c)

(d)

FIGURE 1: Overview of *Aricoris propitia* study sites in Central (a) and Northern ((b)–(d)) Brazil. (a) Cerrado *sensu stricto* area in Alto Paraíso, Goiás; (b) surroundings of Assis Brasil, Acre—note the bordering suburbs, pastures and forest; (c) sandy beach areas along the Xingu river, Porto de Moz, Pará; (d) small home garden in a neighborhood of Belém, Pará—note that within this small space the larvae were able to use three ornamental plant species (see Table 1).

(10°56′S, 69°33′W) (August-September 2006); (3) sandy beach and small-farm cultivation areas along the Xingu river, Porto de Moz, Pará (02°07′S, 52°15′W) (July 2010); (4) house garden in a neighborhood of Belém, Pará (01°25′S, 48°27′W) (several occasions between 2006 and 2009).

2.2. Sampling, Rearing, and Behavioral Observations. Available host-plants in the study sites were visually scanned for the presence of larvae and tending ants (as in [16]). Additionally, some potential host-plants with distinct signs of herbivory and visited by *S. saevissima* ants were excavated in search of larvae and pupae. Plants with immatures (eggs and larvae) were collected for identification, as well as the tending ants. We also recorded the presence of food sources that may promote ant visitation on the plants, such as extrafloral nectaries (EFNs) and/or honeydew-producing hemipterans (HPHs). The immatures of *A. propitia* used for morphological description were collected in the field and reared as follows: eggs were placed in Petri dishes and observed daily until eclosion; newly hatched larvae were reared individually in transparent 250 mL plastic pots under controlled conditions (25 ± 2°C; 12 h L: 12 h D). Branches

of the same host-plant on which each larva was found were offered *ad libitum*, and larvae were checked daily for food replacement and cleaning when necessary. Immatures for morphological analysis were separated, fixed in Dietrich's solution, and then preserved in 70% ethanol. Shed head capsules were collected and preserved for measuring. Voucher specimens of the immature stages were deposited at the Museu de Zoologia "Adão José Cardoso" (ZUEC), Universidade Estadual de Campinas, Campinas, São Paulo, Brazil.

Behavioral interactions between *A. propitia* larvae and tending ants were observed *ad libitum* [17] in the field during the day (ca 10:00–16:00 h), and sometimes at night (ca 18:00–06:00 h), for the population of Porto de Moz. Additional observations on larval ant-organs and their role in the interaction with ants were obtained from larvae reared in plastic pots with their host ants or from larvae maintained in a terrarium together with a captive colony of tending ants (from a population of Belém).

2.3. Morphology. Measurements were taken and general aspects of morphology were observed using a Leica MZ7.5 stereomicroscope equipped with a micrometric scale. Egg

TABLE 1: Summary of recorded host-plants of *Aricoris propitia*, including liquid food source types available for ants (EFNs, extrafloral nectaries; HPHs, honeydew-producing hemipterans) and localities.

Host plants	Sources of liquid food	Localities
Chrysobalanaceae		
Hirtella glandulosa	EFNs, HPHs	Alto Paraíso (Goiás)
Fabaceae		
Senna obtusifolia	EFNs	Belém (Pará)
Malpighiaceae		
Byrsonima sp.	HPHs	Porto de Moz (Pará)
Malvaceae		
Hibiscus rosa-sinensis	HPHs	Belém (Pará)
Verbenaceae		
Aegiphila sp.	EFNs	Assis Brasil (Acre)
Rubiaceae		
Ixora coccinea	HPHs	Belém (Pará)
Simaroubaceae		
Simarouba sp.	EFNs	Alto Paraíso (Goiás)
Turneraceae		
*Turnera ulmifolia**	EFNs	Campinas (São Paulo)

*Lab-accepted host-plant.

size is given as height and diameter. Head capsule width of larvae was considered to be the distance between the most external stemmata; maximum total length for both larvae and pupae corresponded to the distance from head to posterior margin of the tenth abdominal segment in dorsal view (as in [18]). Measurements are given as minimum-maximum values. Scanning electron microscopy (SEM) was conducted using both JEOL JSM-5800 and Carl Zeiss LEO-1430VP microscopes, with samples prepared according to standard techniques (for details, see [19]). Terminology for early stage descriptions follows Downey and Allyn [20] for eggs, Stehr [21] for general morphology of larvae, Mosher [22] for pupae, and DeVries [23] for ant-organs.

3. Results

3.1. Natural History of Aricoris propitia. This butterfly is locally abundant in open areas, where it occurs close to its ant colonies. Adults can be observed flying fast near the ground, perching on the undergrowth where they become almost invisible. Males were observed defending small territories and visiting many wild flowers. Females were seen flying near host-plants infested by host ants (Figure 2(a)), which for all studied populations were ants of the *Solenopsis saevissima* complex. Oviposition occurred in the warmest period of the day, from 11 AM to 2 PM ($n = 15$ oviposition events), a period when ants are more active. Females flew in circles around a host-plant occupied by ants before starting to oviposit (prealighting phase). After landing (postalighting phase), females frequently touched the plant surface with the tip of their abdomen, particularly on ant trails, but

were never attacked by the ants. Eggs were laid singly or in small clusters of two to five eggs (Figure 2(b)). Our host-plant records indicate that the larvae of *A. propitia* are polyphagous using at least seven families of plants, including ornamental (nonnative) species cultivated in urban gardens (see Table 1 and Figure 1(d)). Also, in the laboratory, larvae accepted and developed well on leaves of *Turnera ulmifolia* L. (Turneraceae). All observed host-plants of *A. propitia* provided some source of liquid food that could be potentially used by ants, such as honeydew-producing hemipterans and/or extrafloral nectaries (see Table 1). Other potential host-plants without fire ants or visited by other ant species were also examined at some of the study sites ($n = 51$ at Assis Brasil, $n = 15$ at Alto Paraíso), but no larvae of *A. propitia* were found.

All instars are ant-tended, and even the small first instar is equipped with functional tentacular nectary organs (TNOs). From the second instar on, other ant-organs appear or become functional (Figure 3). Ants antennate the larval body intensely, but especially the anterior region where a row of papilliform setae and the openings of the anterior tentacle organs (ATOs) are located (Figure 3(a)). When everted, these organs provoke clear alterations in ant behavior, such as opening of the jaws and a marked increase in activity and aggressiveness. In the early instars (first to third) the larvae can be found during the day feeding on the host-plant leaves (Figures 2(b)–2(d)). From the fourth instar on, they rest during the day inside underground shelters constructed by ants within the host-plant roots, and that is where pupation occurs. When night falls, the larvae leave the underground shelters and climb up to feed on the host leaves (Figure 2(e)), returning to the shelters by dawn. Large quantities of mature larvae and pupae can be found inside the underground shelters, which are permanently patrolled by tending ants (Figure 2(f)).

3.2. Description of the Immature Stages. The reared immatures from the four sites were very similar and went through five instars. Developmental time is based on material from Alto Paraíso, Goiás, reared on *Turnera ulmifolia* leaves. The egg description and measurements are based on material from Assis Brasil, Acre; the larval and pupal description and measurements are based on material from Porto de Moz and Belém, Pará.

3.2.1. Egg (Figures 2(b) and 4). Duration 6-7 d ($n = 5$). Height 0.30–0.32 mm; diameter 0.54–0.58 mm ($n = 3$). Color whitish-cream when laid, changing to beige before hatching. General spherical shape, with convex upper surface and flattened bottom surface; exochorion with smooth surface and hexagonal cells in lateral view (Figure 4(a)). Slightly depressed micropylar area; annulus present, and rosette surrounded by petal-shaped cells; micropyles at center of the micropylar area (Figure 4(b)). Aeropyles in tiny protuberances in the rib intersections (Figure 4(c)).

3.2.2. First Instar (Figures 2(c) and 5(a)–5(c)). Duration 4-5 d ($n = 2$). Head capsule width 0.24–0.26 mm ($n = 3$), total length 2.2 mm. Dark brown head, prothoracic and anal

Life History of Aricoris propitia (Lepidoptera: Riodinidae)—A Myrmecophilous Butterfly Obligately Associated with Fire Ants

53

FIGURE 2: Life stages of *Aricoris propitia* tended by *Solenopsis saevissima* on *Byrsonima* sp. ((a), (d)–(f)), *Simarouba* sp. ((b), (d) and (e)), and *Hirtella glandulosa* (c). (a) Female at postalighting phase near an ant-tended treehopper aggregation (arrow); (b) eggs (white arrows) and a third instar larva (black arrow); (c) first instar tended by one worker; (d) second instar tended by ants; (e) third instar tended by ants, note that both anterior (black arrow) and tentacle nectary organs (white arrow) are everted; (f) nocturnal fifth (last) instar tended by several ants; (g) nocturnal group of larvae and treehoppers (arrow) tended by ants; (h) diurnal larval group inside a shelter in the host-plant roots.

(a)

(b)

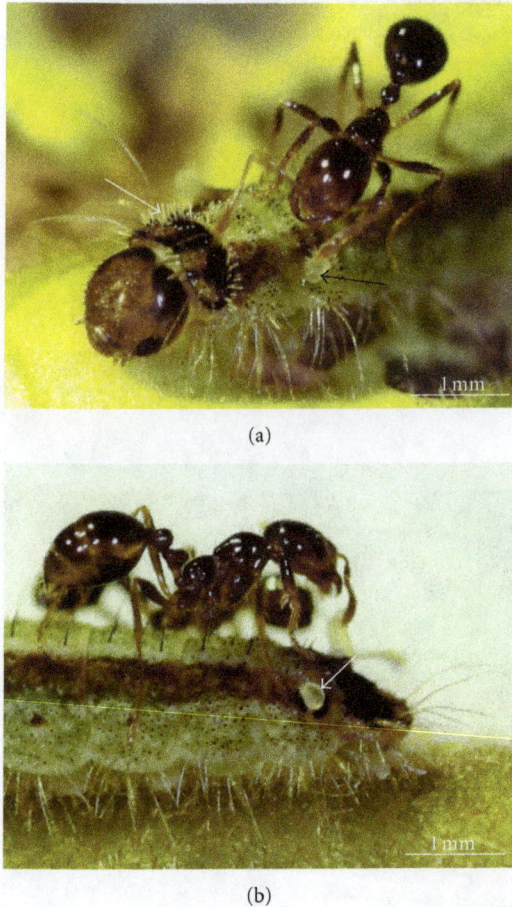

FIGURE 3: Sequence of interactions between *Aricoris propitia* third instar larva and *Solenopsis saevissima* ants. (a) Worker antennating the row of setae on the prothoracic shield (white arrow), note the everted anterior tentacle organ (black arrow); (b) everted nectary tentacle organ (white arrow) after repeated antennation by ant on the A8 segment.

shields; yellowish orange body with beige or translucent setae (Figure 2(c)). Epicranium and frontoclypeus with several setae, pores, and two pairs of perforated cupola organs (PCOs) in the adfrontal areas (Figure 5(a)). Body with long plumose setae in the lateral areas and in the prothoracic and anal shields; the remaining dorsal and subdorsal setae are short and dendritic, and PCOs are associated with these groups of setae. The openings of the anterior tentacle organs (ATOs) are present in the metathoracic segment, but these organs are apparently not functional (Figure 5(b)). Functional tentacle nectary organs (TNOs) are present in the A8 segment (Figure 5(c)).

3.2.3. Second Instar (Figure 2(d)). Duration 5-6 d (n = 2). Head capsule width 0.44 mm (n = 2), total length 3.1 mm. Dark brown head, prothoracic and anal shields; yellowish green body with two longitudinal light brown bands (Figure 2(d)). All ant-organs present, including ATOs, TNOs, PCOs, dendritic setae, and one pair of vibratory papilla on the anterior border of the prothoracic shield. A

dorsal row of papilliform setae is also present on the posterior margin of the prothoracic shield and is maintained in the subsequent instars (Figures 3(a) and 5(d)).

3.2.4. Third Instar (Figures 2(e) and 3). Duration 6 d (n = 2). Head capsule width 0.72–0.84 mm (n = 2), total length 6.2 mm. Brown head; black prothoracic and anal shields with beige spots; green body with two longitudinal brown bands (Figure 2(e)). General morphology is similar to the second instar's, but with more numerous and enlarged setae.

3.2.5. Fourth Instar (Figures 2(g)-2(h) and 5(d)-5(e)). Duration 6 d (n = 2). Head capsule width 1.28–1.30 mm (n = 4), total length 15.2 mm. Brown head; black prothoracic and anal shields with beige and grey spots; variegated body coloring with frosted brown and beige spots (Figures 2(g) and 2(h)). General morphology is similar to preceding instar's, but with more numerous and enlarged setae (Figures 5(d) and 5(e)).

3.2.6. Fifth (Last) Instar (Figures 2(f)–2(h) and 5(f)–5(h)). Duration 6-7 d (n = 2). Head capsule width 1.76–1.87 mm (n = 5), total length 2.1 cm. Coloring is similar to fourth instar (Figures 2(f)–2(h)). Mandibles with eight teeth and six setae (Figure 5(f)). Body covered with several types of setae, including prominent setae on the lateral areas, prothoracic and anal shields; two pairs of prominent dorsal setae in the same position as primary setae on the mesothorax to A8 segments; two types of dendritic setae and several perforated cupola organs (Figures 5(g) and 5(h)). The spiracle on the A1 segment is lateroventral, whereas those on segments A2 to A8 are in a dorsal position.

3.2.7. Pupa (Figure 6). Duration 10–12 d (n = 2). Total length 1.29 cm, width at A1 0.33 cm. Variegated coloring with brown, beige, and dark spots (Figure 6(a)). Tegument is entirely sculptured, with irregular striations and lacking prominent tubercles (Figures 6(b)–6(e)). Prothorax bears dorsal clusters of papilliform setae (Figure 6(a)). Silk girdle crossing the A1 segment near one pair of small tubercles with several associated dendritic setae and PCOs (Figure 6(b)). Body with some small dendritic setae, and PCOs located in clusters on lateral areas close to spiracles (Figures 6(b)– 6(e)); these clusters are absent on the A2 and A7 segments. The intersegmental area between the A4-A5 and A5-A6 abdominal segments features plates and files (Figure 6(f)) that may act as a stridulatory mechanism. The consolidated A9 and A10 segments constitute the ventrally flattened cremaster; with long crochets in a ventral position (Figure 6(g)).

4. Discussion

In general terms, the egg of *Aricoris propitia* resembles those described for other Nymphidiini genera in the Lemoniadina group (such as *Juditha* Hemming, *Lemonias* Hübner, *Synargis* Hübner, and *Thisbe* Hübner), with a semispherical shape, exochorion with hexagonal cells in lateral view, aeropyles in the rib intersections, and micropylar area centered on

Life History of Aricoris propitia (Lepidoptera: Riodinidae)—A Myrmecophilous Butterfly Obligately Associated
with Fire Ants

55

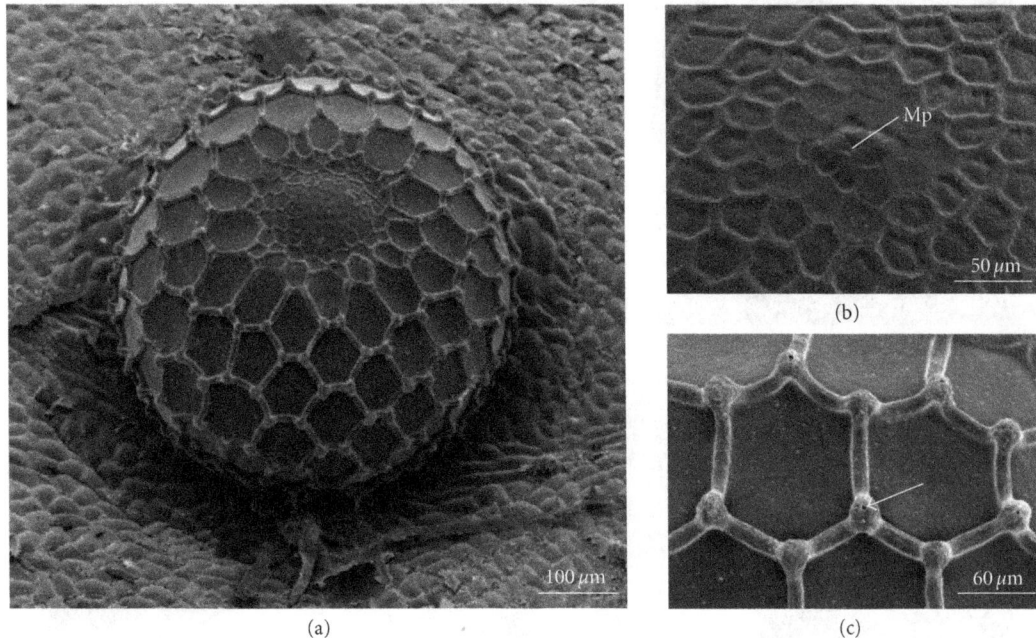

FIGURE 4: Scanning electron microscopy of *Aricoris propitia* egg (material from Assis Brasil, Acre). (a) Lateral view; (b) micropylar area (Mp);
(c) hexagonal cells of the exochorion with aeropyles in the rib intersections (arrow).

the top surface (see [3, 24, 25]). However, it differs in that
the limits of the micropylar area are slightly bounded; this
pattern is shared with other *Aricoris* and *Ariconias* Hall
and Harvey (L.A. Kaminski, unpublished). The first instar
presents some characteristics of myrmecophilous larvae,
namely, conspicuous perforated cupola organs, functional
tentacle nectary organs, and short, dorsally located dendritic
setae (see examples of riodinid first instar larvae in [3, 18,
26]).

Larvae of *A. propitia* present the typical pattern of
Nymphidiini, with the first abdominal spiracle in a ventral
position and vibratory papillae (VPs) on the prothoracic
shield [27]. In addition to the sound producing organs
(VPs), the mature larvae of *A. propitia* feature two other
important types of riodinid ant-organs (see [3, 23, 24]):
the anterior tentacle organs (ATOs) and the tentacle nectary
organs (TNOs). The larvae also present another putative
ant-organ: the row of papilliform setae on the prothorax,
which had already been described for other *Aricoris* species
[4, 8]. Tending ants frequently antennate these papilliform
setae, and usually this palpation is accompanied by eversion
of the ATOs. The way the ants react after ATO eversion
suggests that the ATOs emit a volatile chemical similar to
the ant alarm pheromone, as has been suggested for other
Riodinidae [3, 23, 28]. The chemical compositions of ATO
emissions by myrmecophilous butterflies are still unknown.
In contrast, the chemical ecology of fire ants, including
alarm pheromones and their role in interactions with other
organisms, is relatively well known (e.g., [29]). Thus, the *A.
propitia*/fire ants system may be helpful in answering some
outstanding questions about the functioning of ant-organs
in myrmecophilous butterflies.

The larvae of *A. propitia* can be considered polyphagous
since they feed on at least seven families of host-plants.
Polyphagy in obligate myrmecophilous butterflies, including
Riodinidae, has been regarded as a consequence of ant-
dependent oviposition [3, 12, 25, 30–32], and this seems to
be the case for *A. propitia*. Aphytophagy, on the other hand,
is quite rare in butterfly larvae [33], but it has been suggested
for some species of *Aricoris* [3, 12]. It is believed that the
larvae of these species are able to get food directly from ants,
through regurgitations (trophallaxis) from ant workers or
by preying directly on ant brood. Although *A. propitia* rest
during the day inside underground shelters together with
their tending ants, we do not have evidence that the larvae
get some kind of food from the ants.

All known species of *Aricoris* seem to be engaged in
obligatory associations with their tending ants. To date,
Aricoris domina (Bates) has been associated with *Ectatomma*
Smith [11], and seven *Aricoris* species have been associated
with *Camponotus* Mayr [3, 8–12, 34]. So far, only *Aricoris
hubrichi* (Stichel) and *Aricoris campestris* (Bates) have been
reported to be associated with *Solenopsis* Westwood ants ([4],
A.V.L. Freitas pers. comm.). Both *Aricoris* species are inserted
within the derived "*epulus*-group" sensu Hall and Harvey
[35]. Despite the high species richness and ecological prevalence, symbiotic interactions between butterfly larvae and
Solenopsis are very rare ([36], L.A. Kaminski, unpublished).
Apart from the association with fire ants, several natural
history and morphological features of *A. propitia* are very
similar to those observed for *A. hubrichi* and *A. campestris*,
suggesting an evolutionary relationship among these species.
As the life history of most species within the "*epulus*-group"
is still unknown (see [35]), it is not possible to tell whether

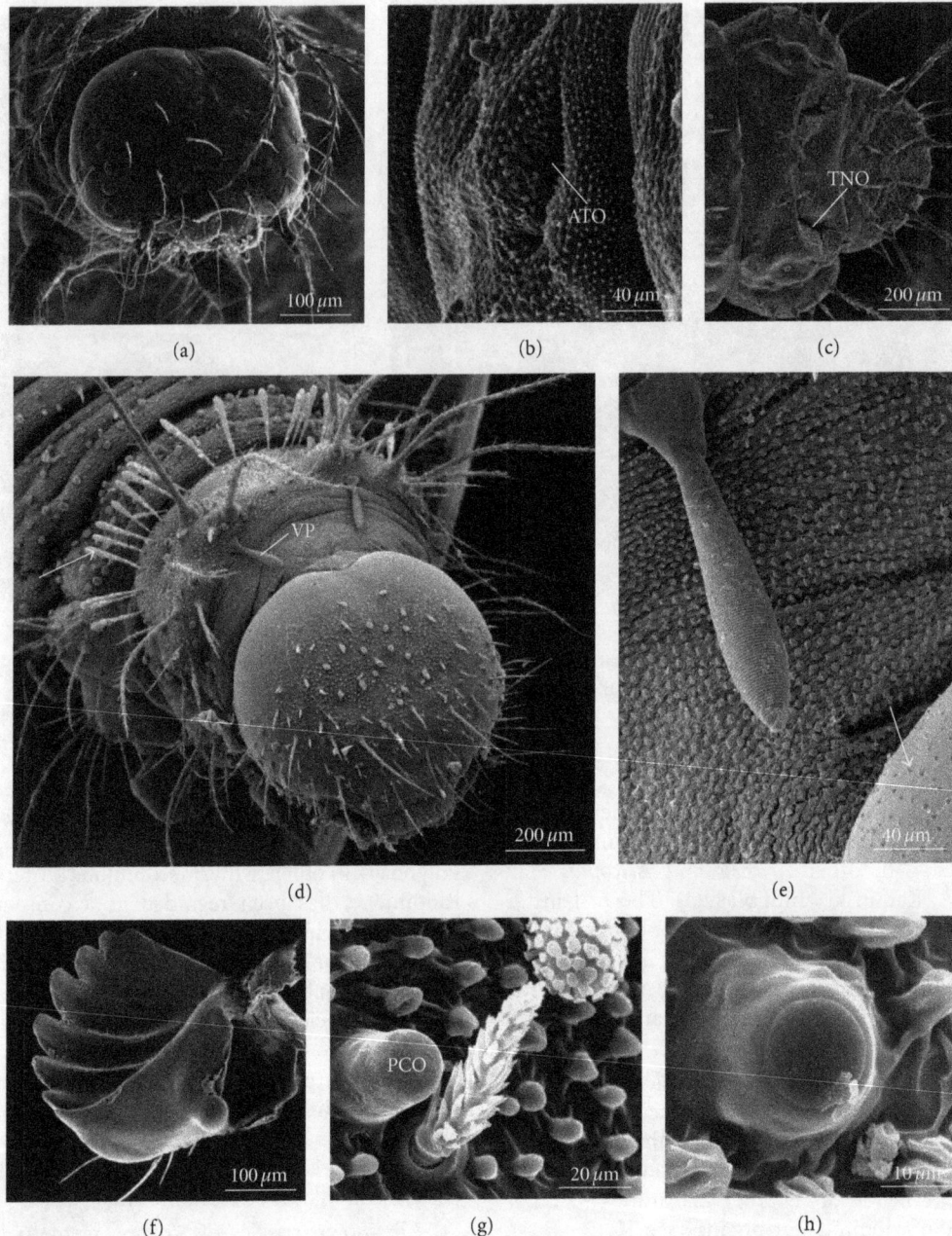

FIGURE 5: Scanning electron microscopy of first ((a)–(c)), fourth ((d) and (e)), and fifth ((g) and (h)) instar of *Aricoris propitia* ((a)–(e) from Assis Brasil, Acre, and ((f)–(h)) from Belém, Pará). (a) Head in frontal view; (b) opening of the anterior tentacle organ (ATO); (c) posterior abdominal segments showing the openings of tentacle nectary organs (TNOs); (d) head and prothorax in dorsofrontal view, note the dorsal row of setae (arrow) and the vibratory papillae (VP); (e) detail of vibratory papilla, note the epicranial granulations (arrow); (f) mandible; (g) two types of dendritic setae and perforated cupola organ (PCO); (h) perforated cupola organ.

interaction with fire ants has a single origin or has arisen more than once in these lineages.

The fire ants are highly dominant organisms and considered one of the most harmful bioinvaders ever known [37]. In their native range, from southern Brazil to Suriname, they are also considered pests in disturbed areas, especially in the Amazon (e.g., [38–40]). Although *A. propitia* occurs naturally in the Amazon (the holotype is from "Amazonas"), continual deforestation over the recent decades—especially

in the "arc of deforestation" (see [41])—could be providing a recent range expansion for this butterfly. Recent studies involving several molecular markers and morphological variation have revealed that *Solenopsis saevissima* belongs to a geographically structured complex of cryptic species [40]. How populations of *A. propitia* respond to ant host structure is an interesting and yet unanswered question. A recent study [42], for example, did not find a direct influence of host ants on the population structure of the obligate myrmecophilous

Life History of Aricoris propitia (Lepidoptera: Riodinidae)—A Myrmecophilous Butterfly Obligately Associated with Fire Ants

57

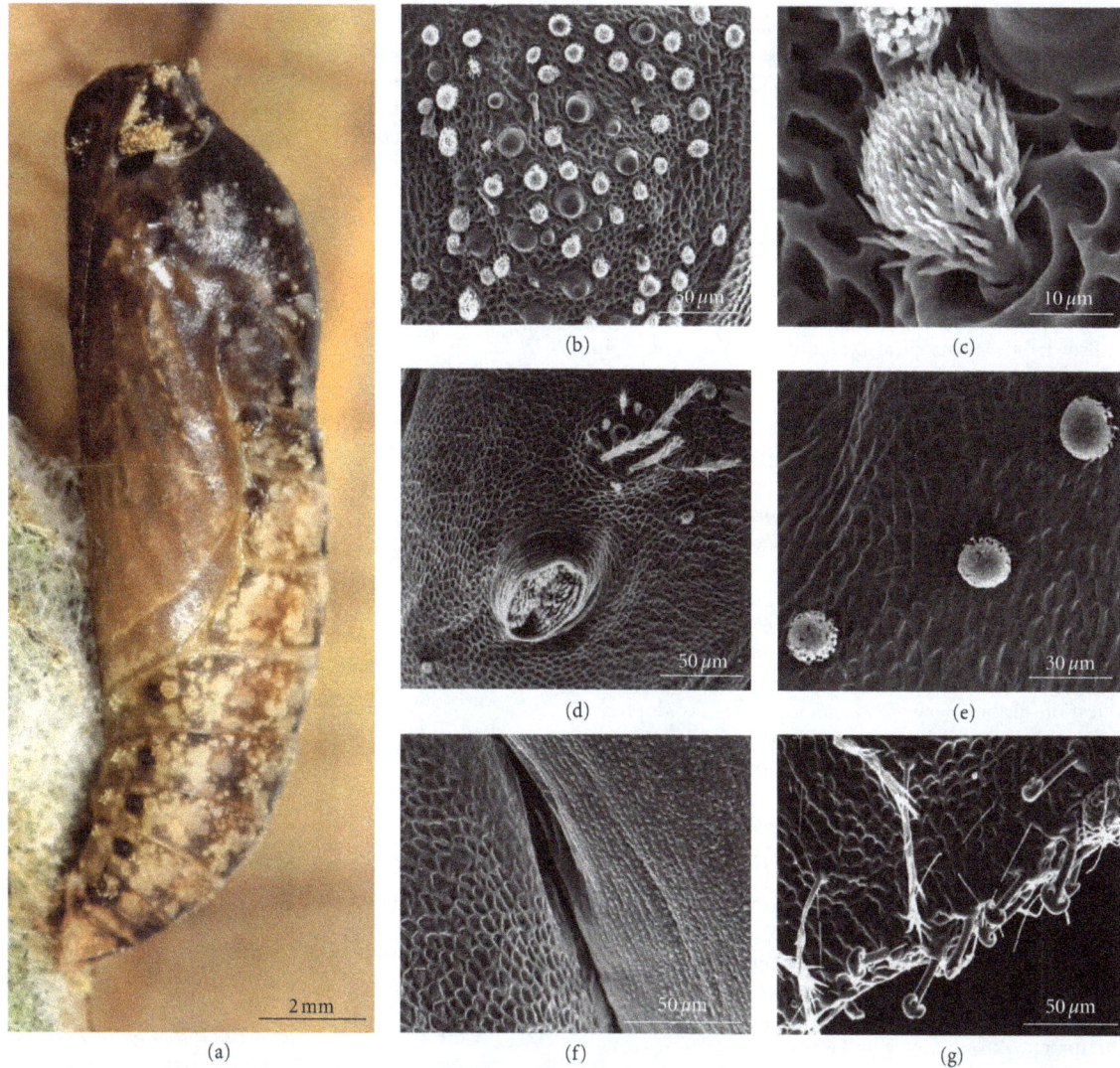

FIGURE 6: Pupa of *Aricoris propitia* in lateral view (a) and details ((b)–(g)) in scanning electron microscopy ((a), from Alto Paraíso and ((b)–(g)), from Belém, Pará). (b) Laterodorsal tubercle on A1 segment with dendritic setae and perforated cupola organs; (c) dendritic setae on A1 segment; (d) spiracle on A5 segment; (e) dendritic setae; (f) detail of putative stridulatory area between A4-A5 segments; (g) detail of cremaster crochet.

butterfly *Jalmenus evagoras* (Donovan) (Lycaenidae), but showed that biogeographical and host-plant aspects have an effect on that structure. *Aricoris propitia* may be a candidate system to elucidate the effects of ant attendance on the diversification of myrmecophilous butterflies.

The system involving *Aricoris propitia* and their tending fire ants presents several features of a model system, including: (1) it is common and widely distributed; (2) it is found in easily accessible environments (open and/or altered areas); (3) it adjusts well to laboratory conditions; (4) it has a short generation time; (5) the larvae accept many host-plant species; (6) the host fire ants have economic importance and various aspects of their biology are well known. Accordingly, we expect that the basic information provided in this work will encourage further studies on this interesting butterfly-ant system.

Acknowledgments

The authors thank Carla Guaitanele, Claudia Bottcher, Hosana Piccardi, Sebastian F. Sendoya, Thais Postali, and JPG Consultoria for their help with the field work; André V. L. Freitas and Paulo S. Oliveira for valuable laboratory support; Curtis Callaghan for confirming butterfly species identification, Fernando Fernández for his help with ant identifications, and Jorge Y. Tamashiro for plant species identifications; Adriana I. Zapata and Ana Gabriela Bieber for their help with Argentinean and German literature, respectively; Curtis Callaghan, Eduardo P. Barbosa, Luísa L. Mota, Pedro P. Rodrigues, and two anonymous reviewers for critically reading the paper. L. A. Kaminski thanks Conselho Nacional de Pesquisa (CNPq 140183/2006-0) and Fundação de Amparo à Pesquisa do Estado de São Paulo

(FAPESP 08/54058-1 and 10/51340-8). This paper is part of the RedeLep "Rede Nacional de Pesquisa e Conservação de Lepidópteros" SISBIOTA-Brasil/CNPq (563332/2010-7) and BIOTA-FAPESP Program (11/50225-3).

References

[1] K. Fiedler, "Systematic, evolutionary, and ecological implications of myrmecophily within the Lycaenidae (Insecta: Lepidoptera: Papilionoidea)," *Bonner Zoologische Monographien*, vol. 31, pp. 1–210, 1991.

[2] K. S. Brown Jr., "Neotropical Lycaenidae—an overview," in *Conservation Biology of Lycaenidae (Butterflies)*, T. R. New, Ed., pp. 45–61, IUCN, Gland, Switzerland, 1993.

[3] P. J. DeVries, *The Butterflies of Costa Rica and Their Natural History. Volume II. Riodinidae*, Princeton University Press, Princeton, NJ, USA, 1997.

[4] C. Bruch, "Orugas mirmecófilas de *Hamearis epulus signatus—Stich*," *Revista de la Sociedad Entomológica Argentina*, vol. 1, pp. 2–9, 1926.

[5] K. Fiedler, "Lycaenid-ant interactions of the *Maculinea* type: tracing their historical roots in a comparative framework," *Journal of Insect Conservation*, vol. 2, no. 1, pp. 3–14, 1998.

[6] Z. Fric, N. Wahlberg, P. Pech, and J. Zrzavý, "Phylogeny and classification of the *Phengaris-Maculinea* clade (Lepidoptera: Lycaenidae): total evidence and phylogenetic species concepts," *Systematic Entomology*, vol. 32, no. 3, pp. 558–567, 2007.

[7] J. A. Thomas, D. J. Simcox, and R. T. Clarke, "Successful conservation of a threatened *Maculinea* butterfly," *Science*, vol. 325, no. 5936, pp. 80–83, 2009.

[8] F. Bourquin, "Notas sobre la metamorfosis de *Hamearis susanea* Orfila, 1953 com oruga mirmecófila," *Revista de la Sociedad Entomológica Argentina*, vol. 16, pp. 83–87, 1953.

[9] K. J. Hayward, "Datos para el studio de la ontogenia de Lepidopteros Argentinos," *Miscelanea*, vol. 31, pp. 1–142, 1969.

[10] F. Schremmer, "Zur Bionomie und Morphologie der myrmekophilen Raupe und Puppe der neotropischen tagfalter-art *Hamearis erostratus* (Lepidoptera: Riodinidae)," *Entomologia Germanica*, vol. 4, pp. 113–121, 1978.

[11] R. K. Robbins and A. Aiello, "Foodplant and oviposition records for Panamanian Lycaenidae and Riodinidae," *Journal of the Lepidopterists' Society*, vol. 36, pp. 65–75, 1982.

[12] P. J. DeVries, I. A. Chacon, and D. Murray, "Toward a better understanding of host use and biodiversity in riodinid butterflies (Lepidoptera)," *Journal of Research on the Lepidoptera*, vol. 31, pp. 103–126, 1994.

[13] C. J. Callaghan and G. Lamas, "Riodinidae," in *Checklist: Part 4A. Hesperioidea—Papilionoidea*, G. Lamas, Ed., pp. 141–170, Association for Tropical Lepidoptera, Gainesville, Fla, USA, 2004.

[14] C. J. Callaghan, "A re-evaluation of the *Aricoris constantius* group with the recognition of three species (Lepidoptera: Riodinidae)," *Zoologia*, vol. 27, pp. 395–402, 2010.

[15] E. O. Emery, K. S. Brown Jr., and C. E. G. Pinheiro, "The butterflies (Lepidoptera, Papilionoidea) of the Distrito Federal, Brazil," *Revista Brasileira de Entomologia*, vol. 50, no. 1, pp. 85–92, 2006.

[16] F. Bodner, G. Brehm, J. Homeier, P. Strutzenberger, and K. Fiedler, "Caterpillars and host plant records for 59 species of geometridae (Lepidoptera) from a montane rainforest in southern ecuador," *Journal of Insect Science*, vol. 10, article 67, pp. 1–22, 2010.

[17] J. Altmann, "Observational study of behavior: sampling methods," *Behaviour*, vol. 49, no. 3-4, pp. 227–267, 1974.

[18] L. A. Kaminski, "Immature stages of *Caria plutargus* (Lepidoptera: Riodinidae), with discussion on the behavioral and morphological defensive traits in nonmyrmecophilous riodinid butterflies," *Annals of the Entomological Society of America*, vol. 101, no. 5, pp. 906–914, 2008.

[19] L. A. Kaminski, D. Rodrigues, and A. V. L. Freitas, "Immature stages of *Parrhasius polibetes* (Lepidoptera: Lycaenidae): host plants, tending ants, natural enemies, and morphology," *Journal of Natural History*. In press.

[20] J. C. Downey and A. C. Allyn, "Eggs of Riodinidae," *Journal of the Lepidopterists' Society*, vol. 34, pp. 133–145, 1980.

[21] F. W. Stehr, "Order Lepidoptera," in *Immature Insects. Vol. I*, F. W. Stehr, Ed., pp. 293–294, Kendall/Hunt Publishing, Dubuque, Iowa, USA, 1987.

[22] E. Mosher, "A classification of the Lepidoptera based on characters of the pupa," *Bulletin of the Illinois State Laboratory of Natural History*, vol. 12, pp. 17–159, 1916.

[23] P. J. DeVries, "The larval ant-organs of *Thisbe irenea* (Lepidoptera: Riodinidae) and their effects upon attending ants," *Zoological Journal of the Linnean Society*, vol. 94, no. 4, pp. 379–393, 1988.

[24] G. N. Ross, "Life-history studies on Mexican butterflies. III. Early stages of *Anatole rossi*, a new myrmecophilous metalmark," *Journal of Research on the Lepidoptera*, vol. 3, pp. 81–94, 1964.

[25] J. P. W. Hall and D. J. Harvey, "A phylogenetic analysis of the Neotropical riodinid butterfly genera *Juditha, Lemonias, Thisbe* and *Uraneis*, with a revision of Juditha (Lepidoptera: Riodinidae: Nymphidiini)," *Systematic Entomology*, vol. 26, no. 4, pp. 453–490, 2001.

[26] K. Nishida, "Description of the immature stages and life history of *Euselasia* (Lepidoptera: Riodinidae) on *Miconia* (Melastomataceae) in Costa Rica," *Zootaxa*, no. 2466, pp. 1–74, 2010.

[27] D. J. Harvey, *The higher classification of the Riodinidae (Lepidoptera)*, Ph.D. dissertation, University of Texas, Austin, Tex, USA, 1987.

[28] G. N. Ross, "Life-history studies on Mexican butterflies. IV. The ecology and ethology of *Anatole rossi*, a myrmecophilous metalmark (Lepidoptera: Riodinidae)," *Annals of the Entomological Society of America*, vol. 59, pp. 985–1004, 1966.

[29] K. Sharma, R. K. Vander Meer, and H. Y. Fadamiro, "Phorid fly, *Pseudacteon tricuspis*, response to alkylpyrazine analogs of a fire ant, *Solenopsis invicta*, alarm pheromone," *Journal of Insect Physiology*, vol. 57, no. 7, pp. 939–944, 2011.

[30] N. E. Pierce and M. A. Elgar, "The influence of ants on host plant selection by *Jalmenus evagora*, a myrmecophilous lycaenid butterfly," *Behavioral Ecology and Sociobiology*, vol. 16, no. 3, pp. 209–222, 1985.

[31] K. Fiedler, "Lycaenid butterflies and plants: is myrmecophily associated with amplified hostplant diversity?" *Ecological Entomology*, vol. 19, no. 1, pp. 79–82, 1994.

[32] L. A. Kaminski, "Polyphagy and obligate myrmecophily in the butterfly *Hallonympha paucipuncta* (Lepidoptera: Riodinidae) in the Neotropical Cerrado savanna," *Biotropica*, vol. 40, no. 3, pp. 390–394, 2008.

[33] N. E. Pierce, "Predatory and parasitic Lepidoptera: carnivores living on plants," *Journal of the Lepidopterists' Society*, vol. 49, no. 4, pp. 412–453, 1995.

[34] G. R. Canals, *Mariposas de Misiones*, L.O.L.A., Buenos Aires, Argentina, 2003.

Life History of Aricoris propitia (Lepidoptera: Riodinidae)—A Myrmecophilous Butterfly Obligately Associated with Fire Ants

59

[35] J. P. W. Hall and D. J. Harvey, "Basal subtribes of the Nymphidiini (Lepidoptera: Riodinidae): phylogeny and myrmecophily," *Cladistics*, vol. 18, no. 6, pp. 539–569, 2002.

[36] K. Fiedler, "Ants that associate with Lycaeninae butterfly larvae: diversity, ecology and biogeography," *Diversity and Distributions*, vol. 7, no. 1-2, pp. 45–60, 2001.

[37] D. A. Holway, L. Lach, A. V. Suarez, N. D. Tsutsui, and T. J. Case, "The causes and consequences of ant invasions," *Annual Review of Ecology and Systematics*, vol. 33, pp. 181–233, 2002.

[38] S. Adams, "Fighting the fire ant: Brazilian trip shows persistence of pest now found across the U.S. South—deforestation in Brazil creates optimum environment for fire ants," *Agricultural Research*, vol. 42, p. 4, 1994.

[39] J. C. Trager, "A revision of the fire ants, *Solenopsis geminata* group (Hymenoptera: Formicidae: Myrmicinae)," *Journal of the New York Entomological Society*, vol. 99, pp. 141–198, 1991.

[40] K. G. Ross, D. Gotzek, M. S. Ascunce, and D. D. Shoemaker, "Species delimitation: a case study in a problematic ant taxon," *Systematic Biology*, vol. 59, no. 2, pp. 162–184, 2010.

[41] L. Durieux, L. A. Toledo Machado, and H. Laurent, "The impact of deforestation on cloud cover over the Amazon arc of deforestation," *Remote Sensing of Environment*, vol. 86, no. 1, pp. 132–140, 2003.

[42] R. Eastwood, N. E. Pierce, R. L. Kitching, and J. M. Hughes, "Do ants enhance diversification in lycaenid butterflies? Phylogeographic evidence from a model myrmecophile, *Jalmenus evagoras*," *Evolution*, vol. 60, no. 2, pp. 315–327, 2006.

8

Diversity of Species and Behavior of Hymenopteran Parasitoids of Ants: A Review

Jean-Paul Lachaud[1,2] and Gabriela Pérez-Lachaud[1]

[1] *Departamento de Entomología Tropical, El Colegio de la Frontera Sur, Avenida Centenario km 5.5, 77014 Chetumal, QRoo, Mexico*
[2] *Centre de Recherches sur la Cognition Animale, CNRS-UMR 5169, Université de Toulouse, UPS, 118 route de Narbonne, 31062 Toulouse Cedex 09, France*

Correspondence should be addressed to Jean-Paul Lachaud, jlachaud@ecosur.mx

Academic Editor: Alain Lenoir

Reports of hymenopterans associated with ants involve more than 500 species, but only a fraction unambiguously pertain to actual parasitoids. In this paper, we attempt to provide an overview of both the diversity of these parasitoid wasps and the diversity of the types of interactions they have formed with their ant hosts. The reliable list of parasitoid wasps using ants as primary hosts includes at least 138 species, reported between 1852 and 2011, distributed among 9 families from 3 superfamilies. These parasitoids exhibit a wide array of biologies and developmental strategies: ecto- or endoparasitism, solitary or gregarious, and idio- or koinobiosis. All castes of ants and all developmental stages, excepting eggs, are possible targets. Some species parasitize adult worker ants while foraging or performing other activities outside the nest; however, in most cases, parasitoids attack ant larvae either inside or outside their nests. Based on their abundance and success in attacking ants, some parasitoid wasps like diapriids and eucharitids seem excellent potential models to explore how parasitoids impact ant colony demography, population biology, and ant community structure. Despite a significant increase in our knowledge of hymenopteran parasitoids of ants, most of them remain to be discovered.

1. Introduction

Ants are distributed all over the world, and their colonies provide both a stable food resource and numerous niches for thousands of other organisms, termed myrmecophiles, that exhibit a diverse array of relationships with their hosts [1–7]. Among myrmecophiles, numerous species of hymenopterans are associated with ants through predation, parasitism on the brood and/or adults, cleptoparasitism, parabiosis, mimetism, true symphily, or indirect parasitism through trophobionts and/or social parasites. However, in most cases, the precise nature of their relationship with their ant hosts remains obscure.

A review of the diversity of parasitoid wasps attacking ants has not been attempted since the work of Schmid-Hempel [7]. In his extensive review of the parasites of social insects, he pointed out the wide variety of hymenopteran parasitoids that attack these insects but, with the exception of the family Eucharitidae (with 33 valid species really involved), his list provided very few other examples (only 10) of true parasitoidism, that is, cases where the attack of the wasp species on ants (adults or brood) has been reliably demonstrated.

Knowledge has increased greatly in the intervening years, and numerous cases of parasitic associations involving wasps and ants have been reported. Moreover, changes in nomenclature and phylogeny have been numerous in the last two decades (see, e.g., [8–15]), and many species names of both the parasitoids and their ant hosts required emendations.

In the present paper, we address only hymenopteran parasitoids and focus strictly on ant-parasitoid wasp associations in which parasitism has been established beyond any doubt, and where ants are proved to be the primary hosts. Therefore, no bethylid species are considered here even though various members of the genera *Pseudisobrachium* and

Dissomphalus are strongly suspected of being parasitoids of ant brood [16–18]. Neither are any species of ceraphronid, dryinid, figitid, platygastrid, proctotrupid, or pteromalid wasps considered although several species belonging to the genera *Ceraphron*, *Conostigmus*, *Gonatopus*, *Kleidotoma*, *Platygaster*, *Exallonyx*, and *Spalangia* are known to be associated with ants, most of them probably as parasitoids [2, 19–24]. All of these species were omitted from the present paper because wasps have not been reliably reared from ants or their brood. Moreover, according to the definition of "parasitoid" which implies the killing of a single host, associations such as those involving numerous sphecid species, particularly those of the genera *Aphilanthops*, *Clypeadon*, and *Tracheliodes*, which are known to specialize with preying on and storing numerous adult ants (of the genus *Formica*, *Pogonomyrmex*, or *Liometopum*, resp.) [25–27] are not dealt with. Likewise, the highly interesting associations of ants with some braconid species such as *Compsobraconoides* sp. [28] and *Trigastrotheca laikipiensis* Quicke [29], which are known to consume various stages of their ant hosts (*Azteca* spp. and *Crematogaster* spp., resp.) during their development, are not covered in the present paper.

In spite of such restrictions, the list of hymenopteran species reliably involved in parasitic associations with ants remains impressive and represents more than a quarter of all of the hymenopteran species known to be associated with ants [30]. Here, we attempt to provide an overview of both the diversity of the species of parasitoid wasps known to attack ants and the diversity of the interactions they have developed with their hosts. By so doing, we also call attention to this little known biodiversity.

2. Checklist of Hymenopteran Parasitoids of Ants

Records of associations of hymenopteran wasps with ants involve more than 500 wasp species [30], but only a fraction have unambiguously been reported as parasitoids. The term parasitoid applies to organisms whose juvenile stages are parasites of a single host individual, eventually sterilizing, killing, or even consuming their host, while the adult parasitoid is free living [31]. With few exceptions, female parasitoid wasps oviposit on or inside the body of their host, typically another arthropod, and all stages of development of the host are susceptible to attack. After hatching, the parasitoid larva feeds on the host's tissues, gradually killing it. A survey of the literature since 1852 and some of our own unpublished results have allowed us to identify at least 138 species (see Table 1 and Supplementary Material available online at doi:10.1155/2012/134746) reported as primary endo- or ectoparasitoids of larvae, pupae, or adult ants. All of these species are included in 3 superfamilies: Chalcidoidea (with 6 families concerned), Ichneumonoidea (2 families), and Diaprioidea (only 1 family) (Table 1). In 2007, Sharkey [12] estimated that there were approximately 115,000 described species of Hymenoptera (perhaps up to 1,000,000 if undescribed species—especially species of parasitoid wasps—were included), and that Chalcidoidea and

Ichneumonoidea were the most species-rich superfamilies among the parasitoid hymenopterans. So, it is not surprising that most of the parasitoid wasps attacking ants belong to these two superfamilies, especially the Chalcidoidea which alone includes more than 70% of all of wasp species parasitizing ants registered until now.

In the following text, we follow Sharkey [12] for the higher-level phylogeny of the order Hymenoptera (see also [15]). The taxonomic validity of the scientific names is in accordance with Bolton [8, 9] for ants, and with different databases available on the web for other hymenopterans: Hymenoptera Name Server (version 1.5) (http://osuc.biosci .ohio-state.edu/hymDB/nomenclator.home_page), Global Name Index (version 0.9.34) (http://gni.globalnames.org/ name_strings), Universal Chalcidoidea Database [32] (http://www.nhm.ac.uk/chalcidoids), and Home of Ichneumonoidea (version 2011) (http://www.ichneumonoidea .name/index.php). Authors of all scientific names are given throughout the text only when they are not already reported in Table 1.

2.1. Diaprioidea. The superfamily Diaprioidea is a monophyletic group, with 4 recognized families [15], and accounts for more than 4000 species around the world in over 210 genera [8, 155–157], almost all in the family Diapriidae. Most diaprioids are primary endoparasitoids of dipterans (eggs, larvae, or pupae), but several species are known to attack Hymenoptera, Homoptera, or Coleoptera, and some are facultative or obligate hyperparasitoids. Some of the species attacking Diptera have been considered as potential biological control agents, but their efficiency has not been demonstrated [157, 158].

2.1.1. Diapriidae. Despite their number, the members of this large family are relatively unknown and less than half of the 4000 species estimated to occur worldwide have been described [8, 156, 159]. Three subfamilies are currently recognized: Ambositrinae, Belytinae, and Diapriinae [15]. Their biologies are diverse, but most species are primary parasitoids of puparia of Diptera [156–160].

Although some diapriids have only occasionally been found in ant nests, a number of species are closely associated with ants (all belonging to the Belytinae and Diapriinae subfamilies). However, there are few behavioral data on host-diapriid myrmecophile interactions (but see [36]). These symphyles are often highly adapted to their hosts, exhibiting morphological and behavioral adaptations to living with ants (extensive morphological mimicry of the host ants—coloration, ocellus regression, convergence in sculpture—, presence of appeasement substances in specialized structures and trichomes, trophallaxis, etc., [161–166]), which presumably aid them in avoiding detection and/or aggression by host ants [34]. The adaptations can include secondary apterism in which the wings of the wasps are assumed to have been bitten off by either the parasite itself or its host (e.g., *Mimopria*, *Bruchopria*, *Lepidopria*, and *Solenopsia*, [156, 161, 164, 167, 168]). Most often, the presence of a diapriid in an ant nest is suspected to be just circumstantial [160] and related to its

TABLE 1: List of parasitic wasps recorded as true primary parasitoids of ants (brood or adult). As all of the eucharitids are true parasitoids of ants, all known associations with ants have been included, but see **. For further details, see text.

Hymenopterous parasitoids		Associated ant host		References
Species	Referred to as	Species	Referred to as	
Diaprioidea: Diapriidae (26)				
Acanthopria sp.	—	Cyphomyrmex salvini Forel	—	[33]
Acanthopria sp.	—	Trachymyrmex cf. zeteki Weber	—	[34]
Acanthopria sp. no. 1	—	Cyphomyrmex transversus Emery	—	[35]
Acanthopria sp. no. 2	—	Cyphomyrmex transversus Emery	—	[35]
Acanthopria sp. no. 3	—	Cyphomyrmex transversus Emery	—	[35]
Acanthopria sp. no. 4	—	Cyphomyrmex transversus Emery	—	[35]
Acanthopria sp. no. 5	—	Cyphomyrmex transversus Emery	—	[35]
Acanthopria sp. no. 6	—	Cyphomyrmex transversus Emery	—	[35]
Acanthopria sp. no. 7	—	Cyphomyrmex transversus Emery	—	[35]
Acanthopria sp. no. 8	—	Cyphomyrmex transversus Emery	—	[35]
Acanthopria sp. 1	—	Cyphomyrmex minutus Mayr	—	[36]
Acanthopria sp. no. 1'	—	Cyphomyrmex rimosus (Spinola)	—	[36]
Acanthopria sp. no. 2'	—	Cyphomyrmex rimosus (Spinola)	—	[36]
Acanthopria sp. no. 3'	—	Cyphomyrmex rimosus (Spinola)	—	[36]
Acanthopria sp. no. 4'	—	Cyphomyrmex rimosus (Spinola)	—	[36]
Mimopriella sp.	—	Cyphomyrmex rimosus (Spinola)	—	[36]
Mimopriella sp. 1	—	Trachymyrmex cf. zeteki Weber	—	[34]
Mimopriella sp. 2	—	Trachymyrmex cf. zeteki Weber	—	[34]
Oxypria sp.	—	Trachymyrmex cf. zeteki Weber	—	[34]
Plagiopria passerai Huggert and Masner	—	Plagiolepis pygmaea (Latr.)	—	[37]
Szelenyiopria lucens (Loiácono)	Gymnopria lucens	Acromyrmex ambiguus (Emery)	—	[38]
Szelenyiopria pampeana (Loiácono)	Gymnopria pampeana	Acromyrmex lobicornis (Emery)	—	[39]
Szelenyiopria sp. 1	—	Trachymyrmex cf. zeteki Weber	—	[34]
Szelenyiopria sp. 2	—	Trachymyrmex cf. zeteki Weber	—	[34]
Trichopria formicans Loiácono	—	Acromyrmex lobicornis (Emery)	—	[40]
Trichopria sp.	—	Acromyrmex lobicornis (Emery)	—	[40]
Chalcidoidea: Chalcididae (2 + 2*)				
Smicromorpha doddi Girault	—	Oecophylla smaragdina (Fabr.)	—	[41, 42]
Smicromorpha keralensis Narendran*	—	Oecophylla smaragdina (Fabr.)	—	[43]
Smicromorpha masneri Darling	—	Oecophylla smaragdina (Fabr.)	—	[44]
Smicromorpha minera Girault *	—	Oecophylla smaragdina (Fabr.)	—	[42]
Chalcidoidea: Encyrtidae (1)				
Blanchardiscus sp. ?pollux Noyes	—	Pachycondyla goeldii (Forel)	—	[45]
Chalcidoidea: Eucharitidae (86 + 7** + 1***)				
Ancylotropus manipurensis	—	Camponotus sp.***	—	[11, 46]
Ancylotropus sp.	—	Odontomachus troglodytes Santschi	—	[11]
Athairocharis vannoorti Heraty	—	Anoplolepis sp.	Anaplolepis sp.	[11]
Austeucharis fasciiventris (Brues)	Psilogaster fasciiventris	Myrmecia gulosa (Fabr.)	—	[47]
Austeucharis implexa (Walker)	—	Myrmecia pilosula F. Smith	—	[11]

TABLE 1: Continued.

Hymenopterous parasitoids		Associated ant host		References
Species	Referred to as	Species	Referred to as	
Austeucharis myrmeciae (Forel)	*Eucharis myrmeciae* Cameron	*Myrmecia forficata* (Fabr.)	—	[48]
Austeucharis sp.	—	*Myrmecia pavida* Clark	*M. atrata* Clark	[49, 50]
	—	*Myrmecia nigriceps* Mayr	*M. nigriceps* Smith	[49, 50]
	—	*Myrmecia pilosula* F. Smith	—	[50]
	Epimetagea sp.	*Myrmecia pyriformis* F. Smith	—	[51]
	—	*Myrmecia tarsata* F. Smith	—	[50]
	—	*Myrmecia vindex* F. Smith	*M. vindex* Forel	[50]
Chalcura affinis (Bingham)	*Rhipipallus affinis*	*Odontomachus ruficeps* F. Smith	*O. ruficeps* subsp. *coriarius* Mayr	[52]
	Chalcuroides versicolor Girault	*Odontomachus* sp.	*Myrmecia* sp.	[53, 54]
Chalcura deprivata (Walker)	—	*Odontomachus haematodus* (L.)	*O. haematodes*	[55]
Chalcura nigricyanea (Girault)	—	*Rhytidoponera metallica* (F. Smith)	*R. metallicum*	[11]
Chalcura polita (Girault)	—	*Rhytidoponera metallica* (F. Smith)	*R. metallicum*	[11]
Chalcura sp.	—	*Formica rufa* L.	—	[56]
Chalcura sp. nr. *polita* (Girault)	—	*Rhytidoponera chalybaea* Emery	—	[11]
Dicoelothorax platycerus Ashmead	—	*Ectatomma brunneum* F. Smith	—	[57]
Dilocantha lachaudii Heraty	—	*Ectatomma tuberculatum* (Olivier)	—	[58, 59]
Eucharis adscendens (Fabr.)	—	*Formica ?cunicularia* Latr.**	*F. glauca* Ruzsky	[60]
	—	*Formica rufa* L.	—	[61]
	—	*Messor barbarus* (L.)**	*Aphaenogaster barbara* L.	[62]
Eucharis bedeli (Cameron)	—	*Cataglyphis bicolor* (Fabr.)***	*C. viaticus*	[63]
	Chalcura bedeli	*Cataglyphis viaticus* (Fabr.)	*Myrmecocystus viaticus*	[64, 65]
	Chalcura bedeli	*Formica rufa* L.***	—	[61, 65]
Eucharis esakii Ishii	*E. scutellaris* Gahan	*Formica japonica* Motschoulski	*F. fusca fusca japonica* Mots.	[66]
	E. scutellaris Gahan	*Formica* sp.	—	[55]
Eucharis microcephala Bouček	—	*Cataglyphis nodus* (Brullé)	*C. bicolor* ssp. *nodus*	[67]
Eucharis punctata Förster	—	*Messor concolor* Santschi**	*M. barbarus* r. *semirufus* v. *concolor* Sm.	[68]
Eucharis rugulosa Gussakovskiy	—	*Cataglyphis* sp.**	—	[60]
Eucharis shestakovi Gussakovskiy	—	*Messor structor* (Latr.)**	—	[69]
Eucharis sp.	—	*Formica neorufibarbis* Emery**	*F. fusca neorufibarbis*	[70]
	—	*Myrmica incompleta* Provancher**	*M. brevinodis* Emery	[70]
Galearia latreillei (Guérin-Méneville)	*Thoracantha bruchi*	*Pogonomyrmex cunicularius* Mayr**	*P. carnivora* Santschi	[11, 71]
Gollumiella longipetiolata Hedqvist	—	*Paratrechina* sp.	—	[72]
Hydrorhoa sp. *striaticeps* Kieffer complex	—	*Camponotus maculatus* (Fabr.)	*C. maculatus* Mayr	[11]
Isomerala coronata (Westwood)	*Isomaralia coronata*	*Ectatomma tuberculatum* (Olivier)	—	[73]
	—	*Ectatomma ruidum* Roger***	—	[11]
Kapala atrata (Walker)	*K. surgens*	*Pachycondyla harpax* (Fabr.)	—	[11]
Kapala cuprea Cameron	—	*Pachycondyla crassinoda* (Latr.)	—	[74]
Kapala floridana (Ashmead)	—	*Pogonomyrmex badius* (Latr.)**	—	[70]
Kapala iridicolor (Cameron)	*K. sulcifacies* (Cameron)	*Ectatomma ruidum* Roger	—	[75, 76]

TABLE 1: Continued.

Hymenopterous parasitoids		Associated ant host		References
Species	Referred to as	Species	Referred to as	
	—	*Gnamptogenys regularis* Mayr	—	[76]
	—	*Gnamptogenys striatula* Mayr	—	[76]
	—	*Gnamptogenys sulcata* (F. Smith)	—	[76]
	—	*Pachycondyla stigma* (Fabr.)	—	[76]
Kapala izapa Carmichael	—	*Ectatomma ruidum* Roger	—	[76]
Kapala sp.	—	*Dinoponera lucida* Emery	—	[77]
	—	*Ectatomma brunneum* F. Smith	—	[78]
	—	*Ectatomma tuberculatum* (Olivier)	—	[79]
	—	*Gnamptogenys sulcata* (F. Smith)	—	[80]
	—	*Gnamptogenys tortuolosa* (F. Smith)	—	[78]
	—	*Hypoponera nitidula* (Emery)	—	[81]
	—	*Odontomachus bauri* Emery	—	[11]
	—	*Odontomachus brunneus* (Patton)	—	[80]
	—	*Odontomachus haematodus* (L.)	—	[77]
	—	*Odontomachus hastatus* (Fabr.)	—	[11]
	—	*Odontomachus insularis* Guérin-Méneville	*O. haematodes insularis pallens* Wheeler	[66]
	—	*Odontomachus laticeps* Roger	—	[80]
	—	*Odontomachus mayi* Mann	—	[78]
	—	*Odontomachus meinerti* Forel	—	[81]
	—	*Odontomachus opaciventris* Forel	—	[80]
	—	*Pachycondyla apicalis* (Latr.)	—	[80]
	—	*Pachycondyla harpax* (Fabr.)	—	[81]
	—	*Pachycondyla stigma* (Fabr.)	—	[81]
	—	*Pachycondyla verenae* (Forel)	—	[78]
	—	*Typhlomyrmex rogenhoferi* Mayr	—	[81]
Kapala terminalis Ashmead	—	*Odontomachus insularis* Guérin-Méneville	*O. haematodes insularis pallens* Wheeler	[66]
Lophyrocera variabilis Torréns, Heraty and Fidalgo	—	*Camponotus* sp.	—	[82]
Mateucharis rugulosa Heraty	—	*Camponotus* sp.	—	[11]
Neolosbanus gemma (Girault)	—	*Hypoponera* sp.	—	[83]
Neolosbanus palgravei (Girault)	—	*Hypoponera* sp.	—	[83]
Obeza floridana (Ashmead)	—	*Camponotus floridanus* (Buckley)	*C. abdominalis floridanus*	[84]
Orasema aenea Gahan	—	*Solenopsis quinquecuspis* Forel	—	[85]
Orasema argentina Gemignani	—	*Pheidole nitidula* Santschi	*P. strobeli misera* Snts.	[71]
Orasema assectator Kerrich	—	*Pheidole* sp.	—	[86, 87]
Orasema coloradensis Wheeler	*O. coloradensis* Ashmead	*Diplorhoptrum validiusculum* (Emery)	*Solenopsis molesta validiuscula*	[70]
	O. coloradensis Gahan	*Formica oreas comptula* Wheeler	—	[88]
	O. coloradensis Gahan	*Formica subnitens* Creighton	—	[88]
	O. coloradensis Ashmead	*Pheidole bicarinata* Mayr	*P. vinelandica* Forel	[70]
Orasema costaricensis Wheeler and Wheeler	—	*Pheidole flavens* Roger	—	[63]

TABLE 1: Continued.

Hymenopterous parasitoids		Associated ant host		References
Species	Referred to as	Species	Referred to as	
	—	*Pheidole vallifica* Forel	—	[89]
Orasema fraudulenta (Reichensperger)	*Psilogaster fraudulentus*	*Pheidole megacephala* (Fabr.)	—	[90]
Orasema minuta Ashmead	—	*Pheidole* nr. *tetra* Creighton	—	[11, 83]
	—	*Temnothorax allardycei* (Mann)	—	[11, 83]
Orasema minutissima Howard	—	*Wasmannia auropunctata* (Roger)	—	[91]
	—	*Wasmannia sigmoidea* (Mayr)	—	[92]
Orasema monomoria Heraty	—	*Monomorium* sp.	—	[93]
Orasema occidentalis Ashmead	—	*Pheidole pilifera* (Roger)	—	[94]
Orasema pireta Heraty	—	*Solenopsis* sp.	—	[85]
Orasema rapo (Walker)	—	*Eciton quadriglume* (Haliday)**	—	[83]
Orasema robertsoni Gahan	—	*Pheidole dentata* Mayr	—	[95]
Orasema salebrosa Heraty	—	*Solenopsis invicta* Buren	—	[85]
	—	*Solenopsis richteri* Forel	—	[96]
Orasema simplex Heraty	—	*Solenopsis invicta* Buren	—	[97]
	—	*Solenopsis macdonaghi* Santschi	—	[85]
	—	*Solenopsis quinquecuspis* Forel	—	[85]
	—	*Solenopsis richteri* Forel	—	[96]
Orasema simulatrix Gahan	—	*Pheidole desertorum* Wheeler	—	[98]
Orasema sixaolae Wheeler and Wheeler	—	*Solenopsis tenuis* Mayr	—	[63]
Orasema sp.	B1 nr. *bakeri*	*Solenopsis geminata* (Fabr.)	—	[83]
	B1 nr. *bakeri*	*Solenopsis xyloni* MacCook	—	[83]
Orasema sp.	B2 nr. *bakeri*	*Pheidole* nr. *californica* Mayr	—	[83]
	B2 nr. *bakeri*	*Pheidole* nr. *clementensis* Gregg	—	[83]
	B2 nr. *bakeri*	*Pheidole* sp.	—	[83]
	B2 nr. *bakeri*	*Tetramorium* sp.	—	[83]
Orasema sp.	C1 nr. *costaricensis*	*Pheidole dentata* Mayr	—	[83]
Orasema sp.	—	*Pheidole bilimeki* Mayr	*P. anastasii* Emery	[99]
Orasema sp.	—	*Pheidole paiute* Gregg	—	[94]
Orasema sp. nr. *bouceki* Heraty	—	*Pheidole* sp.	—	[83]
Orasema sp. *uichancoi*-group	—	*Pheidole* sp.	—	[93]
Orasema susanae Gemignani	—	*Pheidole* nr. *tetra* Creighton	—	[83]
Orasema tolteca Mann	—	*Pheidole hirtula* Forel	*P. vasleti* var. *acohlma*	[100]
Orasema valgius (Walker)	*O. pheidolophaga* Girault	*Pheidole* sp.	—	[53]
Orasema wheeleri Wheeler	*O. wheeleri* Ashmead	*Pheidole ceres* Wheeler	—	[70]
	O. viridis Ashmead	*Pheidole dentata* Mayr	—	[55, 70]
	O. viridis Ashmead	*Pheidole sciophila* Wheeler	—	[55, 70]
	O. viridis Ashmead	*Pheidole tepicana* Pergande	*P. kingi* subsp. *instabilis* Emery	[55, 70]
	O. viridis Ashmead	*Pheidole tepicana* Pergande	*P. carbonaria* Pergande	[55, 70]
Orasema worcesteri (Girault)	*O. doello-juradoi* Gemignani	*Pheidole radoszkowskii* Mayr	*P. nitidula* Emery	[71, 96]
Orasema xanthopus (Cameron)	—	*Solenopsis invicta* Buren	—	[83, 96]
	—	*Solenopsis quinquecuspis* Forel	—	[85]
	—	*Solenopsis richteri* Forel	—	[101]
	—	*Solenopsis saevissima* (F. Smith)	—	[102]

TABLE 1: Continued.

Hymenopterous parasitoids		Associated ant host		References
Species	Referred to as	Species	Referred to as	
Orasemorpha eribotes (Walker)	—	*Pheidole* sp.	—	[54]
Orasemorpha myrmicae (Girault)	—	*Pheidole* sp.	—	[83]
Orasemorpha tridentata (Girault)	*Eucaromorpha wheeleri* Brues	*Pheidole proxima* Mayr	—	[103]
Orasemorpha xeniades (Walker)	—	*Pheidole tasmaniensis* Mayr	—	[83]
Pogonocharis browni Heraty	—	*Gnamptogenys menadensis* (Mayr)	—	[11]
Pseudochalcura gibbosa (Provancher)	—	*Camponotus herculeanus* (L.)	—	[46]
	—	*Camponotus laevigatus* (F. Smith)	—	[104]
	—	*Camponotus novaeboracensis* (Fitch)	*C. ligniperdus* var. *novaeboracensis*	[70]
	—	*Camponotus* sp. ?*vicinus* Mayr	—	[104]
Pseudochalcura nigrocyanea Ashmead	—	*Camponotus* sp.	—	[105]
Pseudochalcura sculpturata Heraty	—	*Camponotus planatus* Roger	—	[11]
Pseudometagea schwarzii (Ashmead)	—	*Lasius neoniger* Emery	—	[106]
Rhipipalloidea madangensis Maeyama, Machida, and Terayama	—	*Camponotus* (*Tanaemyrmex*) sp.	—	[107]
Rhipipalloidea mira Girault	—	*Polyrhachis femorata* F. Smith	—	[11]
Schizaspidia convergens (Walker)	—	*Odontomachus haematodus* (L.)	*O. haematodes*	[55]
Schizaspidia nasua (Walker)	—	*Odontomachus rixosus* F. Smith	—	[11]
Stilbula arenae Girault	—	*Polyrhachis* sp.	*Cyrtomyrma* sp.	[54]
Stilbula cyniformis (Rossi)	*S. cynipiformis*	*Camponotus aethiops* (Latr.)	*C. marginatus* Latr.	[68]
	Schizaspidia tenuicornis	*Camponotus japonicus* Mayr	*C. herculeanus* ssp. *japonicus*	[108]
	Schizaspidia tenuicornis	*Camponotus obscuripes* Mayr	*C. herculeanus* ssp. *ligniperdus* v. *obscuripes*	[66, 108]
	S. cynipiformis	*Camponotus sanctus* Forel	*C. maculatus* r. *sanctus*	[62]
Stilbula polyrhachicida (Wheeler and Wheeler)	*Schizaspidia polyrhachicida*	*Polyrhachis dives* F. Smith	*Polyrhachis* (*Myrmhopla*) *dives*	[109]
Stilbuloida calomyrmecis (Brues)	*Schizaspidia calomyrmecis*	*Calomyrmex purpureus* (Mayr)	—	[103]
Stilbuloida doddi (Bingham)	*Schizaspidia doddi*	*Camponotus* sp.	—	[52]
Timioderus acuminatus Heraty	—	*Pheidole capensis* Mayr	—	[93]
Tricoryna chalcoponerae Brues	—	*Rhytidoponera metallica* (F. Smith)	*Chalcoponera metallica* var. *critulata*	[103]
Tricoryna ectatommae Girault	—	*Rhytidoponera* sp.	*Ectatomma* sp.	[110]
Tricoryna iello (Walker)	—	*Rhytidoponera* sp.	—	[11]
Tricoryna minor (Girault)	—	*Rhytidoponera metallica* (F. Smith)	—	[11]
	—	*Rhytidoponera victoriae* (André)	—	[11]
Tricoryna sp. nr. *alcicornis* (Bouček)	—	*Rhytidoponera violacea* (Forel)	—	[11]
Zulucharis campbelli Heraty	—	*Camponotus* sp.	—	[11]
Chalcidoidea: Eulophidae (5)				
Horismenus floridensis (Schauff and Bouček)	*Alachua floridensis*	*Camponotus atriceps* (F. Smith)	*C. abdominalis* (Fabr.)	[111]
	Alachua floridensis	*Camponotus floridanus* (Buckley)	—	[111]
Horismenus myrmecophagus Hansson, Lachaud, and Pérez-Lachaud	—	*Camponotus* sp. ca. *textor* Forel	—	[112]

TABLE 1: Continued.

Hymenopterous parasitoids		Associated ant host		References
Species	Referred to as	Species	Referred to as	
Myrmokata diparoides Bouček	—	*Crematogaster* sp.	—	[113]
Pediobius marjoriae Kerrich	—	*Lepisotia* sp.	*Acantholepis* sp.	[114]
Unidentified sp. (?*Horismenus*)	nr. *Paracrias*	*Crematogaster acuta* (Fabr.)	—	[109, 112]
Chalcidoidea: Eurytomidae (4)				
Aximopsis affinis (Brues)	*Conoaxima affinis*	*Azteca* sp.	—	[115]
	Conoaxima affinis	*Azteca alfari* Emery	*Azteca alfari* subsp. *lucidula* var. *canalis*	[116]
	Conoaxima affinis	*Azteca pittieri* Forel	—	[117]
Aximopsis aztecicida (Brues)	*Conoaxima aztecicida*	*Azteca alfari* Emery	*Azteca alfaroi*	[115]
	Conoaxima aztecicida	*Azteca constructor* Emery	—	[115]
Aximopsis sp.	*Conoaxima* sp.	*Azteca salti* Wheeler	*Azteca xanthochroa* (Roger) subsp. *salti*	[116]
Aximopsis sp. (?*aztecicida*)	*Conoaxima* sp. (?*aztecicida*)	*Azteca alfari* Emery	—	[118]
	Conoaxima sp. (?*aztecicida*)	*Azteca australis* Wheeler	—	[118]
	Conoaxima sp. (?*aztecicida*)	*Azteca ovaticeps* Forel	—	[119]
	Conoaxima sp. (?*aztecicida*)	*Camponotus balzani* Emery	—	[118]
Chalcidoidea: Perilampidae (1)				
Unidentified sp.	—	*Pachycondyla luteola* (Roger)	—	[119]
Ichneumonoidea: Braconidae (11 + 4*)				
Elasmosoma berolinense Ruthe	—	*Camponotus* spp.	—	[120]
	—	*Camponotus vagus* (Scopoli)	—	[121]
	—	*Formica fusca* L.	—	[48]
	—	*Formica japonica* Motschoulsky	—	[122]
	—	*Formica pratensis* Retzius	—	[123]
	—	*Formica rufa* L.	—	[124–126]
	—	*Formica sanguinea* Latr.	—	[48]
	—	*Formica* spp.	—	[120]
	—	*Lasius niger* (L.)	—	[48, 56]
	—	*Polyergus* sp.	—	[127]
Elasmosoma luxemburgense Wasmann	—	*Formica rufibarbis* Fabr.	—	[128, 129]
Elasmosoma michaeli Shaw	—	*Formica obscuripes* Forel	—	[130]
	E. sp. nr. *pergandei* Ashmead	*Formica obscuriventris clivia* Creighton	—	[131]
Elasmosoma pergandei Ashmead*	—	*Camponotus castaneus* (Latr.)	*C. melleus* (Say)	[132]
	—	*Formica integra* Nylander	—	[126]
	—	*Formica subsericea* Say	—	[126]
Elasmosoma petulans Muesebeck*	—	*Formica integra* Nylander	—	[133]
	—	*Formica opaciventris* Emery	—	[127, 133, 134]
	—	*Formica pergandei* Emery	*F. rubicunda* Emery	[133, 134]
	—	*Formica rubicunda* Emery***	—	[127, 133, 134]
	—	*Formica subintegra* Wheeler	*F. subintegra* Emery	[133]
	—	*Formica subsericea* Say	—	[133]

TABLE 1: Continued.

Hymenopterous parasitoids		Associated ant host		References
Species	Referred to as	Species	Referred to as	
Elasmosoma schwarzi Ashmead*	—	*Formica schaufussi* Mayr	—	[127]
	—	*Polyergus lucidus* Mayr	—	[127]
Elasmosoma vigilans Cockerell	—	*Formica perpilosa* Wheeler	—	[94]
	—	*Formica subpolita* Mayr	—	[135]
Elasmosomites primordialis Brues	—	*Lasius* sp. (?*schiefferdeckeri* Mayr)	—	[136]
Kollasmosoma marikovskii (Tobias)	—	*Formica pratensis* Retzius	—	[137]
Kollasmosoma platamonense (Huddleston)	*Elasmosoma platamonense*	*Cataglyphis bicolor* (Fabr.)	—	[127]
		Messor semirufus (André)	—	[138]
Kollasmosoma sentum van Achterberg and Gómez	—	*Cataglyphis ibericus* (Emery)	—	[129]
Neoneurus auctus (Thomson)	*Euphorus bistigmaticus* Morley	*Formica pratensis* Retzius	—	[139, 140]
	Euphorus bistigmaticus Morley	*Formica rufa* L.	—	[139, 140]
Neoneurus clypeatus (Förster)*	*Elasmosoma viennense* Giraud	*Formica rufa* L.	—	[141]
Neoneurus mantis Shaw	—	*Formica podzolica* Francoeur	—	[142, 143]
Neoneurus vesculus van Achterberg and Gómez	—	*Formica cunicularia* Latr.	—	[129]
Ichneumonoidea: Ichneumonidae (3 + 2*)				
Eurypterna cremieri (de Romand)	*Pachylomma cremieri*	*Formica rufa* L.	—	[144]
	Pachylomma cremieri	*Lasius fuliginosus* (Latr.)	*Formica fuliginosa*	[145–148]
	—	*Lasius niger* (L.)	—	[123]
	Pachylomma cremieri	*Lasius nipponensis* Forel	—	[149]
Ghilaromma fuliginosi (Donisthorpe and Wilkinson)*	*Paxylomma fuliginosi*	*Lasius fuliginosus* (Latr.)	—	[150, 151]
Hybrizon buccatus (Brébisson)	*Pachylomma buccata*	*Formica rufa* L.	*F. rufa* var. *rufo-pratensis*	[152]
	Pachylomma buccata	*Formica rufibarbis* Fabr.	—	[152]
	Pachylomma buccata	*Formica sanguinea* Latr.	—	[152]
	Pachylomma buccata Nees	*Lasius alienus* (Förster)	*Donisthorpea aliena*	[24]
	Pachylomma buccatum	*Lasius brunneus* (Latr.)	—	[144]
	Pachylomma buccata	*Lasius flavus* (Fabr.)	—	[152]
	—	*Lasius grandis* Forel	—	[129]
	Pachylomma buccata	*Lasius niger* (L.)	—	[140]
	Pachylomma buccata Nees	*Myrmica lobicornis* Nylander	—	[24]
	Pachylomma buccata Nees	*Myrmica ruginodis* Nylander	—	[24]
	Pachylomma buccata	*Myrmica scabrinodis* Nylander	—	[153]
	Pachylomma buccata	*Tapinoma erraticum* (Latr.)	—	[152]
Hybrizon rileyi (Ashmead)*	—	*Lasius alienus* (Förster)	—	[154]
Unidentified Hybrizontinae (gen. nov. sp. nov.)	—	*Myrmica kotokui* Forel	—	[149]

*: attack was not observed, but there is strong evidence that all of the species of this genus reported as associated with an ant species are true primary parasitoids of this host.
**: uncertain report of association with the host (e.g., ants of the genera *Pogonomyrmex* and *Messor* do not have cocoons—contrary to what is reported in the original reference—and were probably misidentified), uncertain identification of the ant host (ambiguity between 2 or more species), or wasps not found directly within the nest of the presumed host (e.g., found near a nest—perhaps only by chance—or found on refuse deposit—perhaps as a prey—).
***: erroneous report (misidentification of either the parasitoid or the ant host), or erroneous emendation of the host species.

(a)

(b)

FIGURE 1: Winged females of the diapriid wasp, *Plagiopria passerai* (white pointer) in a nest of the formicine ant *Plagiolepis pygmaea*, just after emergence from queen pupae. Photos courtesy of L. Passera.

search for dipterous hosts, such as *Tetramopria aurocincta* Wasmann found in nests of *Tetramorium caespitum* (L.) [128]. This wasp is in fact a parasitoid of the puparia of *Compsilura concinnata* Meigen (Diptera: Tachinidae), a primary parasite of the lepidopteran *Hyphantria cunea* (Drury) [160]. Occasionally, diapriids enter ant nests for temporary shelter since some species hibernate in the host nest as do *Solenopsia imitatrix* Wasmann and *Lepidopria pedestris* Kieffer in the nests of *Solenopsis fugax* (Latr.) [37, 164].

Only a few diapriids are true parasitoids of ant brood. Ever since the pioneering work of Wasmann in 1899 [128], most diapriids found in ant nests were assumed either to parasitize insect myrmecophiles (dipteran or coleopteran) inside the host nest or, less frequently, to be primary parasitoids of ant larvae. However, the first record of a diapriid positively reared from ant brood was reported just in 1982 by Lachaud and Passera [37], who reared *Plagiopria passerai* from cocoons of queens of the formicine *Plagiolepis pygmaea* (Figures 1(a) and 1(b)). As far as known, diapriid parasitoids attacking ants develop as solitary or gregarious, koinobiont endoparasitoids of the host larvae [34, 36, 38, 169], and worker and/or reproductive immature stages can be parasitized [37, 169, 170]. Ramos-Lacau et al. [35] observed oviposition of *Acanthopria* sp. in young ant larvae under laboratory conditions. Late parasitized larvae are easily recognized by their dark coloration, compared to nonparasitized larvae, due to the developing wasp visible through the cuticle [35, 36, 38]. Worker ants do not discriminate between parasitized and nonparasitized larvae

[35, 38, 169], but adult parasitoids are aggressively attacked by their hosts under laboratory conditions [35, 36].

From the 121 diapriine species in 34 genera that have been collected in association with ants [30], development of immature stages as parasitoids of ant larvae has been demonstrated for only 26 species in 7 genera, most of which are only known at the level of morphospecies (Table 1): 15 species of *Acanthopria*, 3 of *Mimopriella*, 1 of *Oxypria*, 1 of *Plagiopria* (*P. passerai*), 4 of *Szelenyiopria*, and 2 of *Trichopria* (*T. formicans* and *Trichopria* sp.) [34–38, 169, 170]. The ant hosts of these diapriines belong to 8 species in only 4 genera: the myrmicine fungus-growing ants *Cyphomyrmex*, *Trachymyrmex*, and *Acromyrmex* and the formicine *Plagiolepis*. Fifteen species of Belytinae belonging to 11 genera have also been reported from ant nests [30, 171–173], but none has been reliably reared from the ants, and their actual relationship with their hosts remains unknown.

In some cases, the rate of parasitism can reach high levels. Two recent studies have provided important details of the biology of diapriids and have also investigated their impact on ant-host populations. Fernández-Marín et al. [36] found that between 27 and 70% of the colonies of 2 species of *Cyphomyrmex* were parasitized by one species in Puerto Rico and by up to 4 concurrent morphospecies of diapriids in Panama. Similarly, the work of Pérez-Ortega et al. [34] showed that another fungus-growing ant, *Trachymyrmex* cf. *zeteki*, was attacked by a diverse community of diapriids in Panama, with a mean intensity of larval parasitism per ant colony of 33.9%, and a prevalence across all ant populations of 27.2% (global data for all 6 diapriid morphospecies present at the study site).

2.2. Chalcidoidea. The superfamily Chalcidoidea is considered as one of the most abundant, species-rich, and biologically diverse groups of insects with 23,000 species described and a conservative estimation of about 400,000 to 500,000 species in over 2040 genera distributed in 19 families [32, 174–178]. Though some species are phytophagous, most Chalcidoidea are parasitoids of other insects, and numerous species are currently used as biological control agents against insect pests.

2.2.1. Chalcididae. Chalcididae is a moderate-sized family with more than 1450 species and over 85 genera. Chalcids are primary parasitoids of Lepidoptera or, to a much lesser extent, of Coleoptera, Diptera, Hymenoptera, and Neuroptera, and various species are hyperparasitoids of other hymenopterous parasitoids [179]. Most often they parasitize host larvae or pupae, but a few species can parasitize eggs.

Very few species, like *Epitranus chilkaensis* (Mani) (referred to as *Anacryptus chilkaensis*) found with the formicine *Camponotus compressus* (Fabr.) in the Barkuda Island (India) [180], are known to be associated with ants [179, 181], but true parasitoidism has rarely been documented. Only species of the genus *Smicromorpha* seem to be specialized as parasitoids of the larvae of the green ant, *Oecophylla smaragdina*. The only unquestionable (see [44]) record of parasitoidism is that of Dodd in the early 20th century, describing *Smicromorpha doddi* in North

Queensland (Australia) parasitizing larvae of this weaver ant, "*depositing eggs upon them when the workers are using their silk-spinning larvae for the purpose of binding the leaves together when building a new nest*" [41]. No other example of true parasitoidism has ever been quoted for the genus *Smicromorpha* but, more recently, adults of another species of this genus, *S. masneri*, were reported emerging from *O. smaragdina* nests collected in Vietnam and maintained in controlled green-house conditions in the USA, which strongly suggests that these wasps are also primary parasitoids of weaver ants [44]. Moreover, two other species, *S. keralensis* [43] and *S. minera* [42], have been observed hovering over nests of *O. smaragdina* in India and Australia, respectively, a behavior likely to be related to parasitism of ants (see below under Braconidae and Ichneumonidae). For such reasons, all these members of the genus *Smicromorpha* can reasonably be suspected of being true parasitoids of the larvae of this ant host and were included in our list (Table 1).

2.2.2. Encyrtidae.
Encyrtidae is a large family of parasitic wasps, currently including more than 460 genera and 3700 species, and is one of the key chalcidoid families for the biological control of insect pests [178, 182, 183]. Most encyrtids are primary endoparasitoids of immatures or, less commonly, adults of Coccidae and Pseudococcidae; others are hyperparasitic through other hymenopterous parasitoids, and some can attack insects in other orders, mites, ticks, or spiders [184, 185]. Some species are polyembryonic, a single egg multiplying clonally in the host, producing large numbers of identical adult wasps.

At least 25 species of encyrtid wasps representing 16 genera are known to be indirectly associated with ants through primary parasitism of the trophobionts they exploit and protect [32]; for example, the species *Anagyrus ananatis* Gahan is indirectly associated with the ant *Pheidole megacephala* through the trophobiotic Pseudococcidae present in their nest [186]. However, very few encyrtids have been reported as directly associated with ants. Apart from *Taftia prodeniae* Ashmead, which was found to exhibit a phoretic association (wasps were found clinging to the ant's antennae) with the dolichoderine ant *Dolichoderus thoracicus* (F. Smith) (referred to as *D. bituberculatus* (Mayr)) [187], and an unidentified species recently reported from a refuse deposit of the ecitonine ant *Eciton burchellii* [188], only *Holcencyrtus wheeleri* (Ashmead) (referred to as *Pheidoloxenus wheeleri*), found in nests of the myrmicine ants *Pheidole tepicana* Pergande (referred to as *P. instabilis*) [70] and *P. ceres* Wheeler (referred to as *P. ceres* var. *tepaneca* Wheeler) [100], has been suspected of being "*probably also entoparasitic on these ants or their progeny during its larval stages*" [1]. However, the parasitic relationship was never proved. Only very recently a Neotropical, gregarious endoparasitoid species, *Blanchardiscus* sp. (?*pollux*) (determination by J. S. Noyes), was recorded from French Guiana attacking pupae of the ponerine ant *Pachycondyla goeldii* [45] and thus constitutes the first true case of parasitism on ants for this family. However, no information has yet been published, and the exact identification of the species still needs to be confirmed.

2.2.3. Eucharitidae.
This is a small family but the largest and most diverse group of hymenopteran parasitoids attacking ants since all of its members, where the host is known, parasitize ant brood [11, 66, 72, 78, 83, 189–191]. Fifty-three genera and more than 470 species are currently described and distributed in three subfamilies: Oraseminae, Eucharitinae, and Gollumielinae.

All of the species have a highly modified life cycle [63, 66, 76, 83, 108]. Like the Perilampidae [191] and the ichneumonid species *Euceros frigidus* [192], but unlike most parasitic wasp species, eucharitid females deposit their eggs away from the host nest, in or on plant tissue (leaves and buds) [72, 189] (Figures 2(a) and 2(b)), and the very active, minute (less than 0.13 mm), strongly sclerotized first-instar larva is termed a "planidium" (Supplementary material 2 available online at doi:10.1155/2012/134746). It is responsible for gaining access to the host ant brood by using various phoretic behaviors including either attachment to an intermediate host (as in some orasemine species [11, 72, 83, 86, 88, 93] and, possibly, in *Gollumiella antennata* (Gahan) ([190] but see [72]) or, more generally, to foraging ant workers. On occasions (as is apparently the case for *Pseudochalcura gibbosa* and *Gollumiella longipetiolata*), attractive substances are suspected to be present in or on the eggs [46, 72]. Within the nest, the planidium attaches itself to an ant larva (Figures 2(c) and 2(d)): Eucharitine planidia attach externally to the host larva, whereas orasemine and gollumielline planidia partially burrow into the host larva, in the thoracic region just posterior to the head capsule [11, 70, 72]. All of the Eucharitidae develop as koinobiont, larval-pupal ectoparasitoids. At molting of the host larva, the planidium migrates to the ventral region, just under the legs (Figure 2(e)), of the newly formed ant pupa for further development which is only completed when the host pupates [76, 83, 93, 189] (Supplementary material 3 available online at doi:10.1155/2012/134746). In general, only one parasitoid develops per host but, occasionally, more than one adult eucharitid can develop in a single host (superparasitism) (Figure 2(f)) [72, 83], especially when larger brood (sexual brood) is parasitized [193, 194], and one exceptional case of multiparasitism involving two different species from two different eucharitid genera (*Dilocantha lachaudii* and *Isomerala coronata*) has even been reported from a single pupa of the ectatommine ant *Ectatomma tuberculatum* [79]. In almost all of the cases, adults emerge among ant brood (but see [77]), and, even if in some cases they are well treated within the nest by their hosts (as is the case for *Orasema coloradensis* which is transported, cared for, and even fed by the workers of *Pheidole bicarinata* [70]), they have to leave the host nest to reproduce. Ants show only moderate aggression to newly emerged eucharitids [58, 70, 75, 106, 189, 195, 196], suggesting passive or active chemical mimicry of the host ants [58, 75, 195]. If the parasitoid wasps do not exit their host nest by themselves, ant workers transport them outside (Figure 2(g)) as if they were refuse [58, 77, 196], ultimately enhancing wasp dispersal. Parasitism is very variable and localized in time and space [106, 193, 194]. A very high local prevalence may lead to only a low impact at the regional scale, suggesting that these parasitoids do not

FIGURE 2: Life cycle of a typical eucharitid wasp. (a) Female *Dilocantha lachaudii* ovipositing on *Lantana camara* L. (Verbenaceae). (b) *D. lachaudii* female with eggs scattered on leaf surface. (c) Planidium (white pointer) attached upon an *Ectatomma tuberculatum* larva. Insert: SEM picture of a planidium. (d) Two *D. lachaudii* swollen planidia (white pointers) feeding upon an *E. tuberculatum* larva. (e) 2nd instar larva (white pointer) relocated after host pupation. (f) Two *D. lachaudii* pupae from a single host pupa. The host cocoon has been removed. (g) *E. tuberculatum* worker transporting a recently emerged *D. lachaudii* female. Photos: J.-P. Lachaud and G. Pérez-Lachaud.

have a major influence on the dynamics of their ant host population [194].

According to Heraty [11], the hypothesized phylogeny of Eucharitidae is highly correlated with the subfamilies of their ant hosts and responsible for differences in behavior related with egg placement, activity of the planidium, and access to the ant host. Oraseminae (*Orasema*, *Orasemorpha*, and *Timioderus*) primarily attack myrmicine ants (numerous species of *Pheidole* and *Solenopsis*, and some species of *Diplorhoptrum*, *Monomorium*, *Temnothorax*, *Tetramorium*, and *Wasmannia*, see Table 1), and exceptionally formicines (*Formica subnitens* and *F. oreas comptula* in the case of

O. coloradensis, [88]) or ecitonines (*Eciton quadriglume* in the questionable case of *O. rapo*, [11, 83]). For Eucharitinae, the only two host records for the tribe Psilocharitini (*Neolosbanus*) concern the ponerine genus *Hypoponera* [83], while the numerous members of the tribe Eucharitini are essentially parasitic on medium to large ponerines (*Pachycondyla, Odontomachus*, and *Dinoponera*) and ectatommines (*Ectatomma, Gnamptogenys, Typhlomyrmex*, and *Rhytidoponera*), but also on myrmeciines (*Myrmecia*) and numerous formicines (*Anoplolepis, Calomyrmex, Camponotus, Cataglyphis, Formica, Lasius*, and *Polyrhachis*); without exception, all of the scarce records of associations of eucharitines with myrmicine ants (*Messor, Myrmica*, and *Pogonomyrmex*) are highly doubtful (Table 1). Finally, the only host record for the Gollumiellinae concerns a formicine (*Paratrechina*).

The hosts of most eucharitid genera seem to be restricted to only one or a few closely related ant genera and, for a long time, all species were considered as host-specific parasitoids, at least at the host genus level [83]. However, recent results [76, 78, 79] raised questions concerning the degree of host specificity in eucharitids and about the factors that determine the association of these parasitoids and their hosts. Results in the guild of eucharitid parasitoids associated with ponerine ant species in southeastern Mexico and French Guiana suggest that some eucharitid wasps tend to be oligophagous in their host choice: some eucharitid species can attack different hosts from different genera and different subfamilies such as *Kapala iridicolor*, which parasitizes one species of *Ectatomma*, two of *Gnamptogenys*, and one of *Pachycondyla* [76, 78]. Furthermore, concurrent parasitism has been reported for *Ectatomma tuberculatum*, which is simultaneously parasitized by *Dilocantha lachaudii*, *Isomerala coronata*, and *Kapala* sp. [79], or for *E. ruidum* parasitized by two *Kapala* species, *K. iridicolor*, and *K. izapa* [76, 193].

2.2.4. Eulophidae. The family Eulophidae is the largest of the Chalcidoidea with up to 4470 species in 297 genera. The majority of the species are primary parasitoids attacking a large variety of insects (mainly Lepidoptera and Coleoptera, but also Diptera, Thysanoptera, and Hymenoptera), and occasionally mites or spiders. Many species are facultative or obligate hyperparasitoids of other Hymenoptera, and some are even phytophagous. Entomophagous larvae can develop as koino- or idiobionts, gregarious or solitary, and ecto- or endoparasitoids, and according to the species, eulophids can attack eggs, larvae, pupae, or even the adults of their hosts [197].

Despite the large number of species in this family, parasitization of ants is uncommon among Eulophidae, and only few associations involving eulophid wasps and ant hosts have been reported to date. Almost all are from genera belonging to the subfamily Entedoninae. Three concern species indirectly associated with ants as they parasitize insects living in ant nests: *Pediobius acraconae* Kerrich which has been reported [114] from a last instar larva of the pyralid lepidopteran *Acracona remipedalis* Karsch found in a nest of *Crematogaster depressa* (Latr.) or *C. africana* Mayr in Nigeria, and both *Microdonophagus woodleyi* Schauff in Panama

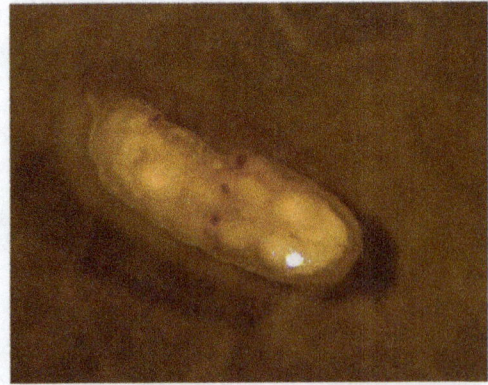

FIGURE 3: Larva of the neotropical weaver ant *Camponotus* sp. ca. *textor* parasitized by the gregarious endoparasitoid *Horismenus myrmecophagus* (Eulophidae). Several wasp larvae can be observed through the host cuticle. Photo: G. Pérez-Lachaud.

and *Horismenus microdonophagus* Hansson et al. in Mexico, which parasitize larvae of *Microdon* sp. syrphid flies living in nests of the dolichoderine *Technomyrmex fulvus* (Wheeler) (referred to as *Tapinoma fulvum*) [198] and of the formicine *Camponotus* sp. ca. *textor* [112], respectively. Three other species (two Entedoninae and a Tetrastichinae) have been reported associated with ant nests, but direct parasitism on the ant brood was not clearly established in any of these cases: *Myrmobomyia malayana* Gumovsky and Bouček with nests of an ant species of the genus *Dolichoderus* in Malaysia [199], an unidentified species of *Horismenus* from the bivouac and refuse deposits of the army ant *Eciton burchellii* [188], and an unidentified species of *Tetrastichus* from a nest of the formicine *Myrmecocystus mexicanus* Wesmael in Nevada [94].

In fact, only five species are known as true primary parasitoids of ants (Table 1). An unidentified gregarious parasitoid, apparently closely related to the genus *Paracrias* (according to Gahan in [109]), possibly *Horismenus* sp. [112], was recorded parasitizing larvae of the myrmicine *Crematogaster acuta* in Guyana, the prepupae of another unidentified species of *Crematogaster* were parasitized by *Myrmokata diparoides* [113] in Cameroon, *Pediobius marjoriae* was reared from cocoons of the formicine ant *Lepisiota* sp. in Uganda [114], and two species of *Horismenus*, *H. floridensis* and *H. myrmecophagus*, were found parasitizing the pupae of *Camponotus atriceps* and *C. floridanus* in Florida [111], and of the weaver ant *Camponotus* sp. ca. *textor* in Mexico [112], respectively. In the latter two cases, *Horismenus* larvae develop as gregarious endoparasitoids of the ant larvae (Figure 3), and large numbers of parasitoid individuals can develop from the same host: up to 21 for *H. floridensis* and between 4 and 12 for *H. myrmecophagus*. Finally, two other cases deserve to be added to this list since two other ant species have recently been found parasitized by eulophids: the ponerine ant *Pachycondyla crenata* (Roger) in Mexico and an unidentified species of *Camponotus* (*Dendromyrmex*) in French Guiana [112]; however, the identity of the parasitoids has not been confirmed yet.

2.2.5. Eurytomidae.

Eurytomidae is a moderate-sized family with 90 genera and at least 1400 nominal species [13, 32, 200, 201]. Eurytomid wasps exhibit a wide range of biologies, but most of the larvae are endophytic either as seed or plant stem eaters or as parasitoids of gall formers or other phytophagous insects. Most species are primary or secondary parasitoids, attacking eggs, larvae, or pupae of various arthropods (Diptera, Coleoptera, Hymenoptera, Lepidoptera, Orthoptera, and Araneae).

A few species have been reported as indirectly associated with ants, like *Eurytoma rosae* Nees von Esenbeck found with *Lasius flavus* and *Eurytoma* sp. found with *Formica* (?) *rufibarbis* (misidentified as *Polyergus rufibarbis*) [20], but most probably these eurytomids only fed on the gall-forming cynipid larvae and/or on the gall tissue on *Rosa* spp. which are visited by these ant species, without any direct relationship with the ants. Recently, various adults of a new genus and species, *Camponotophilus delvarei* Gates, were found within nests of the weaver ant *Camponotus* sp. ca. *textor* [202], but the exact nature of their relationship with the ants remains unclear. As a matter of fact, only 3 or 4 species from the single genus *Aximopsis* (see Table 1) have been reported from Guatemala, Costa Rica, Guyana, Colombia, and Peru as parasitoids of queens of various species of dolichoderine ants (*Azteca alfari*, *A. australis*, *A. constructor*, *A. pitieri*, *A. ovaticeps*, and *A. salti*) and one formicine (*Camponotus balzani*), all of which colonize *Cecropia* spp. internode chambers by chewing a hole through a prostoma and entering the internode. The parasitoids attack only founding queens and feed on their host, while the internode chamber is sealed with parenchyma scraped from the internal stem walls [115, 116, 118]; there is never more than one wasp larva or pupa per foundress ant [117]. Queen parasitization was thought to occur before they entered their dwellings (Bailey, in [115]); however, as suggested by Davidson and Fisher [119], the location of the ant host may occur through searching for host plants since female *Aximopsis* were observed to visit various seedlings, where they inspected newly sealed prostoma. This fact has been confirmed recently. A picture of an *A. affinis* female ovipositing through a prostoma into an *Azteca* queen at La Selva Biological Station, Costa Rica, was provided by Weng et al. [203] (their Figure 16). In this site, among the internodes that harbored *Azteca* ants, 43% contained dead queens, of which 13% contained *A. affinis* [203].

2.2.6. Perilampidae.

Perilampidae is a small family closely related to the Eucharitidae, composed of up to 270 species from 15 genera. A feature shared with Eucharitidae is that the first-instar larva, the "planidium", is responsible for gaining access to the host, rather than the egg-laying female [191]. Most species are hyperparasitoids on ichneumonid wasps or tachinid flies which are primary parasitoids of Hymenoptera or Lepidoptera, or parasitoids of wood-boring platypodid and anobiid beetles, and some species can attack Orthoptera, Neuroptera, or Hymenoptera [190, 204].

Association of perilampids with ants seems extremely casual. The only report deals with an unidentified species from Peru found parasitizing cocoons of the ponerine ant *Pachycondyla luteola*, inhabiting internode chambers of a *Cecropia*, with as many as nine perilampid wasps emerging from a single pupa of this ant [119]. However, no other details were ever published, and the species apparently remained undescribed.

2.3. Ichneumonoidea.

The superfamily Ichneumonoidea, with only two extant families, accounts for more than 40,000 species around the world, and there are estimated to be approximately 100,000 species [205–207]. Most are primary ecto- or endoparasitoids, idio- or koinobionts, especially attacking immature stages of a wide variety of insects and arachnids, and more occasionally adults. Some members use many different insects as hosts, and others are very specific in host choice. Various ichneumonoids are successfully employed as biological control agents in controlling insect pests such as flies or beetles.

2.3.1. Braconidae.

This is a very large family with 48 subfamilies, more than 1050 genera and about 17,600 described species worldwide and exhibiting a variety of biologies [207–209]. The total number of species is estimated to be 40–50,000. Many braconids parasitize nymphal stages of Hemiptera, Isoptera, and Psocoptera; a few genera also parasitize adult Coleoptera and Hymenoptera [209]. Two major lineages occur within the Braconidae: (a) the cyclostome braconids, most of which are idiobiont ectoparasitoids of concealed Lepidoptera and Coleoptera larvae although many are koinobiont endoparasitoids of Diptera and Hemiptera, and (b) the noncyclostome braconids which are all endoparasitoids, and most generally koinobionts, typically attacking an early instar of their hosts (see [210] for a comprehensive overview of their biology).

Numerous braconid species have been reported in association with ants. Some, such as *Compsobraconoides* sp. [28] and *Trigastrotheca laikipiensis* [29], are predatory on several developmental stages of ants. Others, such as *Aclitus sappaphis* Takada and Shiga found in nests of *Pheidole fervida* Smith [211, 212], *Paralipsis enervis* (Nees von Esenbeck) found with *Lasius niger* [213], or *P. eikoae* (Yasumatsu) found with *L. japonicus* Santschi (referred to as *L. niger* (L.)) and *L. sakagamii* Yamauchi and Hayashida [212, 214], are in fact primary parasitoids of root aphids and can only be considered as indirectly associated with the aphid-attending ants; however, they have developed highly sophisticated relationships with their hosts involving chemical mimicry and chemical and tactile communication to obtain regurgitated food (trophallaxis).

For several other species, the exact nature of the association with the ant host has not been clearly established, but at least 15 euphorine species can be considered as true parasitoids of adult ants even if direct evidence of oviposition has been obtained for only 11 of them (see Table 1). All of these parasitoids are grouped in three extant genera, *Elasmosoma*, *Kollasmosoma*, and *Neoneurus*, and one fossil genus, *Elasmosomites*, all belonging to the tribe Neoneurini. Evidence from Eocene Baltic amber, as demonstrated from an individual of *Elasmosomites primordialis* emerging from the abdomen of a *Lasius* worker (Figure 4(a)), indicates that the parasitoid association between neoneurine braconids and ants has been in existence for at least 40 million years [136]. Although oviposition into the abdomen of adult worker ants has been reported on several occasions [56, 120, 121, 126, 127, 140], detailed descriptions were rare and, until recently, restricted to only two species. In the case of *N. mantis* attacking *Formica podzolica*, Shaw [142, 143] gave interesting information both on the "perching" behavior displayed by the parasitoid females in their ambush strategy to locate their hosts and on the attack sequence which is completed in less than 1 s and is characterized by a reduction of the usual braconid oviposition sequence, the first two steps (antennation of the host and ovipositor probing) being entirely lost in favor of speed. For *E. michaeli*, Poinar [131] not only described the attack behavior, exclusively focused on major workers of *Formica obscuriventris clivia* (Figure 4(b)), but also provided invaluable information on the altered behavior of parasitized ants, on the development of the immature stages, and on cocoon formation and adult emergence. Immature stages of Neoneurini parasitoids attacking adult ants develop as koinobiont endoparasitoids in the abdomen of workers, and fully developed larvae leave the host to pupate in the soil [131].

Very recently, slow motion video recordings were used to describe the oviposition behavior in adult ants for 3 other species [129], and we refer the reader to their excellent films, which show the variability in oviposition behavior within the tribe. Neoneurini wasps parasitize worker ants in the vicinity of the nest entrance(s), or while foraging. Females of *Elasmosoma luxemburgense* hover over the nest entrance of *Formica rufibarbis* and attack workers from behind, grasping the ant abdomen with the three pairs of legs involved, and probably ovipositing through the anus. The whole behavioral sequence (alighting, grasping, ovipositor insertion, and takeoff) lasted a mean of 0.73 s. The ants were aware of these attacks, turning around and chasing the wasps with open mandibles ([129] doi: 10.3897/zookeys.125.1754.app1). Females of *Kollasmosoma sentum* attack workers of *Cataglyphis iberica* in the vicinity of nest entrances, or when carrying prey and walking more slowly than usual. Attacks usually occurred during the brief stops characterizing *Cataglyphis* workers walks. The wasps were extremely fast and attacked the ants from behind. Oviposition took place in both the dorsal and ventral surfaces of the ant's gaster, likely through intersegmental membranes. Wasps adjusted their alighting strategies according to the direction of their own approach to the targeted ant, and to the position of the ant's gaster (horizontal or vertical position, distinctive for the genus *Cataglyphis*), and accomplished extraordinary pirouettes. The whole oviposition behavior lasted only 0.05 s on average. The ants were often aware of the presence of the parasitoids, aggressively turning around with open mandibles, or extending their hind or middle legs to hit them ([129] doi: 10.3897/zookeys.125.1754.app2). Finally, *N. vesculus* females alight and probably oviposit in the mesosoma of *Formica cunicularia* workers. As for *N. mantis* [142, 143], they were observed ambushing or hovering over the nest entrance. Females preferentially attacked ants while at a vertical

(a)

(b)

FIGURE 4: (a) *Elasmosomites primordialis* larva (white pointer) emerging from the abdomen of a *Lasius* worker in Baltic amber. Photo courtesy of G. Poinar Jr. (see [136]). (b) *Elasmosoma michaeli* larva leaving its *Formica obscuriventris clivia* host to pupate in the soil. Photo courtesy of G. Poinar Jr.

position (going up a tree trunk, e.g.). The wasps approached the ants from behind, alighted, held the ant's thorax with their raptorial fore legs, bent their abdomen towards the postero-lower part of the ant's thorax, and oviposited. The ovipositor is thought to be inserted near the posterior coxal cavities. The whole oviposition behavior lasted a mean of 2.02 s ([128] doi: 10.3897/zookeys.125.1754.app3).

With few exceptions, neoneurine wasps have been found in association with formicine ants [129, 207, 215, 216]. It is thought that formic acid used by these ants could serve also as a kairomonal stimulant to host-seeking hymenopterous parasitoids [120, 127, 129]. Far less is known about the fate of parasitized ants. According to Poinar [131], *Formica* ants parasitized by *E. michaeli* form an assembly along the edge of their superficial nest when the parasitoid larvae are about to leave the host to pupate. This behavioral modification is thought to increase the survival of adult wasps.

Several morphological and behavioral adaptations, apart from rapidity of attack, contribute to the success of these wasps in parasitizing aggressive adult ants: for example, the vestigial tarsal claws and enlarged pulvilli (suction like disks, [130, 131, 217]) of *Elasmosoma* spp., or the raptorial fore legs of *Neoneurus* spp., enable wasps to grasp and hold the ant firmly while ovipositing. Likewise, the peculiar ventral spine of *K. sentum* females, located on the fifth sternite, could help to fix the wasp's position during oviposition, when the body of the wasp goes back tending to the vertical position, and fore legs detach from the ant's cuticle. Finally, the longitudinal disposition of *K. sentum* females's tarsi on the ant metasoma, one over the other, enables the necessary rotation of the body to adjust itself to the position of the ant's gaster, before oviposition. The wasp rotates counterclockwise if the right tarsus is placed over the left one; and if the left tarsus is placed over the right one, the rotation is clockwise.

2.3.2. Ichneumonidae. Ichneumonidae is the largest family in the Hymenoptera with about 23,330 described species worldwide in 46 subfamilies and 1207 genera; the total number of species is estimated to be more than 60,000 [207, 218, 219]. Most of the members of this large family are parasites of holometabolous insects, but a few species parasitize spiders (egg sacs, spiderlings, or adults) or egg sacs of pseudoscorpions. Many ichneumonids are hyperparasitoids of other ichneumonoids or of tachinid flies, and some species are egg-larval parasitoids, laying an egg in the host egg but consuming the host in its larval stage [218, 219].

Various species of the genus *Gelis* (all of them initially referred to as *Pezomachus*) and a few others of the genera *Agrothereutes, Aptesis, Pleolophus,* and *Thaumatogelis* have been reported by various authors to be associated with ants of the genera *Lasius, Formica, Myrmica, Temnothorax,* and *Solenopsis* [24, 56, 220–222]. However, no information is available on the exact relationship with their ant host, except that in some cases (such as *Pleolopus micropterus* (Gravenhost) (referred to as *Pezomachus micropterus*) and *T. vulpinus* (Gravenhorst) (referred to as *Pezomachus vulpinus*)), they were clearly reported as "*found in the nest of* Formica rufa, *not reared from cocoons*" [220]. Until now, true ichneumonid parasitism on ants has been demonstrated only for 3 species, all belonging to the subfamily Hybrizontinae and very likely to the same tribe Hybrizontini. The most ancient report dates back to 1852 [145] and concerns *Eurypterna cremieri* described as hovering over a nest of *Lasius fuliginosus* in Germany. This behavior, suspected to be related to the search of an appropriate host, was later confirmed by different authors not only for the same host species in France and Italy [146–148] but also for three other species of ants in the genera *Lasius* and *Formica* in France, England, and Japan [123, 144, 149]. In the early 20th century, Cobeli [148] described how four females of *E. cremieri* were hovering over trails of *L. fuliginosus,* while ants were moving their nest to another nest site, inspecting each ant worker that was transporting a larva. The female parasitoids quickly drew closer to the larva, and folding up the abdomen touched it, presumably depositing an egg. Such behavior was only observed with ants transporting a larva and did not trigger any reaction from the workers. In spite of the interesting information supplied, this report passed more or less unnoticed until 2010 when the parasitic nature of this behavior could be confirmed (and even photographed) concerning *Lasius nipponensis* transporting brood between two nests [149]. Only workers carrying something in their mandibles were tracked by *E. cremieri* females hovering about 2 cm above them. And only those carrying a larva were attacked after a sudden dive of the wasp which gripped the targeted larva with the tarsi of its fore and middle legs, bent its abdomen down, exerted its ovipositor, and oviposited in the larva before flying away in search of a new host. The complete sequence lasted less than 1 s and elicited some brief excitement from the worker ant. Dissection of a stung ant larva showed that a wasp egg was present in the somatic cavity. Another undescribed Hybrizontinae species (gen. nov. sp. nov.) was similarly reported by the same authors as hovering over workers of the slow moving ant *Myrmica kotokui* which were holding something in their mandibles. As for *E. cremieri,* only those carrying a larva were more closely inspected and were attacked in a similar manner as previously described, but in that case, the complete attack sequence lasted longer (3-4 s), and oviposition itself took at least 1 s. A third case of ant larval parasitism has very recently been confirmed and involves *Hybrizon buccatus* females. This species had been frequently reported in association with (or hovering over) different ant species from various genera (*Myrmica, Lasius, Formica,* and *Tapinoma,* see Table 1) [24, 140, 144, 146, 152, 153] and was reared from nests of *Lasius alienus* where the ichneumonid naked pupae had been found among ant-host cocoons [150]. But it was not until 2011 that the oviposition into larvae transported by *Lasius grandis* workers could be observed and filmed during brood transfer between two nest entrances [129]. Only final instar larvae were attacked, in a very similar way to that previously described for *E. cremieri,* and the complete sequence lasted between 0.40 and 0.58 s. Chemical and/or visual cues are likely to be involved in the location of the ants' trail since *H. buccatus* females have been observed continuously hovering over the trail for a period of time, even in the absence of ants. Finally, considering both the hovering behavior as a reliable evidence of parasitism and the fact that all three ichneumonid parasitoids known until now to attack ants are restricted to the Hybrizontinae, two other cases are likely to be added to our list: *Ghilaromma fuliginosi* and *H. rileyi* which have been reported swarming and hovering over the nests of *Lasius fuliginosus* [150, 151] or attracted to a disturbed nest of *L. alienus* [154], respectively. However, in both cases, direct oviposition into ant larvae or adults needs to be confirmed.

3. Conclusions

Since the last paper on parasites of social insects by Schmid-Hempel [7], the number of reliable records of parasitoid wasps attacking ants and their brood has grown dramatically from about 43 species to at least 138 belonging to 9 hymenopteran families. Furthermore, the knowledge of the biology and behavior of those wasps and the nature of their interactions with ants has significantly progressed, though many gaps still remain. Most likely, hymenopterous parasitoids of ants are more abundant than suggested by our list of reliable records, and future studies focusing on the immature stages of ants under close scrutiny would certainly increase this list substantially.

All castes of ants and all developmental stages, excepting eggs, are the target of parasitoid wasps. For example, neoneurine braconids parasitize adult worker ants while foraging or performing other activities outside the nest [129, 131, 143], while eurytomids of the genus *Aximopsis* attack adult queens at the very moment of nest foundation [115, 116, 118, 119]. However, in most cases, ant larvae are the target of parasitoid attacks, either inside or outside their nests. Larvae can be parasitized outside the protective walls of the nest during transportation when ants move from one nest to another as for some euphorine braconids and hybrizontine ichneumonids [129, 149], or while being employed to fix or build a new nest as occurred for the green weaver ant larvae attacked by the chalcidid *Smicromorpha* [41]. Most often, ant larvae are attacked inside the nest, notwithstanding the pugnacious character of ants. For eucharitid and perilampid wasps, planidia are transported by phoresis into the targeted nest where they actively search for a larval host. The extremely small size of the planidia is assumed to facilitate both entrance into the host colony and initial parasitization [195], but in most other parasitoid wasps (diapriids, encyrtids, entedonine eulophids, and some eurytomids), it has been assumed that it is the female that searches for a host nest, enters it, and oviposits on or in the larval host. So far, however, how the females gain access into the ant nest and complete the oviposition process has never been described, and the initial stages of development of these parasitoids are in most cases unknown (but see [35, 131]).

Hymenopterous parasitoids attacking ants exhibit a wide array of biologies and developmental strategies: ecto- or endoparasitism, solitary or gregarious, and idio- or koinobiosis. Besides, the behavioral strategies evolved to cope with ant aggression or to exploit the communication system of ants are also impressive. Most of these parasitoids belong to families with species using a wide range of insects or arthropods as primary hosts, and in many cases of recorded associations between parasitic wasps and ants [20, 23, 112, 114, 128, 160, 186, 198], the primary host of the parasitoids is not the ant but another insect species present in the ant nests. Such indirect association through parasitism of trophobionts or other myrmecophiles suggests that a possible path to the parasitization of ants by hymenopterous parasitoids could have evolved as a shift from the initial primary host (Diptera, Coleoptera, or other insect myrmecophiles) to the ant host larvae through a gradual process of association and integration with the ant hosts. Such a hypothesis proposed for diapriids by Huggert and Masner [160] and widened by Hanson et al. [223] to hymenopterous parasitoids in general might apply for numerous families, and a supporting example has recently been suggested among eulophids [112]. However, other evolutionary paths are likely to be involved in the case of eucharitids and perilampids and those species that attack adult ants and deserve further study.

Despite a significant increase in our knowledge of hymenopterous parasitoids of ants in the last 15 years, the remark of Schmid-Hempel [7] concerning parasitism in social insects in general: "*the existing knowledge is bound to be a massive underestimation, since the true abundance and distribution of parasites remain to be discovered*" is still, more than ever, a topical subject. Most hymenopterous parasitoids attacking ants remain to be discovered. Moreover, despite the presumed importance of some of them as natural enemies of ants, few quantitative data are available on the impact of these natural enemies on their hosts (see [224]). Based on their abundance and success in attacking ant hosts [36, 83, 193, 194], some parasitoid wasps like, for example, diapriids and eucharitids, seem excellent potential models to explore how parasitoids impact ant colony demography, population biology, and ant community structure, and further studies focusing on these issues will certainly contribute to deepening our knowledge on this important group of parasites.

Acknowledgments

The authors thank M. W. Gates, C. Hansson, J. M. Heraty, M. Loiácono, and T. C. Narendran for making available various bibliographic references, and L. Passera and G. Poinar Jr. for kindly supplying their original pictures of ant attack by *Plagiopria passerai*, *Elasmosomites primordialis*, and *Elasmosoma michaeli*. They are also indebted to T. C. Narendran, G. Poinar Jr., and an anonymous reviewer for constructive comments and suggestions on a previous version of this paper, and to Peter Winterton for grammatical improvement.

References

[1] W. M. Wheeler, *Ants, their Structure, Development and Behavior*, Mac Millan, The Columbia University Press, NY, USA, 1910.

[2] H. S. J. K. Donisthorpe, *The Guests of British Ants: Their Habits and Life-Histories*, G. Routledge and Sons, London, UK, 1927.

[3] E. O. Wilson, *The Insect Societies*, The Belknap Press of Harvard University Press, Cambridge, Mass, USA, 1971.

[4] D. H. Kistner, "Social and evolutionary significance of social symbionts," in *Social Insects*, H. R. Hermann, Ed., vol. 1, pp. 339–413, Academic Press, New York, NY, USA, 1979.

[5] D. H. Kistner, "The social insects' bestiary," in *Social Insects*, H. R. Hermann, Ed., vol. 3, pp. 1–244, Academic Press, New York, NY, USA, 1982.

[6] B. Hölldobler and E. O. Wilson, *The Ants*, Springer, Berlin, Germany, 1990.

[7] P. Schmid-Hempel, "Parasites in social insects," in *Monographs in Behavior and Ecology*, J. R. Krebs and T. Clutton-Brock, Eds., Princeton University Press, Princeton, NJ, USA, 1998.

[8] N. F. Johnson, "Catalog of world species of Proctotrupoidea, exclusive of Platygastridae (Hymenoptera)," *Memoirs of the American Entomological Institute*, vol. 51, pp. 1–825, 1992.

[9] B. Bolton, *A New General Catalog of the Ants of the World*, Harvard University Press, Cambridge, Mass, USA, 1995.

[10] B. Bolton, "Synopsis and classification of Formicidae," *Memoirs of the American Entomological Institute*, vol. 71, pp. 1–370, 2003.

[11] J. M. Heraty, "A revision of the genera of Eucharitidae (Hymenoptera: Chalcidoidea) of the world," *Memoirs of the American Entomological Institute*, vol. 68, pp. 1–367, 2002.

[12] M. J. Sharkey, "Phylogeny and classification of Hymenoptera," *Zootaxa*, no. 1668, pp. 521–548, 2007.

[13] H. Lotfalizadeh, G. Delvare, and J. Y. Rasplus, "Phylogenetic analysis of Eurytominae (Chalcidoidea: Eurytomidae) based on morphological characters," *Zoological Journal of the Linnean Society*, vol. 151, no. 3, pp. 441–510, 2007.

[14] J. Heraty, F. Ronquist, J. M. Carpenter et al., "Evolution of the hymenopteran megaradiation," *Molecular Phylogenetics and Evolution*, vol. 60, no. 1, pp. 73–88, 2011.

[15] M. J. Sharkey, J. M. Carpenter, L. Vilhelmsen et al., "Phylogenetic relationships among superfamilies of Hymenoptera," *Cladistics*, vol. 28, no. 1, pp. 80–112, 2012.

[16] H. E. Evans, "A revision of the genus *Pseudisobrachium* in the North and Central America (Hymenoptera, Bethylidae)," *Bulletin of the Museum of Comparative Zoology*, vol. 126, no. 2, pp. 211–318, 1961.

[17] C. Bruch, "Nuevas capturas de insectos mirmecófilos," *Physis, Revista de la Sociedad Argentina de Ciencias Naturales*, vol. 3, pp. 458–463, 1917.

[18] H. Eidmann, "Die gäste und gastverhältnisse der blattschneiderameise *Atta sexdens* L.," *Zeitschrift für Morphologie und Ökologie der Tiere*, vol. 32, no. 3, pp. 391–462, 1937.

[19] J. J. Kieffer, "Nouveaux proctotrypides myrmécophiles," *Bulletin de la Société d'Histoire Naturelle de Metz*, vol. 23, pp. 31–58, 1904.

[20] J. J. Kieffer, "Ueber neue myrmekophile Hymenopteren," *Berliner Entomologische Zeitschrift*, vol. 50, pp. 1–10, 1905.

[21] J. J. Kieffer, *Species des Hyménoptères d'Europe et d'Algérie*, vol. 10, Librairie Scientifique A. Hermann & Fils, Paris, France, 1908.

[22] J. J. Kieffer, "Description de nouveaux microhyménoptères," *Brotéria*, vol. 11, pp. 169–198, 1913.

[23] J. J. Kieffer, *Hymenoptera. Diapriidae*, Das Tierreich, Verlag von R. Friedländer und Sohn, Berlin, Germany, 1916.

[24] H. S. J. K. Donisthorpe, *British Ants, Their Life-History & Classification*, William Brendon and Son Limited, Plymouth, UK, 1915.

[25] H. E. Evans, "A review of nesting behavior of digger wasps of the genus *Aphilanthops*, with special attention to the mechanics of prey carriage," *Behaviour*, vol. 19, no. 3, pp. 239–260, 1962.

[26] H. E. Evans, "Ecological-behavioral studies of the wasps of Jackson Hole, Wyoming," *Bulletin of the Museum of Comparative Zoology*, vol. 140, no. 7, pp. 451–511, 1970.

[27] R. M. Bohart and A. S. Menke, *Sphecid Wasps of the World: A Generic Revision*, University of California Press, Berkeley, Calif, USA, 1976.

[28] D. W. Yu and D. L. J. Quicke, "*Compsobraconoides* (Braconidae: Braconinae), the first hymenopteran ectoparasitoid of adult *Azteca* ants (Hymenoptera: Formicidae)," *Journal of Hymenoptera Research*, vol. 6, no. 2, pp. 419–421, 1997.

[29] D. L. J. Quicke and M. L. Stanton, "*Trigastrotheca laikipiensis* sp. nov. (Hymenoptera: Braconidae): a new species of brood parasitic wasp that attacks foundress queens of three coexisting acacia-ant species in Kenya," *Journal of Hymenoptera Research*, vol. 14, no. 2, pp. 182–190, 2005.

[30] J.-P. Lachaud and G. Pérez-Lachaud, In prep.

[31] O. M. Reuter, *Lebensgewohnheiten und Instinkte der Insekten bis zum Erwachen der Sozialen Instinkte*, Friendländer und Sohn, Berlin, Germany, 1913.

[32] J. S. Noyes, "Universal Chalcidoidea Database," 2011, http://www.nhm.ac.uk/chalcidoids.

[33] J. Longino, "*Cyphomyrmex salvini* Forel 1899," 2004, http://www.discoverlife.org/mp/20q?search=Cyphomyrmex+salvini.

[34] B. Pérez-Ortega, H. Fernández-Marín, M. S. Loiácono, P. Galgani, and W. T. Wcislo, "Biological notes on a fungus-growing ant, *Trachymyrmex* cf. *zeteki* (Hymenoptera, Formicidae, Attini) attacked by a diverse community of parasitoid wasps (Hymenoptera, Diapriidae)," *Insectes Sociaux*, vol. 57, no. 3, pp. 317–322, 2010.

[35] L. S. Ramos-Lacau, J. H. C. Delabie, O. C. Bueno et al., "Estratégia comportamental de *Acanthopria* Ashmead (Hymenoptera: Diapriidae), parasitóide de *Cyphomyrmex transversus* Emery (Hymenoptera: Formicidae)," *Biológico, São Paulo*, vol. 69, supplement 2, pp. 451–454, 2007.

[36] H. Fernández-Marín, J. K. Zimmerman, and W. T. Wcislo, "*Acanthopria* and *Mimopriella* parasitoid wasps (Diapriidae) attack *Cyphomyrmex* fungus-growing ants (Formicidae, Attini)," *Naturwissenschaften*, vol. 93, no. 1, pp. 17–21, 2006.

[37] J.-P. Lachaud and L. Passera, "Données sur la biologie de trois Diapriidae myrmécophiles: *Plagiopria passerai* Masner, *Solenopsia imitatrix* Wasmann et *Lepidopria pedestris* Kieffer," *Insectes Sociaux*, vol. 29, no. 4, pp. 561–568, 1982.

[38] M. S. Loiácono, "Un nuevo diáprido (Hymenoptera) parasitoide de larvas de *Acromyrmex ambiguus* (Emery) (Hymenoptera, Formicidae) en el Uruguay," *Revista de la Sociedad Entomológica Argentina*, vol. 44, no. 2, pp. 129–136, 1987.

[39] M. S. Loiácono, "Diaprinos asociados a la hormiga *Camponotus rufipes* (Hymenoptera: Diapriidae)," *Revista de la Sociedad Entomológica Argentina*, vol. 59, no. 1–4, pp. 198–200, 2000.

[40] M. S. Loiácono, C. B. Margaría, E. M. Quirán, and B. M. Corró Molas, "Diápridos (Hymenoptera) parasitoides de larvas de la hormiga cortadora *Acromyrmex lobicornis* (Hymenoptera: Formicidae) en la Argentina," *Revista de la Sociedad Entomológica Argentina*, vol. 59, no. 1–4, pp. 7–15, 2000.

[41] A. A. Girault, "Some chalcidoid Hymenoptera from North Queensland," *Archiv für Naturgeschichte*, vol. 79, pp. 70–90, 1913.

[42] I. D. Naumann, "A revision of the Indo-Australian Smicromorphinae (Hymenoptera: Chalcididae)," *Memoirs of the Queensland Museum*, vol. 22, pp. 169–187, 1986.

[43] T. C. Narendran, "A new species and a new record of the interesting genus *Smicromorpha* Girault (Hymenoptera: Chalcididae) from Oriental region," *Journal of Bombay Natural History Society*, vol. 75, pp. 908–911, 1979.

[44] D. C. Darling, "A new species of *Smicromorpha* (Hymenoptera, Chalcididae) from Vietnam, with notes on the host association of the genus," *ZooKeys*, vol. 20, pp. 155–163, 2009.

[45] G. Pérez-Lachaud and J.-P. Lachaud, In prep.

[46] J. M. Heraty and K. N. Barber, "Biology of *Obeza floridana* (Ashmead) and *Pseudochalcura gibbosa* (Provancher) (Hymenoptera: Eucharitidae)," *Proceedings of the Entomological Society of Washington*, vol. 92, no. 2, pp. 248–258, 1990.

[47] C. T. Brues, "A new chalcid-fly parasitic on the Australian bull-dog ant," *Annals of the Entomological Society of America*, vol. 12, pp. 13–23, 1919.

[48] A. Forel, "Un parasite de la *Myrmecia forficata* Fabr.," *Annales de la Société Entomologique de Belgique*, vol. 34, pp. 8–10, 1890.

[49] G. P. Browning, *Taxonomy of Myrmecia Fabricius (Hymenoptera: Formicidae)*, Ph.D. thesis, University of Adelaide, Adelaide, Australia, 1987.

[50] R. W. Taylor, P. Jaisson, I. D. Naumann, and S. O. Shattuck, "Notes on the biology of Australian bulldog ants (*Myrmecia*) and their chalcidoid parasites of the genus *Austeucharis* Bouček (Hymenoptera: Formicidae: Myrmeciinae: Eucharitidae: Eucharitinae)," *Sociobiology*, vol. 23, no. 2, pp. 109–114, 1993.

[51] J. M. Barnett, *Ecology and general biology of five sympatric species of Myrmecia (Hymenoptera: Formicidae)*, M.S. thesis, Monash University, Melbourne, Australia, 1976.

[52] F. P. Dodd, "Notes upon some remarkable parasitic insects from North Queensland," *Transactions of the Entomological Society of London*, pp. 119–124, 1906.

[53] A. A. Girault, "New genera and species of chalcidoid Hymenoptera in the South Australian Museum," *Transactions of the Royal Society of South Australia*, vol. 37, pp. 67–115, 1913.

[54] Z. Bouček, *Australasian Chalcidoidea (Hymenoptera). A Biosystematic Revision of Genera of Fourteen Families, with a Reclassification of Species*, C. A. B. International, Wallingford, UK, 1988.

[55] A. B. Gahan, "A contribution to the knowledge of the Eucharidae (Hymenoptera: Chalcidoidea)," *Proceedings of the United States National Museum*, vol. 88, pp. 425–458, 1940.

[56] E. Wasmann, *Kritisches Verzeichniss der Myrmekophilen und Termitophilen Arthropoden. Mit Angabe der Lebensweise und mit Beschreibung neuer Arten*, Verlag von Felix L. Dames, Berlin, Germany, 1894.

[57] J. Torréns and J. M. Heraty, "Description of the species of *Dicoelothorax* Ashmead (Chalcidoidea, Eucharitidae) and biology of *D. platycerus* Ashmead *ZooKeys*.," vol. 165, pp. 33–46, 2012.

[58] J.-P. Lachaud, G. Pérez-Lachaud, and J. M. Heraty, "Parasites associated with the ponerine ant *Ectatomma tuberculatum* (Hymenoptera: Formicidae): First host record for the genus *Dilocantha* (Hymenoptera: Eucharitidae)," *The Florida Entomologist*, vol. 81, no. 4, pp. 570–574, 1998.

[59] J. M. Heraty, "The genus *Dilocantha* (Hymenoptera: Eucharitidae)," *Proceedings of the Entomological Society of Washington*, vol. 100, no. 1, pp. 72–87, 1998.

[60] H. Lotfalizadeh, "New distribution records for Eucharitidae (Hym.: Chalcidoidea) in Iran," *North-Western Journal of Zoology*, vol. 4, no. 1, pp. 134–138, 2008.

[61] F. Ruschka, "Die europäisch-mediterranen Eucharidinae und Perilampinae. (Hym. Chalc.). [Der Chalcididenstudien IV. und V. Teil.]," *Deutsche Entomologische Zeitschrift*, vol. 41, pp. 82–96, 1924.

[62] J. Fahringer and F. Tölg, "Beiträge zur Kenntnis˜der Lebensweise und Entwicklungsgeschichte einiger Hautflüger," *Verhandlungen des Naturforschenden Vereines in Brünn*, vol. 50, pp. 242–269, 1912.

[63] G. C. Wheeler and E. W. Wheeler, "New hymenopterous parasites of ants (Chalcidoidea: Eucharidae)," *Annals of the Entomological Society of America*, vol. 30, pp. 163–175, 1937.

[64] P. Cameron, "Hymenopterological notices," *Memoirs and Proceedings of the Manchester Literary and Philosophical Society*, vol. 4, pp. 182–194, 1891.

[65] M. L. Bedel, "Communications. Note sur un Hyménoptère parasite des Fourmis et sur l'état actuel des connaissances relatives aux Arthropodes myrmécophiles et termitophiles," *Bulletin de la Société Entomologique de France*, pp. 35–36, 1895.

[66] C. P. Clausen, "The habits of the Eucharidae," *Psyche*, vol. 48, pp. 57–69, 1941.

[67] Z. Bouček, "A contribution to the knowledge of the Chalcididae, Leucospididae and Eucharitidae (Hymenoptera, Chalcidoidea) of the Near East," *Bulletin of the Research Council of Israel*, vol. 5, no. 3-4, pp. 227–259, 1956.

[68] J. Fahringer, "Beiträge zur kenntnis der lebensweise einiger Chalcididen," *Zeitschrift für Wissenschaftliche Insektenbiologie*, vol. 17, pp. 41–47, 1922.

[69] I. Andriescu, "Beitrag zur kenntnis der Eucharitiden Rumäniens (Chalcidoidea, Hym. Insecta)," in *Lucrarile Sesiunii Stiintifice a Statiunii de Cercetari Marine*, S. Carausu and P. Jitariu, Eds., pp. 225–241, Publications of University "Alexandru Ioan Cuza", Agigea, Iasi, 1968.

[70] W. M. Wheeler, "The polymorphism of ants, with an account of some singular abnormalities due to parasitism," *Bulletin of the American Museum of Natural History*, vol. 23, pp. 1–93, 1907.

[71] E. V. Gemignani, "La familia "Eucharidae" (Hymenoptera: Chalcidoidea) en la República Argentina," *Anales del Museo Nacional de Historia Natural*, vol. 37, pp. 477–493, 1933.

[72] J. Heraty, D. Hawks, J. S. Kostecki, and A. Carmichael, "Phylogeny and behaviour of the Gollumiellinae, a new subfamily of the ant-parasitic Eucharitidae (Hymenoptera: Chalcidoidea)," *Systematic Entomology*, vol. 29, no. 4, pp. 544–559, 2004.

[73] O. F. Cook, "The social organization and breeding habits of the cotton-protecting Kelep of Guatemala," Technical Series No. 10, pp. 1–55, United States Department of Agriculture, 1905.

[74] J. G. Myers, "Descriptions and records of parasitic Hymenoptera from British Guiana and the West Indies," *Bulletin of Entomological Research*, vol. 22, pp. 267–277, 1931.

[75] R. W. Howard, G. Pérez-Lachaud, and J.-P. Lachaud, "Cuticular hydrocarbons of *Kapala sulcifacies* (Hymenoptera: Eucharitidae) and its host, the ponerine ant *Ectatomma ruidum* (Hymenoptera: Formicidae)," *Annals of the Entomological Society of America*, vol. 94, no. 5, pp. 707–716, 2001.

[76] G. Pérez-Lachaud, J. M. Heraty, A. Carmichael, and J.-P. Lachaud, "Biology and behavior of *Kapala* (Hymenoptera: Eucharitidae) attacking *Ectatomma*, *Gnamptogenys*, and *Pachycondyla* (Formicidae: Ectatomminae and Ponerinae) in Chiapas, Mexico," *Annals of the Entomological Society of America*, vol. 99, no. 3, pp. 567–576, 2006.

[77] S. C. Buys, R. Cassaro, and D. Salomon, "Biological observations on *Kapala* Cameron 1884 (Hymenoptera Eucharitidae) in parasitic association with *Dinoponera lucida* Emery 1901 (Hymenoptera Formicidae) in Brazil," *Tropical Zoology*, vol. 23, no. 1, pp. 29–34, 2010.

[78] J.-P. Lachaud, P. Cerdan, and G. Pérez-Lachaud, "Poneromorph ants associated with parasitoid wasps of the genus *Kapala* Cameron (Hymenoptera: Eucharitidae) in French Guiana," *Psyche*, vol. 2012, Article ID 393486, 6 pages, 2012.

[79] G. Pérez-Lachaud, J. A. López-Méndez, and J.-P. Lachaud, "Eucharitid parasitism of the Neotropical ant *Ectatomma tuberculatum*: parasitoid co-occurrence, seasonal variation, and multiparasitism," *Biotropica*, vol. 38, no. 4, pp. 574–576, 2006.

[80] J.-P. Lachaud and G. Pérez-Lachaud, "Fourmis ponérines associées aux parasitoïdes du genre *Kapala* Cameron (Hymenoptera, Eucharitidae)," *Actes des Colloques Insectes Sociaux*, vol. 14, pp. 101–105, 2001.

[81] A. De la Mora and S. M. Philpott, "Wood-nesting ants and their parasites in forests and coffee agroecosystems," *Environmental Entomology*, vol. 39, no. 5, pp. 1473–1481, 2010.

[82] J. Torréns, J. M. Heraty, and P. Fidalgo, "Biology and description of a new species of *Lophyrocera* Cameron (Hymenoptera: Eucharitidae) from Argentina," *Zootaxa*, no. 1871, pp. 56–62, 2008.

[83] J. M. Heraty, "Classification and evolution of the Oraseminae in the Old World, with revisions of two closely related genera of Eucharitinae (Hym: Eucharitidae)," *Life Sciences Contributions (Royal Ontario Museum)*, vol. 157, pp. 1–174, 1994.

[84] L. R. Davis Jr. and D. P. Jouvenaz, "*Obeza floridana*, a parasitoid of *Camponotus abdominalis floridanus* from Florida (Hymenoptera: Eucharitidae, Formicidae)," *The Florida Entomologist*, vol. 73, no. 2, pp. 335–337, 1990.

[85] L. Varone, J. M. Heraty, and L. A. Calcaterra, "Distribution, abundance and persistence of species of *Orasema* (Hym: Eucharitidae) parasitic on fire ants in South America," *Biological Control*, vol. 55, no. 1, pp. 72–78, 2010.

[86] G. M. Das, "Preliminary studies on the biology of *Orasema assectator* Kerrich (Hym., Eucharitidae), parasitic on *Pheidole* and causing damage to leaves of tea in Assam," *Bulletin of Entomological Research*, vol. 54, pp. 373–379, 1963.

[87] G. J. Kerrich, "Descriptions of two species of Eucharitidae damaging tea, with comparative notes on other species (Hym., Chalcidoidea)," *Bulletin of Entomological Research*, vol. 54, pp. 365–372, 1963.

[88] J. B. Johnson, T. D. Miller, J. M. Heraty, and F. W. Merickel, "Observations on the biology of two species of *Orasema* (Hym.: Eucharitidae)," *Proceedings of the Entomological Society of Washington*, vol. 88, no. 3, pp. 542–549, 1986.

[89] W. W. Kempf, "A study of some Neotropical ants of genus *Pheidole* Westwood. I. (Hymenoptera: Formicidae)," *Studia Entomologica*, vol. 15, pp. 449–464, 1972.

[90] A. Reichensperger, "Zur kenntnis von myrmecophilen aus abessinien," *I. Zoologische Jahrbücher*, vol. 35, pp. 185–218, 1913.

[91] W. M. Mann, "Some myrmecophilous insects from Cuba," *Psyche*, vol. 25, pp. 104–106, 1918.

[92] J. M. Heraty, "Biology and importance of two eucharitid parasites of *Wasmannia* and *Solenopsis*," in *Exotic Ants: Biology, Impact and Control of Introduced Species*, D. Williams, Ed., pp. 104–120, Westview Press, Boulder, Colo, USA, 1994.

[93] J. M. Heraty, "Phylogenetic relationships of Oraseminae (Hymenoptera: Eucharitidae)," *Annals of the Entomological Society of America*, vol. 93, no. 3, pp. 374–390, 2000.

[94] G. C. Wheeler and J. N. Wheeler, *The Ants of Nevada*, Natural History Museum of Los Angeles County, 1986.

[95] A. F. Van Pelt, "*Orasema* in nests of *Pheidole dentata* Mayr (Hymenoptera: Formicidae)," *Entomological News*, vol. 61, no. 6, pp. 161–163, 1950.

[96] J. M. Heraty, D. P. Wojcik, and D. P. Jouvenaz, "Species of *Orasema* parasitic on the *Solenopsis saevissima*-complex in South America (Hymenoptera: Eucharitidae, Formicidae)," *Journal of Hymenoptera Research*, vol. 2, no. 1, pp. 169–182, 1993.

[97] L. Varone and J. Briano, "Bionomics of *Orasema simplex* (Hymenoptera: Eucharitidae), a parasitoid of *Solenopsis* fire ants (Hymenoptera: Formicidae) in Argentina," *Biological Control*, vol. 48, no. 2, pp. 204–209, 2009.

[98] B. G. Carey, *Behavioral ecology of Orasema (Hymenoptera: Eucharitidae) and the mechanism of indirect parasitism of ants*, M.S. thesis, University of California, Riverside, Calif, USA, 2000.

[99] M. A. Naves, "A monograph of the genus *Pheidole* in Florida (Hymenoptera: Formicidae)," *Insecta Mundi*, vol. 1, no. 2, pp. 53–90, 1985.

[100] W. M. Mann, "Some myrmecophilous insects from Mexico," *Psyche*, vol. 21, pp. 171–184, 1914.

[101] A. Silveira-Guido, P. San-Martin, C. Crisci-Pisano, and J. Carbonnell-Bruhn, "Investigations on the biology and biological control of the fire ant, *Solenopsis saevissima richteri* Forel in Uruguay. Third report," Departamento de Sanidad Vegetal, Facultad de Agronomía, Universidad de la República, Montevideo, Uruguay, pp. 1–67, 1964.

[102] M. A. Pesquero and A. M. Penteado-Dias, "New records of *Orasema xanthopus* (Hymenoptera: Eucharitidae) and *Solenopsis daguerrei* (Hymenoptera: Formicidae) from Brazil," *Brazilian Journal of Biology*, vol. 64, no. 3, p. 737, 2004.

[103] C. T. Brues, "Some new eucharid parasites of Australian ants," *Bulletin of the Brooklyn Entomological Society*, vol. 29, pp. 201–207, 1934.

[104] J. M. Heraty, "*Pseudochalcura* (Hymenoptera: Eucharitidae): a New World genus parasitic upon ants," *Systematic Entomology*, vol. 11, pp. 183–212, 1986.

[105] J. M. Heraty, J. M. Heraty, and J. Torréns, "A new species of *Pseudochalcura* (Hymenoptera, Eucharitidae), with a review of antennal morphology from a phylogenetic perspective," *ZooKeys*, vol. 20, pp. 215–231, 2009.

[106] G. L. Ayre, "*Pseudometagea schwarzii* (Ashm.) (Eucharitidae: Hymenoptera), a parasite of *Lasius neoniger* Emery (Formicidae: Hymenoptera)," *The Canadian Journal of Zoology*, vol. 40, pp. 157–164, 1962.

[107] T. Maeyama, M. Machida, and M. Terayama, "The ant-parasitic genus *Rhipipalloidea* Girault (Hymenoptera: Eucharitidae), with a description of a new species," *Australian Journal of Entomology*, vol. 38, no. 4, pp. 305–309, 1999.

[108] C. P. Clausen, "The biology of *Schizaspidia tenuicornis* Ashm., a eucharid parasite of *Camponotus*," *Annals of the Entomological Society of America*, vol. 16, pp. 195–217, 1923.

[109] G. C. Wheeler and E. H. Wheeler, "A new species of *Schizaspidia* (Eucharidae) with notes on a eulophid ant parasite," *Psyche*, vol. 31, pp. 49–56, 1924.

[110] A. A. Girault, "Australian Hymenoptera Chalcidoidea–X," *Memoirs of the Queensland Museum*, vol. 4, pp. 225–237, 1915.

[111] M. E. Schauff and Z. Bouček, "*Alachua floridensis*, a new genus and species of Entedoninae (Hymenoptera, Eulophidae) parasitic on the Florida carpenter ant, *Camponotus abdominalis* (Formicidae)," *Proceedings of the Entomological Society of Washington*, vol. 89, pp. 660–664, 1987.

[112] C. Hansson, J.-P. Lachaud, and G. Pérez-Lachaud, "Entedoninae wasps (Hymenoptera, Chalcidoidea, Eulophidae) associated with ants (Hymenoptera, Formicidae) in tropical America, with new species and notes on their biology," *ZooKeys*, vol. 134, pp. 65–82, 2011.

[113] Z. Bouček, "Descriptions of new eulophid parasites (Hym., Chalcidoidea) from Africa and the Canary Islands," *Bulletin of Entomological Research*, vol. 62, no. 2, pp. 199–205, 1972.

[114] G. J. Kerrich, "A revision of the tropical and subtropical species of the eulophid genus *Pediobius* Walker (Hymenoptera: Chalcidoidea)," *Bulletin of the British Museum of Natural History*, vol. 29, no. 3, pp. 113–199, 1973.

[115] C. T. Brues, "*Conoaxima*, a new genus of the hymenopterous family Eurytomidae, with a description of its larva and pupa," *Psyche*, vol. 29, pp. 153–158, 1922.

[116] W. M. Wheeler, "Studies of Neotropical ant-plants and their ants," *Bulletin of the Museum of Comparative Zoology of Harvard College*, vol. 90, no. 1, pp. 1–262, 1942.

[117] M. D. Trager, *Ant occupancy and anti-herbivore defense of Cordia alliodora, a Neotropical myrmecophyte*, M.S. thesis, University of Florida, Fla, USA, 2005.

[118] D. W. Yu and D. W. Davidson, "Experimental studies of species-specificity in *Cecropia*-ant relationships," *Ecological Monographs*, vol. 67, no. 3, pp. 273–294, 1997.

[119] D. W. Davidson and B. F. Fisher, "Symbiosis of ants with *Cecropia* as a function of light regime," in *Ant-Plant Interactions*, C. R. Huxley and D. F. Cutler, Eds., pp. 289–309, Oxford University Press, Oxford, UK, 1991.

[120] A. Forel, "Les fourmis de la Suisse. Systématique. Notices anatomiques et physiologiques. Architecture. Distribution géographique. Nouvelles expériences et observations de moeurs," *Neue Denkschriften der Allgemeinen Schweizerischen Gesellschaft für di Gesammten Naturwissenschaften*, vol. 26, pp. 1–452, 1874.

[121] A. Panis, "Les braconides parasitoïdes de fourmis et observations biologiques sur *Elasmosoma berolinense* Ruthe (Hymenoptera: Braconidae) et les ouvrières parasitées de *Camponotus vagus* Scopoli (Hymenoptera: Formicidae)," *Bulletin Mensuel de la Société Linnéenne de Lyon*, vol. 76, no. 4, pp. 57–62, 2007.

[122] C. Watanabe, "On two hymenopterous guests of ants in Japan," *Insecta Matsumurana*, vol. 9, pp. 90–94, 1935.

[123] O. Schmiedeknecht, "Die schlupfwespen (Ichneumonidea)," in *Die Insekten Mitteleuropas insbesondere Deutschlands. Zweiter Band. Hymenopteren Zweiter Teil*, C. Schröder, Ed., pp. 113–256, Franckh'sche Verlagshanlung, Stuttgart, Germany, 1914.

[124] J. F. Ruthe, "Beiträge zur kenntniss der braconiden," *Berliner Entomologische Zeitschrift*, vol. 2, pp. 1–10, 1858.

[125] J. Giraud, "Note sur l'*Elasmosoma Berolinense* et description d'une espèce nouvelle (*viennense*) du même genre," *Annales de la Société Entomologique de France*, pp. 299–302, 1871.

[126] E. Olivier, "Notes entomologiques," *Bulletin des Séances et Bulletin Bibliographique de la Société Entomologique de France*, no. 1, pp. LXX–LXXI, 1893.

[127] T. Huddleston, "A revision of *Elasmosoma* Ruthe (Hymenoptera, Braconidae) with two new species from Mongolia," *Annales Historico-Naturales Musei Nationalis Hungarici*, vol. 68, pp. 215–225, 1976.

[128] E. Wasmann, "Die psychischen Fähigkeiten der Ameisen," *Zoologica*, vol. 11, pp. 1–133, 1899, 1909.

[129] J.-M. Gómez Durán and C. van Achterberg, "Oviposition behaviour of four ant parasitoids (Hymenoptera, Braconidae, Euphorinae, Neoneurini and Ichneumonidae, Hybrizontinae), with the description of three new European species," *ZooKeys*, vol. 125, pp. 59–106, 2011.

[130] S. R. Shaw, "A new species of *Elasmosoma* Ruthe (Hymenoptera: Braconidae: Neoneurinae) from the northwestern United States associated with the western thatching ants, *Formica obscuripes* Forel and *Formica obscuriventris clivia* Creighton (Hymenoptera: Formicidae)," *Proceedings of the Entomological Society of Washington*, vol. 109, no. 1, pp. 1–8, 2007.

[131] G. Poinar Jr., "Behaviour and development of *Elasmosoma* sp. (Neoneurinae: Braconidae: Hymenoptera), an endoparasite

[132] W. H. Ashmead, "Discovery of the genus *Elasmosoma*, Ruthe, in America," *Proceedings of the Entomological Society of Washington*, vol. 3, pp. 280–284, 1895.

[133] G. A. Coovert, "The ants of Ohio (Hymenoptera: Formicidae)," *Ohio Biological Survey Bulletin New Series*, vol. 15, no. 2, pp. 1–196, 2005.

[134] C. F. W. Muesebeck, "A new ant parasite (Hymenoptera, Braconidae)," *Bulletin of the Brooklyn Entomological Society*, vol. 36, pp. 200–201, 1941.

[135] T. D. A. Cockerell, "A new braconid of the genus *Elasmosoma*," *Proceedings of the Entomological Society of Washington*, vol. 10, pp. 168–169, 1909.

[136] G. Poinar Jr. and J. C. Miller, "First fossil record of endoparasitism of adult ants (Formicidaes: Hymenoptera) by Braconidae (Hymenoptera)," *Annals of the Entomological Society of America*, vol. 95, no. 1, pp. 41–43, 2002.

[137] V. I. Tobias, "Obzor naezdnikov-brakonid (Hymenoptera) fauny SSSR," *Trudy Vsesoyuznogo Entomologicheskogo Obshchestva*, vol. 54, pp. 156–268, 1971.

[138] C. van Achterberg and Q. Argaman, "*Kollasmosoma* gen. nov. and a key to the genera of the subfamily Neoneurinae (Hymenoptera: Braconidae)," *Zoologische Mededelingen Leiden*, vol. 67, pp. 63–74, 1993.

[139] C. Morley, "Notes on Braconidae, X.: On the Pachylommatinae, with descriptions of new species," *Entomologist's Monthly Magazine*, vol. 45, pp. 209–214, 1909.

[140] H. S. J. K. Donisthorpe, "Myrmecophilous notes for 1909," *The Entomologist's Record and Journal of Variation*, vol. 22, no. 1, pp. 15–17, 1910.

[141] J. Fahringer, "Opuscula braconologica. Band 4. Palaearktischen Region. Lieferung 1-3," *Opuscula braconologica* (1935), pp. 1–276. Fritz Wagner, Wien, 1936.

[142] S. R. Shaw, "Seven new North American species of *Neoneurus* (Hymenoptera: Braconidae)," *Proceedings of the Entomological Society of Washington*, vol. 94, no. 1, pp. 26–47, 1992.

[143] S. R. Shaw, "Observations on the ovipositional behavior of *Neoneurus mantis*, an ant-associated parasitoid from Wyoming (Hymenoptera: Braconidae)," *Journal of Insect Behavior*, vol. 6, no. 5, pp. 649–658, 1993.

[144] J. de Gaulle, "Catalogue systématique & biologique des Hyménoptères de France, suite," *La Feuille des Jeunes Naturalistes*, vol. 37, no. 441, pp. 185–189, 1907.

[145] J. T. C. Ratzeburg, *Die Ichneumonen der Forstinsecten in Forstlicher und Entomologischer Beziehung. Ein Anhang zur Abbildung und Beschreibung der Forstinsecten. Dritter Band*, Nicolai'schen Buchhandlung, Berlin, Germany, 1852.

[146] J. Giraud, "Description de quelques hyménoptères nouveaux ou rares," *Verhandlungen der Zoologisch-Botanischen Gesellschaft Wien*, vol. 7, pp. 163–184, 1857.

[147] R. Cobelli, "L'ibernazione delle Formiche," *Verhandlungen der Zoologisch-Botanischen Gesellschaft Wien*, vol. 53, pp. 369–380, 1903.

[148] R. Cobelli, "Il *Pachylomma cremieri* de Romand ed il *Lasius fuliginosus* Latr.," *Verhandlungen der Zoologisch-Botanischen Gesellschaft Wien*, vol. 56, pp. 475–477, 1906.

[149] T. Komatsu and K. Konishi, "Parasitic behaviors of two ant parasitoid wasps (Ichneumonidae: Hybrizontinae)," *Sociobiology*, vol. 56, no. 3, pp. 575–584, 2010.

[150] H. S. J. K. Donisthorpe and D. S. Wilkinson, "Notes on the genus *Paxylomma* (Hym. Brac.), with the description of

of *Formica* ants (Formicidae: Hymenoptera)," *Parasitology*, vol. 128, no. 5, pp. 521–531, 2004.

a new species taken in Britain," *Transactions of the Entomological Society of London*, vol. 78, no. 1, pp. 87–93, 1930.

[151] C. Watanabe, "Notes on Paxylommatinae with review of Japanese species (Hymenoptera, Braconidae)," *Kontyû, Tokyo*, vol. 52, no. 4, pp. 553–556, 1984.

[152] H. S. J. K. Donisthorpe, "Myrmecophilous notes for 1913," *The Entomologist's Record and Journal of Variation*, vol. 26, no. 2, pp. 37–45, 1914.

[153] T. A. Marshall, "I. A monograph of British Braconidae. Part VIII," *Transactions of the Entomological Society of London*, vol. 47, pp. 1–79, 1897.

[154] P. M. Marsh, "Notes on the genus *Hybrizon* in North America (Hymenoptera: Paxylommatidae)," *Proceedings of the Entomologial Society of Washington*, vol. 91, no. 1, pp. 29–34, 1989.

[155] J. W. Early, L. Masner, I. D. Naumann, and A. D. Austin, "Maamingidae, a new family of proctotrupoid wasp (Insecta: Hymenoptera) from New Zealand," *Invertebrate Taxonomy*, vol. 15, no. 3, pp. 341–352, 2001.

[156] L. Masner and J. L. García, "The genera of Diapriinae (Hymenoptera: Diapriidae) in the New World," *Bulletin of the American Museum of Natural History*, no. 268, pp. 1–138, 2002.

[157] L. Musetti and N. F. Johnson, "Revision of the New World species of the genus *Monomachus* Klug (Hymenoptera: Proctotrupoidea, Monomachidae)," *The Canadian Entomologist*, vol. 136, no. 4, pp. 501–552, 2004.

[158] E. L. Aguiar-Menezes, E. B. Menezes, and M. S. Loiácono, "First record of *Coptera haywardi* Loiácono (Hymenoptera: Diapriidae) as a parasitoid of fruit-infesting Tephritidae (Diptera) in Brazil," *Neotropical Entomology*, vol. 32, no. 2, pp. 355–358, 2003.

[159] L. Masner, "Superfamily Proctotrupoidea," in *Hymenoptera of the World: An Identification Guide to Families*, H. Goulet and J. T. Huber, Eds., pp. 537–557, Research Branch, Agriculture Canada Publications, Ottawa, Canada, 1993.

[160] L. Huggert and L. Masner, "A review of myrmecophilic-symphilic diapriid wasps in the Holartic realm, with descriptions of new taxa and a key to genera (Hymenoptera: Proctotrupoidea: Diapriidae)," *Contributions of the American Entomological Institute*, vol. 20, pp. 63–89, 1983.

[161] K. Hölldobler, "Zur Biologie der diebischen Zwergameise (*Solenopsis fugax*) und ihrer Gäste," *Biologisches Zentralblatt*, vol. 48, no. 3, pp. 129–142, 1928.

[162] M. W. Wing, "A new genus and species of myrmecophilous Diapriidae with taxonomic and biological notes on related forms," *Transactions of the Royal Entomological Society of London*, vol. 102, no. 3, pp. 195–210, 1951.

[163] J.-P. Lachaud, "Les communications tactiles interspécifiques chez les diapriides myrmécophiles *Lepidopria pedestris* Kieffer et *Solenopsia imitatrix* Wasmann et leur hôte *Diplorhoptrum fugax* Latr. (*Solenopsis fugax* Latr.)," *Biologie-Écologie Méditerranéenne*, vol. 7, no. 3, pp. 183–184, 1980.

[164] J.-P. Lachaud, *Étude des relations hôte-myrmécophile entre les Diapriidae Lepidopria pedestris Kieffer et Solenopsia imitatrix Wasmann et la fourmi Diplorhoptrum fugax Latreille*, Ph.D. dissertation, Université Paul-Sabatier, Toulouse, France, 1981.

[165] J.-P. Lachaud, "Les glandes tégumentaires chez deux espèces de Diapriidae: aspects structuraux et ultrastructuraux," *Bulletin Intérieur de la Section Française de l'UIEIS, Toulouse, France*, pp. 83–85, 1981.

[166] J.-P. Lachaud, "Estudio sobre las relaciones trofalácticas entre *Lepidopria pedestris* Kieffer (Hymenoptera, Diapriidae) y su huésped *Diplorhoptrum fugax* Latreille (Hymenoptera, Formicidae)," *Folia Entomológica Mexicana*, vol. 54, pp. 46–47, 1982.

[167] C. Ferrière, "Notes sur un Diapriide (Hyménoptère), hôte de *Solenopsis fugax* Latr.," *Konowia (Vienna)*, vol. 6, pp. 282–286, 1927.

[168] L. Masner, "A revision of ecitophilous diapriid-genus *Mimopria* Holmgren (Hym., Proctotrupoidea)," *Insectes Sociaux*, vol. 6, no. 4, pp. 361–367, 1959.

[169] L. S. Ramos-Lacau, *Bioecologia comparada de duas espécies de Cyphomyrmex Mayr (Formicidae: Myrmicinae)*, Ph.D. dissertation, Universidade Estadual Paulista, Rio Claro, SP, Brasil, 2006.

[170] M. S. Loiácono and C. B. Margaría, "A note on *Szelenyiopria pampeana* (Loiácono) n. comb., parasitoid wasps (Hymenoptera: Diapriidae) attacking the fungus growing ant, *Acromyrmex lobicornis* Emery (Hymenoptera: Formicidae: Attini) in La Pampa, Argentina," *Zootaxa*, no. 2105, pp. 63–65, 2009.

[171] G. E. J. Nixon, "A new British proctotrupid of the subfamily Belytinae," *The Entomologist's Record*, vol. 43, pp. 83–84, 1931.

[172] G. E. J. Nixon, *Hymenoptera, Proctotrupoidea, Diapriidae Subfamily Belytinae*, Handbooks for the Identification of British Insects, Royal Entomological Society of London, London, UK, 1957.

[173] L. Huggert, "*Cryptoserphus* and Belytinae wasps (Hymenoptera, Proctotrupoidea) parasiting fungus- and soil-inhabiting Diptera," *Notulae Entomologicae*, vol. 59, pp. 139–144, 1979.

[174] J. S. Noyes, "A word on chalcidoid classification," *Chalcid Forum*, vol. 13, pp. 6–7, 1990.

[175] E. E. Grissell and M. E. Schauff, "Chalcidoidea," in *Annotated Keys to the Genera of Nearctic Chalcidoidea (Hymenoptera)*, G. A. P. Gibson, J. T. Huber, and J. B. Woolley, Eds., pp. 45–116, National Research Council of Canada Press, Ontario, Canada, 1997.

[176] G. A. P. Gibson, J. M. Heraty, and J. B. Woolley, "Phylogenetics and classification of Chalcidoidea and Mymarommatoidea – a review of current concepts (Hymenoptera, Apocrita)," *Zoologica Scripta*, vol. 28, no. 1-2, pp. 87–124, 1999.

[177] J. B. Munro, J. M. Heraty, R. A. Burks et al., "A molecular phylogeny of the Chalcidoidea (Hymenoptera)," *PLoS ONE*, vol. 6, no. 11, article e27023, 2011.

[178] J. S. Noyes, "Encyrtidae of Costa Rica (Hymenoptera: Chalcidoidea) 1," *Memoirs of the American Entomological Institute*, vol. 62, pp. 1–354, 2000.

[179] Z. Bouček, "The New World genera of Chalcididae," *Memoirs of the American Entomological Institute*, vol. 53, pp. 49–118, 1992.

[180] M. S. Mani, "Chalcids (parasitic Hymenoptera) from India," *Record of the Indian Museum*, vol. 38, pp. 125–129, 1936.

[181] P. Dessart, "Matériel typique des Microhymenoptera myrmécophiles de la collection Wasmann déposé au Muséum Wasmannianum à Maastricht (Pays-Bas)," *Publicatiës van het Natuurhistorisch Genootschap in Limburg*, vol. 24, no. 1-2, pp. 1–94, 1975.

[182] J. S. Noyes and M. Hayat, *Oriental Mealybug Parasitoids of the Anagyrini (Hymenoptera: Encyrtidae)*, CAB International, Wallingford, UK, 1994.

[183] V. A. Trjapitzin, "A review of encyrtid wasps (Hymenoptera, Chalcidoidea, Encyrtidae) of Macaronesia," *Entomological Review*, vol. 88, no. 2, pp. 218–232, 2008.

[184] T. Tachikawa, "Hosts of encyrtid genera in the World (Hymenoptera: Chalcidoidea)," *Memoirs of the College of Agriculture, Ehime University*, vol. 25, no. 2, pp. 85–110, 1981.

[185] J. S. Noyes, J. B. Woolley, and G. Zolnerowich, "Encyrtidae," in *Annotated Keys to the Genera of Nearctic Chalcidoidea (Hymenoptera)*, G. A. P. Gibson, J. T. Huber, and J.B. Woolley, Eds., pp. 170–320, National Research Council of Canada Press, Ontario, Canada, 1997.

[186] H. González-Hernández, M. W. Johnson, and N. J. Reimer, "Impact of *Pheidole megacephala* (F.) (Hymenoptera: Formicidae) on the biological control of *Dysmicoccus brevipes* (Cockerell) (Homoptera: Pseudococcidae)," *Biological Control*, vol. 15, no. 2, pp. 145–152, 1999.

[187] W. Roepke, "Some additional remarks concerning Mr. Girault's descriptions of new Javanese chalcid flies," *Treubia*, vol. 1, p. 60, 1919.

[188] C. W. Rettenmeyer, M. E. Rettenmeyer, J. Joseph, and S. M. Berghoff, "The largest animal association centered on one species: the army ant *Eciton burchellii* and its more than 300 associates," *Insectes Sociaux*, vol. 58, no. 3, pp. 281–292, 2011.

[189] C. P. Clausen, "The immature stages of the Eucharidae," *Proceedings of the Entomological Society of Washington*, vol. 42, no. 8, pp. 161–170, 1940.

[190] C. P. Clausen, "The oviposition habits of the Eucharidae (Hymenoptera)," *Journal of the Washington Academy of Sciences*, vol. 30, no. 12, pp. 504–516, 1940.

[191] J. M. Heraty and D. C. Darling, "Comparative morphology of the planidial larvae of Eucharitidae and Perilampidae (Hymenoptera: Chalcidoidea)," *Systematic Entomology*, vol. 9, pp. 309–328, 1984.

[192] H. A. Tripp, "The biology of a hyperparasite, *Euceros frigidus* Cress. (Ichneumonidae) and description of the planidial stage," *The Canadian Entomologist*, vol. 93, no. 1, pp. 40–58, 1961.

[193] J.-P. Lachaud and G. Pérez-Lachaud, "Impact of natural parasitism by two eucharitid wasps on a potential biocontrol agent ant in southeastern Mexico," *Biological Control*, vol. 48, no. 1, pp. 92–99, 2009.

[194] G. Pérez-Lachaud, J. A. López-Méndez, G. Beugnon, P. Winterton, and J.-P. Lachaud, "High prevalence but relatively low impact of two eucharitid parasitoids attacking the Neotropical ant *Ectatomma tuberculatum* (Olivier)," *Biological Control*, vol. 52, no. 2, pp. 131–139, 2010.

[195] R. K. Vander Meer, D. P. Jouvenaz, and D. P. Wojcik, "Chemical mimicry in a parasitoid (Hymenoptera: Eucharitidae) of fire ants (Hymenoptera: Formicidae)," *Journal of Chemical Ecology*, vol. 15, no. 8, pp. 2247–2261, 1989.

[196] G. Pérez-Lachaud and J.-P. Lachaud, "Comportement de transport de parasitoïdes Eucharitidae par leur hôte: mimétisme chimique et effet de la taille de l'objet à transporter," *Actes du Colloque Annuel de la Section Française de l'UIEIS*, p. 32, 2007.

[197] N. Gauthier, J. Lasalle, D. L. J. Quicke, and H. C. J. Godfray, "Phylogeny of Eulophidae (Hymenoptera: Chalcidoidea), with a reclassification of Eulophinae and the recognition that Elasmidae are derived eulophids," *Systematic Entomology*, vol. 25, no. 4, pp. 521–539, 2000.

[198] M. E. Schauff, "*Microdonophagus*, a new entedontine genus (Hymenoptera, Eulophidae) from Panama," *Proceedings of the Entomological Society of Washington*, vol. 88, no. 1, pp. 167–173, 1986.

[199] A. Gumovsky and Z. Bouček, "A new genus of Entedoninae from Malaysia, associated with ant nests (Hymenoptera, Eulophidae)," *Entomological Problems*, vol. 35, no. 1, pp. 39–42, 2005.

[200] M. W. Gates, M. A. Metz, and M. E. Schauff, "The circumscription of the generic concept of *Aximopsis* Ashmead (Hymenoptera: Chalcidoidea: Eurytomidae) with the description of seven new species," *Zootaxa*, no. 1273, pp. 9–54, 2006.

[201] M. V. Gates, "Species revision and generic systematics of world Rileyinae (Hymenoptera: Eurytomidae)," *University of California Publications in Entomology*, vol. 127, pp. 1–332, 2008.

[202] M. W. Gates and G. Pérez-Lachaud, "Description of *Camponotophilus delvarei*, gen. n. and sp. n. (Hymenoptera: Chalcidoidea: Eurytomidae), with discussion of diagnostic characters," *Proceedings of the Entomological Society of Washington*, vol. 114, no. 1, 2012, In press.

[203] J.-L. Weng, K. Nishida, P. Hanson, and L. LaPierre, "Biology of *Lissoderes* Champion (Coleoptera, Curculionidae) in *Cecropia* saplings inhabited by *Azteca* ants," *Journal of Natural History*, vol. 41, no. 25–28, pp. 1679–1695, 2007.

[204] G. A. P. Gibson, "Superfamilies Mymarommatoidea and Chalcidoidea," in *Hymenoptera of the World: An Identification Guide to Families*, H. Goulet and J. T. Huber, Eds., pp. 570–655, Research Branch, Agriculture Canada Publications, Ottawa, Canada, 1993.

[205] P. M. Marsh and R. W. Carlson, "Superfamily Ichneumonoidea," in *Catalog of Hymenoptera in America North of Mexico, vol. 1, Symphita and Apocrita (Parasitica)*, K. V. Krombein, P. D. Hurd Jr., D. R. Smith, and B. D. Burks, Eds., pp. 143–144, Smithonian Institution Press, Washington, DC, USA, 1979.

[206] D. B. Wahl and M. J. Sharkey, "Superfamily Ichneumonoidea," in *Hymenoptera of the World: An Identification Guide to Families*, H. Goulet and J. T. Huber, Eds., pp. 358–362, Research Branch, Agriculture Canada Publications, Ottawa, Canada, 1993.

[207] D. S. Yu, C. van Achterberg, and K. Horstmann, "World Ichneumonoidea 2004. Taxonomy, biology, morphology and distribution," Taxapad 2005 (Scientific names for information management), Interactive catalogue on DVD/CDROM, Vancouver, Canada, 2005.

[208] D. L. J. Quicke and C. van Achterberg, "Phylogeny of the subfamilies of the family Braconidae (Hymenoptera: Ichneumonoidea)," *Zoologische Verhandelingen Leiden*, vol. 258, pp. 1–95, 1990.

[209] M. J. Sharkey, "Family Braconidae," in *Hymenoptera of the World: An Identification Guide to Families*, H. Goulet and J. T. Huber, Eds., pp. 362–395, Research Branch, Agriculture Canada Publications, Ottawa, Canada, 1993.

[210] M. R. Shaw and T. Huddleston, "Classification and biology of braconid wasps (Hymenoptera: Braconidae)," in *Handbooks for the Identification of British Insects*, W. R. Dowling and R. R. Askew, Eds., vol. 7, part 11, pp. 1–126, Royal Entomological Society of London, London, UK, 1991.

[211] H. Maneval, "Observations sur un Aphidiidae (Hym.) myrmécophile. Description du genre et de l'espèce," *Bulletin Mensuel de la Société Linnéenne de Lyon*, vol. 9, pp. 9–14, 1940.

[212] H. Takada and Y. Hashimoto, "Association of the root aphid parasitoids *Aclitus sappaphis* and *Paralipsis eikoae* (Hymenoptera, Aphidiidae) with the aphid-attending ants *Pheidole fervida* and *Lasius niger* (Hymenoptera, Formicidae)," *Kontyû, Tokyo*, vol. 53, no. 1, pp. 150–160, 1985.

[213] W. Völkl, C. Liepert, R. Birnbach, G. Hübner, and K. Dettner, "Chemical and tactile communication between the root aphid parasitoid *Paralipsis enervis* and trophobiotic ants: consequences for parasitoid survival," *Experientia*, vol. 52, no. 7, pp. 731–738, 1996.

[214] T. Akino and R. Yamaoka, "Chemical mimicry in the root aphid parasitoid *Pavalipsis eikoae* Yasumatsu (Hymenoptera: Aphidiidae) of the aphid-attending ant *Lasius sakagamii* Yamauchi & Hayashida (Hymenoptera: Formicidae)," *Chemoecology*, vol. 8, no. 4, pp. 153–161, 1998.

[215] R. D. Shenefelt, "Braconidae 1. Hybrizoninae, Euphorinae, Cosmophorinae, Neoneurinae, Macrocentrinae," in *Hymenopterum Catalogus*, C. Ferrière and J. van der Vecht, Eds., pars 4, pp. 1–176, Junk, The Hague, The Netherlands, 1969.

[216] P. M. Marsh, "Family Braconidae," in *Catalog of Hymenoptera in America North of Mexico, vol. 1, Symphita and Apocrita (Parasitica)*, K. V. Krombein, P. D. Hurd Jr., D. R. Smith, and B. D. Burks, Eds., pp. 144–295, Smithonian Institution Press, Washington, DC, USA, 1979.

[217] S. R. Shaw, "A phylogenetic study of the subfamilies Meteorinae and Euphorinae (Hymenoptera: Braconidae)," *Entomography*, vol. 3, pp. 277–370, 1985.

[218] R. W. Carlson, "Family Ichneumonidae," in *Catalog of Hymenoptera in America North of Mexico, vol. 1, Symphita and Apocrita (Parasitica)*, K. V. Krombein, P. D. Hurd Jr., D. R. Smith, and B. D. Burks, Eds., pp. 315–740, Smithonian Institution Press, Washington, DC, USA, 1979.

[219] D. B. Wahl, "Family Ichneumonidea," in *Hymenoptera of the World: An Identification Guide to Families*, H. Goulet and J. T. Huber, Eds., pp. 395–442, Research Branch, Agriculture Canada Publications, Ottawa, Canada, 1993.

[220] F. Smith, "Hymenoptera. Observations on the effects of the late unfavourable season on hymenopterous insects; notes on the economy of certain species, on the capture of others of extreme rarity, and on species new to the British fauna," *The Entomologist's Annual for 1861*, pp. 33–45, 1861.

[221] C. T. Brues, "Descriptions of new ant-like and myrmecophilous Hymenoptera," *Transactions of the American Entomological Society*, vol. 29, pp. 119–128, 1903.

[222] C. Morley, *Ichneumonologia Britannica. ii. The Ichneumons of Great Britain. Cryptinae*, J. H. Keys, Plymouth, UK, 1907.

[223] P. E. Hanson, M. J. West-Eberhard, and I. D. Gauld, "Interspecific interactions of nesting Hymenoptera," in *The Hymenoptera of Costa Rica*, P. E. Hanson and I. D. Gauld, Eds., pp. 76–88, Oxford University Press, Oxford, UK, 1995.

[224] D. H. Feener Jr., "Is the assembly of ant communities mediated by parasitoids?" *Oikos*, vol. 90, no. 1, pp. 79–88, 2000.

9

First Record of *Lenomyrmex inusitatus* (Formicidae: Myrmicinae) in Ecuador and Description of the Queen

T. Delsinne[1] and F. Fernández[2]

[1] Biological Evaluation Section, Royal Belgian Institute of Natural Sciences, 29 rue Vautier, 1000 Brussels, Belgium
[2] Instituto de Ciencias Naturales, Universidad Nacional de Colombia, Apartado 7495, Bogotá D.C., Colombia

Correspondence should be addressed to T. Delsinne, thibaut.delsinne@sciencesnaturelles.be

Academic Editor: Jacques H. C. Delabie

The rarely collected ant *Lenomyrmex inusitatus* Fernández 2001 is recorded for the first time in Ecuador. The queen is described. The new record is the southernmost limit of distribution for the genus. A key to the workers of the six *Lenomyrmex* species and a key for the known queens are provided.

1. Introduction

The myrmicine ant genus *Lenomyrmex* Fernández and Palacio 1999 includes six species rarely collected from Costa Rica to Ecuador [1–3]. The genus is characterised by elongate mandibles bearing a series of minute peg-like denticles that arise behind the masticatory margin, by frontal lobes that are poorly expanded laterally, by large and deep antennal fossae, and by pedunculate petiole, with a poorly defined node [1]. The fact that *Lenomyrmex* possesses both primitive (e.g., promesonotal suture well developed) and derived (e.g., specialized morphology of the mandibles) characters makes ascertaining its correct phylogenetic position challenging [1, 2, 4]. The genus was tentatively placed in its own tribe, Lenomyrmecini [5], but its position within the Myrmicinae remains to be determined [5]. Preliminary results of a phylogenetic analysis (Ant-AToL project, http://www.antweb.org/atol.jsp) indicate that *Lenomyrmex* falls within a clade of predominantly New World ants that includes the tribes Attini, Cephalotini, Dacetini, and the genus *Pheidole* (T. Schultz and P. Ward, comm. pers.).

The worker of *Lenomyrmex inusitatus* Fernández 2001 is distinguished from other *Lenomyrmex* workers by smooth and shiny mesosoma with well-developed propodeal spines and by the foveolate-striate sculpture covering all the dorsal surface of its head [2]. *L. inusitatus* has an unusual distribution since it is the single *Lenomyrmex* species recorded east of the Andes [2]. Nevertheless, it was previously only known from the type locality ("Territorio Kofanes", Nariño, Colombia). Here, the species is recorded for the first time in the Eastern Cordillera of the South-Ecuadorian Andes.

Among *Lenomyrmex* species, the queen caste has been described only for *L. mandibularis* Fernández and Palacio 1999 and *L. wardi* Fernández and Palacio 1999. In this paper, we provide the first record and a description of the queen of *L. inusitatus*.

2. Materials and Methods

The sampling of *Lenomyrmex* in the Ecuadorian Andes is part of a rainfall exclusion experiment [6] and was based on the Winkler extraction method. The leaf litter inside a 0.25 or 0.5-m² quadrat was collected and sifted and its fauna was extracted during 48 h. All specimens were collected close to the Podocarpus National Park, within the "Copalinga" property, at 1420 m (Zamora-Chinchipe province, Ecuador). Vegetation corresponds to an evergreen lower montane forest [7]. Mean annual precipitation is about 2100 mm. Mean temperature in the leaf litter from December 2009 to May 2010 was 18.5°C (min–max: 15.7–22.2°C).

(a)

(b)

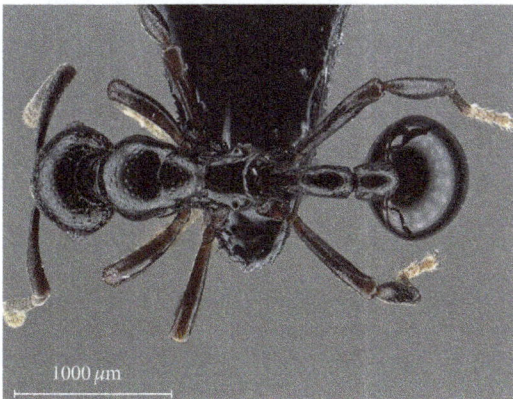

(c)

FIGURE 1: Worker (specimen number 4042619) of *Lenomyrmex inusitatus* Fernández 2001: in (a) frontal, (b) lateral, and (c) dorsal views. Note the predominantly smooth and shiny mesosoma, with no erect hairs (b, c) and the foveolate head, with median longitudinal striae (a).

A worker (no. 4042619, from sample no. 40426) and a queen (no. 4042602, from the same sample) have been imaged (Figures 1 and 2, resp.) and are available at http://projects.biodiversity.be/ants.

Measurements and Indices. All measurements are in millimeters. The abbreviations are as follows:

HL: Head length, measured in full face view, from the anterior margin of the medial lobe of the clypeus to the posterior border of the head (excluding the mandibles).

HW: Head width, the maximum width of the head measured in full face view, excluding the compound eyes.

ML: Mandible length, the maximum length of the mandible measured in dorsal view, from the anteriormost portion of the head to the apex of closed mandibles.

EL: Eye length, the maximum diameter of the eye in frontal view.

SL: Scape length, excluding the basal condyle and the neck.

WL: Weber's length, measured diagonally in lateral view from the anterior edge of the pronotum to the posterior edge of the propodeal lobe.

PL: Petiole length, the axial distance from the dorsal corner of the posterior peduncle to the nearest edge of the propodeal lobe.

PW: Petiole width, the maximum transverse distance across the node measured in dorsal view.

PPL: Postpetiole length, the axial distance from the base of the node in front to the tip of the posterior peduncle measured in lateral view.

PPW: Postpetiole width, the maximum transverse distance across the postpetiole in dorsal view.

GL: Gaster length, in lateral view, from the anterior edge of the first tergum to the posterior edge of the last visible tergum.

GW: Gaster width, in dorsal view, the maximum transverse distance across the gaster.

TL: Total length measured in lateral view (ML + HL + WL + PL + PPL + GL).

OI: Ocular index, EL/HW × 100.

CI: Cephalic index, HW/HL × 100.

SI: Scape index, SL/HL × 100.

Queens and workers have been deposited at the Royal Belgian Institute of Natural Sciences, Brussels, Belgium, (RBINS), the Laboratorio de Entomología—Universidad Técnica Particular de Loja, Loja, Ecuador (UTPL), and the Museo de Insectos, Instituto de Ciencias Naturales—Museo de Historia Natural, Universidad Nacional de Colombia, Santafé de Bogotá D.C., Colombia (ICN).

3. Results (Tables 1 and 2)

3.1. Material Examined. A total of 34 workers and two dealated queens of *Lenomyrmex inusitatus* were collected. The worker (Figure 1) corresponds to the description of the holotype [2], except that it is slightly smaller.

TABLE 1: Key to the workers of the six described *Lenomyrmex* species.

1. Mesosoma predominantly smooth and shiny, with no erect hairs	2
–Mesosoma with conspicuous sculpture and at least a pair of erect hairs	3
2(1). Propodeum without spines; head only foveolate (SW Colombia)	*foveolatus*
–Propodeum with a pair of acute and well-defined spines; head foveolate, with median longitudinal striae (Cordillera Oriental of the Andes in S Colombia and S Ecuador)	*inusitatus*
3(1). Dorsum of head and petiole with longitudinal conspicuous costae; erect hairs of antennal scape as long as or longer than maximum diameter of scape; body ferruginous yellow (W Panama)	*costatus*
–Dorsum of head densely rugo-reticulate; sculpture of the petiole variable, rugulate to rugo-reticulate or longitudinally striate but never costate; erect hairs of antennal scape not longer than maximum diameter of the scape; body brownish black or dark red brown	4
4(3). Length of propodeal spines approximately equal to distance between their bases; mesopleuron with some irregular longitudinal striae, but mostly smooth and shiny; metapleuron with irregular longitudinal striae; HL > 0.80 mm; mesosoma with only two suberect hairs on the pronotum (SW Colombia)	*mandibularis*
–Length of propodeal spines variable, either shorter or longer than distance between their bases; metapleuron and subsequent portion of mesopleuron with fine transverse rugulae or rugo-reticulate, without smooth areas; HL < 0.80 mm; mesosoma with numerous erect to suberect hairs	5
5(4). Propodeal spines shorter than distance between their bases; eyes with six or seven facets in maximum diameter; petiolar node protruding over the peduncle and well defined; postpetiolar dorsum with longitudinal striae (NW Ecuador, SW Colombia)	*wardi*
–Propodeal spines longer than distance between their bases; eyes with about nine facets in maximum diameter; petiolar node undifferentiated from the peduncle; postpetiolar dorsum smooth and polished (Costa Rica)	*colwelli*

TABLE 2: Key for the known queens of *Lenomyrmex*.

1. Head foveolate, with median longitudinal striae; mesosoma predominantly smooth and shiny, with sparse punctures on pronotum, mesopleuron, metapleuron, and propodeum, scutellum and axillae foveolate, mesoscutum foveolate-striate, no erect hairs	*inusitatus*
–Head densely rugo-reticulate; mesosoma covered by sculpture, mesopleuron, scutellum, and propodeal dorsum with striae, axillae rugo-reticulate, mesoscutum rugulose, erect hairs	2
2(1). Propodeal spines approximately equal in length to distance between their bases; integument predominantly shiny; HL > 0.80	*mandibularis*
–Propodeal spines notably shorter than distance between their bases; integument predominantly opaque; HL < 0.80	*wardi*

Workers. ECUADOR: Zamora-Chinchipe province: Zamora: Bombuscaro: Copalinga property; Lat: −4.083; Long: −78.967; 26.IV-01.V.2010; collected by Delsinne T. and Arias Penna T.; 34 workers in 23 Winkler samples (number of specimens/Winkler sample: 1–4); sample codes: 40343, 40367, 40369, 40374, 40375, 40382, 40387, 40391, 40395, 40417, 40418, 40424, 40426, 40428, 40437, 40439, 40440, 40446, 40449, 40453, 40455, 40457, 40459, 40461; RBINS, UTPL, ICN.

Worker Measurements (no. 4042619). TL 4.23, HL 0.74, HW 0.64, ML 0.41, SL 0.60, EL 0.16, WL 1.15, PL 0.62, PW 0.20, PPL 0.30, PPW 0.24, GL 1.11, GW 0.76, CI 86, OI 24, SI 81.

Queens. ECUADOR: Same data as workers; two queens in two Winkler samples; sample codes: 40426 and 40343; RBINS, UTPL.

Queen Measurements (no. 4042602). TL 4.34, HL 0.75, HW 0.65, ML 0.41, SL 0.59, EL 0.20, WL 1.16, PL 0.64, PW 0.21, PPL 0.27, PPW 0.24, GL 1.11, GW 0.78, CI 86, OI 31, SI 79.

Queen Diagnosis (Figure 2). The queen is similar to the worker [2] but differing in the following characters: anterior margin of clypeus mostly convex, with a slight median notch or concavity; compound eyes bigger, with 11-12 facets in maximum diameter; three ocelli present; mesosoma robust; dorsum of pronotum smooth and shiny, with sparse punctures; mesoscutum foveolate, with longitudinal striae; scutellum and axillae foveolate, with smooth and shiny interspaces; dorsum of propodeum completely smooth and polished; propodeal spines long and stout but shorter than distance between their bases; mesopleuron with anepisternum clearly separated from katepisternum by a suture; lateral face of pronotum, anepisternum, katepisternum, metapleuron, and

(a)

(b)

(c)

FIGURE 2: Queen (specimen number 4042602) of *Lenomyrmex inusitatus* Fernández 2001: in (a) frontal, (b) lateral, and (c) dorsal views. Note the predominantly smooth and shiny mesosoma, with mesoscutum foveolate-striate and without erect hairs (b, c) and the foveolate head, with median longitudinal striae (a).

lateral face of propodeum mostly smooth and shiny, with some sparse punctures; punctures of lateral and dorsal faces of petiole and postpetiole more defined and deeper than in workers; short and appressed pilosity more abundant on mesosoma than in workers.

4. Discussion

Lenomyrmex inusitatus is, with *L. wardi* and *L. foveolatus*, the third *Lenomyrmex* species collected in Ecuador [1, 8]. To our knowledge, the new record represents only the tenth locality known for the entire genus and constitutes its southernmost limit of distribution. The range of the species and of the genus increases nearly 510 km and 415 km to the South, respectively. Although data remain insufficient to understand the biogeography of *Lenomyrmex*, it is interesting to note that the new record confirms the presence of *L. inusitatus* on the Eastern side of the Cordillera Oriental of the Andes.

Lenomyrmex species were collected from elevations close to sea level to 1800 m but seem to be mainly restricted to mid-elevations, that is, 1100–1500 m ([1–3], this study). The degree of queen-worker dimorphism is weak, suggesting small colony sizes and absence of claustral independent colony foundation [9]. *Lenomyrmex* ants seem always locally rare and it is in fact the first time that up to 34 workers have been collected within a relatively small area (400 m^2). A thorough inspection of the dead wood laying on the ground and of soil samples failed to uncover any nest of *L. inusitatus*. This and the fact that both workers and dealate queens were extracted from the leaf litter (Winkler method) may indicate that this species nests and forages in the leaf litter. The unusual morphology of the mandibles suggests that *Lenomyrmex* is a specialist predator on an unknown prey. This habit is possibly linked to its apparent rarity and restricted elevational distribution. More data are needed to accurately determine the biology and biogeography of these interesting ants.

N.B. After submitting the paper, two additional workers were found within a soil sample, at slightly higher elevation (1500 m), within the "Copalinga" property. The two workers were maintained alive during six days. They moved relatively slowly and feigned death when disturbed. They did not feed on any offered food items (alive and dead termites, millipedes, mites, various insect parts, sugar/water, tuna, biscuits). The information for these specimens are ECUADOR: Zamora-Chinchipe province: Zamora: Bombuscaro: Copalinga property; Lat: −4.082; Long: −78.968; 13.IV.2011; collected by Delsinne T. and Arias Penna T.; two workers in one soil sample (= a thorough visual search for ants for twenty person-minutes from a 15 × 15 × 15-cm core of soil); specimen codes: 4649901 and 4649902; RBINS.

Acknowledgments

The authors thank C. Vits and B. de Roover from "Copalinga" for allowing them to sample ants within their property, J. Bendix, the "Deutsche Forschungsgemeinschaft" (DFG)-Research Unit 816, and the team of the "Estación Científica San Francisco" for allowing and extensively facilitating their work, I. Bachy, Y. Laurent, and M. Leponce for ant digitization, J. Peña and T. M. Arias Penna for assistance with fieldwork, and two anonymous referees for helpful suggestions on the paper. This research was funded by the Belgian Science Policy (BELSPO) and was carried out in the framework of EDIT (European Distributed Institute of Taxonomy). In accordance with section 8.6 of the ICZNs International Code of Zoological Nomenclature, printed copies of the edition of Psyche containing this article are deposited at the following six publicly accessible libraries: Green Library (Stanford University), Bayerische Staatsbibliothek, Library—ECORC (Agriculture & Agri-Food Canada), Library—Bibliotheek (Royal Belgium Institute of Natural Sciences), Koebenhavns Universitetsbibliotek, University of Hawaii Library.

References

[1] F. Fernández C. and E. E. Palacio G., "*Lenomyrmex*, an enigmatic new ant genus from the Neotropical Region (Hymenoptera: Formicidae: Myrmicinae)," *Systematic Entomology*, vol. 24, no. 1, pp. 7–16, 1999.

[2] F. C. Fernández, "Hormigas de Colombia. IX: Nueva especie de *Lenomyrmex* (Formicidae: Myrmicinae)," *Revista Colombiana de Entomología*, vol. 27, pp. 201–204, 2001.

[3] J. T. Longino, "New species and nomenclatural changes for the Costa Rican ant fauna (Hymenoptera: Formicidae)," *Myrmecologische Nachrichten*, vol. 8, pp. 131–143, 2006.

[4] F. Fernández, "Subfamilia myrmicinae," in *Introducción a Las Hormigas de la Región Neotropical*, F. Fernández, Ed., pp. 307–330, Instituto de Investigación de Recursos Biológicos Alexander von Humbold, Bogotá, Colombia, 2003.

[5] B. Bolton, "Synopsis and classification of Formicidae," *Memoirs of the American Entomological Institute*, vol. 71, pp. 1–370, 2003.

[6] T. Delsinne, T. M. Arias Penna, and M. Leponce, "Effects of experimental rainfall exclusion on a diverse ant assemblage along an elevational gradient in the Ecuadorian Andes," in *Proceedings of the 5th International conference of the International Biogeography Society*, p. 97, Crete, Greece, 2011.

[7] J. Homeier, F. A. Werner, S. R. Gradstein, S.-W. Breckle, and M. Richter, "Potential vegetation and floristic composition of Andean forests in South Ecuador, with a focus on the RBSF," in *Gradients in a Tropical Mountain Ecosystem of Ecuador*, E. Beck, J. Bendix, I. Kottke, F. Makeschin, and R. Mosandl, Eds., vol. 198, pp. 87–100, Ecological Studies, 2008.

[8] D. Donoso and G. Ramón, "Composition of a high diversity leaf litter ant community (Hymenoptera: Formicidae) from an Ecuadorian pre-montane rainforest," *Annales de la Société Entomologique de France (numéro spécial)*, vol. 45, pp. 487–499, 2009.

[9] C. Peeters and M. Molet, "Colonial reproduction and life histories," in *Ant Ecology*, L. Lach, C. L. Parr, and K. L. Abbott, Eds., pp. 159–193, Oxford University Press, New York, NY, USA, 2010.

10

Flight Dynamics and Abundance of *Ips sexdentatus* (Coleoptera: Curculionidae: Scolytinae) in Different Sawmills from Northern Spain: Differences between Local *Pinus radiata* (Pinales: Pinaceae) and Southern France Incoming *P. pinaster* Timber

Sergio López and Arturo Goldarazena

Neiker-Basque Institute of Agricultural Research and Development, Arkaute, 01080 Vitoria, Spain

Correspondence should be addressed to Arturo Goldarazena, agoldarazena@neiker.net

Academic Editor: Kleber Del-Claro

In January 2009, the windstorm "Klaus" struck the southern part of France, affecting 37.9 million m³ of maritime pine *Pinus pinaster* Aiton (Pinales: Pinaceae). This breeding plant material favored the outbreak of *Ips sexdentatus* (Börner) (Coleoptera: Curculionidae: Scolytinae). As much of this timber is imported to the Basque Country (northern Spain), a potential risk to conifer stands is generated, due to the emergence of the incoming beetles. Thus, flight dynamics and beetle abundance were compared in different sawmills, according to the timber species (either local *P. radiata* D. Don or imported *P. pinaster*). A maximum flight peak of *I. sexdentatus* was observed in mid-June in *P. pinaster* importing sawmills, whereas a second lighter peak occurred in September. In contrast, only a maximum peak in mid-June was observed in *P. radiata* inhabiting beetles, being significantly smaller than in local *P. pinaster* trading sawmills. In addition, significant differences were found between imported *P. pinaster* and *P. radiata* regarding the number of insects beneath the bark. The development of IPM strategies for controlling *I. sexdentatus* populations is recommended, due to the insect abundance found in *P. pinaster* imported timber.

1. Introduction

Bark beetles (Coleoptera: Curculionidae: Scolytinae) are an insect group that contains at least 6,000 species from 181 genera around the world [1]. Bark beetles are considered as important agents of forest succession and initiate the sequence of nutrient cycling in infested tree material [2]. However, it is well known that some species are among the most destructive insects of coniferous forests, representing a continuous threat [1, 3]. Although bark beetles tend to colonize dead or weakened trees, it is well reported that some species can attack healthy trees under epidemic conditions. Frequently, improper forestry management or adverse abiotic and climatic conditions (e.g., storms, fires, and droughts) act as precursors by providing breeding substrate that unleashes population outbreaks for these bark beetles species [4–6]. For instance, the storms "Vivian/Wiebke"

in February/March 1990 and "Lothar" in December 1999 triggered the propagation of *Ips typographus* (L.) in Centre Europe [7]. Recently, "Klaus" named windstorm affected 37.9 million m³ of maritime pine *Pinus pinaster* Aiton (Pinales: Pinaceae) in Aquitaine (southern France) during January 2009 [8]. As a consequence a great amount of windthrown timber was left as suitable breeding material for the six-toothed beetle *I. sexdentatus* (Börner) Figure 1. Despite its preference for weak, decaying or dead trees, the six-toothed beetle can attack healthy trees under outbreak conditions. Much of this timber from Landes region is imported to many sawmill and timber-processing industries located at the Basque Country (northern Spain), due to its low cost. The long-time storage of such infested logs could put into risk the local forestry management, since new emerging *I. sexdentatus* would disperse beyond sawmills and attack the adjacent Monterey pine (*P. radiata* D. Don) stands,

(a)

(b)

(c)

FIGURE 1: *Ips sexdentatus* (Börner) (Coleoptera: Curculionidae: Scolytinae), lateral (a) and dorsal views (b), and detail of the elytral declivity of male (left) and female (right). Note the fusion at the base of the 3rd and 4th teeth in male (white arrow).

TABLE 1: Sampling sawmills located at the Basque Country (northern Spain). *Pinus* L. species (Pinales: Pinaceae) is also indicated within each row.

Locality	Province	Latitude and longitude	Timber
Amezketa	Guipuzcoa	43° 02′ N, 02° 04′ W	*P. pinaster* Aiton
Tolosa	Guipuzcoa	43° 07′ N, 02° 04′ W	*P. pinaster*
Aia	Guipuzcoa	43° 15′ N, 02° 09′ W	*P. pinaster*
Berrobi	Guipuzcoa	43° 08′ N, 02° 01′ W	*P. radiata* D. Don
Zalla	Biscay	43° 12′ N, 03° 08′ W	*P. radiata*
Legutiano	Alava	42° 58′ N, 02° 38′ W	*P. radiata*

from the Basque Country, according to different timber species (either *P. radiata* or imported *P. pinaster*). Secondly, in order to evaluate the infestation level of maritime pine, the density of beetles was evaluated, through direct observation on debarked logs. These primary objectives would allow inferring the significance and risk of importing maritime pine to the Basque Country.

2. Material and Methods

Monitoring trapping took place from 1st April to 31th October 2011. Six different commercial sawmills were chosen. Three of them use *P. radiata* planted in the Basque Country as primary resource, whereas the other three import maritime pine timber from Landes region (southwestern France). The locations of sampling sites are provided in Table 1.

Two eight-unit Lindgren multiple funnel traps (Econex S.L., Murcia, Spain) were placed in each sawmill. Each trap was hung with the top of the trap at 2 m above the ground and the distance between traps was at least 50 m. One trap was unbaited, as a blank control, whereas the other trap was baited with a synthetic *I. sexdentatus*-specific pheromone (a mixture of ipsdienol (212.9 mg), *cis*-verbenol (60.8 mg), and ipsenol (13.6 mg), SEDQ, Barcelona, Spain). Baits were replaced every two months. Fifty mL of propylene glycol were added to each trap cup to kill and preserve captured insects. Not only *I. sexdentatus*, but also other bark beetles species and other accidentally trapped beetles were collected. Samples were removed every fifteen days and taken to the laboratory. Voucher specimens have been deposited at the Entomology Collection of the NEIKER-Basque Institute for Agricultural Research and Development, Arkaute, Basque Country, Spain.

In order to determine what *Pinus* species showed the largest density of *I. sexdentatus*, sections of 70 cm × 30 cm of seven randomly chosen logs (from both *P. radiata* and *P. pinaster*) were peeled off every week from 2nd May to 31th July in each sawmill. Debarking was made with the aid of a chisel. All *I. sexdentatus* present in the galleries beneath the bark were collected. The number of galleries was also recorded.

Data of mean catches of flying beetles caught in baited traps were subjected to a two-way ANOVA analysis (with pine species and date considered as factors). Subsequent Tukey *post-hoc* tests at a significance level of $\alpha = 0.05$

which is the most common tree species planted in the Basque Country [9].

Ips sexdentatus is a Palearctic species distributed throughout Europe which is capable of breeding in many coniferous genera, including *Pinus* L., *Picea* A. Dietr. (Pinaceae), *Larix* Mill. (Pinaceae), and *Abies* Mill. (Pinaceae) [10, 11]. Concerning the Basque Country, it has been trapped in both *P. radiata* and *P. sylvestris* L. stands [12]. It is associated with several species of ophiostomatoid fungi (Sordariomycetes: Ophiostomatales) [13, 14], which are involved in many tree diseases and sapstain [15]. Not only with blue-staining fungi, but also the association with the fungus *Fusarium circinatum* Niremberg and O'Donnell (= *F. subglutinans* f. sp. *pini* Correll et al. (Hypocreales: Nectriaceae), causal agent of the pitch canker disease, has been detected in *P. radiata* inhabiting populations in the Basque Country [16].

Thus, the aim of the current work was to determine the flight dynamics of *I. sexdentatus* in different sawmills

Flight Dynamics and Abundance of Ips sexdentatus (Coleoptera: Curculionidae: Scolytinae) in Different Sawmills
from Northern Spain: Differences between Local Pinus radiata (Pinales: Pinaceae)....

91

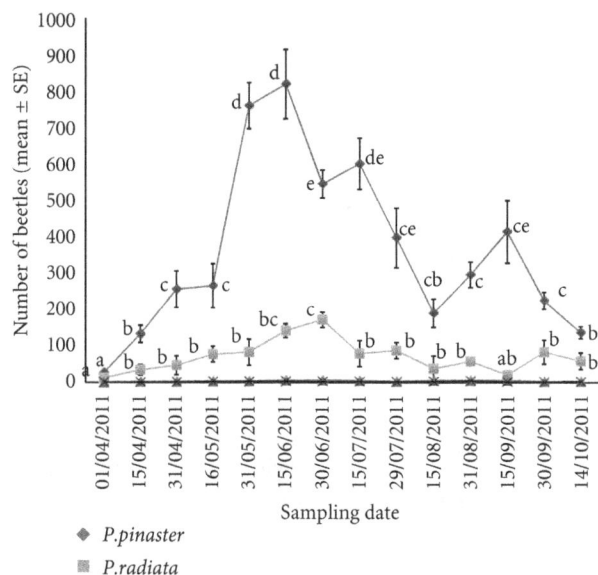

FIGURE 2: Number (mean ± SE) of *Ips sexdentatus* (Börner) (Coleoptera: Curculionidae: Scolytinae) captured in *Pinus pinaster* Aiton (Pinales: Pinaceae) (dark grey) and *P. radiata* D. Don (light grey) sawmills from 1 April to 14 October 2011. Dates within each *Pinus* L. species with different letters are significantly different at a significance level of $\alpha = 0.05$. Control catches in both cases were insignificant to perform any statistical analysis.

were applied to compare mean catches between dates within each *Pinus* species. Concerning density data, Student's *t*-test was used to compare mean number of galleries and beetle collections in different *Pinus* species for each month. A square root transformation was used to normalize the data and correct the heteroscedasticity. All the analyses were performed with the statistical software SPSS 2004 SYSTAT statistical package (version 13.0, SPSS, Chicago).

3. Results

A total of 15,184 specimens of *I. sexdentatus* were trapped in *P. pinaster* importing sawmills, whereas 2,774 were captured in *P. radiata* sawmills. As expected, pheromone-baited traps caught significant more insects in *P. pinaster* sawmills when compared with captures in *P. radiata* sawmills ($F = 108.927$, df = 1, $P < 0.001$). An interaction between sampling dates and *Pinus* species was found ($F = 7.2440$, df = 13, $P < 0.001$). A maximum flight peak was observed from the end of May to middle June for maritime pine, whereas a slighter peak occurred on September (Figure 2). Regarding *P. radiata* sawmills, a significant peak was observed only at the end of June. No statistical differences were observed in catches of other accidentally trapped insects.

Significant differences were found between the mean number of beetles and galleries under the bark during the three months. Maritime pine sections showed significant more galleries (May: $t = 4.152$, df = 12, $P = 0.002$; June: $t = 5.928$, df = 12, $P < 0.001$; July: $t = 5.063$, df = 12, $P < 0.001$) (Figure 3(a)) and beetles (May: $t = 9.367$, df =

12, $P < 0.001$; June: $t = 8.538$, df = 12, $P < 0.001$; July: $t = 7.900$, df = 12, $P < 0.001$) (Figure 3(b)) than in local *P. radiata*.

In addition, many other bark and ambrosia beetles species were accidentally captured in pheromone-baited traps. Table 2 details the different bark and ambrosia beetles caught per locality, along with other xylophagous species (Coleoptera: Cerambycidae) and bark beetle predators (Coleoptera: Cleridae).

4. Discussion

Current work demonstrates that maritime pine timber imported from France to commercial sawmills is highly infested compared to *P. radiata* timber, according to observed differences in the amount of insects caught in both field trapping and log debarking.

The six-toothed beetle has two generations per year, with adult flight periods from April to May and July to August. However, *I. sexdentatus* can undergo a third generation in Mediterranean regions of Europe [17]. Our results are consistent with other studies. Similar maximum flight peaks have been observed in *Picea orientalis* (L.) Link (in Turkey) and *Pinus sylvestris* (in Romania) stands [18, 19]. In contrast, *I. sexdentatus* showed three different peak flights in *P. pinaster* stands at the province of Leon (northern Spain), with the maximum peak occurring in September [20]. It has been suggested that this latter increase might be due to a strong increasing of the population during that season or a seasonal pheromone production, as it occurs in *I. pini* (Say) [21].

Ips sexdentatus is a polygamous species in which male is the pioneer sex which initiates the host seeking process. Afterwards, up to 2–5 females join each male within the gallery systems [10]. Galleries are star shaped, with a central nuptial chamber built by the male and in which mating occurs. Females bore egg galleries, which radiate outwards from the nuptial chamber. All the observed galleries in the current study had more than two arms.

Among accidentally trapped bark beetles species, it is worth noting the find of a female exemplar of the small spruce bark beetle *Polygraphus poligraphus* (L.), which would represent the first record for the Iberian Peninsula. *Polygraphus poligraphus* inhabits *Picea abies* (L.) H. Karst. and *P. obovata* Ledeb. [11], rarely breeding in *Pinus sylvestris* and *P. strobus* L. [10, 11, 22]. This unique specimen was trapped in the sawmill located at Berrobi, in which *P. radiata* timber is used. In addition, its distribution area is supposed to extend from Central Europe to Northern Europe and Siberia [11], being absent in the Mediterranean region [10]. Thus, the presence of this insect in the sampling area should be clearly stated.

Moreover, two species of *Monochamus* Dejean (Coleoptera: Cerambycidae) were also trapped, mainly in two *P. pinaster* trading sawmills: *M. sutor* (L.) and *M. galloprovincialis* (Olivier). The latter shows special relevance, as it is known to be the vector of the pine wood nematode, *Bursaphelenchus xylophilus* (Steiner and Buhrer) Nickle (Aphelenchida, Parasitaphelenchidae), causal agent of the pine wilt

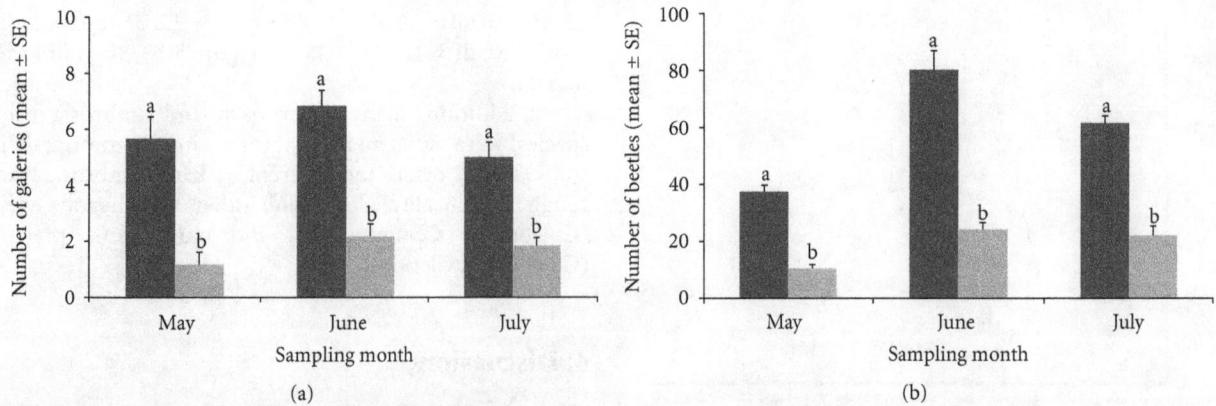

FIGURE 3: Number (mean ± SE) of (a) galleries and (b) *Ips sexdentatus* (Börner) (Coleoptera: Curculionidae: Scolytinae) found under the bark of *Pinus pinaster* Aiton (Pinales: Pinaceae) (dark grey) and *Pinus radiata* D. Don (Pinales: Pinaceae) (light grey) logs from May to July (*n* = 28). Means within each month with different letters are significantly different at a significance level of α = 0.05.

TABLE 2: Total number of accidentally trapped species of bark and ambrosia beetles (Curculionidae: Scolytinae), cerambycid (Cerambycidae) and checkered beetles (Cleridae). Species within family/subfamily are sorted by alphabetical order.

Species/Locality	Amezketa	Tolosa	Aia	Berrobi	Zalla	Legutiano
Coleoptera: Curculionidae: Scolytinae						
Dryocoetes autographus (Ratzeburg)	5	0	0	1	0	0
Dryocoetes villosus (F.)	9	0	3	2	0	0
Gnathotrichus materiarius (Fitch)	249	7	112	23	22	23
Hylastes ater (Paykull)	22	0	0	3	0	22
Hylurgops palliatus (Gyllenhal)	2	0	1	0	0	0
Hylurgus ligniperda (F.)	137	20	33	3	1	3
Kissophagus hederae (Schmitt)	3	0	0	0	0	0
Orthotomicus erosus (Wollaston)	111	14	57	2	26	95
Orthotomicus laricis (F.)	86	0	18	7	12	1
Pityogenes calcaratus (Eichhoff)	11	1	0	0	9	54
Polygraphus poligraphus (L.)*	0	0	0	1	0	0
Xyleborinus saxeseni (Ratzeburg)	0	0	0	0	0	16
Xyleborus eurygraphus (Ratzeburg)	35	0	11	0	0	0
Xyleborus dryographus (Ratzeburg)	1	0	7	1	0	0
Xylosandrus germanus (Blandford)	1	0	4	2	0	0
Coleoptera: Cerambycidae						
Monochamus galloprovincialis (Olivier)	12	0	2	0	0	0
Monochamus sutor (L.)	28	0	4	0	0	1
Coleoptera: Cleridae						
Allonyx quadrimaculatus (Schaller)	2	0	0	0	0	0
Clerus mutillarius F.	1	0	0	0	0	0
Thanasimus formicarius (L.)	337	157	33	43	25	45

*Indicates first record for the Iberian Peninsula.

disease in different countries, including in Europe (Portugal and Spain) [23–26]. The kairomonal attraction to bark beetle pheromone components has been previously reported in some long-horned beetles, including *M. galloprovincialis* in Spain, another North American species of the genus [27–29].

The checkered beetle *Thanasimus formicarius* (L.) (Coleoptera: Cleridae) was the most common predator found in traps (527 individuals in *P. pinaster* sawmills and 113 in *P. radiata* sawmills). This insect is a common predator of European conifer bark beetles [30], and it is capable of locating their preys by detecting bark beetle produced-pheromones as kairomonal signals [31]. Moreover, it has been reported that they recognize conifer volatiles and even volatiles from angiosperm trees that act as nonhost volatiles to conifer bark beetles [32]. *Allonyx quadrimaculatus* (Schaller) is also considered as a predator of *Tomicus piniperda* L. [33], although there are not concrete studies about the mechanisms involved in prey detection.

As in other species of the genus, management programs should be focused on minimizing attacks on living trees, the sanitation of infested trees and the establishment of a trapping system [7]. The use of semiochemicals with antiaggregative effects should be considered as a useful management tool for trees protection. (1*S*, 4*S*)-(-)-Verbenone (4,6,6-trimethylbicyclo-[3.1.1]hept-3-en-2-one, hereafter (-)-verbenone), has been demonstrated to be capable of disrupting the pheromone-mediated attraction of *I. sexdentatus* [16, 34]. Romón et al. [16] detected a significant negative dose-dependent relationship between different (-)-verbenone release rates (0.01, 0.2, 1.8, and 3.1 mg/24 h) and catches of *I. sexdentatus* in a *P. radiata* stand. Etxebeste and Pajares [34] also found significant reduction in catches when testing (-)-verbenone at 2 and 40 mg/day in a mixed pine stand (ca. 40-year-old *P. pinaster* with younger ca. 30-year-old *P. sylvestris*). In addition, the spyroketal *trans*-7-methyl-1,6-dioxaspiro[4.5]decane (commonly known as *trans*-conophthorin) has also shown promising results. There are evidences of its electrophysiological detection by *I. sexdentatus* [35],and the antiaggregative effect is supported by field assays, although with some disparities. Despite Jactel et al. did not find any significant reduction in trap catches when testing *trans*-conophthorin at 5 mg/day [35], a 16-time lower release rate (i.e., 0.3 mg/day) is capable of reducing the response of *I. sexdentatus* to aggregation pheromone [34]. Moreover, *trans*-conophthorin seems to achieve stronger effects when combined either with (-)-verbenone or NHV alcohols [34, 35]. Thus, taken into account these results, we suggest the development of "push-pull" strategies [36], using pheromone-baited traps inside the park (to favor insect mass trapping) and blends of disruptant semiochemicals at the edges of close pine stands, in order to repel incoming beetles. Long-time buildup of logs should also be not recommended. Future field studies are needed to evaluate the impact of these incoming *I. sexdentatus* populations upon local conifer stands.

Acknowledgments

The authors would like to express thanks to the Basque Country Forest Confederation for financial support and the owners of sampling sawmills for allowing our experimental work. They would like to thank to members of NEIKER for technical assistance. The present study was funded by The Government of the Basque Country.

References

[1] M. P. Ayres and M. J. Lombardero, "Assessing the consequences of global change for forest disturbance from herbivores and pathogens," *Science of the Total Environment*, vol. 262, no. 3, pp. 263–286, 2000.

[2] R. W. Stark, "Generalized ecology and life cycle of bark beetles," in *Bark Beetles in North American Conifers: A System for the Study of Evolutionary Biology*, J. B. Mitton and K. B. Sturgeon, Eds., pp. 21–45, University of Texas Press, Austin, Tex, USA, 1982.

[3] S. L. Wood, "The bark and ambrosia beetles of North and Central America, a taxonomic monograph," *Great Basin Naturalist Memoirs*, vol. 6, pp. 1–1359, 1982.

[4] M. Peltonen, "Windthrows and dead-standing trees as bark beetle breeding material at forest-clearcut edge," *Scandinavian Journal of Forest Research*, vol. 14, no. 6, pp. 505–511, 1999.

[5] M. Eriksson, A. Pouttu, and H. Roininen, "The influence of windthrow area and timber characteristics on colonization of wind-felled spruces by *Ips typographus* (L.)," *Forest Ecology and Management*, vol. 216, no. 1–3, pp. 105–116, 2005.

[6] M. M. Fernández, "Colonization of fire-damaged trees by *Ips sexdentatus* (Boerner) as related to the percentage of burnt crown," *Entomologica Fennica*, vol. 17, no. 4, pp. 381–386, 2006.

[7] B. Wermelinger, "Ecology and management of the spruce bark beetle *Ips typographus*—a review of recent research," *Forest Ecology and Management*, vol. 202, no. 1–3, pp. 67–82, 2004.

[8] Inventaire Forestier Nationale, "Tempête Klaus du 24 janvier 2009: estimations pour l'ensemble de la zone évaluée," 2011, http://www.ifn.fr/spip/spip.php?article618.

[9] M. Michel, *El pino radiata* (Pinus radiata D.Don) *en la historia de la Comunidad Autónoma de Euskadi. Análisis de un proceso de forestalismo intensivo [Ph.D. dissertation]*, University of Madrid, Madrid, Spain, 2004.

[10] A. Balachowsky, *Faune de France, Volume 50: Coléoptères Scolytides*, Fédération Française des Sociétés de Sciences:, Paris, France, 1949.

[11] A. Pfeffer, *Zentral- und westpalaearktische Borken- und Kenkäfer*, Pro Entomologia, c/o Naturhistorisches Museum Basel:, Basel, Switzerland, 1995.

[12] S. López, P. Romón, J. C. Iturrondobeitia, and A. Goldarazena, *Conifer Bark Beetles of the Basque Country: Practical Guide for Their Identification and Control*, Servicio Central de Publicaciones del País Vasco, Vitoria, Spain, 2007.

[13] P. Romón, X. Zhou, J. C. Iturrondobeitia, M. J. Wingfield, and A. Goldarazena, "Ophiostoma species (Ascomycetes: Ophiostomatales) associated with bark beetles (Coleoptera: Scolytinae) colonizing *Pinus radiata* in northern Spain," *Canadian Journal of Microbiology*, vol. 53, no. 6, pp. 756–767, 2007.

[14] A. Bueno, J. J. Díez, and M. M. Fernández, "Ophiostomatoid fungi transported by *Ips sexdentatus* (Coleoptera; Scolytidae) in *Pinus pinaster* in NW Spain," *Silva Fennica*, vol. 44, no. 3, pp. 387–397, 2010.

[15] T. Kirisits, "Fungal associates of European bark beetles with special emphasis to ophiostomatoid fungi," in *Bark and Wood Boring Insects in Living Trees in Europe, A Synthesis*, F. Lieutier, K. R. Day, A. Battisti, J. C. Grégoire, and H. F. Evans, Eds., pp. 181–235, Kluwer Academic Press, Dordrech, The Netherlands, 2004.

[16] P. Romón, J. C. Iturrondobeitia, K. Gibson, B. S. Lindgren, and A. Goldarazena, "Quantitative association of bark beetles with pitch canker fungus and effects of verbenone on their semiochemical communication in Monterey Pine Forests in Northern Spain," *Environmental Entomology*, vol. 36, no. 4, pp. 743–750, 2007.

[17] J. F. Abgrall and A. Soutrenon, *La Forêt et ses Ennemis*, Centre National du Machinisme Agricole du Genie Rural des Eaux et des Forets, Paris, France, 1991.

[18] G. Isaia, A. Manea, and M. Paraschiv, "Study on the effect of pheromones on the bark beetles of the Scots pine," *Bulletin of the Transilvania University of Brasov*, vol. 3, no. 52, pp. 67–72, 2010.

[19] G. E. Ozcan, M. Eroglu, and H. A. Akinci, "Use of pheromone-baited traps for monitoring *Ips sexdentatus* (Boerner) (Coleoptera: Curculionidae) in oriental spruce stand," *African Journal of Biotechnology*, vol. 10, no. 72, pp. 16351–16360, 2011.

[20] J. M. Sierra and A. B. Martín, "Pheromone-baited traps effectiveness in the massive capture of *Ips sexdentatus* Boern. (Coleoptera: Scolytidae), bark beetle of pines," *Boletín Sanidad Vegetal, Plagas*, vol. 30, pp. 745–752, 2004.

[21] G. N. Lanier, M. C. Birch, R. F. Schmitz, and M. M. Furniss, "Pheromones of *Ips pini* (Coleoptera: Scolytidae): variation in response among three populations," *Canadian Entomologist*, vol. 104, no. 12, 1917.

[22] B. Lekander, B. Bejer-Peterson, E. Kangas, and A. Bakke A, "The distribution of bark beetles in the Nordic countries," *Acta Entomologica Fennica*, vol. 32, pp. 1–36, 1977.

[23] M. J. Wingfield, R. A. Blanchette, T. H. Nichols, and K. Robbins, "The pine wood nematode: a comparison of the situation in the United States and Japan," *Canadian Journal of Forest Research*, vol. 12, pp. 71–75, 1982.

[24] Y. Mamiya, "Pathology of the pine wilt disease caused by *Bursaphelenchus xylophilus*," *Annual Review of Phytopathology*, vol. 21, pp. 201–220, 1983.

[25] M. M. Mota, H. Braasch, M. A. Bravo et al., "First report of *Bursaphelenchus xylophilus* in Portugal and in Europe," *Nematology*, vol. 1, no. 7-8, pp. 727–734, 1999.

[26] A. Abelleira, A. Picoaga, J. P. Mansilla, and O. Aguin, "Detection of *Bursaphelenchus xylophilus*, causal agent of Pine Wilt Disease on *Pinus pinaster* in Northwestern Spain," *Plant Disease*, vol. 95, no. 6, p. 776, 2011.

[27] J. A. Pajares, F. Ibeas, J. J. Díez, and D. Gallego, "Attractive reponses by *Monochamus galloprovincialis* (Col., Cerambycidae) to host and bark bettle semiochemicals," *Journal of Applied Entomology*, vol. 128, no. 9-10, pp. 633–638, 2004.

[28] D. R. Miller and C. Asaro, "Ipsenol and ipsdienol attract *Monochamus titillator* (Coleoptera: Cerambycidae) and associated large pine woodborers in Southeastern United States," *Journal of Economic Entomology*, vol. 98, no. 6, pp. 2033–2040, 2005.

[29] D. R. Miller, C. Asaro, C. Crowe, and D. Duerr, "Bark Beetle pheromones and pine volatiles: attractant kairomone lure blend for Longhorn Beetles (Cerambycidae) in pine stands of the southeastern United States," *Journal of Economic Entomology*, vol. 104, no. 4, pp. 1245–1257, 2011.

[30] R. Gauß, "Der Ameisenbuntkäfer *Thanasimus (Clerus) formicarius* Latr. als Borkenkäferfeind," in *Die Grosse Borkenkäferkalamität in Südwestdeutschland 1944-1951*, G. Wellenstein and G. Ringingen, Eds., pp. 417–442, Selbstverlag der Forstschutzstelle Südwest, Ringingen, Germany, 1954.

[31] A. Bakke and T. Kvamme, "Kairomone response in *Thanasimus* predators to pheromone components of *Ips typographus*," *Journal of Chemical Ecology*, vol. 7, no. 2, pp. 305–312, 1981.

[32] Q. H. Zhang and F. Schlyter, "Inhibition of predator attraction to kairomones by non-host plant volatiles for herbivores: a bypass-trophic signal," *PloS one*, vol. 5, no. 6, Article ID e11063, 2010.

[33] F. Herard and G. Mercadier, "Natural enemies of *Tomicus piniperda* and *Ips acuminatus* (col, scolytidae) on *Pinus sylvestris* near orléans, france: temporal occurrence and relative abundance, and notes on eight predatory species," *Entomophaga*, vol. 41, no. 2, pp. 183–210, 1996.

[34] I. Etxebeste and J. A. Pajares, "Verbenone protects pine trees from colonization by the six-toothed pine bark beetle, *Ips sexdentatus* Boern. (Col.: Scolytinae)," *Journal of Applied Entomology*, vol. 135, no. 4, pp. 258–268, 2011.

[35] H. Jactel, I. van Halder, P. Menassieu, Q. H. Zhang, and F. Schlyter, "Non-host volatiles disrupt the response of the stenographer bark beetle, *Ips sexdentatus* (Coleoptera: Scolytidae), to pheromone-baited traps and maritime pine logs," *Integrated Pest Management Reviews*, vol. 6, no. 3-4, pp. 197–207, 2001.

[36] B. S. Lindgren and J. H. Borden, "Displacement and aggregation of mountain pine beetles, *Dendroctonus ponderosae* (Coleoptera: Scolytidae), in response to their antiaggregation and aggregation pheromones," *Canadian Journal of Forest Research*, vol. 23, no. 2, pp. 286–290, 1993.

11

Evolution of the Heme Peroxidases of Culicidae (Diptera)

Austin L. Hughes

Department of Biological Sciences, University of South Carolina, 715 Sumter Street, Coker Life Sciences Building, Columbia, SC 29208, USA

Correspondence should be addressed to Austin L. Hughes, austin@biol.sc.edu

Academic Editor: Zainulabeuddin Syed

Phylogenetic analysis of heme peroxidases (HPXs) of Culicidae and other insects revealed six highly conserved ancient HPX lineages, each of which originated by gene duplication prior to the most recent common ancestor (MRCA) of Hemimetabola and Holmetabola. In addition, culicid HPX7 and HPX12 arose by gene duplication after the MRCA of Culicidae and Drosophilidae, while HPX2 orthologs were not found in any other order analyzed except Diptera. Within Diptera, HPX2, HPX7, and HPX12 were relatively poorly conserved at the amino acid level in comparison to the six ancient lineages. The genome of *Anopheles gambiae* included genes ecoding five proteins (HPX10, HPX11, HPX13, HXP14, and HPX15) without ortholgs in other genomes analyzed. Overall, gene expression patterns did not seem to reflect phylogenetic relationships, but genes that evolved rapidly at the amino acid sequence level tended to have divergent expression patterns as well. The uniquely high level of duplication of HPXs in *A. gambiae* may have played a role in coevolution with malaria parasites.

1. Introduction

The production of nitric oxide (NO) is an important immune defense mechanism against cellular microorganisms in insects and other invertebrates [1]. Nitric oxide synthase (NOS), encoded by a single gene in insect genomes sequenced to date [2], is the major enzyme involved in NO production, but the full pathway of NO production is only beginning to be understood. In the mosquito *Anopheles gambiae* (Diptera: Culicidae) when infected by malaria parasites (Apicomplexa: *Plasmodium*), a heme peroxidase (HPX2) and NADPH oxidase (NOX5) were found to play a crucial role in potentiating NO in antiparasite defense [3].

NOX5 is represented by a single ortholog in insects, but HPX2 is a member of a multigene heme peroxidase family in *A. gambiae* and other insects [3–5]. Other mosquito heme peroxidases (HPXs) of known function include those expressed in the salivary glands of female *A. gambiae* and *A. albimanus* [6, 7] and one involved in the catalysis of protein-crosslinking in the chorion of *Aedes aegypti* eggs [8]. In another member of Diptera, the fruit fly *Drosophila melanogaster*, heme peroxidases have been implicated in chorion assembly and other developmental processes [9, 10].

For nearly 5000 orthologous genes, Waterhouse and colleagues [4] compared the amino acid sequence distance between *D. melanogaster* and *A. gambiae* with that between *D. melanogaster* and *Ae. aegypti*. On average genes with known immune function showed a greater level of amino acid sequence divergence than other genes, but certain immune-related proteins were well conserved [4]. HPXs included in these analyses likewise included a number of more conserved and less conserved proteins.

Here I use a phylogenetic analysis in order to establish orthologous sets of HPX genes in Culicidae. Unlike previously published analyses of insect HPXs [4, 5], the present phylogenetic analysis includes HPXs from Hemimetabola, making it possible to time gene duplication events in the HPX family relative to the time of the most recent common ancestor (TMRCA) of Hemimetabola and Holometabola (including Diptera). Using sets of orthologs established by the phylogenetic analysis, I analyze the patterns of amino acid sequence conservation of HPXs within and between two families of Diptera: (1) Culicidae (mosquitoes), represented by three species with completely sequenced genomes (*A. gambiae*, *Ae. aegypti*, and *Culex quinquefasciatus*) and (2) Drosophilidae (fruit flies), represented by *D. melanogaster* and *D. grimshawi*. In addition, using data from the MozAtlas *A. gambiae* gene expression database [11], I compare across-tissue expression patterns with phylogenetic relationships and with patterns of amino acid sequence conservation.

2. Methods

2.1. Sequences and Alignment. The phylogenetic analysis was based on 81 heme peroxidase (HPX) sequences from 10 insect species with completely or nearly completely sequenced genomes, belonging to five orders: (1) the pea aphid *Acyrthosiphon pisum* (order Hemiptera); (2) the body louse *Pediculus humanus* (order Phthiraptera); (3) the red flour beetle *Tribolium confusum* (order Coleoptera); (4) the jewel wasp *Nasonia vitripennis* (order Hymenoptera); (5) the honeybee *Apis mellifera* (Hymenoptera); (6) the yellow fever mosquito *Aedes aegypti* (Diptera); (7) the African malaria mosquito *Anopheles gambiae* (Diptera); (8) the Southern house mosquito *Culex quinquefasciatus* (Diptera); (9) the fruit fly *Drosophila melanogaster* (Diptera); (10) the Hawaiian fruit fly *Drosophila grimshawi* (Diptera). Of the five orders represented, two (Hemiptera and Phthiraptera) belong to Hemimetabola, while the rest belong to Holometabola. Preliminary analyses including species from the order Lepidoptera and additional species from Hymenoptera produced results similar to those reported (not shown). However, since the main focus of the present analyses was on HPXs of Diptera, only the above 11 species were used in the final analysis for ease of presentation. Nomenclature of individual proteins and their putative orthologs was based on names provided by VectorBase (http://www.vectorbase.org/) for *A. gambiae*, where those were available.

Amino acid sequences were aligned by the CLUSTAL W algorithm in the MEGA 5 program [12]. In all comparisons among aligned sequences, amino acid sites at which the alignment postulated a gap in any sequence were excluded from all comparisons. Following Waterhouse and colleagues [4], the DBLOX protein of *A. gambiae*, which has an internally duplicated structure, was divided into N-terminal and C-terminal segments, each of which was aligned separately with the other HPXs. The same approach was applied to 8 additional sequences related to DBLOX, which were found to share the same internally duplicated structure. The phylogenetic analysis was thus applied to 90 aligned sequences, including both N-terminal and C-terminal segments of the 9 internally duplicated sequences.

Phylogenetic analyses were conducted using the MEGA program, version 5.05 [12], by the maximum likelihood (ML) method. The Model test function in MEGA was used to choose models by the Bayes Information Criterion (BIC). The reliability of branching patterns was tested by bootstrapping (1000 samples). Relative rate tests were conducted by Tajima's [13] method in MEGA.

Expression data for *A. gambiae* were downloaded from the MozAtlas database [11]. Expression data for the following tissue types were included separately for adult males and adult (blood-fed) females: (1) the head, including the brain;

(2) the salivary gland; (2) the midgut; (3) the Malpighian tubules; (4) the thorax excluding the gut, Mapighian tubules, and the gonads. In addition, there were data for the following sex-specific tissues: (1) ovary in females; (2) testis in males; (3) accessory glands in males. For each of these tissues, the mean expression value from MozAtlas was standardized with gene, and the genes were clustered by the McQuitty algorithm using the Manhattan distance in Minitab version 15.0 (http://www.minitab.com/).

3. Results

3.1. Phylogenetic Analyses. A phylogenetic tree of insect HPXs was constructed by the ML method using the WAG+G+I model at 301 aligned amino acid sites (Figure 1). In the tree there were seven major clusters each defining a lineage which included one or more sequences from Diptera and one or more sequences from Hemimetabola; these seven clusters are here designated HPX1, HPX4, HPX5, HPX6, DBLOX-N, DBLOX-C, and CG42331 (Figure 1). The first four lineages were named according to the nomenclature used in VectorBase for the *Anopheles gambiae* sequence. The DBLOX-N and DBLOX-C clusters were named, respectively, for the N-terminal and C-terminal portions of *A. gambiae* DBLOX (Figure 1). The CG42331 cluster was named for the *Drosophila melanogaster* sequenced included in that cluster (Figure 1). Each of the seven clusters was supported by an internal branch that received at least 86% bootstrap support; and five of the seven clusters received 99% percent or greater bootstrap support (Figure 1).

Each of the seven major clusters included, in addition to sequences from the three mosquito species, apparent ortholgs from the two *Drosophila* species (Figure 1). Furthermore, each of the seven clusters included at least one sequence from Hemimetabola (*Acyrthosiphon pisum* or *Pediculus humanus*); this topology supported the hypothesis that each of these seven major groups of HPX orthologs arose by gene duplications that occurred prior to the most recent common ancestor (MRCA) of Hemimetabola and Holometabola. Since each of the two portions of DBLOX formed a cluster including sequences from Hemimetabola (Figure 1), the phylogenetic analysis supported the hypothesis that the internal duplication of DBLOX occurred prior to the MRCA of Hemimetabola and Holometabola.

Outside the seven major clusters, *A. gambiae* HPX sequences showed very different patterns of relatedness. These patterns were examined further by an additional phylogenetic analysis of the 32 sequences corresponding to the subtree of the original phylogeny (Figure 2) that included the HPX1 cluster and the other sequences that fell outside the seven major clusters (Figure 2). The phylogeny was constructed on the basis of the WAG+G+I model at 402 aligned sites; and the topology of the phylogenetic tree (Figure 2) was broadly similar to that of the corresponding portion of the original tree (Figure 1).

In both trees, *A. gambiae* HPX2 formed a cluster with apparent orthologs from the other mosquito species and from *Drosophila* (Figures 1 and 2). However, there were

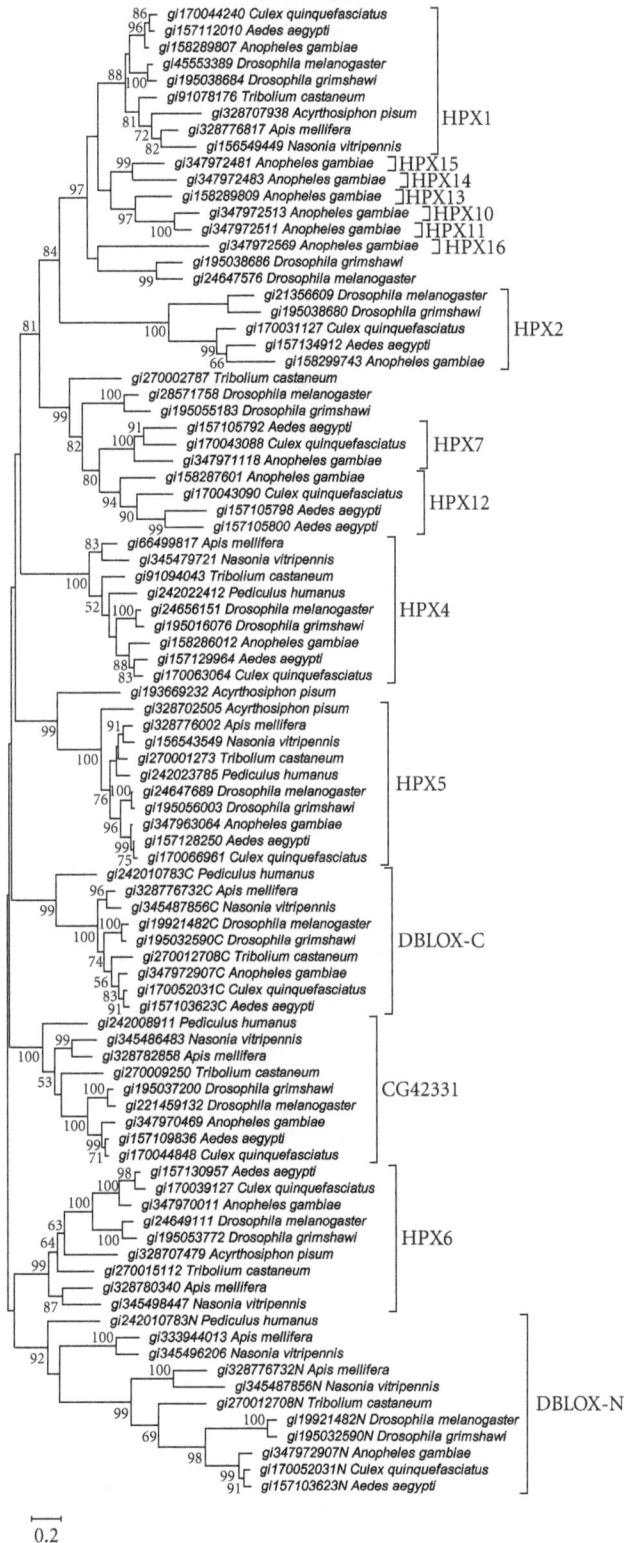

FIGURE 1: Maximum likelihood tree of insect heme peroxidases, based on the WAG+G+I model at 301 aligned amino acid sites. Numbers on the branches represent the percentage of 1000 bootstrap pseudosamples supporting the branch; only values ≥50% are shown.

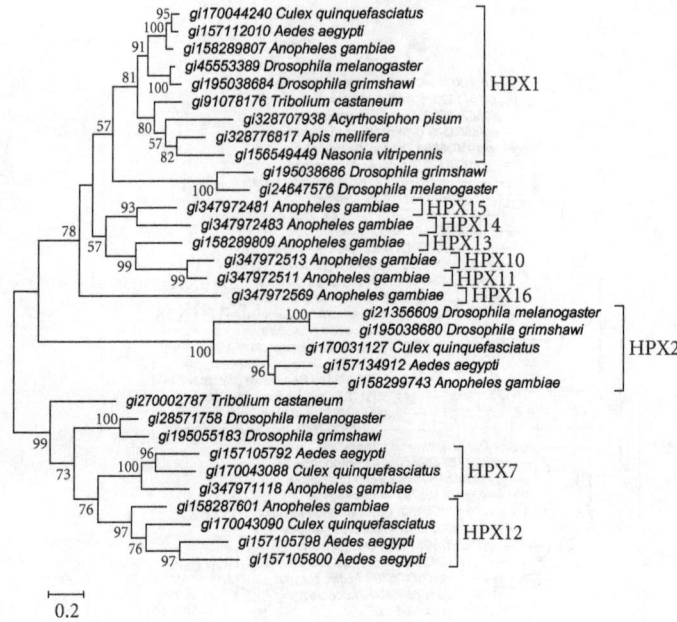

FIGURE 2: Maximum likelihood tree of select insect heme peroxidases, based on the WAG+G+I model at 402 aligned amino acid sites. Numbers on the branches represent the percentage of 1000 bootstrap pseudosamples supporting the branch; only values ≥50% are shown.

no apparent orthologs of HPX2 from outside of Diptera (Figures 1 and 2). In both trees, there was a cluster including *A. gambiae* HPX10, HPX11, HPX13, HXP14, and HPX15, with no apparent orthologs from any of the other genomes analyzed (Figures 1 and 2). Bootstrap support for the latter cluster was weak in both trees (Figures 1 and 2). However, there was high bootstrap support for the clustering of *A. gambiae* HPX10, HPX11, and HPX13 (99%; Figure 2); and for clustering of *A. gambiae* HPX 10 and HPX11 (99%, Figure 2). There was moderately high support for clustering of *A. gambiae* HPX4 and HPX15 (93%, Figure 2).

Culicid HPX7 and HPX12 clustered together with sequences from the two *Drosophila* species and one sequence from *Triboilum castaneum*; the latter cluster received 99% bootstrap support in both phylogenetic analyses (Figures 1 and 2). Because the *Drosophila* sequence fell outside the cluster of culicid HPX7 and HPX12 in both phylogenetic analyses (Figures 1 and 2), the topology is consistent with the hypothesis that the gene duplication giving rise to culicid HPX7 and HPX12 occurred after the MRCA of Culicidae and Drosophilidae.

3.2. Sequence Conservation. In order to obtain evidence regarding conservation of amino acid sequences in different HPXs, aligned amino acid sets were compared in 10 HPX orthologous groups represented in both Cuclicidae and Drosophilidae (Table 1). Comparisons were made among the three mosquito species, between the two *Drosophila* species, and between the two families (Table 1). HPX5 showed the strongest conservation in all comparisons (Table 1). HPX2 was much less conserved; in fact, of the 9 other orthologs, 6 were significantly more conserved that HPX2 in all comparisons (Table 1). On the other hand, HPX12, unique

to Culicidae (Figure 1), was significantly less conserved than HPX2 in comparisons within Culicidae (Table 1). HPX7 was significantly less conserved than HPX2 in the comparison between *Ae. aegypti* and *C. quinquefasciatus*, and HPX7 was less conserved than HPX2, though not significantly so, in the comparison among the three culicid species (Table 1). CG42331 did not differ significantly from HPX2 in the level of conservation within Culicidae, but CG42331 was significantly more conserved than HPX2 within Drosophilidae and in both families (Table 1).

For each of the orthologous groups listed in Table 1, relative rate tests were used to compare the rates of amino acid evolution in *A. gambiae* sequences with that in *Ae. aegypti* orthologs, using *D. melanogaster* orthologs as an outgroup. In no case was there a significant rate difference between *A. gambiae* and *Ae. aegypti* (not shown). Similar analyses likewise showed no significant differences in the rate of amino acid sequence evolution between *A. gambiae* and *C. quinquefasciatus*.

3.3. Expression Pattern. When the 15 *A. gambiae* genes were clustered on the basis of adult male and female expression pattern (Figure 3), clustering showed little relationship with phylogenetic relationships (Figures 1 and 2). For example, the closely related pair of genes HPX14 and HPX15 did not show very similar expression patterns, nor did the closely related pair of genes HPX10 and HPX11 (Figure 3). Likewise, the closely related pair HPX2 and HPX7 did not show similar expression patterns (Figure 3). A cluster of genes very similar in terms of expression pattern (HPX1, DBLOX, HPX14, HPX, and HPX10; Figure 3) shared low levels of expression across all tissues analyzed. By contrast, CG42331 had the most divergent expression pattern (Figure 3), with

TABLE 1: Percentages of amino acid residues in heme peroxidases that are conserved among sets of dipteran taxa.

Protein	No. aligned residues	Set of taxa compared			
		Ae. aegypti and *C. quinquesfasciatus*	*Ae. aegypti, C. quinquesfasciatus,* and *A. gambiae*	*D. melanogaster* and *D. grimshawi*	Culicidae and Drosophilidae
HPX1	672	87.5%***	79.8%***	85.3%***	56.3%***
HPX2	609	68.1%	50.6%	61.9%	24.8%
HPX4	868	85.5%***	74.3%***	88.1%***	54.0%***
HPX5	746	93.4%***	91.4%***	96.0%***	78.8%***
HPX6	800	82.0%*	64.9%***	78.8%***	41.4%***
HPX7	654	58.3%**	46.0%	67.1%[1]	18.7%[1]
HPX12	654	41.4%***	35.5%***		
CG42331	1132	74.9%	54.8%	72.1%***	34.0%**
DBLOX-N	585	86.8%***	72.5%***	83.8%***	39.5%***
DBLOX-C	775	92.5%***	86.5%***	92.3%***	66.3%***

Chi-square tests of the hypothesis that the proportion conserved is equal to that for HPX2 in the same set of taxa (Bonferroni-corrected): *$P < 0.05$; **$P < 0.01$; ***$P < 0.001$.
[1]Homologous to both HPX7 and HPX12 (see Figure 1).

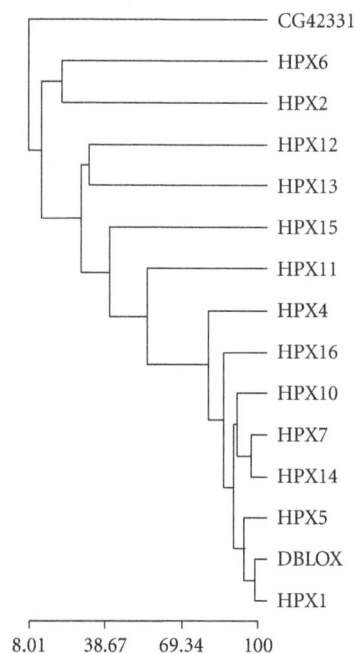

FIGURE 3: Hierarchical clustering of gene expression of 15 *Anopheles gambiae* HPXs across 11 tissue types. The scale on the left represents expression pattern similarity score.

particularly high levels in both male and female thorax. Next most divergent were HPX2 (with high levels in male thorax and testis and in the head in both sexes) and HPX6 (with high levels in the head in both sexes).

4. Discussion

Phylogenetic analysis of the heme peroxidases (HPXs) of Culicidae showed support for the existence of six separate lineages that originated by gene duplication events that occurred prior to the MRCA of Hemimetabola and Holometabola, which is believed to have occurred in the late Carboniferous period, 318–300 million years ago [14]: HPX1, HPX4, HPX5, HPX6, CG42331, and DBLOX. Likewise the results supported the hypothesis that the internal duplication of DBLOX occurred before the MRCA of Hemimetabola and Holometabola.

In addition to these ancient HPX lineages, culicid genomes include certain members of the HPX family that were apparently duplicated more recently than the six ancient lineages. Culicid HPX7 and HPX12 clustered with a *Tribolium castaneum* sequence, implying an origin before the MRCA of Diptera and Coleoptera. The duplication giving rise to HPX7 and HPX12 appeared to have occurred independently in the culicid lineage after the MRCA of Culicidae and Drosophilidae. However, since HPX7 and HPX12 were found in *Anopheles gambiae, Aedes aegypti*, and *Culex quinquefasciatus*, the duplication occurred prior to the MRCA of those three genera.

Neither of the two available genomes of Hemimetabola was found to contain orthologs of HPX7, HPX12, HPX2, HPX10, HPX11, HPX13, HPX14, HPX15, or HPX16. The absence of these genes from Hemimetabola suggests the possibility that these genes have arisen in the Holometabola after the MRCA of Hemimetabola and Holometabola. However, it may be that future analysis of additional genomes of Hemimetabola will discover orthologs of these genes. Therefore the conclusion that they are unique to Holometabola remains tentative.

HPX2, which is known to be involved in resistance to malaria parasite infection in *A. gambiae* [3], did not have orthologs in any of the other orders included in the present analysis. In addition, the genome of *A. gambiae* included six HPXs (HPX10, HPX11, HPX13, HPX14, HPX15, and HPX16) with no reported orthologs in any other species. Thus, extensive HPX duplication appears to be a unique trait of *A. gambiae* among the mosquito genomes analyzed.

Members of the six ancient families generally showed a greater level of amino acid sequence conservation in Diptera than did sets of orthologs not found outside Holometabola (HPX2, HPX7, and HPX12). Of the latter, HPX2 was somewhat more conserved in Culicidae than was either HPX7 or HPX12. Because of the presence of relatively poorly conserved HPXs and the unique paralogs, the HPX family of Culicidae and of *A. gambiae* in particular seem to have undergone a degree of rapid evolution unusual for insect HPXs. By contrast, CG42331 was the one ancient lineage that was relatively nonconserved within Culicidae.

Overall, gene expression patterns did not reflect phylogenetic relationships, with closely related gene pairs often showing marked differences in expression pattern. On the other hand, there was some association between lack of amino acid sequence conservation and divergent patterns of gene expression (Table 1 and Figure 3). For example, CG42331, HPX2, and HPX12 represented relatively unconserved genes whose patterns of gene expression were dissimilar to those of other HPX genes, as indicated by the fact that these three genes clustered apart from the other genes in the cluster analysis of gene expression patterns (Figure 3).

It might be proposed that the evolutionary pattern of HPXs reflects coevolution of the insect host with parasitic microorganisms, particularly malaria parasites in the case of *A. gambiae*. However, there was no evidence of an increased rate of amino acid evolution of *A. gambiae* HPXs in comparison to orthologs in the other culicid species. Therefore coevolution with malaria parasites seems not to have enhanced the rate of amino acid replacement of the HPXs of *A. gambiae*. On the other hand, the extensive duplication of HPXs, unique in the present data to *A. gambiae*, may have played a role in coevolution with malaria parasites. An increased understanding of the functions of individual HPXs, including more detailed information on tissue expression, particularly in response to infection by malaria parasites, will provide evidence to test the latter hypothesis.

References

[1] A. Rivero, "Nitric oxide: an antiparasitic molecule of invertebrates," *Trends in Parasitology*, vol. 22, no. 5, pp. 219–225, 2006.

[2] N. M. Gerardo, B. Altincicek, C. Anselme et al., "Immunity and other defenses in pea aphids, Acyrthosiphon pisum," *Genome Biology*, vol. 11, no. 2, article r21, 2010.

[3] G. de A. Oliveira, J. Lieberman, and C. Barillas-Mury, "Epithelial nitration by a peroxoidase/NOX5 system mediates mosquito antiplasmodial immunity," *Science*, vol. 335, pp. 856–859, 2012.

[4] R. M. Waterhouse, E. V. Kriventseva, S. Meister et al., "Evolutionary dynamics of immune-related genes and pathways in disease-vector mosquitoes," *Science*, vol. 316, no. 5832, pp. 1738–1743, 2007.

[5] G.-Q. Shi, Q.-Y. Yu, and Z. Zhang, "Annotation and evolution of the antioxidant genes in the silkworm, Bombyx mori," *Archives of Insect Biochemistry and Physiology*, vol. 79, no. 2, pp. 87–103, 2012.

[6] I. M. B. Francischetti, J. G. Valenzuela, V. M. Pham, M. K. Garfield, and J. M. C. Ribeiro, "Toward a catalog for the transcripts and proteins (sialome) from the salivary gland of the malaria vector Anopheles gambiae," *Journal of Experimental Biology*, vol. 205, no. 16, pp. 2429–2451, 2002.

[7] J. M. C. Ribeiro and J. G. Valenzuela, "Purification and cloning of the salivary peroxidase/catechol oxidase of the mosquito Anopheles albimanus," *Journal of Experimental Biology*, vol. 202, no. 7, pp. 809–816, 1999.

[8] J. Li, S. R. Kim, and J. Li, "Molecular characterization of a novel peroxidase involved in Aedes aegypti chorion protein crosslinking," *Insect Biochemistry and Molecular Biology*, vol. 34, no. 11, pp. 1195–1203, 2004.

[9] M. Fakhouri, M. Elalayli, D. Sherling et al., "Minor proteins and enzymes of the Drosophila eggshell matrix," *Developmental Biology*, vol. 293, no. 1, pp. 127–141, 2006.

[10] R. E. Nelson, L. I. Fessler, Y. Takagi et al., "Peroxidasin: a novel enzyme-matrix protein of Drosophila development," *EMBO Journal*, vol. 13, no. 15, pp. 3438–3447, 1994.

[11] D. A. Baker, T. Nolan, B. Fischer, A. Pinder, A. Crisanti, and S. Russell, "A comprehensive gene expression atlas of sex- and tissue-specificity in the malaria vector, Anopheles gambiae," *BMC Genomics*, vol. 12, article 296, 2011.

[12] K. Tamura, D. Peterson, N. Peterson, G. Stecher, M. Nei, and S. Kumar, "MEGA5: molecular evolutionary genetics analysis using maximum likelihood, evolutionary distance, and maximum parsimony methods," *Molecular Biology and Evolution*, vol. 28, pp. 2731–2739, 2011.

[13] F. Tajima, "Simple methods for testing the molecular evolutionary clock hypothesis," *Genetics*, vol. 135, no. 2, pp. 599–607, 1993.

[14] B. M. Wiegmann, M. D. Trautwein, J. W. Kim et al., "Single-copy nuclear genes resolve the phylogeny of the holometabolous insects," *BMC Biology*, vol. 7, article 34, 2009.

12

Mechanisms of Odor Coding in Coniferous Bark Beetles: From Neuron to Behavior and Application

Martin N. Andersson

Department of Biology, Lund University, 223 62 Lund, Sweden

Correspondence should be addressed to Martin N. Andersson, martin_n.andersson@biol.lu.se

Academic Editor: John A. Byers

Coniferous bark beetles (Coleoptera: Curculionidae: Scolytinae) locate their hosts by means of olfactory signals, such as pheromone, host, and nonhost compounds. Behavioral responses to these volatiles are well documented. However, apart from the olfactory receptor neurons (ORNs) detecting pheromones, information on the peripheral olfactory physiology has for a long time been limited. Recently, however, comprehensive studies on the ORNs of the spruce bark beetle, *Ips typographus*, were conducted. Several new classes of ORNs were described and odor encoding mechanisms were investigated. In particular, links between behavioral responses and ORN responses were established, allowing for a more profound understanding of bark beetle olfaction. This paper reviews the physiology of bark beetle ORNs. Special focus is on *I. typographus*, for which the available physiological data can be put into a behavioral context. In addition, some recent field studies and possible applications, related to the physiological studies, are summarized and discussed.

1. Introduction

Bark beetles (Coleoptera: Curculionidae: Scolytinae) constitute some of the most destructive pests of coniferous trees throughout the world, destroying forests of great economic value. Currently, the large-scale outbreak of the mountain pine beetle, *Dendroctonus ponderosae*, in North America has resulted in the loss of hundreds of millions m³ timber and turned the forests into major sources of carbon release [1]. In Europe and parts of Asia [2, 3], the European spruce bark beetle, *Ips typographus* (Figure 1), is considered the most destructive bark beetle of coniferous forests [4, 5].

Bark beetles, like most insects, locate their hosts mainly by means of olfactory signals. It is clear that they utilize both attractants and antiattractants that emanate from host and nonhost plants, as well as from conspecific and heterospecific bark beetle individuals [3, 6–10]. The odor molecules are transported downwind from their source of release as an odor plume with a complex structure [11–13]. Molecules are picked up by olfactory receptors (ORs) or ionotropic receptors (IRs) [14], located mainly in the antennae and maxillary palps. Specifically, the ORs are present in the cell membrane of olfactory receptor neuron (ORN) dendrites that, in turn, are housed within olfactory sensilla [15]. The ORs are encoded by a large and diverse family of olfactory receptor genes [16]. Each ORN is generally thought to express only one member from this family in addition to the widely expressed coreceptor, Orco [17]. IRs act in combinations of up to three subunits that are comprised of odor-specific receptors and one or two broadly expressed coreceptors [14]. These receptors are expressed in neurons that do not express ORs. When an odor molecule binds to a receptor, the ORN sends a neuronal signal to the primary olfactory center of the brain, the antennal lobe. Typically, the signal that is generated by an ORN is an increase in the firing frequency of action potentials (excitation), but some odorants may instead cause a decrease in firing activity (inhibition). ORNs can be divided into classes based on their odor response profiles. Often, ORNs are fairly specific and activated by only one or a few compounds, but some appear to have a broader tuning. In addition, each compound often activates more than one type of ORN, and thus, the odor input is thought to be constructed as a combinatorial code [18].

FIGURE 1: The European spruce bark beetle, *Ips typographus*. Photo: Göran Birgersson.

In contrast to the well-studied chemical ecology of bark beetles, until recently, little was known about the physiological responses of individual bark beetle ORNs. Mainly in the 1980s, Single-sensillum recordings (SSR) were carried out, primarily identifying classes of ORNs that responded to various pheromone compounds. Some decades later, comprehensive studies on *I. typographus* have characterized additional ORNs that respond also to host and nonhost plant compounds [19] and have provided novel insights into potential odor coding mechanisms in insects in general [8]. This review summarizes the results from early and recent studies on the physiology of ORNs in conifer-feeding bark beetles. Particular focus is on *I. typographus*, for which a sufficient amount of information has emerged in order to bridge the physiological data with previously recorded behavioral responses to several semiochemicals. In addition, some recent behavioral studies with connections to olfactory physiology are summarized and possible applications discussed. First, however, a brief overview of the semiochemicals that are used by *I. typographus* in host selection is presented.

2. Host Selection by *I. typographus*

The male is the initial host seeking, or "pioneering," sex of *I. typographus*. Once a male has located a suitable host material to colonize, it releases an aggregation pheromone, a mixture of (4S)-*cis*-verbenol and 2-methyl-3-buten-2-ol [20], which attracts individuals of both sexes. Although the olfactory-mediated host location behavior of *I. typographus* has been extensively studied, it is not known how the pioneering males locate a suitable host tree, as no primary attraction (in the absence of pheromone) to spruce volatiles has been demonstrated. However, spruce volatiles may modulate the pheromone response [9] or possibly attract beetles to a suitable habitat [21]. It is also possible that pioneering beetles land randomly on trees and assess host quality upon contact [22]. However, apart from the few pioneering males, the aggregation pheromone attracts the majority of individuals to the host.

The attraction to the pheromone is modulated by other semiochemicals that appear in later attack phases. Verbenone

and ipsenol are two such compounds that are believed to be used as cues to avoid heavily attacked trees [23]. In addition, volatiles that are particularly abundant in nonhost angiosperm plants (so called nonhost volatiles, NHV), such as green leaf volatiles (GLVs) [24] and compounds from the bark, such as C8-alcohols and *trans*-conophthorin [25, 26], have inhibitory effects on pheromone attraction. Combining these compounds with verbenone produces a strong synergistic effect and a potent antiattractant blend [27]. Possibly, the individual constituents in the synergistic blend represent different levels in the host selection sequence [6]. The GLVs that are common to broad-leaved plants may represent a signal of a nonhost dominated habitat. More specific plant volatiles, such as *trans*-conophthorin, may indicate nonhosts at the tree species level [7], whereas the antiattractive pheromone components may signal unsuitability of individual spruce trees.

3. Olfactory Receptor Neurons of *I. typographus* and Other Bark Beetles

Many compounds that are either attractants or antiattractants for conifer bark beetles have been identified [3, 6, 7]. Single-sensillum recordings from the ORNs of several bark beetle species have shown that many of the behaviorally active compounds elicit responses in different classes of neurons (Table 1). It is obvious that, except for *I. typographus* and the ambrosia beetle *Trypodendron lineatum*, more is known about ORN responses to pheromone components than about responses to plant odors (Table 1). In addition, several of the tested compounds (i.e., ipsdienol, ipsenol, verbenone, *cis/trans*-verbenol, *exo*-brecicomin, and α-pinene) elicit strong responses in the majority of species studied. For more details on ORN specificity, sensitivity, and abundance in each species, the reader is referred to the cited literature.

Olfactory sensilla of *I. typographus* are present in three areas (or bands) on the antenna (Figure 2(a)) [41]. Andersson et al. [19] screened 150 olfactory sensilla for responses to an odor panel comprised of similar numbers of synthetic pheromone, host, and nonhost compounds. Strong excitatory responses were obtained from 106 ORNs; 45 responded specifically to various bark beetle pheromone compounds, 37 to host compounds, and 24 to antiattractive nonhost volatiles (NHVs). Based on response spectra, the 106 ORNs were grouped into 17 different classes (Figure 3). Additionally, 26 neurons (divided into 12 ORN classes) responded only weakly to any test odorant, indicating that the most potent compounds for these ORNs were lacking. In addition to the ORN classes described by Andersson et al. [19], three other classes, responding specifically to (+)-*trans*-verbenol, phenylethanol, or camphor plus pino-camphone, respectively, had been identified previously (Table 1) [28]. Furthermore, the majority of the ORN classes responding to pheromone compounds was found in both studies. Many ORN classes have been subjected to dose-response trials that indicated that the ORNs, in general, are highly sensitive and specific for only one or a few structurally related pheromone or plant compounds (Figures 4(a)–4(c)) [19, 28]. Response thresholds for the best ligand(s) were normally found around

TABLE 1: Compounds from different ecological sources that elicit strong responses in olfactory receptor neurons in eight species of Scolytinae.

Species	Beetle-produced compounds	Host compounds	Nonhost compounds	References
Ips typographus	(+)-ipsdienol (−)-ipsdienol (−)-ipsenol (−)-*cis*-verbenol (+)-*trans*-verbenol (−)-verbenone 2-methyl-3-buten-2-ol Amitinol Phenylethanol *exo*-brevicomin (A)* (±)-chalcogran (A)	Myrcene Campher (B) Pino-camphone (B) *p*-cymene 3-carene 1,8-cineole (+)-α-pinene (C) (−)-α-pinene (C)	Pine bark extract Birch bark extract 1-hexanol (D) E2-hexenol (D) Z3-hexenol (D) 1-octen-3-ol 3-octanol (S,S)-*trans*-conophthorin (A)	[19, 28–31]
Ips pini	(+)-ipsdienol (−)-ipsdienol (±)-ipsenol *cis*-verbenol *trans*-verbenol Verbenone	Linalool Camphor Myrcene		[29, 32–34]
Ips paraconfusus	(+)-ipsdienol (−)-ipsdienol (±)-ipsenol			[33]
Dendroctonus pseudotsugae	Frontalin 3-methyl-2-cyclohexenone 3-methyl-2-cyclohexenol 1-methyl-2-cyclohexenol *trans*-verbenol *cis*-verbenol Verbenone Ipsenol	α-pinene Limonene		[35, 36]
Dendroctonus frontalis	(−)-frontalin *exo*-brevicomin *endo*-brevicomin Verbenone *trans*-verbenol	α-pinene 3-carene		[37, 38]
Dendroctonus micans	(+)-ipsdienol *exo*-brevicomin			[31]
Trypodendron lineatum	(+)-lineatin Phenylethanol	Ethanol Methanol Butanol α-pinene β-pinene Spruce bark extract	Pine bark extract Birch bark extract	[39]
Tomicus destruens		Compounds in pine extract	Benzyl alcohol	[40]

*Compounds that elicit responses of similar strength in the same ORN class are indicated by the same capital letter. Odorants eliciting secondary responses are omitted for clarity.

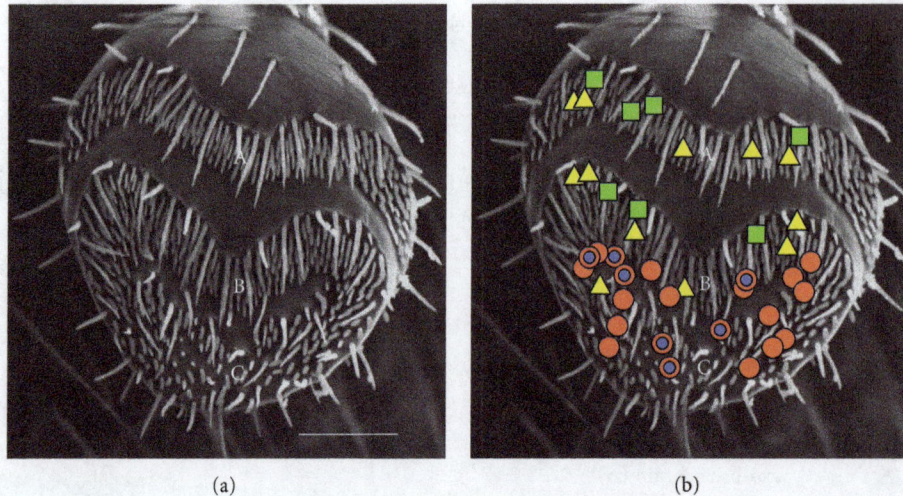

FIGURE 2: (a) Olfactory sensilla are present in three areas (A, B, and C) on the antennal club of *Ips typographus*. (b) Spatial distribution patterns of four classes of olfactory receptor neurons (ORNs). ORNs responding to green leaf volatile alcohols (nonhost) = green squares, myrcene (host) = yellow triangles, *cis*-verbenol (pheromone) = red circles, 1,8-cineole (host) = blue small circles (from [19], with permission from the publisher). Scale bar = 50 μm.

the 1 ng dose on the filter paper using paraffin oil as solvent [19]. A high specificity, not only among pheromone ORNs, but also among those for plant compounds, seems to be a general rule also in other bark beetle species (see especially [32, 35]).

The ORNs of *I. typographus* are not randomly distributed on the antenna. Instead, ORNs from a particular class are generally found either in both the proximal and medial bands of sensilla, or exclusively in the distal area (Figure 2(b)) [19]. This distribution pattern seems to correspond to the distribution of the two morphological types of single-walled sensilla previously identified [41].

It is common in insects that the pheromone ORNs are numerous on the antenna and that the most common ORN type is tuned to the major (most abundant) component [42, 43]. In *I. typographus*, the most recurrent ORN class was tuned to (−)-*cis*-verbenol [19]. In contrast, there were only few cells specific for 2-methyl-3-buten-2-ol (MB) (Figure 3) [19, 28], an essential pheromone component which is produced, and behaviorally active, in much larger quantities [20, 44]. This suggests that the pattern might be reversed in the bark beetle. However, the MB cells were found in a restricted area on the antenna [19], that is, on the borderline between the medial band and distal area of sensilla, which could have resulted in this cell type being underrepresented among the sampled sensilla. Alternatively, as MB is highly volatile, the low number of cells could be the result of the compound being lost from the stimulus cartridge upon stimulation. Indeed, photoionization detector measurements showed that the airborne amount of MB released from the stimulus pipette drops dramatically upon stimulation (Figure 5) [45]. However, the insect ORN still responded vigorously despite the low concentration, rendering this explanation unlikely. In contrast to Andersson et al. [19], Tømmerås [28] found that the ORNs tuned to ipsdienol were the most common ones

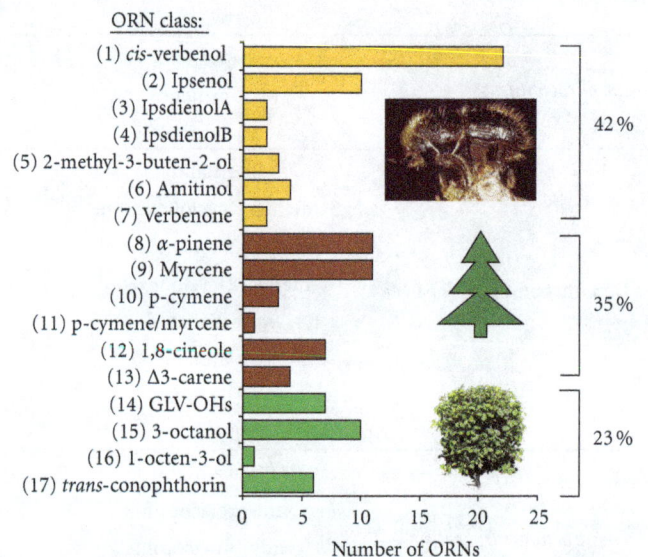

FIGURE 3: Number of olfactory receptor neurons (ORNs) of 17 strongly responding classes of *Ips typographus* (data from [19]). ORN classes are labeled according to which compound(s) elicited the strongest response. As pure enantiomers were not tested, it is likely that the ipdienol ORN classes A and B correspond to the ORNs responding to (+)- and (−)-ipdienol, respectively [29]. Orange = bark beetle pheromone compounds, brown = conifer compounds, green = nonhost volatiles. GLV-OHs = green leaf volatile alcohols.

in *I. typographus*. This discrepancy may also be explained by the nonrandom localization of ORNs on the antenna (i.e., neurons for *cis*-verbenol are abundant only in the distal part of the antennae, Figure 2(b)).

Although no primary attraction has been demonstrated, the high frequency of ORNs tuned to conifer-related

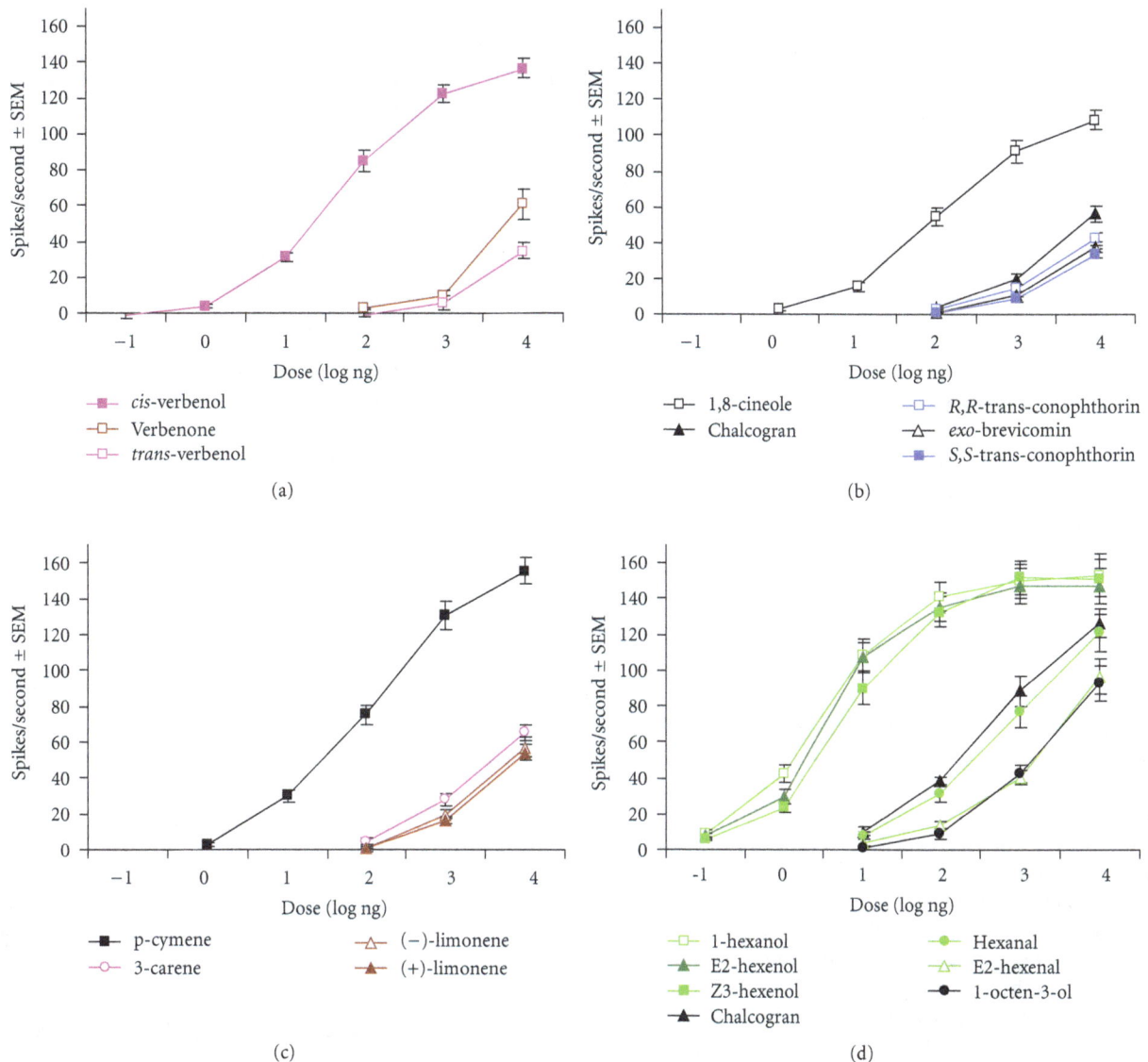

FIGURE 4: Dose-response curves from four receptor neuron classes of *Ips typographus*, demonstrating specific primary responses to (a) the pheromone component *cis*-verbenol, the spruce compounds (b) 1,8-cineole and (c) p-cymene. (d) Indiscriminate response to the three green leaf volatiles 1-hexanol, *E*2-hexenol, and *Z*3-hexenol (modified from [19], with permission from the publisher).

monoterpenes (Table 1, Figure 3) suggests that host kairomone is relevant for host location by *I. typographus*. As mentioned previously, these compounds may serve as habitat-scale attractants [21], or as modulators of pheromone attraction [8, 9]. Perhaps the most striking finding from the bark beetle SSR studies is that almost 25% of the strongly responding ORNs were specifically tuned to antiattractive NHV (Table 1, Figure 3) [19]. This indicates that insects may devote a lot of olfactory capacity to the detection of compounds from sources that they avoid. Similar results have not been found in any other insect studied so far, however, it is likely that many other bark beetles that show strong GC-EAD responses to NHV also have a large proportion of ORNs tuned to such compounds [7, 46–48].

4. Discrimination of Enantiomers

Most bark beetle pheromone compounds are chiral. Attraction is typically evoked by only one of the enantiomers, while the other sometimes inhibits attraction (e.g., [20, 33]). The enantiospecific behavioral response is reflected in the specificity of the ORNs detecting the compounds (Table 1). For instance, ORNs that are specific for either the (+)- or the (−)-enantiomer of ipsdienol, ipsenol, verbenone, or *cis*- and *trans*-verbenol have been identified in several *Ips* species [28, 33]. Other examples are the ORNs in the Southern pine beetle, *D. frontalis* [37], and in the Douglas-fir beetle, *D. pseudotsugae* [36], that discriminate between the (+)- and (−)-enantiomer of frontalin (Table 1).

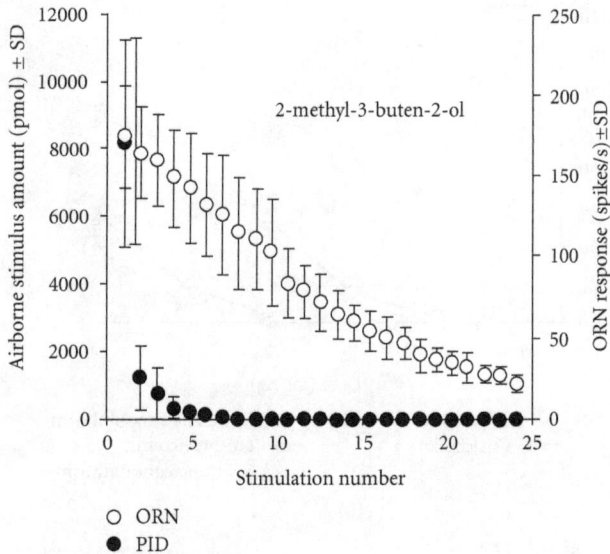

FIGURE 5: Response of *Ips typographus* olfactory receptor neurons (ORNs) and a photoionization detector (PID) to successive stimulations with 2-methyl-3-buten-2-ol ($N = 4$) (modified from [61]).

In general, the sensitivity of the pheromone ORNs seems to be 10–100-fold higher for the enantiomer that they are tuned to, as compared to the other [28, 33]. In addition, there seems to be a correspondence between the attraction to a specific enantiomer, and the frequency of ORNs on the antenna that responds to it. For instance, *I. pini* is attracted by (−)-ipsdienol and has more of its ORNs tuned to (−)-ipsdienol than to (+)-ipsdienol. Similarly, *I. paraconfusus*, which is attracted by (+)-ipsdienol, has most of its ipsdienol ORNs tuned to the (+)-enantiomer [33].

Enantiospecific responses to plant compounds have been recorded in *I. typographus* [19]. The neuron class that responded most strongly to the nonhost volatile *trans*-conophthorin (Table 1) was >100-fold more sensitive to the (5S,7S)-enantiomer than to the (5R,7R)-enantiomer. In fact, other structurally related compounds (racemic *exo*-brevicomin and chalcogran) elicited stronger responses in this ORN than did the (5R,7R)-enantiomer of *trans*-conophthorin [19]. In another class of ORN, the naturally occurring (−)-1-octen-3-ol elicited a slightly stronger response than the racemic mixture, indicating that the (−)-enantiomer is the key ligand. In contrast, the neuron that is tuned to α-pinene responded similarly to both enantiomers (Table 1) [19].

5. Olfactory Receptor Neuron Responses and Behavior

The results that have been obtained from single sensillum recordings [19, 28] indicate that behavioral responses of *I. typographus* to several compounds can likely be explained by the responses of the ORNs.

Several volatiles from nonhost plants were previously shown to inhibit pheromone attraction of *I. typographus*

[24–26]. The three GLVs: 1-hexanol, *E*2-hexenol, and *Z*3-hexenol all reduced pheromone attraction to a similar extent. However, combining the three did not produce a stronger inhibition of attraction, a phenomenon defined as redundancy [27]. Interestingly, the only ORN that was sensitive to any of these volatiles had a more or less identical sensitivity to all three of them (Table 1, Figure 4(d)) [19]. Thus, it appears as if the bark beetle cannot differentiate between the compounds at the physiological level, which agrees well with their behavioral redundancy. In contrast, the compounds verbenone and *trans*-conophthorin that synergize the inhibition are detected by different ORNs [19, 28]. Interestingly, the pheromone component, chalcogran, of the sympatric *Pityogenes chalcographus*, was primarily detected, by *I. typographus*, by the same neuron as *trans*-conophthorin (Table 1). Chalcogran also inhibits pheromone attraction of *I. typographus* [49].

Most insects house their ORNs for pheromone compounds in sensilla that are distinct from the ones that detect plant compounds (e.g., [50, 51]). However, in some sensilla in *I. typographus*, the ORN for the aggregation pheromone component *cis*-verbenol (cV) is colocalized with an ORN that responds to the host plant compound 1,8-cineole (Ci) [8, 19] (Figure 2(b)). This lack of segregation between ORNs detecting pheromones and plant volatiles may suggest that host finding in bark beetles is an integrated process that involves both pheromones and plant volatiles. When the ORN for Ci responded, the colocalized cV cell was inhibited, indicative of interactions between ORNs in the periphery. In addition, Ci was found to be particularly abundant in heavily attacked spruce trees and the compound strongly reduced pheromone attraction (88% reduction in trap catch) in the field [8]. Possibly, Ci is a signal of an unsuitable (crowded) host or a well-defended tree.

6. Peripheral Modulation of ORN Responses

Colocalization of insect ORNs in the same sensillum is thought to improve coincidence detection, which increases the insect's spatiotemporal resolution of odor signals [52] and improves ratio detection of ecologically relevant odor mixtures [53]. In addition, the presence of two or more neurons in the same sensillum may provide opportunities for signal modulation in the periphery. Indeed, in the Douglas-fir beetle, *Dendroctonus pseudotsugae*, two ORNs, each specific for one of the two pheromone components 3-methyl-2-cyclohexenone or 3-methyl-2-cyclohexenol, are colocalized. When either one of the ORNs responded to its specific ligand, the spontaneous activity of the other ORN was reduced. This observation indicated reciprocal interactions, either directly between the two neurons, or between the two ligands and their respective receptors [35]. In addition, when another ORN type that responded to limonene (10 ng dose) was challenged with a binary mixture of limonene and 3-methyl-2-cyclohexenol (10 : 1000 ng), the response to limonene was completely shut down [35].

In *I. typographus*, not all cV neurons (large amplitude A-cell) are colocalized with the neuron for Ci (small amplitude B-cell) (Figure 6(a)). These cV neurons are instead found

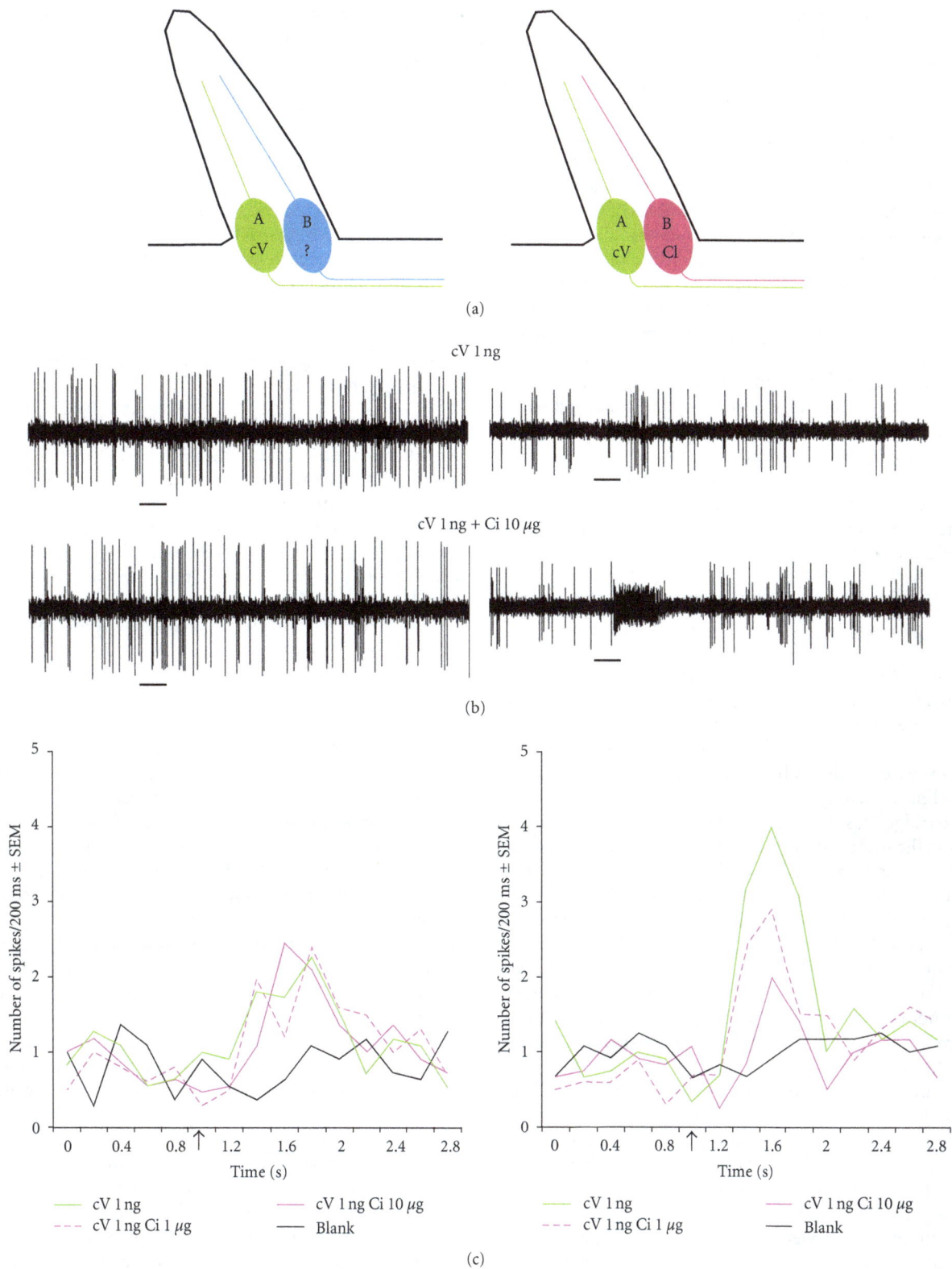

FIGURE 6: (a) Schematic drawing of the two types of sensilla in *Ips typographus* containing the A cell (large amplitude spikes) for *cis*-verbenol (cV), accompanied either by a nonresponsive (*left column*) B cell (small amplitude), or a B cell for 1,8-cineole (Ci) (*right column*). (b) Responses of both sensillum types to 1 ng cV (*upper traces*), and a binary mixture of 1 ng cV and 10 μg Ci (*lower traces*). Note the inhibition of the cV response during the response to Ci in the B cell. Black horizontal bars indicate the 0.5 s stimulation period. (c) Detailed response curves to cV and binary cV : Ci mixtures showing a Ci dose-dependent inhibition of the cV response only in sensilla that also contain the Ci cell ($N = 10$–12). Arrows indicate the onset of the 0.5 s stimulation period (modified from [8], with permission from the publisher).

together with another ORN type that does not respond to any odorant tested so far [19]. The Ci inhibited the cV cell only in sensilla in which the two neurons were colocalized, implying that the inhibition might be due to interactions between the ORNs. To test this hypothesis, Andersson et al. [8] recorded both types of cV sensilla (with or without the Ci cell) and tested responses to binary cV/Ci mixtures. They found that not only the spontaneous activity but also the ORN response to the lowest cV dose (1 ng) was inhibited by simultaneous stimulation with high doses of Ci (1–10 μg). This inhibition occurred only in sensilla that also contained the Ci cell (Figures 6(b)-6(c)). In addition, the response to the higher cV dose (10 ng) was more strongly inhibited in sensilla where the Ci cell was colocalized. Thus, it seems plausible that the two ORNs interact, possibly by means of passive electrical interactions [54]. However, if or to which extent the reduction in pheromone trap catches by the presence of Ci [8] can be explained by the inhibition of the cV ORN remains unknown, as the excitatory input from the two ORNs provides the means also for central integration [55]. It seems like similar inhibitory interactions between colocalized ORNs occur also in other insects [51, 56, 57], but the phenomenon has so far only been systematically addressed in bark beetles.

7. Difficulties in Comparing ORN Responses to Compounds with Different Volatility

In most SSR studies, odor stimuli are prepared based on a known amount (e.g., in nano- or microgram) of compound applied to a piece of filter paper that is positioned inside a Pasteur pipette odor cartridge. Upon stimulation, the headspace in the cartridge is blown over the insect preparation. Depending on compound, solvent, and how many times the cartridge has been used, the quantity of molecules reaching the insect can be highly variable and seriously affect the ORN response [58, 59]. Indeed, different stimulation regimes, compound doses, and solvents (mostly hexane and paraffin oil) have been used in the various bark beetle SSR studies (Table 1), making it difficult to directly compare the sensitivity and specificity of ORNs characterized in different species or studies. Furthermore, the physical parameters of the odor-delivery system also affect the integrity of an airborne odor stimulus [60], which may further increase the variability among responses.

Airborne amounts of different compounds have been measured with a photoionization detector [45]. A huge variation among compounds was observed. For the most volatile compounds, such as 2-methyl-3-buten-2-ol (Figure 5), ca 80% of the headspace in the odor cartridge was lost at the first puff, even though paraffin oil was used as solvent. Airborne amounts of heavier compounds, such as linalool, were reduced by only ca 50% after 50 reiterative stimulations. In addition, compounds that were dissolved in pentane were released at a much higher rate than compounds in paraffin [45].

The large variation between compounds, solvents, and successive stimulations could easily bias electrophysiological responses in insects. This was verified by reanalyzing the response of the 3-octanol ORN of *I. typographus* [19] to two C8-alcohols (3-octanol and 1-octen-3-ol) and two C6-alcohols (Z3-hexenol and 1-hexanol) using both fresh (not used) and "old" (used 10 times) stimulus pipettes [45]. The ORN response to fresh pipettes was clearly different from the response to the "old" pipettes. In particular the response to the C6-alcohols was clearly lower when old pipettes were used. In fact, the difference in response was so large that it falsely implied that recordings were made from two distinct ORN classes. Such a finding suggests that it is absolutely necessary to use very strict experimental protocols for electrophysiological recordings, and that it sometimes is required to measure airborne odor amounts, especially when compounds of different volatility are used as stimuli [45].

8. Odor Coding in Bark Beetles Compared to Other Insects

In insects in general, neurons that detect pheromone constituents have a narrow tuning. Bark beetles are no exception as the ORNs that respond to aggregation pheromone compounds are, in most cases, sensitive to only one compound. The tuning width of insect ORNs detecting plant volatiles seems to range from narrow to broad, although ORN specificity is strongly correlated to the stimulus concentration tested [18]. Most of the ORNs for plant volatiles in *I. typographus* are narrowly tuned [8]. However, some show more indiscriminate responses, such as the GLV neuron that had similar sensitivity to 1-hexanol, E2-hexenol, and Z3-hexenol. This is in contrast to the highly specific GLV neurons that have been described in, for instance, scarab beetles [50, 51], and in the Colorado potato beetle [62]. The difference may be related to the fact that these other species feed on angiosperms, which presumably requires a better resolution of angiosperm dominated volatiles (i.e., GLVs) than what is needed for a conifer specialist. Many of the bark beetle ORNs are highly selective for specific enantiomers, both in terms of pheromones and plant compounds. However, this feature is not unique for bark beetles; highly enantioselective neurons have been characterized also in other insects [63–65]. In contrast, no other insect studied so far has a comparable frequency of ORNs tuned to antiattractants as the one found in *I. typographus* [19].

The co-localization of ORNs for pheromone and plant compounds in *I. typographus* is not commonly found in insects. This special type of ORN pairing may be related to the fact that host colonization in bark beetles often involves both pheromone and plant-produced compounds [8]. In addition, the colocalized neurons for plant and pheromone compounds also interacted by inhibiting their neighboring neuron while responding [8]. It is difficult to say whether a similar interaction occurs also in other insects, since it has not been systematically addressed elsewhere. However, inhibition of the spontaneous activity of the large-spiking cell when the small-spiking cell responds seems to be a common phenomenon [35, 51, 56, 57], indicating that the same type of modulation could be present. Indirect evidence for ORN interactions was found previously in the honeybee [66]. The 18–35 ORNs that are housed within honeybee

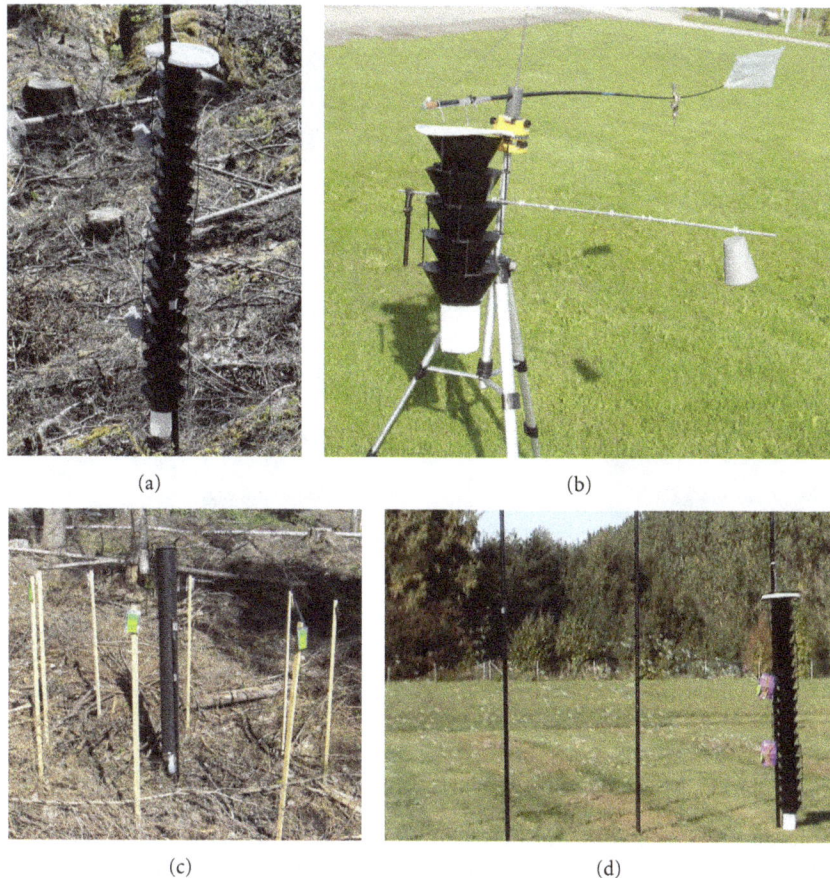

(a)

(b)

(c)

(d)

FIGURE 7: (a) Lindgren funnel traps (19-funnel size) were used in the vertical spacing tests with *Ips typographus*. Dispensers positioned under grey cups. (b) A Lindgren trap (5-funnel size) was attached to a wind vane in the horizontal spacing tests to ensure constant distance between plumes. (c) Pipe trap surrounded by eight nonhost volatile dispensers in the antiattractant background tests. (d) Soap bubble visualization of vertical plume overlap at a spacing distance of 48 cm. Distance between black poles = 1 m (modified from [61]).

sensilla placodea seemed to respond to odors in a coordinated manner, indicating that the individual ORNs do not act as independent response units. However, in that study, it was not possible to keep track of the individual ORNs.

Taken together, odor coding in bark beetles is, in general, similar to odor coding in other insects, but it also exhibits some rare features. The coding principle seems to be consistent with the "combinatorial code" theory, but the olfactory input travels mainly through highly specific channels.

9. Detection and Behavior in Odor-Diverse Habitats

Activation of an ORN by an attractant may cause an upwind flight by the insect towards the odor source. However, if repulsive compounds simultaneously trigger other ORNs to fire, the upwind flight may be aborted. Thus, in environments with a high "semiochemical diversity" [27] where odor plumes from different sources intermix, localization of host plants may be hampered by the presence of odors from nonhosts [67, 68]. Thus, for bark beetles, it may be possible to reduce the risk of attacks by making the environment more semiochemically diverse. Homogenous mixing of

odor plumes from different sources is, however, contradicted by the partitioning of plumes into "odor packages" (or filaments) that are interspersed with pockets of "clean air" [11, 12]. This, in turn, is thought to facilitate plume discrimination by insects.

Placing an NHV mixture inside a pheromone trap, that is, next to the pheromone bait, greatly reduces trap catch of *I. typographus* [27]. However, to test the "semiochemical diversity hypothesis," pheromone trap catches in the presence of NHV at different vertical and horizontal distances from the pheromone dispenser (Figures 7(a)-7(b)), were investigated [69]. Trap catches in response to separated pheromone components (*cis*-verbenol and 2-methyl-3-buten-2-ol) were also tested (in the absence of NHV) to further investigate responses to separated baits in general. In addition, the response of the beetle was compared to the response of the Egyptian cotton leaf worm, *Spodoptera littoralis* (Lepidoptera: Noctuidae), to separated sex pheromone components and to separated pheromone and behavioral antagonist. In both species, increased spacing between pheromone and antiattractants led to increased trap catch, whereas, as expected, increased spacing between pheromone components had the opposite effect. However,

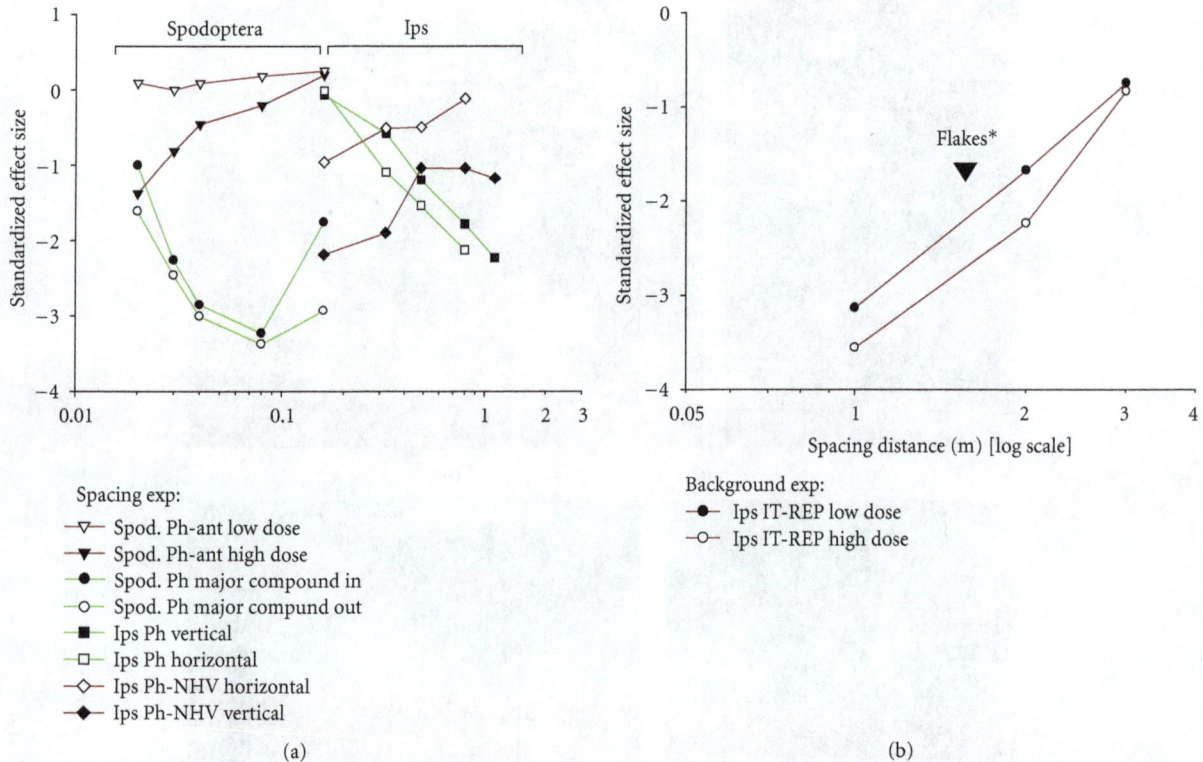

(a)   (b)

FIGURE 8: (a) The effect of spacing between attractant and antiattractant sources on trap catches of *Ips typographus* and *Spodoptera littoralis*, illustrated by measures of effect size (*Hedges'* unbiased *g*). The effect size provides a measure of a biological treatment effect by scaling the difference between the treatment and the control means, with the pooled standard deviation for those means. Effect sizes further from zero than 0.8 are regarded as strong effects. In all experiments, the pheromone bait alone (zero distance between components) was the control. The zero cm spacing distance in experiments involving antiattractants is omitted for clarity. (b) Effect sizes in the *Ips* antiattractant background experiments using nonhost volatile dispensers at eight positions, or flakes around the trap. *Flakes were evenly distributed on the ground 0–2 m from the trap. Thus, this treatment is "not to scale" on the *x*-axis. Ph = pheromone; Ant = *Spodoptera* pheromone antagonist; NHV = nonhost volatiles; IT-REP = semicommercial *Ips typographus* repellent dispenser (from [69], with permission from the publisher).

the two species differed greatly with respect to the spacing distances that affected their trap catch (Figure 8(a)). While beetle trap catches were affected by separation of some decimeters, trap catches of the moth were affected by separation distances of just a few centimeters [69]. In each species, the spacing distances affecting trap catch did not differ between the pheromone component spacing and the pheromone/antiattractant spacing experiments [69].

The bark beetle pheromone/NHV spacing experiments indicated antiattractive effects of NHV up to a distance of >1 m [69]. To further investigate potential effects of NHV at even longer distances, pheromone attraction was studied in the presence of a synthetic background of NHV, either created by eight NHV point sources positioned in a ring (with 1, 2, or 3 m radius) around a central pheromone trap (Figure 7(c)), or by ca 6000 small (ca 3 × 3 mm) NHV impregnated flakes [70] on the ground around a pheromone trap [69]. With the eight NHV sources, bark beetle attraction was reduced up to the 2 m spacing distance, and there was still a tendency for reduced attraction at the 3 m distance (Figure 8(b)). Similar to the eight point sources, the NHV flakes also reduced pheromone attraction [69]. The active spacing distances are in accordance with the "active

inhibitory range" of NHV of at least 2 m that was estimated previously [27]. The pheromone dose used by Andersson et al. [69] was comparable to that released from a mass-attacked tree, which is a very strong signal. Thus, it is striking that volatiles from nonhost plants can inhibit attraction when they are released a few meters away from the pheromone source. This indicates that avoiding not only nonhost species, but also nonhost habitats, likely improves bark beetle fitness.

The different spacing distances that affected trap catches of the beetle and the moth may reflect differences in the size of the natural odor sources (and plumes) the insects orient to [69]. While a male moth orients towards a single calling female, bark beetles may orient to large patches of trees with hundreds of calling males. Furthermore, the moth sex pheromone communication system is highly specialized. A male moth flies towards a calling female for mating only, whereas the bark beetle aggregation pheromone can be used as a signal of mates, food, and oviposition sites. Thus, the different selection pressures that operate on these systems have likely resulted in different degrees of specialization. The ORNs for pheromone compounds in moths are housed in specific sensilla (*trichodea*), distinct from the ones that detect plant odors [43]. In contrast, *I. typographus* groups the *cis*-verbenol pheromone ORN together in the same sensilla as the ORN for the plant compound 1,8-cineole, although the ORNs themselves are specific in their response [8].

Similar to *I. typographus*, studies on *Dendroctonus* bark beetles indicated synergistic interactions between pheromone components when two baits were separated by several meters [71, 72]. The sharp response of *S. littoralis* to odor source spacing has been observed previously in other moths [73–75]. The most extreme example is provided by Fadamiro et al. [52], who found that 1 mm separation between pheromone and antagonist was sufficient to restore upwind flight to the pheromone by male *Helicoverpa zea*. It was hypothesized that coincident detection of pheromone components and antagonists, achieved by colocalization of the ORNs, was the reason for this amazing ability of the males. Furthermore, synchronous detection of pheromone compounds was shown to improve the temporal spiking pattern by projection neurons in the antennal lobe of *Manduca sexta* moths [76]. Thus, it is clear that coincidence detection is of great importance in the pheromone system of moths. Soap bubble generators were used to visualize plume overlap at the different spacing distances used for *I. typographus* (Figure 7(d)) [69]. The simulations indicated that filaments from different plumes are more likely to overlap and, thus, to be detected coincidently, when the sources are close to each other. Therefore, the lower sensitivity of *I. typographus* to small-scale spatial separation of odor sources might indicate that coincidence detection is of less importance for bark beetles than for moths [69].

10. Applications

Conifer pest insect infestations are typically less common in diversified habitats [67], which in part may be due to the presence of antiattractive NHV. The finding that NHV, from a distance of at least 2 m (see also [27]), can reduce attraction to a pheromone dose comparable to that released from a mass-attacked tree suggests a potential for NHV in forest protection. However, pheromone attraction was not completely shut down so it is more likely that, instead of counteracting ongoing mass attacks, synthetic or natural NHV sources may reduce the risk of spruces being attacked in the first place. Indeed, spruces were previously protected by NHV dispensers attached to every second tree, demonstrating a protective effect of ca 2 m [77]. In another study, groups of ten trees were all protected by 20 NHV dispensers, and bark beetle attacks were diverted to trees >15 m away [78].

In addition, the spruce compound 1,8-cineole that strongly reduced pheromone attraction should be further tested in combination with the other active semiochemicals for possible improvement of antiattractant blends. It is possible that the repression of the ORN for *cis*-verbenol by 1,8-cineole, adds another inhibitory mechanism by distorting the "perceived blend ratio" of the aggregation pheromone. If so, it is likely that a more effective antiattractant blend can be obtained than the one that is comprised of GLV alcohols, C8-alcohols, *trans*-conophthorin, and verbenone [27].

11. Conclusions and Future Directions

The recent advances in bark beetle olfactory physiology have provided a connection between the physiological and behavioral responses of *I. typographus* to ecologically relevant compounds. This connection has allowed for a deeper understanding about how bark beetles (and possibly insects in general) may encode, and respond to, the odor environment. However, there are still several ORNs of *I. typographus* (and other species) for which odor ligands have not been identified, meaning that there is yet more to be learned about its olfactory physiology. Identification of active compounds should be achieved by GC-coupled SSR and by testing headspace collections from, for instance, attacked and unattacked or resistant host trees.

At the molecular level, Andersson and collaborators [61, 79] recently sequenced the antennal transcriptome of *I. typographus*, leading to identification of gene sequences for 40 different candidate olfactory receptors (ORs). The amino acid sequences of the receptors were compared, in a sequence similarity tree, with receptors that were previously identified from the genome of the flour beetle, *Tribolium castaneum*. Many of the *Ips* ORs formed a bark beetle-specific branch, indicating an extension of OR function. Possibly, these receptors detect conifer-related volatiles or pheromones that are especially relevant for bark beetles. The other ORs of *Ips* were grouped together with ORs of *T. castaneum*, which may indicate conserved functionality of some sets of ORs within Coleoptera. Functional studies to reveal which compounds the ORs of *Ips* bind will be the next step in the study. Such studies will hopefully extend the connection from behavior, through physiology, all the way to the level of the receptor and gene.

The identification of the bark beetle ORs paves the way for the development of potential novel management strategies in the future. If the receptors for pheromone components and antiattractive NHV can be identified, it might be possible to identify ligands that pharmacologically block the pheromone receptors or hyperstimulate [80] receptors for nonhost volatiles. If such compounds are found, they might be dispensed in the forest to disrupt bark beetle pheromone communication and host tree localization.

One hypothesis why insect colocalize specific ORNs in the same sensilla is that it allows for improved spatiotemporal resolution of odor stimuli [52]. This hypothesis could be tested by comparing trap catches of *I. typographus* in response to spacing between pheromone and 1,8-cineole (ORNs for *cis*-verbenol and 1,8-cineole co-localized), with trap catches in response to spacing between pheromone and verbenone, the latter compound being detected by an ORN that is never colocalized with an aggregation pheromone neuron. Predictably, the beetle should be more "sensitive" to small-scale spacing between pheromone and 1,8-cineole than to spacing between pheromone and verbenone.

In order to put the sensory physiology into a more natural context, a portable single sensillum recording device [81] should be used in the field. The sensillum that contains the ORNs for *cis*-verbenol and 1,8-cineole could be used as a biological detector for measurements of plume filament overlap. Such measurements would reveal whether filaments from overlapping plumes are detected coincidently or not. It would also provide some indirect clue if beetles temporally integrate filaments from different plumes to a larger degree than moths, which could explain the difference in response to spacing in the two types of insects.

Acknowledgments

The author is grateful to Professor Fredrik Schlyter (SLU, Alnarp) for useful criticism on a previous manuscript draft. Funding was provided by FORMAS Project no. 230-2005-1778 "Semiochemical diversity and insect dynamics," the Linnaeus program "Insect chemical ecology, ethology, and evolution (ICE3)," and by the Department of Plant Protection Biology, SLU, Alnarp.

References

[1] W. A. Kurz, C. C. Dymond, G. Stinson et al., "Mountain pine beetle and forest carbon feedback to climate change," *Nature*, vol. 452, no. 7190, pp. 987–990, 2008.

[2] L. Stauffer, F. Lakatos, and G. M. Hewitt, "Phylogeography and postglacial colonization routes of *Ips typographus* L. (Coleoptera, Scolytidae)," *Molecular Ecology*, vol. 8, no. 5, pp. 763–773, 1999.

[3] J. A. Byers, "Chemical ecology of bark beetles in a complex olfactory landscape," in *Bark and Wood Boring Insects in Living Trees in Europe, a Synthesis*, F. Lieutier, K. R. Day, A. Battisti, J.-C. Grégoire, and H. F. Evans, Eds., pp. 89–134, Kluwer Academic Publishers, Dordrecht, The Netherlands, 2004.

[4] B. Økland and O. N. Bjørnstad, "Synchrony and geographical variation of the spruce bark beetle (*Ips typographus*) during a non-epidemic period," *Population Ecology*, vol. 45, no. 3, pp. 213–219, 2003.

[5] B. Økland and O. N. Bjørnstad, "A resource-depletion model of forest insect outbreaks," *Ecology*, vol. 87, no. 2, pp. 283–290, 2008.

[6] F. Schlyter and G. A. Birgersson, "Forest beetles," in *Pheromones of Non-Lepidopteran Insects Associated with Agricultural Plants*, J. Hardie and A. K. Minks, Eds., pp. 113–148, CAB International, Oxford, UK, 1999.

[7] Q.-H. Zhang and F. Schlyter, "Olfactory recognition and behavioural avoidance of angiosperm nonhost volatiles by conifer-inhabiting bark beetles," *Agricultural and Forest Entomology*, vol. 6, no. 1, pp. 1–19, 2004.

[8] M. N. Andersson, M. C. Larsson, M. Blaženec, R. Jakuš, Q.-H. Zhang, and F. Schlyter, "Peripheral modulation of pheromone response by inhibitory host compound in a beetle," *Journal of Experimental Biology*, vol. 213, no. 19, pp. 3332–3339, 2010.

[9] N. Erbilgin, P. Krokene, T. Kvamme, and E. Christiansen, "A host monoterpene influences *Ips typographus* (Coleoptera: Curculionidae, Scolytinae) responses to its aggregation pheromone," *Agricultural and Forest Entomology*, vol. 9, no. 2, pp. 135–140, 2007.

[10] J. A. Byers and Q.-H. Zhang, "Chemical ecology of bark beetles in regard to search and selection of host trees," in *Recent Advances in Entomological Research*, T.-X. Liu and L. Kang, Eds., pp. 91–113, Higher Education Press, Beijing, China, 2010.

[11] N. J. Vickers, "Mechanisms of animal navigation in odor plumes," *Biological Bulletin*, vol. 198, no. 2, pp. 203–212, 2000.

[12] R. T. Cardé and M. A. Willis, "Navigational strategies used by insects to find distant, wind-borne sources of odor," *Journal of Chemical Ecology*, vol. 34, no. 7, pp. 854–866, 2008.

[13] J. Murlis, M. A. Willis, and R.T. Cardé, "Spatial and temporal structures of pheromone plumes in fields and forests," *Physiological Entomology*, vol. 25, pp. 211–222, 2000.

[14] L. Abuin, B. Bargeton, M. H. Ulbrich, E. Y. Isacoff, S. Kellenberger, and R. Benton, "Functional architecture of olfactory ionotropic glutamate receptors," *Neuron*, vol. 69, no. 1, pp. 44–60, 2011.

[15] L. B. Vosshall and R. F. Stocker, "Molecular architecture of smell and taste in *Drosophila*," *Annual Review of Neuroscience*, vol. 30, pp. 505–533, 2007.

[16] M. de Bruyne and T. C. Baker, "Odor detection in insects: volatile codes," *Journal of Chemical Ecology*, vol. 34, no. 7, pp. 882–897, 2008.

[17] L. B. Vosshall and B. S. Hansson, "A unified nomenclature system for the insect olfactory coreceptor," *Chemical Senses*, vol. 36, no. 6, pp. 497–498, 2011.

[18] E. A. Hallem and J. R. Carlson, "Coding of odors by a receptor repertoire," *Cell*, vol. 125, no. 1, pp. 143–160, 2006.

[19] M. N. Andersson, M. C. Larsson, and F. Schlyter, "Specificity and redundancy in the olfactory system of the bark beetle *Ips typographus*: single-cell responses to ecologically relevant odors," *Journal of Insect Physiology*, vol. 55, no. 6, pp. 556–567, 2009.

[20] F. Schlyter, G. Birgersson, J. A. Byers, J. Löfqvist, and G. Bergström, "Field response of spruce bark beetle, *Ips typographus*, to aggregation pheromone candidates," *Journal of Chemical Ecology*, vol. 13, no. 4, pp. 701–716, 1987.

[21] M. Saint-Germain, C. M. Buddle, and P. Drapeau, "Primary attraction and random landing in host-selection by wood-feeding insects: a matter of scale?" *Agricultural and Forest Entomology*, vol. 9, no. 3, pp. 227–235, 2008.

[22] J. A. Byers, "An encounter rate model of bark beetle populations searching at random for susceptible host trees," *Ecological Modelling*, vol. 91, no. 1-3, pp. 57–66, 1996.

[23] F. Schlyter, G. Birgersson, and A. Leufvén, "Inhibition of attraction to aggregation pheromone by verbenone and ipsenol - Density regulation mechanisms in bark beetle *Ips typographus*," *Journal of Chemical Ecology*, vol. 15, no. 8, pp. 2263–2277, 1989.

[24] Q.-H. Zhang, F. Schlyter, and P. Anderson, "Green leaf volatiles interrupt pheromone response of spruce bark beetle, *Ips typographus*," *Journal of Chemical Ecology*, vol. 25, no. 12, pp. 2847–2861, 1999.

[25] Q.-H. Zhang, F. Schlyter, and G. Birgersson, "Bark volatiles from nonhost angiosperm trees of spruce bark beetle, *Ips typographus* (L.) (Coleoptera: Scolytidae): chemical and electrophysiological analysis," *Chemoecology*, vol. 10, no. 2, pp. 69–80, 2000.

[26] Q.-H. Zhang, T. Tolasch, F. Schlyter, and W. Francke, "Enantiospecific antennal response of bark beetles to spiroacetal (*E*)-conophthorin," *Journal of Chemical Ecology*, vol. 28, no. 9, pp. 1839–1852, 2002.

[27] Q.-H. Zhang and F. Schlyter, "Redundancy, synergism, and active inhibitory range of non-host volatiles in reducing pheromone attraction in European spruce bark beetle *Ips typographus*," *Oikos*, vol. 101, no. 2, pp. 299–310, 2003.

[28] B. Å. Tømmerås, "Specialization of the olfactory receptor cells in the bark beetle *Ips typographus* and its predator *Thanasimus formicarius* to bark beetle pheromones and host tree volatiles," *Journal of Comparative Physiology A*, vol. 157, no. 3, pp. 335–341, 1985.

[29] H. Mustaparta, B. Å. Tømmerås, P. Baeckström, J. M. Bakke, and G. Ohloff, "Ipsdienol-specific receptor cells in bark beetles: structure-activity relationships of various analogues and of deuterium-labelled ipsdienol," *Journal of Comparative Physiology A*, vol. 154, no. 4, pp. 591–595, 1984.

[30] B. Å. Tømmerås and H. Mustaparta, "Chemoreception of host volatiles in the bark beetle *Ips typographus*," *Journal of Comparative Physiology A*, vol. 161, no. 5, pp. 705–710, 1987.

[31] B. Å. Tømmerås, H. Mustaparta, and J. C. Gregoire, "Receptor cells in *Ips typographus* and *Dendroctonus micans* specific to pheromones of the reciprocal genus," *Journal of Chemical Ecology*, vol. 10, no. 5, pp. 759–770, 1984.

[32] H. Mustaparta, M. E. Angst, and G. N. Lanier, "Specialization of olfactory cells to insect-and host-produced volatiles in the bark beetle *Ips pini* (say)," *Journal of Chemical Ecology*, vol. 5, no. 1, pp. 109–123, 1979.

[33] H. Mustaparta, M. E. Angst, and G. N. Lanier, "Receptor discrimination of enantiomers of the aggregation pheromone ipsdienol, in two species of *Ips*," *Journal of Chemical Ecology*, vol. 6, no. 3, pp. 689–701, 1980.

[34] H. Mustaparta, M. E. Angst, and G. N. Lanier, "Responses of single receptor cells in the pine engraver beetle, *Ips pini* (SAY) (Coleoptera: Scolytidae) to its aggregation pheromone, ipsdienol, and the aggregation inhibitor, ipsenol," *Journal of Comparative Physiology. A*, vol. 121, no. 3, pp. 343–347, 1977.

[35] J. C. Dickens, T. L. Payne, L. C. Ryker, and J. A. Rudinsky, "Single cell responses of the Douglas-fir beetle, *Dendroctonus pseudotsugae* hopkins (Coleoptera: Scolytidae), to pheromones and host odors," *Journal of Chemical Ecology*, vol. 10, no. 4, pp. 583–600, 1984.

[36] J. C. Dickens, T. L. Payne, L. C. Ryker, and J. A. Rudinsky, "Multiple acceptors for pheromonal enantiomers on single olfactory cells in the Douglas-fir beetle, *Dendroctonus pseudotsugae* Hopk. (Coleoptera: Scolytidae)," *Journal of Chemical Ecology*, vol. 11, no. 10, pp. 1359–1370, 1985.

[37] T. L. Payne, J. V. Richerson, J. C. Dickens et al., "Southern pine beetle: olfactory receptor and behavior discrimination of enantiomers of the attractant pheromone frontalin," *Journal of Chemical Ecology*, vol. 8, no. 5, pp. 873–881, 1982.

[38] J. C. Dickens and T. L. Payne, "Bark beetle olfaction: pheromone receptor system in *Dendroctonus frontalis*," *Journal of Insect Physiology*, vol. 23, no. 4, pp. 481–489, 1977.

[39] B. Å. Tømmerås and H. Mustaparta, "Single cell responses to pheromones, host and non-host volatiles in the ambrosia beetle *Trypodendron lineatum*," *Entomologia Experimentalis et Applicata*, vol. 52, no. 2, pp. 141–148, 1989.

[40] A. Guerrero, J. Feixas, J. Pajares, L. J. Wadhams, J. A. Pickett, and C. M. Woodcock, "Semiochemically induced inhibition of behaviour of *Tomicus destruens* (Woll.) (Coleoptera: Scolytidae)," *Naturwissenschaften*, vol. 84, no. 4, pp. 155–157, 1997.

[41] E. Hallberg, "Sensory organs in *Ips typographus* (Insecta: Coleoptera)—Fine structure of antennal sensilla," *Protoplasma*, vol. 111, no. 3, pp. 206–214, 1982.

[42] T. C. Baker, S. A. Ochieng, A. A. Cossé et al., "A comparison of responses from olfactory receptor neurons of *Heliothis subflexa* and *Heliothis virescens* to components of their sex pheromone," *Journal of Comparative Physiology A*, vol. 190, no. 2, pp. 155–165, 2004.

[43] H. Ljungberg, P. Anderson, and B. S. Hansson, "Physiology and morphology of pheromone-specific sensilla on the antennae of male and female *Spodoptera littoralis* (Lepidoptera: Noctuidae)," *Journal of Insect Physiology*, vol. 39, no. 3, pp. 253–260, 1993.

[44] G. Birgersson, F. Schlyter, J. Löfqvist, and G. Bergström, "Quantitative variation of pheromone components in the spruce bark beetle *Ips typographus* from different attack phases," *Journal of Chemical Ecology*, vol. 10, no. 7, pp. 1029–1055, 1984.

[45] M. N. Andersson and F. Schlyter, "What reaches the antenna? How to calibrate odor flux and ligand-receptor affinities," *Chemical Senses*. In press.

[46] I. M. Wilson, J. H. Borden, R. Gries, and G. Gries, "Green leaf volatiles as antiaggregants for the mountain pine beetle, *Dendroctonus ponderosae* Hopkins (Coleoptera: Scolytidae)," *Journal of Chemical Ecology*, vol. 22, no. 10, pp. 1861–1875, 1996.

[47] Q.-H. Zhang, N. Erbilgin, and S. J. Seybold, "GC-EAD responses to semiochemicals by eight beetles in the subcortical community associated with Monterey pine trees in coastal California: similarities and disparities across three trophic levels," *Chemoecology*, vol. 18, no. 4, pp. 243–254, 2008.

[48] D. P. W. Huber, R. Gries, J. H. Borden, and H. D. Pierce, "A survey of antennal responses by five species of coniferophagous bark beetles (Coleoptera: Scolytidae) to bark volatiles of six species of angiosperm trees," *Chemoecology*, vol. 10, no. 3, pp. 103–113, 2000.

[49] J. A. Byers, "Avoidance of competition by spruce bark beetles, *Ips typographus* and *Pityogenes chalcographus*," *Experientia*, vol. 49, no. 3, pp. 272–275, 1993.

[50] B. S. Hansson, M. C. Larsson, and W. S. Leal, "Green leaf volatile-detecting olfactory receptor neurones display very high sensitivity and specificity in a scarab beetle," *Physiological Entomology*, vol. 24, no. 2, pp. 121–126, 1999.

[51] M. C. Larsson, W. S. Leal, and B. S. Hansson, "Olfactory receptor neurons detecting plant odours and male volatiles in *Anomala cuprea* beetles (Coleoptera: Scarabaeidae)," *Journal of Insect Physiology*, vol. 47, no. 9, pp. 1065–1076, 2001.

[52] H. Y. Fadamiro, A. A. Cossé, and T. C. Baker, "Fine-scale resolution of closely spaced pheromone and antagonist filaments by flying male *Helicoverpa zea*," *Journal of Comparative Physiology, A*, vol. 185, no. 2, pp. 131–141, 1999.

[53] T. J. A. Bruce, L. J. Wadhams, and C. M. Woodcock, "Insect host location: a volatile situation," *Trends in Plant Science*, vol. 10, no. 6, pp. 269–274, 2005.

[54] A. Vermeulen and J. P. Rospars, "Why are insect olfactory receptor neurons grouped into sensilla? The teachings of a model investigating the effects of the electrical interaction between neurons on the transepithelial potential and the neuronal transmembrane potential," *European Biophysics Journal*, vol. 33, no. 7, pp. 633–643, 2004.

[55] A. F. Silbering, R. Okada, K. Ito, and C. G. Galizia, "Olfactory information processing in the *Drosophila* antennal lobe: anything goes?" *Journal of Neuroscience*, vol. 28, no. 49, pp. 13075–13087, 2008.

[56] M. M. Blight, J. A. Pickett, L. J. Wadhams, and C. M. Woodcock, "Antennal perception of oilseed rape, *Brassica napus* (Brassicaceae), volatiles by the cabbage seed weevil *Ceutorhynchus assimilis* (Coleoptera, Curculionidae)," *Journal of Chemical Ecology*, vol. 21, no. 11, pp. 1649–1664, 1995.

[57] E. A. Hallem, M. G. Ho, and J. R. Carlson, "The molecular basis of odor coding in the *Drosophila* antenna," *Cell*, vol. 117, no. 7, pp. 965–979, 2004.

[58] J. E. Cometto-Muñiz, W. S. Cain, and M. H. Abraham, "Quantification of chemical vapors in chemosensory research," *Chemical Senses*, vol. 28, no. 6, pp. 467–477, 2003.

[59] T. Tsukatani, T. Miwa, M. Furukawa, and R. M. Costanzo, "Detection thresholds for phenyl ethyl alcohol using serial dilutions in different solvents," *Chemical Senses*, vol. 28, no. 1, pp. 25–32, 2003.

[60] R. S. Vetter, A. E. Sage, K. A. Justus, R. T. Cardé, and C. G. Galizia, "Temporal integrity of an airborne odor stimulus is greatly affected by physical aspects of the odor delivery system," *Chemical Senses*, vol. 31, no. 4, pp. 359–369, 2006.

[61] M. N. Andersson, *Olfaction in the spruce bark beetle*, Ips typographus: *receptor, neuron and habitat*, Ph.D. dissertation, Swedish University of Agricultural Sciences, Alnarp, Sweden, 2011.

[62] R. De Jong and J. H. Visser, "Integration of olfactory information in the Colorado potato beetle brain," *Brain Research*, vol. 447, no. 1, pp. 10–17, 1988.

[63] K. Hansen, "Discrimination and production of disparlure enantiomers by the gypsy moth and the nun moth," *Physiological Entomology*, vol. 9, pp. 9–18, 1984.

[64] G. P. Svensson and M. C. Larsson, "Enantiomeric specificity in a pheromone-kairomone system of two threatened saproxylic beetles, *Osmoderma eremita* and *Elater ferrugineus*," *Journal of Chemical Ecology*, vol. 34, no. 2, pp. 189–197, 2008.

[65] H. Wojtasek, B. S. Hansson, and W. S. Leal, "Attracted or repelled? A matter of two neurons, one pheromone binding protein, and a chiral center," *Biochemical and Biophysical Research Communications*, vol. 250, no. 2, pp. 217–222, 1998.

[66] W. M. Getz and R. P. Akers, "Honeybee olfactory sensilla behave as integrated processing units," *Behavioral and Neural Biology*, vol. 61, no. 2, pp. 191–195, 1994.

[67] H. Jactel and E. G. Brockerhoff, "Tree diversity reduces herbivory by forest insects," *Ecology Letters*, vol. 10, no. 9, pp. 835–848, 2007.

[68] H. Jactel, G. Birgersson, S. Andersson, and F. Schlyter, "Nonhost volatiles mediate associational resistance to the pine processionary moth," *Oecologia*, vol. 166, no. 3, pp. 703–711, 2011.

[69] M. N. Andersson, M. Binyameen, M. M. Sadek, and F. Schlyter, "Attraction modulated by spacing of pheromone components and anti-attractants in a bark beetle and a moth," *Journal of Chemical Ecology*, vol. 37, no. 8, pp. 899–911, 2011.

[70] N. E. Gillette, J. D. Stein, D. R. Owen et al., "Verbenone-releasing flakes protect individual *Pinus contorta* trees from attack by *Dendroctonus ponderosae* and *Dendroctonus valens* (Coleoptera: Curculionidae, Scolytinae)," *Agricultural and Forest Entomology*, vol. 8, no. 3, pp. 243–251, 2006.

[71] J. A. Byers, "Interactions of pheromone component odor plumes of western pine beetle," *Journal of Chemical Ecology*, vol. 13, no. 12, pp. 2143–2157, 1987.

[72] B. T. Sullivan and K. Mori, "Spatial displacement of release point can enhance activity of an attractant pheromone synergist of a bark beetle," *Journal of Chemical Ecology*, vol. 35, no. 10, pp. 1222–1233, 2009.

[73] G. H. L. Rothschild, "Problems in defining synergists and inhibitors of the Oriental Fruit Moth pheromone by field experimentation," *Entomologia Experimentalis et Applicata*, vol. 17, no. 2, pp. 294–302, 1974.

[74] P. Witzgall and E. Priesner, "Wind-tunnel study on attraction inhibitor in male *Coleophora laricella* Hbn. (Lepidoptera: Coleophoridae)," *Journal of Chemical Ecology*, vol. 17, no. 7, pp. 1355–1362, 1991.

[75] M. Coracini, M. Bengtsson, L. Cichon, and P. Witzgall, "Codling moth males do not discriminate between pheromone and a pheromone/antagonist blend during upwind flight," *Naturwissenschaften*, vol. 90, no. 9, pp. 419–423, 2003.

[76] T. A. Christensen and J. G. Hildebrand, "Coincident stimulation with pheromone components improves temporal pattern resolution in central olfactory neurons," *Journal of Neurophysiology*, vol. 77, no. 2, pp. 775–781, 1997.

[77] R. Jakuš, F. Schlyter, Q.-H. Zhang et al., "Overview of development of an anti-attractant based technology for spruce protection against *Ips typographus*: from past failures to future success," *Journal of Pest Science*, vol. 76, no. 4, pp. 89–99, 2003.

[78] C. Schiebe, M. Blaženec, R. Jakuš, C. R. Unelius, and F. Schlyter, "Semiochemical diversity diverts bark beetle attacks from Norway spruce edges," *Journal of Applied Entomology*, vol. 135, no. 10, pp. 726–737, 2011.

[79] M. N. Andersson, J. M. Bengtsson, E. Grosse-Wilde et al., "Olfactory receptors in *Ips typographus*. Transcriptome from antenna analysed and preliminary compared to *Dendroctonus ponderosae* and *Tribolium castaneum* (Coleoptera: Curculionidae & Tenebrionidae)," in *Genetics of Bark Beetles and Associated Microorganisms*, F. Lakatos, B. Mészáros, and C. Stauffer, Eds., p. 15, Sopron, Hungary, 2011.

[80] N. Triballeau, E. van Name, G. Laslier et al., "High-potency olfactory receptor agonists discovered by virtual high-throughput screening: molecular probes for receptor structure and olfactory function," *Neuron*, vol. 60, no. 5, pp. 767–774, 2008.

[81] J. N. C. van der Pers and A. K. Minks, "Pheromone monitoring in the field using single sensillum recording," *Entomologia Experimentalis et Applicata*, vol. 68, no. 3, pp. 237–245, 1993.

13

Incorporating a Sorghum Habitat for Enhancing Lady Beetles (Coleoptera: Coccinellidae) in Cotton

P. G. Tillman[1] and T. E. Cottrell[2]

[1] USDA, ARS, Crop Protection and Management Research Laboratory, P.O. Box 748, Tifton, GA 31793, USA
[2] USDA, ARS, Southeastern Fruit & Tree Nut Research Laboratory, 21 Dunbar Road, Byron, GA 31008, USA

Correspondence should be addressed to P. G. Tillman, glynn.tillman@ars.usda.gov

Academic Editor: Ai-Ping Liang

Lady beetles (Coleoptera: Coccinellidae) prey on insect pests in cotton. The objective of this 2 yr on-farm study was to document the impact of a grain sorghum trap crop on the density of Coccinellidae on nearby cotton. *Scymnus* spp., *Coccinella septempunctata* (L.), *Hippodamia convergens* Guérin-Méneville, *Harmonia axyridis* (Pallas), *Coleomegilla maculata* (De Geer), *Cycloneda munda* (Say), and *Olla v-nigrum* (Mulsant) were found in sorghum over both years. Lady beetle compositions in sorghum and cotton and in yellow pyramidal traps were similar. For both years, density of lady beetles generally was higher on cotton with sorghum than on control cotton. Our results indicate that sorghum was a source of lady beetles in cotton, and thus incorporation of a sorghum habitat in farmscapes with cotton has great potential to enhance biocontrol of insect pests in cotton.

1. Introduction

Lady beetles (Coleoptera: Coccinellidae) have a significant impact on aphids (Hemiptera: Aphididae) [1–4], including the cotton aphid (*Aphis gossypii* Glover) attacking cotton (*Gossypium hirsutum* L.) [5] and the corn leaf aphid (*Rhopalosiphum maidis* (Fitch) and greenbug (*Schizaphis graminum* (Rondani) attacking grain sorghum (*Sorghum bicolor* (L.) Moench spp. *bicolor*) [6–8]. In the southeastern USA, cotton fields commonly are closely associated with other agronomic crops, especially corn (*Zea mays* L.) and peanut (*Arachis hypogea* L.), and in these farmscapes polysphagous pest species are known to move from corn and peanut into cotton to find newly available, suitable food, or oviposition sites [9]. As part of a larger pest management strategy, strips of grain sorghum planted between a source crop and cotton have proved useful as a trap crop to reduce pest movement, especially stink bugs (Hemiptera: Pentatomidae), into cotton [10, 11]. Additionally, a grain sorghum trap crop is beneficial to natural enemies by hosting the corn leaf aphid and greenbug [12]. Thus, grain sorghum, when planted adjacent to cotton, can perform as a trap crop for stink bugs and possibly as a source of natural enemies moving into cotton. In fact, many species of Coccinellidae are commonly found inhabiting grain sorghum: *Harmonia axyridis* (Pallas), *Hippodamia convergens* Guérin-Méneville, *H. sinuata* Mulsant, *H. parenthesis* (Say), *Coccinella septempunctata* L., *Coleomegilla maculata* (De Geer), *Cycloneda munda* (Say), *Scymnus* spp., *Olla v-nigrum* (Mulsant), *Exochomus* sp., and *Psyllobora vigintimaculata* (Say) [13–16]. These same species colonize cotton [4, 16, 17], and their presence within a grain sorghum trap crop may lead to these predators moving into cotton and facilitating insect pest management.

The objective of this study was to document the impact of a grain sorghum trap crop on the density of Coccinellidae on nearby cotton. Two treatments were used: (1) cotton fields without sorghum and (2) cotton fields bordered on one side by a strip of grain sorghum. Within the grain sorghum, Coccinellidae were not only sampled on plants but also from yellow pyramid traps that predominantly served in the larger pest management scheme to kill stink bugs in sorghum.

2. Materials and Methods

2.1. Study Sites. Six cotton fields, ranging from 5 to 18 ha in size, were sampled each year, 2006 and 2007, in Irwin County GA (Table 1). Recommended agricultural practices

TABLE 1: Planting date (PD) and variety for cotton (Ct) with sorghum trap crops, control cotton, and sorghum (So) in 2006 and 2007.

Year	Treatment	Rep	Crop	Variety[a]	PD
2006	Cotton w/trap crop	1	Ct	DP 555	4/28
	Cotton w/trap crop	2	Ct	DP 555	5/4
	Cotton w/trap crop	3	Ct	DP 555	5/10
	Sorghum trap crop	1–3	So	DK 54	4/14
	Control cotton	1	Ct	DP 555	5/1
	Control cotton	2	Ct	DP 555	5/4
	Control cotton	3	Ct	DP 555	5/26
2007	Cotton w/trap crop	1	Ct	DP 555	5/9
	Cotton w/trap crop	2	Ct	DP 555	6/7
	Cotton w/trap crop	3	Ct	DP 555	6/11
	Sorghum trap crop	1–3	So	DK 54	6/13
	Control cotton	1	Ct	DP 555	5/11
	Control cotton	2	Ct	DP 555	5/11
	Control cotton	3	Ct	DP 555	6/11

[a] Seed companies; DK: DeKalb; DP: Deltapine.

for production of sorghum [18] and cotton [19] were followed. Row width was 0.91 m for each crop, and rows for each crop were parallel to each other.

2.2. Yellow Pyramidal Traps. These traps consisted of a 2.84-liter clear plastic polyethylene terephthalate jar (United States Plastic Corp., Lima, OH, USA) on top of a 1.22 m-tall yellow pyramidal base [20, 21]. An insecticidal ear tag (Saber Extra, Coppers Animal Health Inc., Kansas City, KS, USA) was placed in the plastic jar at the beginning of a test to prevent escape of captured specimens. Active ingredients in the ear tag were lambda-cyhalothrin (10%) and piperonyl butoxide (13%). As part of the larger strategy to reduce pest movement into cotton, stink bug attraction to the traps was enhanced by placing *Euschistus* spp. stink bug lures (40 μL of the *Euschistus* spp. pheromone, methyl (*E, Z*)-2,4-decadienoate (CAS registry no. 4493-42-9) (Degussa AG Fine Chemicals, Marl, Germany, loaded onto rubber septa) in traps and replacing lures weekly. Insects from weekly collections were taken to the laboratory for identification.

2.3. Experimental Design. Two treatments were used each year: control cotton (without a sorghum trap crop) and cotton bordered by a sorghum trap crop and yellow pyramidal traps within the trap crop. At the beginning of the study, six commercial cotton fields were selected in Irwin County, Georgia, and each treatment was assigned randomly to three cotton fields similar to a completely randomized design. For the sorghum trap crop treatment, sorghum was planted in a strip (4 rows) along one edge of the cotton field; row 1 of sorghum was adjacent to a peanut field and row 4 was adjacent to the cotton field. Then 25–28 yellow pyramidal traps (depending on field width) were placed 12 m apart in row 1 of sorghum.

2.4. Insect Sampling. Each year of the study, crops and yellow pyramidal traps were examined weekly for the presence of lady beetles; from the week of 5 July to the week of 16 August in 2006 and from the week of 19 July to the week of 23 August in 2007. Due to time constraints of sampling these large fields, not all farmscapes were sampled on the same day of the week, but crops and/or yellow pyramidal traps within a field were sampled on the same day. For each sorghum sample, all plant parts within a 1.83 m length of row were visually checked for all lady beetles. For each cotton sample, all plants within a 1.83 m length of row were shaken over a drop cloth, and the aerial parts of all plants were visually checked thoroughly for all lady beetles. Voucher specimens are stored in the USDA-ARS, Crop Protection and Management Research Laboratory in Tifton, GA, USA.

For sampling purposes, the edge of a cotton field adjacent to a peanut field was labeled as side A, and in a clockwise direction the other 3 sides of a field were labeled as sides B, C, and D. Each year, samples were obtained from within the cotton field at 3 distances from the edge of side A (i.e., at rows 1, 2, and 5), and at 6 interior locations along the length of the field (i.e., rows 16, 33, 100, 167, 233, and 300 from the edge of the field on side A). In both years, the 300-row samples were not close to the edge of side C; 24–31 m from side C in 2006 and 61 m from side C in 2007. For sides B–D, samples were taken from 2 edge locations, rows 1 and 5 from the edge of the field. The number of samples from each field on each date was as follows: 9 from each row on side A, 3 from each row on sides B–D, and 6 from each interior location. For both years, the 4-row strip of sorghum was sampled by taking 9 samples from each of the 4 rows.

2.5. Statistical Analysis. Lady beetle species compositions in sorghum strips, cotton fields, and yellow pyramidal traps were similar for both years, and then data for the two years were combined. Means were obtained for number of lady beetle adults per sample for sorghum and yellow pyramidal traps using PROC MEANS [22]. The number of lady beetle adults per sample in cotton with sorghum trap crops and control cotton was compared using *t*-tests. In 2007, one cotton field with a sorghum trap crop was excluded from data analysis on week 6 due to an insecticide application after sampling on week 5. One control cotton field was excluded from data analysis on weeks 5 and 6 due to an insecticide application after sampling on week 4.

3. Results and Discussion

Scymnus spp., *C. septempunctata*, *H. convergens*, *H. axyridis*, *C. maculata*, *C. munda*, and *O. v-nigrum* were found in crops and yellow pyramidal traps over both years in Georgia. The corn leaf aphid was observed feeding on sorghum mainly during the vegetative stage, and the greenbug was observed mainly feeding in sorghum grain heads. Cotton aphids were present on cotton for much of the growing season, but they were mainly observed on this crop early in the season (late June-early July).

Scymnus spp. and *C. septempunctata* were the predominant species in sorghum; however, *C. septempunctata* and

TABLE 2: Percentage composition (within columns) of lady beetle species in sorghum trap crops, yellow pyramidal traps, cotton with sorghum trap crops, and control cotton.

Species	Percentage in sorghum trap crops ($n = 1789$)	Percentage in yellow pyramidal traps ($n = 20,313$)	Percentage in cotton w/sorghum trap crops ($n = 4804$)	Percentage in control cotton ($n = 2879$)
Scymnus spp.	51.9	16.5	28.6	43.0
C. septempunctata	33.9	38.1	13.7	14.8
H. convergens	6.3	6.3	31.7	22.6
H. axyridis	5.4	32.1	24.6	18.3
C. maculata	2.0	5.7	1.1	0.9
O. v-nigrum	0.4	0.1	0.2	0.2
C. munda	0.1	1.2	0.1	0.2

H. axyridis were the most abundant coccinellids captured in the yellow pyramidal traps (Table 2). Species composition was similar for cotton with or without (i.e., control) a trap crop with the predominant species being *Scymnus* spp., *C. septempunctata*, *H. convergens*, and *H. axyridis*. The lady beetle species in sorghum and cotton have been previously reported to colonize these crops [4, 13–17].

It was not surprising that yellow pyramidal traps (baited with an aggregation pheromone for *Euschistus* spp. stink bugs) captured adult lady beetles. Captures of lady beetles in this yellow trap, with or without the stink bug pheromone, are common (T.E.C., personal observation), and yellow sticky cards have been used in previous studies to sample adult Coccinellidae [23–27]. The similarity in lady beetle captures in traps and those sampled on sorghum may indicate that the yellow trap itself does not attract lady beetles from significant distances; lady beetle capture in the trap was likely facilitated by the attractiveness of the surrounding sorghum. Nevertheless, modifying the yellow pyramidal traps (intended to attract and kill stink bugs) to reduce lady beetle capture could conserve these predators in sorghum.

In 2006, lady beetle density remained relatively low in flowering and milking sorghum and then peaked during the soft dough stage of seed development (Figure 1). Generally, corn leaf aphids are first observed on sorghum when plants have three to five leaves, and then their numbers increase on vegetative sorghum until declining around the boot or early bloom stages [7]. Greenbugs also are present on sorghum during the three to five leaf stage, but their numbers do not increase until after plants have about 10 leaves, with peak abundance at the half bloom or soft dough stage and then declining as sorghum matures [7]. After week 4, density of adult lady beetles began an overall decline in sorghum, a likely result of prey depletion on sorghum. Apparently, as prey were depleted on sorghum, beetles moved from sorghum to yellow pyramidal traps during week 5 but their capture in these traps dropped precipitously thereafter (Figure 1). Density of lady beetles increased slightly on sorghum during the hard dough stage (i.e., when 75% of the grain dry weight has accumulated) and then declined as sorghum heads matured. In cotton, lady beetles first appeared in relatively low numbers in early July and peaked on cotton in late July-early August. Lady beetle density was significantly

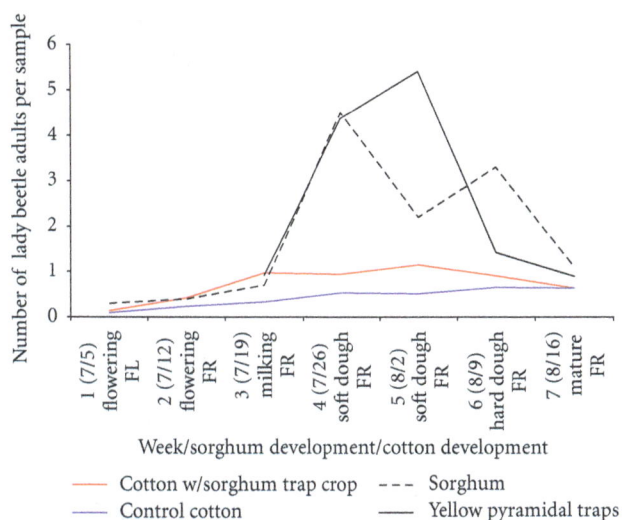

FIGURE 1: Mean number of lady beetle adults per sample in cotton with a sorghum trap crop, control cotton, sorghum, and yellow pyramidal traps in 2006. FL: flowers; FR: fruit. Number of lady beetles in yellow pyramidal traps divided by 10. Date refers to middle of sampling week.

higher on cotton with sorghum trap crops than on control cotton during weeks 2 through 6 (Table 3). Altogether, these results indicate that sorghum was a source of adult lady beetles moving into cotton fields. Because cotton aphids were observed on cotton early in the season, lady beetles were likely responding to populations of cotton aphids in cotton. Aphidophagous lady beetles, though, can be generalist predators; therefore, when they moved into fruiting cotton, they were likely also preying on other pest insects that feed on cotton fruit such as lepidopteran pests and stink bug eggs [28, 29].

In 2007, lady beetle abundance on sorghum followed a similar pattern as seen during 2006. Beetles first moved to flowering sorghum, and density peaked when sorghum heads reached the soft dough stage (Figure 2). Lady beetle density was significantly higher on cotton with sorghum trap crops than on control cotton during weeks 1 through 5 (Table 3). As above, these results suggest that sorghum can serve as a source of lady beetles dispersing to cotton.

TABLE 3: Number (mean ± SE) of lady beetle adults per 1.83 m of row in cotton with sorghum trap crops and control cotton in 2006 and 2007.

Year	Week	Cotton w/sorghum trap crop	Control cotton	$\|t\|$	df	P
	1	0.15 ± 0.02	0.106 ± 0.022	1.45	937	0.1465
	2	0.431 ± 0.039	0.246 ± 0.04	3.2	937	0.0014
	3	0.982 ± 0.076	0.349 ± 0.038	7.47	1132	0.0001
2006	4	0.952 ± 0.075	0.545 ± 0.052	4.51	1132	0.0001
	5	1.153 ± 0.094	0.52 ± 0.057	5.79	1132	0.0001
	6	0.918 ± 0.072	0.66 ± 0.053	2.91	1132	0.0037
	7	0.65 ± 0.059	0.648 ± 0.065	0.02	1327	0.9819
	1	0.611 ± 0.051	0.302 ± 0.039	4.6	503	0.0001
	2	0.447 ± 0.045	0.309 ± 0.036	2.4	536	0.0168
2007	3	0.732 ± 0.058	0.411 ± 0.045	4.39	563	0.0001
	4	0.637 ± 0.059	0.487 ± 0.049	1.97	413	0.049
	5	0.696 ± 0.075	0.479 ± 0.066	2.03	476	0.0432
	6	0.412 ± 0.047	0.338 ± 0.05	1.07	341	0.2845

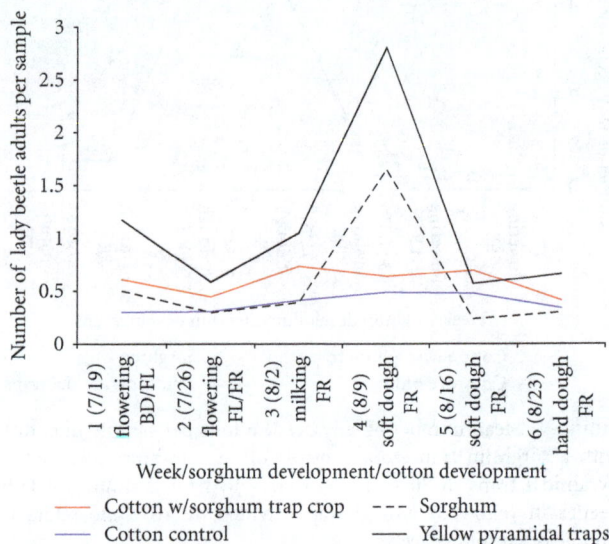

FIGURE 2: Mean number of lady beetle adults per sample in cotton with a sorghum trap crop, control cotton, sorghum, and yellow pyramidal traps in 2007. BD: buds; FL: flowers; FR: fruit. Number of lady beetles in yellow pyramidal traps divided by 10. Date refers to middle of sampling week.

In the current study, results suggest that adult lady beetles dispersed from sorghum into cotton. Previous studies also have demonstrated sorghum as a source of lady beetles moving into cotton. For example, populations of insect predators, including *H. convergens*, increased when feeding on greenbug in grain sorghum fields adjacent to cotton in Arizona [30, 31]. These predators dispersed into cotton as sorghum matured and the greenbug population declined. Cage studies also indicate that adult lady beetles disperse from sorghum into cotton in response to crop phenology and prey abundance [32]. In the current study, lady beetle density similarly declined as sorghum matured. In a study in Texas, as greenbug and corn leaf aphid numbers increased in sorghum, predators, including the predominant *Hippodamia* spp.

predators, also increased [33]. They reported that predator levels in cotton began to increase at about the same time that predator density began to decrease in sorghum indicating that predators dispersed from sorghum into cotton. In fact, fluorescent dust marking demonstrated predator dispersal from sorghum into cotton. In another study using rubidium to mark predators in sorghum and cotton, *H. convergens* and *Scymnus loewii* Mulsant were documented to move from sorghum into cotton [32].

In these previous studies, adult lady beetles dispersed from sorghum into cotton in response to sorghum senescence and prey decline, but in our study, lady beetles continuously moved from sorghum to cotton throughout development of fruit in cotton. Although the reason for lady beetles continuously moving from sorghum into cotton was not determined, it was likely due to resource availability (e.g., prey, pollen, and extrafloral nectaries) in cotton. Perhaps, planting sorghum earlier would result in relaying lady beetles from senescing sorghum into cotton as documented in a 3 yr relay intercropping study in Texas [34]. There, the intercrops acted as a reservoir for predators, including lady beetles, during the noncotton season. These intercrops "relayed" the aphid predators from canola and wheat in the winter to sorghum in the spring and finally to cotton in the summer. Of the intercrop species tested, predator numbers were highest in sorghum. Average aphid abundance was lower in relay intercropped cotton than in isolated cotton, and average predator numbers were higher in relay intercropped cotton than in isolated cotton. Predators appeared in higher numbers earlier in the summer in relay intercropped cotton than in isolated cotton suggesting that this management strategy aids early colonization of predators in cotton, thereby inhibiting increase of the cotton aphid. In a 2 yr study in Texas, a sorghum relay strip-crop system enhanced numbers of predators, including lady beetles, and suppressed cotton aphid abundance in cotton [35]. It can be concluded from these two studies and the current study that incorporating a source crop for lady beetles in a cotton field can be a successful management tactic for control of cotton aphids. Also, a multifunctional habitat of sorghum to detract stink bugs from feeding and ovipositing on cash crops, and using pheromone traps to capture and kill stink bugs has great potential for suppressing stink bugs in cotton while preserving lady beetles.

Lady beetles in this study were present in cotton fields with or without sorghum indicating that these natural enemies disperse into cotton from other plants. Peanut fields were adjacent to all the cotton fields, but early-season host plants such as corn and rye were also prevalent in these agricultural landscapes. Because each of these crops harbors lady beetles [29, 36, 37], they likely contributed lady beetles to sorghum and cotton. Nevertheless, placement of a strip of sorghum along the cotton field edge near peanut enhanced abundance of lady beetles in cotton. Possible explanations for this enhancement include providing newly abundant prey during senescence of corn and rye, providing new or preferred prey for adults developing in peanut, concentrating abundant prey, and thus lady beetles, next to cotton fields. Regardless of the mechanisms involved, a habitat of sorghum can be utilized in conserving biocontrol of these natural enemies in cotton.

Acknowledgments

The authors thank Kristie Graham (USDA, ARS, Crop Protection and Management Research Laboratory, Tifton, GA, USA) and Ann Amis (USDA, ARS, Southeastern Fruit and Tree Nut Research Laboratory) for their technical assistance.

References

[1] T. J. Kring, F. E. Gilstrap, and G. J. Michels Jr., "Role of indigenous coccinellids in regulating greenbugs (Homoptera: Aphididae) on Texas grain sorghum," *Journal of Economic Entomology*, vol. 78, no. 1, pp. 269–273, 1985.

[2] M. E. Rice and G. E. Wilde, "Experimental evaluation of predators and parasitoids in suppressing greenbugs (Homoptera: Aphididae) in sorghum and wheat," *Environmental Entomology*, vol. 17, no. 5, pp. 836–841, 1988.

[3] P. W. Kidd and D. R. Rummel, "Effect of insect predators and a pyrethroid insecticide on cotton aphid, *Aphis gossypii* glover, population density," *Southwestern Entomologist*, vol. 22, no. 4, pp. 381–393, 1997.

[4] M. L. Wells, R. M. McPherson, J. R. Ruberson, and G. A. Herzog, "Coccinellids in cotton: population response to pesticide application and feeding response to cotton aphids (Homoptera: Aphididae)," *Environmental Entomology*, vol. 30, no. 4, pp. 785–793, 2001.

[5] M. R. Abney, J. R. Ruberson, G. A. Herzog, T. J. Kring, D. C. Steinkraus, and P. M. Roberts, "Rise and fall of cotton aphid (Hemiptera: Aphididae) populations in southeastern cotton production systems," *Journal of Economic Entomology*, vol. 101, no. 1, pp. 23–35, 2008.

[6] W. R. Young and G. L. Teetes, "Sorghum entomology," *Annual Review of Entomology*, vol. 22, pp. 193–218, 1977.

[7] T. L. Archer, J. C. ves Losada, and E. D. Bynum Jr., "Influence of planting date on abundance of foliage-feeding insects and mites associated with sorghum," *Journal of Agricultural Entomology*, vol. 7, no. 3, pp. 221–232, 1990.

[8] G. D. Buntin, *Sorghum Insect Pests and their Management, Bulletin 1283*, Cooperative Extension Service, College of Agriculture & Environmental Science, University of Georgia, Athens, Ga, USA, 2009.

[9] P. G. Tillman, "Influence of corn on stink bugs (Heteroptera: Pentatomidae) in subsequent crops," *Environmental Entomology*, vol. 40, no. 5, pp. 1159–1176, 2011.

[10] P. G. Tillman, "Sorghum as a trap crop for *Nezara viridula* L. (Heteroptera: Pentatomidae) in cotton in the Southern United States," *Environmental Entomology*, vol. 35, no. 3, pp. 771–783, 2006.

[11] P. G. Tillman and T. E. Cottrell, "Case study: trap crop with pheromone traps for suppressing *Euschistus servus* (Heteroptera: Pentatomidae) in cotton," *Psyche*. In press.

[12] T. J. Kring and F. E. Gilstrap, "Beneficial role of corn leaf aphid, *Rhopalosiphum maidis* (Fitch) (Homoptera: Aphididae), in maintaining *Hippodamia* spp. (Coleoptera: Coccinellidae) in grain sorghum," *Crop Protection*, vol. 5, no. 2, pp. 125–128, 1986.

[13] G. J. Michels Jr. and R. W. Behle, "Evaluation of sampling methods for lady beetles (Coleoptera: Coccinellidae) in grain sorghum," *Journal of Economic Entomology*, vol. 85, no. 6, pp. 2251–2257, 1992.

[14] G. J. Michels Jr., N. C. Elliott, R. L. Romero, and T. D. Johnson, "Sampling aphidophagous coccinellidae in grain sorghum," *Southwestern Entomologist*, vol. 21, no. 3, pp. 237–246, 1996.

[15] G. J. Michels JR. and J. H. Matis, "Corn leaf aphid, *Rhopalosiphum maidis* (Hemiptera: Aphididae), is a key to greenbug, *Schizaphis graminum* (Hemiptera: Aphididae), biological control in grain sorghum, *Sorghum bicolor*," *European Journal of Entomology*, vol. 105, no. 3, pp. 513–520, 2008.

[16] M. W. Phoofolo, K. L. Giles, and N. C. Elliott, "Effects of relay-intercropping sorghum with winter wheat, alfalfa, and cotton on lady beetle (coleoptera: Coccinellidae) abundance and species composition," *Environmental Entomology*, vol. 39, no. 3, pp. 763–774, 2010.

[17] H. E. Conway and T. J. Kring, "Coccinellids associated with the cotton aphid (Homoptera: Aphididae) in northeast Arkansas cotton," *Journal of Entomological Science*, vol. 45, no. 2, pp. 129–139, 2010.

[18] Auburn University, *Production Guide for Grain Sorghum, ANR-502*, Alabama Cooperative Extension System, Auburn University, Auburn, Ala, USA, 1997.

[19] University of Georgia, *Georgia Cotton Production Guide, CSS-04-01*, Cooperative Extension Service, College of Agriculture & Environmental Science, University of Georgia, Athens, Ga, USA, 2004.

[20] R. F. Mizell and W. L. Tedders, "A new monitoring method for detection of the stink bug complex in pecan orchards," *Proceedings of the Southeastern Pecan Growers Association*, vol. 88, pp. 36–40, 1995.

[21] T. E. Cottrell and D. L. Horton, "Trap capture of brown and dusky stink bugs (Hemiptera: Pentatomidae) as affected by pheromone dosage in dispensers and dispenser source," *Journal of Entomological Science*, vol. 46, no. 2, pp. 135–147, 2011.

[22] SAS Institute, *SAS/STAT User's Guide, Version 9.2*, SAS Institute, Cary, NC, USA, 2008.

[23] K. M. Maredia, S. H. Gage, D. A. Landis, and T. M. Wirth, "Visual response of *Coccinella septempunctata* (L.), *Hippodamia parenthesis* (Say), (Coleoptera: Coccinellidae), and *Chrysoperla carnea* (Stephens), (Neuroptera: Chrysopidae) to colors," *Biological Control*, vol. 2, no. 3, pp. 253–256, 1992.

[24] M. P. Hoffmann, M. S. Orfanedes, L. H. Pedersen, J. J. Kirkwyland, E. R. Hoebeke, and R. Ayyappath, "Survey of lady beetles (Coleoptera: Coccinellidae) in sweet corn using yellow sticky cards," *Journal of Entomological Science*, vol. 32, no. 3, pp. 358–369, 1997.

[25] R. K. Mensah, "Yellow traps can be used to monitor populations of *Coccinella transversalis* (F.) and *Adalia bipunctata* (L.) (Coleoptera: Coccinellidae) in Cotton crops," *Australian Journal of Entomology*, vol. 36, no. 4, pp. 377–381, 1997.

[26] M. Colunga-Garcia and S. H. Gage, "Arrival, establishment and habitat use of the multicolored Asian ladybeetle in a Michigan landscape," *Environmental Entomology*, vol. 27, no. 6, pp. 1574–1580, 1998.

[27] E. J. Stephens and J. E. Losey, "Comparison of sticky cards, visual and sweep sampling of coccinellid populations in Alfalfa," *Environmental Entomology*, vol. 33, no. 3, pp. 535–539, 2004.

[28] W. A. Frank and J. E. Slosser, *An Illustrated Guide to the Predaceous Insects of the Northern Texas Plains*, Texas Agricultural Experiment Station, Publication MP-1718, 1996.

[29] P. G. Tillman, "Natural biological control of stink bug (Heteroptera: Pentatomidae) eggs in corn, peanut, and cotton farmscapes in Georgia," *Environmental Entomology*, vol. 40, no. 2, pp. 303–314, 2011.

[30] R. E. Fye, "Grain sorghum – a source of insect predators for insects on cotton," *Progress on Agriculture in Arizona*, vol. 23, pp. 12–13, 1971.

[31] R. E. Fye and R. L. Carranza, "Movement of insect predators from grain sorghum to cotton," *Environmental Entomology*, vol. 1, no. 6, pp. 790–791, 1972.

[32] J. R. Prasifka, K. M. Heinz, and R. R. Minzenmayer, "Timing, magnitude, rates, and putative causes of predator movement between cotton and grain sorghum fields," *Environmental Entomology*, vol. 33, no. 2, pp. 282–290, 2004.

[33] E. G. Lopez and G. L. Teetes, "Selected predators of aphids in grain sorghum and their relation to cotton," *Journal of Economic Entomology*, vol. 69, no. 2, pp. 198–204, 1976.

[34] M. N. Parajulee, R. Montandon, and J. E. Slosser, "Relay intercropping to enhance abundance of insect predators of cotton aphid (*Aphis gossypii* Glover) in Texas cotton," *International Journal of Pest Management*, vol. 43, no. 3, pp. 227–232, 1997.

[35] M. N. Parajulee and J. E. Slosser, "Evaluation of potential relay strip crops for predator enhancement in Texas cotton," *International Journal of Pest Management*, vol. 45, no. 4, pp. 275–286, 1999.

[36] G. Tillman, H. Schomberg, S. Phatak et al., "Influence of cover crops on insect pests and predators in conservation tillage cotton," *Journal of Economic Entomology*, vol. 97, no. 4, pp. 1217–1232, 2004.

[37] G. Tillman, "Populations of stink bugs (Heteroptera: Pentatomidae) and their natural enemies in peanuts," *Journal of Entomological Science*, vol. 43, no. 2, pp. 191–207, 2008.

14

Effect of Crude Plant Extracts on Mushroom Mite, *Luciaphorus* sp. (Acari: Pygmephoridae)

Prapassorn Bussaman,[1] **Chirayu Sa-uth,**[1]
Paweena Rattanasena,[1] **and Angsumarn Chandrapatya**[2]

[1] *Department of Biotechnology, Faculty of Technology, Mahasarakham University, Maha Sarakham 44150, Thailand*
[2] *Department of Entomology, Faculty of Agriculture, Kasetsart University, Bangkok 10900, Thailand*

Correspondence should be addressed to Prapassorn Bussaman, prapassorn_b@yahoo.com

Academic Editor: Kabkaew Sukontason

The use of plant extracts for controlling agricultural pests has become increasingly popular in the recent years. Mushroom mite, *Luciaphorus* sp., is a destructive pest of several mushroom species and has been reported to cause severe loss of mushroom cultivation in many settings. The efficacies of 23 rhizome and leaf extracts were evaluated against female adults of *Luciaphorus* sp. At 3 days after treatment, the rhizome extracts derived from *Curcuma xanthorrhiza* Roxb. and *Zingiber montanum* (Koenig) Link ex Dietr. were found to have very strong acaricidal activities, resulting in 100% mite mortality, followed by *Curcuma longa* Linn. (98.89%), *Zingiber zerumbet* (L.) Smith. (97.78%), *Kaempferia parviflora* Wall. Ex Baker (88.89%), and *Zingiber officinale* Roscoe. (84.44%). The leaf extracts of *Ocimum sanctum* Linn. and *Melissa officinalis* L. also resulted in 100% mite mortality 3 days after treatment, while the other leaf extracts induced mite mortality only below 70%. The results suggested that rhizome extracts of *C. xanthorrhiza* and *Z. montanum* and leaf extracts of *O. sanctum* and *M. officinalis* have a great potential for future development as natural acaricides for controlling *Luciaphorus* sp.

1. Introduction

Luciaphorus sp. (Acari: Pygmephoridae) is considered as one of the most destructive pests of mushroom cultivation in Thailand. This pygmephorid mite is responsible for the severe production losses of *Lentinus squarrosulus* (Mont.) Singer, *L. polychrous* Lev., *Auricularia auricula-judae* (Bull.:Fr.) Wettst. and *Flammulina velutipes* (Curt.:Fr.) Karst. mushrooms in the Northeast of Thailand [1]. Despite that, little is known about the effective measures for controlling this mite and routine horticultural hygiene is the only procedure to alleviate the problem. To make the situation worse, desperate mushroom growers in Thailand use a large amount of carbamate and organophosphate insecticides and even some harmful solvents to manage this mite; however, this results in very limited success [2].

As a consequence, this mite becomes rapidly resistant and more harmful miticides have to be applied. The use of toxic miticides raises the concerns because of their effects on environments, human safety, and nontarget organisms. Hence, the use of nontoxic natural products for controlling this agricultural pest has been proposed. There are several higher plants that are rich in natural substances, especially the secondary metabolites, such as terpenes, steroids, alkaloids, phenolics, and cardiac glycosides, and can be used as nonharmful, environmentally friendly agents for insect control. Indeed, the use of natural compounds derived from plant extracts has been suggested as alternative treatments for insect and mite controls due to their multiple modes of action, including repellence, feeding and oviposition deterrence, toxicity, and growth regulatory activity [3–6]. Moreover, plant-based pesticides are often found to contain a mixture of active substances which can delay or prevent resistance development [7]. Therefore, in this study, the acaricidal activities of 23 plant extracts were determined against the mushroom mite, *Luciaphorus* sp.

2. Materials and Methods

2.1. Mushroom and Mite Culture. *Lentinus squarrosulus* Mont. mushroom culture was obtained from the Mushroom Growers Society of Thailand. The mycelium was freshly sub cultured on 90 mm plastic Petri dish plates containing potato dextrose agar (PDA, Sigma) and grown at 25°C.

Luciaphorus sp. mites were collected from infested *L. squarrosulus* composts obtained from Rapeephan mushroom farm in Khon Kaen province in the Northeast of Thailand. A pair of male and female mites was maintained at 28°C using fresh *L. squarrosulus* spawn that was grown on sawdust and sorghum grains in a glass bottle. The offspring that were in-house bred inside this glass bottle were used for all the experiments.

2.2. Preparation of Plant Extracts. Leaves and rhizomes of 23 plants were collected locally from Mahasarakham province in the Northeast of Thailand (Table 1). Plant materials were cut into small pieces and dried in hot air oven at 45°C for 3 days.

The dried plants were separately ground into powders using a small grinder and stored at 4°C in polypropylene bags. For extraction, 100 g of each powdered plant materials and 300 mL of 80% ethyl alcohol were added into sterile 2L Erlenmeyer flask, and the flask was agitated at 100 rpm for 24 h. After filtering through a Buchner funnel and Whatman No. 1 filter paper, the extracts were concentrated under low pressure using rotary evaporator. The crude extracts were reconstituted to have the concentration of 20% (w/v) using 80% ethyl alcohol (v/v, in distilled water) and stored at 4°C in glass vials to be used as stock plant extracts. For the tests, these stock plant extracts were dissolved in distilled water containing 0.05% Tween 80 to have the concentration of 5% (w/v).

2.3. Bioassay. For evaluation of each plant extract, 100 adult female mites were transferred to a 50 mm Petri dish plate containing mushroom mycelial culture grown on PDA medium, and the plate was then sprayed with 500 μL of each plant extracts prepared at the concentration of 5% (w/v). The same volumes of the sterilized distilled water (DW) and 0.005% Omite (commercial miticide) were used as control groups. The experiments for each plant extracts were performed in triplicates. All plates were incubated in the growth chamber at 28°C and 85% RH in the dark. The mortality of mites was recorded every day for 5 days after application with plant extracts.

2.4. Statistical Analysis. Data on the percentages of mite mortality due to application with plant extracts were arcsine-transformed and subjected to analysis of variance using the general linear models procedure (SAS Institute, Cary, NC, USA). Significant differences between the treatments were determined using the LSD test at $P < 0.05$.

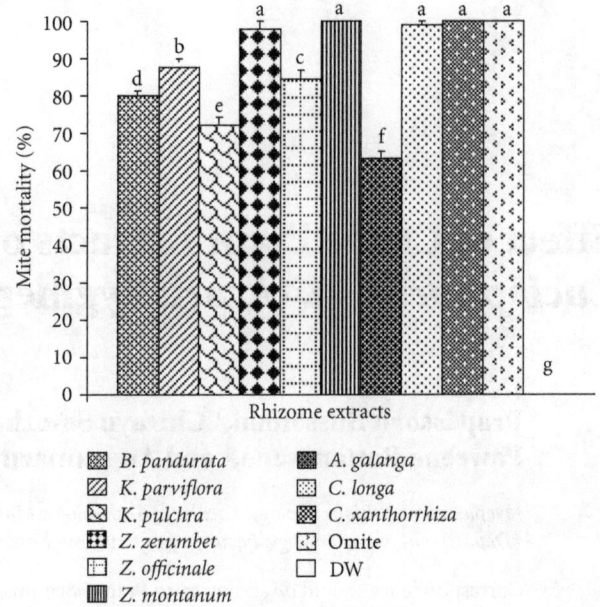

FIGURE 1: The mortality rates of adult female *Luciaphorus* sp. after being treated with 5% rhizome extracts at 3 days after application. Bars (mean ± SE) with same letter(s) are not significantly different as determined by LSD test at $P < 0.05$.

3. Results

3.1. Acaricidal Activities of Rhizome Extracts. In this study, all rhizome extracts were shown to have acaricidal activities against *Luciaphorus* sp., and the percentages of mite mortality increased progressively and reached the plateau after 3 days of applications (Figure 1). On day 3, the significantly high levels of mortality rates were caused by the rhizome extracts of *C. xanthorrhiza* (100%), *Z. montanum* (100%), *C. longa* (98.89%), and *Z. zerumbet* (97.78%), followed by *K. parviflora* (88.89%), *Z. officinale* (84.44%), *B. pandurata* (80.00%), *K. pulchra* (72.22%), and *A. galanga* (63.33%) (Figure 1). Interestingly, on day 1, *K. parviflora*, *Z. officinale*, *C. longa*, and *C. xanthorrhiza* extracts resulted in mortality rates at over 70% which were significantly higher than the other treatments (data not shown). However, on day 2, mite mortality rates in almost all treatments were over 70% with the exception of *A. galanga* (56.67%) and *K. pulchra* (67.78%) (data not shown).

3.2. Acaricidal Activity of Leaf Extracts. The levels of mite mortality after applications with leaf extracts also reached maximum on day 3 (Figure 2). On day 3, the leaf extracts of *O. sanctum* and *M. officinalis* resulted in maximum mortality (100%), but the other treatments were shown to result in mortality at levels below 70% (Figure 2). This was not unexpected because only the applications with the leaf extracts of *O. sanctum* and *M. officinalis* caused over 70% of mortality on day 1 (data not shown). Also, on day 2, mortality rates in all treatments increased and the leaf extracts of *O. sanctum* and *M. officinalis* still resulted in mite mortality at the levels significantly higher than the rest,

TABLE 1: Plants and their parts used for evaluation of acaricidal activities against *Luciaphorus* sp.

Scientific name	Family	Common name	Parts
Boesenbergia pandurata (Roxb) Schltr.	Zingiberaceae	Fingerroot	Rhizome
Kaempferia parviflora Wall. Ex Baker	Zingiberaceae	Belamcanda chinensis	Rhizome
Kaempferia pulchra (Ridl.) Ridl.	Zingiberaceae	Peacock ginger, resurrection lily	Rhizome
Zingiber zerumbet (L.) Smith.	Zingiberaceae	Wild ginger, Martinique ginger	Rhizome
Zingiber officinale Roscoe.	Zingiberaceae	Ginger	Rhizome
Zingiber montanum (Koenig) Link ex Dietr.	Zingiberaceae	Phlai, cassumunar	Rhizome
Alpinia galanga (L.) Swartz.	Zingiberaceae	Kha, galingale, galanga	Rhizome
Curcuma longa Linn.	Zingiberaceae	Turmeric	Rhizome
Curcuma xanthorrhiza Roxb.	Zingiberaceae	Curcuma	Rhizome
Cymbopogon citratus Stapf.	Gramineae	Takhrai, lemongrass	Leaf
Citrus hystrix DC.	Rutaceae	Leech lime	Leaf
Ocimum basilicum Linn.	Labiatae	Ho-ra-pa, sweet-basil, common basil	Leaf
Ocimum canum Linn.	Labiatae	Hairy basil	Leaf
Ocimum sanctum Linn.	Malvaceae	Holy basil, sacred basil	Leaf
Moringa oleifera Lam.	Moringaceae	Horse radish tree	Leaf
Annona squamosa Linn.	Annonaceae	Sugar apple	Leaf
Psidium guajava Linn.	Myrtaceae	Guava	Leaf
Eucalyptus camaldulensis Dehnh.	Myrtaceae	Red river gum, Murray red gum, red gum	Leaf
Artocarpus heterophyllus Lam.	Moraceae	Jackfruit tree	Leaf
Piper sarmentosum Roxb. Ex Hunter.	Piperaceae	Cha-plu	Leaf
Murraya paniculata (L.) Jack.	Rutaceae	Orange jessamine, satin-wood,	Leaf
Melissa officinalis L.	Lamiaceae	Kitchen mint, marsh mint	Leaf
Cassia siamea (Lam.) Irwin et Barnaby	Fabaceae	Kassod tree, siamese senna, Thai copperpod, siamese cassia	Leaf

accounting for 97.78% and 94.44%, respectively (data not shown).

4. Discussion

Several plants have been found to contain bioactive compounds with a variety of biological actions against insects and mites, including repellent, antifeedant, anti-ovipositional, toxic, chemosterilant, and growth regulatory activities [4, 8]. Therefore, botanical insecticides have long been recommended as attractive alternatives to synthetic chemical insecticides for pest management because these chemicals pose little threat to the environment or to human health [9]. For example, the crude foliar extracts of five subfamilies of Australian Lamiaceae, including Ajugoideae, Scutellarioideae, Chloanthoideae, Viticoideae, and Nepetoideae, were found to have contact toxicity against the polyphagous mite (*Tetranychus urticae* Koch) [10]. This *T. urticae* could also be inhibited by the essential oil in crude foliar extract of sandalwood (*Santalum austrocaledonicum*), resulting in 87.2 ± 2.9% mortality and 89.3% reduction of the total number of eggs on leaf disks treated with this oil [11]. Piperoctadecalidine, which is the alkaloid isolated from *Piper longum* Linn., was also found to have activities against *T. urticae* at LD$_{50}$ of 246 ppm [12]. Moreover, Aslan et al. [13] reported that essential oil vapours from *Satureja hortensis* Linn., *Ocimum basilicum* Linn, and *Thymus vulgaris* Linn. had potential against *T. urticae*, but the essential oil obtained from *S. hortensis* was the most effective at 1.563 µL/L air

dose by causing 100% mortality of *T. urticae* after 4 days of treatment.

In recent years, many studies have also been conducted to investigate the activities of plant extracts or essential oils against carmine spider mite (*Tetranychus cinnabarinus* Boisd. Tunc) and Hawthorn red spider mite (*Tetranychus viennensis* Zacher). The chloroform extract of *Kochia scoparia* Linn. was shown to have rapid acaricidal activities against *T. urticae*, *T. cinnabarinus*, and *T. viennensis*, resulting in the highest mortality at 92.58, 88.88, and 84.47%, respectively, within 24 h after treatment [14]. Also, toxicity against *T. cinnabarinus* and *T. viennensis* could be quickly induced by the petroleum ether extract of *Juglans regia* Linn., resulting in mortality rates at 81.58 and 78.58%, respectively, within 24 h [7].

Furthermore, the complete 100% mortality of *T. cinnabarinus* was found to be induced by the essential oils of *Cuminum cyminum* Linn., *Pimpinella anisum* Linn., and *Origanum syriacum* var. *bevaii* (Holmes) as fumigants in greenhouse experiments [15]. This complete mortality could also be produced by using the acetone parallel extract of *Artemisia annua* Linn. leaves collected in July [16]. In addition, Zhang et al. [17] reported that benzene extracts derived from *C. longa* Linn. had LC$_{50}$ against *T. cinnabarinus* at 99.3 ppm after 72 h. The high mortality rates of *T. cinnabarinus* could be induced by methanol extracts of *Gliricidia sepium* (Jacq) Kunth ex Steud. (100%) and *Lippia origanoides* Kunth (96.6%) when used at the concentration of 20% [18]. Additionally, Sertkaya et al. [8] evaluated the

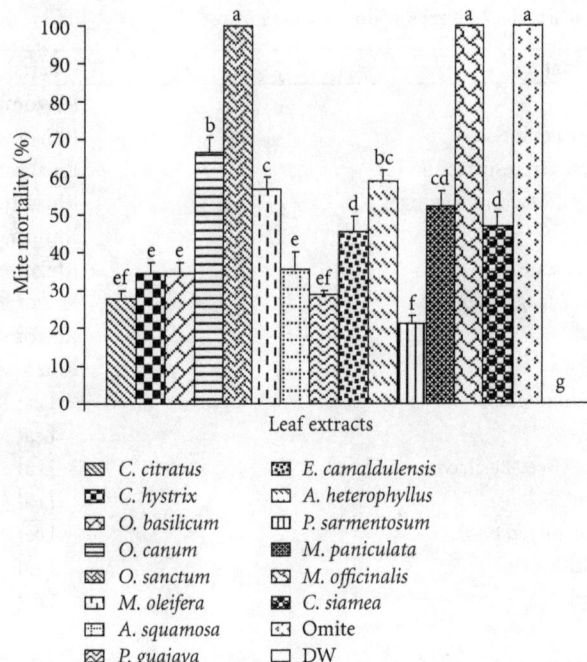

FIGURE 2: The mortality rates of adult female *Luciaphorus* sp. after being treated with 5% leaf extracts at 3 days after application. Bars (mean ± SE) with same letter(s) are not significantly different as determined by LSD test at *P* < 0.05.

efficacy of essential oils derived from medicinal plants against *T. cinnabarinus* and showed that thyme (*Thymbra spicata* Linn. subsp. *spicata*), oregano (*Origanum onites* Linn.), mint (*Mentha spicata* Linn.), and lavender (*Lavandula stoechas* Linn. subsp. *stoechas*) essential oils had LC$_{50}$ values of 0.53, 0.69, 1.83, and 2.92 ppm, respectively. Moreover, the acetone extract of *Aloe vera* Linn. leaves was shown to have acaricidal activity against female *T. cinnabarinus* at 3 days after treatment with LC$_{50}$ value of 90 ppm [6].

Other insect pests were also found to be inhibited by plant extracts. According to the results of Liu et al. [19], the ethanol extracts of *Eupatorium adenophorum* Spreng. (0.1% w/v) could cause mortality of citrus red mite (*Panonychus citri* (McGregor)) at 71.10 and 73.53% after 12 and 24 h, respectively. Also, the activities against *P. citri* of the ethanol extracts derived from *Boenninghausenia sessilicarpa* H. Lev., *Laggera pterodonta* (DC.) Benth., *Humulus scandens* (Lour) Merr., and *Rabdosia* were reported with LC$_{50}$ values of 0.9241, 0.9827, 0.9905, and 1.0196 mg/mL, respectively [20]. In addition, applications with aqueous extracts of *Acorus calamus* Linn., *Xanthium strumarium* Linn., *Polygonum hydropiper* Linn., and *Clerodendron infortunatum* (Gaertn.) could lead to more than 50% mortality of *Oligonychus coffeae* (Nietner) [21]. Moreover, 3% methanolic extracts of *Ocimum tenuiflorum* Linn. and *Cassia alata* Linn. exhibited acaricidal activities against *Tetranychus neocaledonicus* Andre. and resulted in the mortality at 93.3 and 97.0%, respectively [22]. On the other hand, 3% aqueous extracts of *C. alata* and *O. tenuiflorum* could lead to mortality of *T. neocaledonicus* at 75% and 82.2%, respectively, after exposure

for 3 days. In addition, the volatile oils of *Citrus reticulata* Blanco. and *C. longa* Linn. could cause mortality of *Sitophilus oryzae* Linn. as high as 100 and 90%, respectively [23]. The essential oils of *Ocimum basilicum* Linn., *Coriandrum sativum* Linn., *Eucalyptus globulus* Labill, *Mentha piperita* Linn. and *Satureja hortensis* Linn. were toxic against poultry red mite (*Dermanyssus gallinae* (De Geer)), and, when using the *in vitro* direct contact method, these essential oils at the dose of 0.6 mg/cm could result in mortality rates over 80% after 24 h of contract [24]. Furthermore, *Eucalyptus citriodora* Hook extract was found to be effective against *D. gallinae*, resulting in 85% mortality over a 24 h exposure period in contact toxicity tests [25].

In this study, the rhizome extracts of *C. xanthorrhiza* and *Z. montanum* and the leaf extracts of *O. sanctum* and *M. officinalis* at the dose of 5% (w/v) were found to be highly effective against female adults of *Luciaphorus* sp. The results revealed that the rhizome extracts were likely to have more potent acaricidal activities than those derived from leaves. The acaricidal activities of plant extracts against *Luciaphorus* sp. mites have been previously described. The essential oils derived from lemon grass (*Cymbopogon citratus* Stapf.) and citronella grass (*Cymbopogon nardus* Rendle) were shown to be effective against *Luciaphorus perniciosus* Rack., and the median effective concentration (EC$_{50}$) was 18.15 and 19.66 ppm, respectively [26]. In addition, the essential oils of *Litsea cubeba* Pers. were effective against *L. perniciosus* by contact and fumigation methods with LD$_{50}$ values equivalent to 0.932 and 0.166 ppm, respectively [27].

In conclusion, the results in this study suggest the possibility of developing plant extracts derived from the rhizomes of *C. xanthorrhiza* and *Z. montanum* and the leaves of *O. sanctum* and *M. officinalis* for controlling *Luciaphorus* mites. The effective concentration and mode of action of these plant extracts against *Luciaphorus* sp. remain to be determined for the future development of highly potent products to be used in the real settings.

Acknowledgments

This work was financially supported by the Thailand Research Fund under Grant number RTA 4880006 and Mahasarakham University. The authors also thank the Department of Biotechnology, Faculty of Technology, Mahasarakham University, Thailand, for providing laboratory equipments and facility.

References

[1] P. Bussaman, A. Chandrapatya, R. W. Sermswan, and P. S. Grewal, "Morphology, biology and behavior of the genus *Pygmephorus* (Acari: Heterostigmata), a new parasite of economic edible mushroom," in *Proceedings of the 22th International Congress of Entomology*, Brisbane, Australia, 2004.

[2] P. Bussaman, R. W. Sermswan, and P. S. Grewal, "Toxicity of the entomopathogenic bacteria *Photorhabdus* and *Xenorhabdus* to the mushroom mite (Luciaphorus sp.; Acari: Pygmephoridae)," *Biocontrol Science and Technology*, vol. 16, no. 3, pp. 245–256, 2006.

[3] R. Jbilou, A. Ennabili, and F. Sayah, "Insecticidal activity of four medicinal plant extracts against *Tribolium castaneum* (Herbst) (Coleoptera: Tenebrionidae)," *African Journal of Biotechnology*, vol. 5, no. 10, pp. 936–940, 2006.

[4] R. N. Singh and B. Saratchandra, "The development of botanical products with special reference to seri-ecosystem," *Caspian Journal of Environmental Science*, vol. 3, no. 1, pp. 1–8, 2005.

[5] A. Gokce, L. L. Stelinski, M. E. Whalon, and L. J. Gut, "Toxicity and antifeedant activity of selected plant extracts against larval obliquebanded leafroller, *Choristoneura rosaceana* (Harris)," *The Open Entomology Journal*, vol. 3, pp. 30–36, 2009.

[6] J. Wei, W. Ding, Y. G. Zhao, and P. Vanichpakorn, "Acaricidal activity of *Aloe vera* L. leaf extracts against *Tetranychus cinnabarinus* (Boisduval) (Acarina: Tetranychidae)," *Journal of Asia-Pacific Entomology*, vol. 14, no. 3, pp. 353–356, 2011.

[7] Y. N. Wang, G. L. Shi, L. L. Zhao et al., "Acaricidal activity of juglans regia leaf extracts on *Tetranychus viennensis* and *Tetranychus cinnabarinus* (Acaris Tetranychidae)," *Journal of Economic Entomology*, vol. 100, no. 4, pp. 1298–1303, 2007.

[8] E. Sertkaya, K. Kaya, and S. Soylu, "Acaricidal activities of the essential oils from several medicinal plants against the carmine spider mite (*Tetranychus cinnabarinus* Boisd.) (Acarina: Tetranychidae)," *Industrial Crops and Products*, vol. 31, no. 1, pp. 107–112, 2010.

[9] M. B. Isman, "Botanical insecticides, deterrents, and repellents in modern agriculture and an increasingly regulated world," *Annual Review of Entomology*, vol. 51, pp. 45–66, 2006.

[10] H. L. Rasikari, D. N. Leach, P. G. Waterman et al., "Acaricidal and cytotoxic activities of extracts from selected genera of Australian lamiaceae," *Journal of Economic Entomology*, vol. 98, no. 4, pp. 1259–1266, 2005.

[11] H. S. Roh, E. G. Lim, J. Kim, and C. G. Park, "Acaricidal and oviposition deterring effects of santalol identified in sandalwood oil against two-spotted spider mite, *Tetranychus urticae* Koch (Acari: Tetranychidae)," *Journal of Pest Science*, vol. 84, no. 4, pp. 495–501, 2011.

[12] B. S. Park, S. E. Lee, W. S. Choi, C. Y. Jeong, C. Song, and K. Y. Cho, "Insecticidal and acaricidal activity of pipernonaline and piperoctadecalidine derived from dried fruits of *Piper longum* L.," *Crop Protection*, vol. 21, no. 3, pp. 249–251, 2002.

[13] I. Aslan, H. Özbek, Ö. Çalmasur, and F. Sahln, "Toxicity of essential oil vapours to two greenhouse pests, *Tetranychus urticae* Koch and *Bemisia tabaci* Genn.," *Industrial Crops and Products*, vol. 19, no. 2, pp. 167–173, 2004.

[14] G. L. Shi, L. L. Zhao, S. Q. Liu, H. Cao, S. R. Clarke, and J. H. Sun, "Acaricidal activities of extracts of *Kochia scoparia* against *Tetranychus urticae*, *Tetranychus cinnabarinus*, and *Tetranychus viennensis* (Acari: Tetranychidae)," *Journal of Economic Entomology*, vol. 99, no. 3, pp. 858–863, 2006.

[15] I. Tunc and S. Sahinkaya, "Sensitivity of two greenhouse pests to vapours of essential oils," *Entomologia Experimentalis et Applicata*, vol. 86, no. 2, pp. 183–187, 1998.

[16] Y. Q. Zhang, W. Ding, Z. M. Zhao, J. Wu, and Y. H. Fan, "Studies on acaricidal bioactivities of *Artemisia annua* L. extracts against *Tetranychus cinnabarinus* Bois. (Acari: Tetranychidae)," *Agricultural Sciences in China*, vol. 7, no. 5, pp. 577–584, 2008.

[17] Y. Q. Zhang, W. Ding, Z. M. Zhao, J. J. Wang, and H. J. Liao, "Research on acaricidal bioactivities of turmeric, *Curcuma longa*," *Acta Phytophylacica Sinica*, vol. 31, pp. 390–394, 2004.

[18] A. Sivira, M. E. Sanabria, N. Valera, and C. Vásquez, "Toxicity of ethanolic extracts from *Lippia origanoides* and *Gliricidia sepium* to *Tetranychus cinnabarinus* (Boisduval) (Acari: Tetranychidae)," *Neotropical Entomology*, vol. 40, no. 3, pp. 375–379, 2011.

[19] Y. P. Liu, P. Gao, W. G. Pan, F. Y. Xu, and S. G. Liu, "Effect of several plant extracts on *Tetranychus urticae* and *Panoychus cotri*," *Journal of Sichuan University*, vol. 41, pp. 212–215, 2004.

[20] H. Z. Yang, J. H. Hu, Q. Li et al., "Primary studies of acarocidal activity of twenty plants extracts against *Panonychus citri*," *Southwest China Journal of Agricultural Sciences*, vol. 20, no. 5, pp. 1012–1015, 2007.

[21] M. Sarmah, A. Rahman, A. K. Phukan, and G. Gurusubramanian, "Effect of aqueous plant extracts on tea red spider mite, *Oligonychus coffeae*, Nietner (tetranychidae: acarina) and *Stethorus gilvifrons* mulsant," *African Journal of Biotechnology*, vol. 8, no. 3, pp. 417–423, 2009.

[22] I. Roy, G. Aditya, and G. K. Saha, "Preliminary assessment of selected botanicals in the control of *Tetranychus neocaledonicus* Andre' (Acari: Tetranychidae)," *Proceedings of Zoological Society*, vol. 64, no. 2, pp. 124–127, 2011.

[23] B. Chayengia, P. Patgiri, Z. Rahman, and S. Sarma, "Efficacy of different plant products against *Sitophilus oryzae* (Linn.) (Coleoptera: Curculionidae) infestation on stored rice," *Journal of Biopesticides*, vol. 3, no. 3, pp. 604–609, 2010.

[24] C. Magdaş, M. Cernea, H. Baciu, and E. Şuteu, "Acaricidal effect of eleven essential oils against the poultry red mite *Dermanyssus gallinae* (Acari: Dermanyssidae)," *Scientica Parasitologica*, vol. 11, no. 2, pp. 71–75, 2010.

[25] D. R. George, D. Masic, O. A. E. Sparagano, and J. H. Guy, "Variation in chemical composition and acaricidal activity against *Dermanyssus gallinae* of four eucalyptus essential oils," *Experimental and Applied Acarology*, vol. 48, no. 1-2, pp. 43–50, 2009.

[26] J. Pumnuan, A. Insung, and P. Rongpol, "Effectiveness of medical plant essential oils on pregnant female of *Luciaphorus perniciosus* Rack. (Acari: Pygmephoridae)," *Asian Journal of Food and Agro-Industry*, vol. 2, pp. S410–S414, 2009.

[27] J. Pumnuan, A. Chandrapatya, and A. Insung, "Acaricidal activities of plant essential oils from three plants on the mushroom mite, *Luciaphorus perniciosus* Rack (Acari: Pygmephoridae)," *Pakistan Journal of Zoology*, vol. 42, no. 3, pp. 247–252, 2010.

15

Spider-Ant Associations: An Updated Review of Myrmecomorphy, Myrmecophily, and Myrmecophagy in Spiders

Paula E. Cushing

Department of Zoology, Denver Museum of Nature & Science, 2001 Colorado Boulevard, Denver, CO 80205, USA

Correspondence should be addressed to Paula E. Cushing, paula.cushing@dmns.org

Academic Editor: Jean Paul Lachaud

This paper provides a summary of the extensive theoretical and empirical work that has been carried out in recent years testing the adaptational significance of various spider-ant associations. Hundreds of species of spiders have evolved close relationships with ants and can be classified as myrmecomorphs, myrmecophiles, or myrmecophages. Myrmecomorphs are Batesian mimics. Their close morphological and behavioral resemblance to ants confers strong survival advantages against visually hunting predators. Some species of spiders have become integrated into the ant society as myrmecophiles or symbionts. These spider myrmecophiles gain protection against their own predators, live in an environment with a stable climate, and are typically surrounded by abundant food resources. The adaptations by which this integration is made possible are poorly known, although it is hypothesized that most spider myrmecophiles are chemical mimics and some are even phoretic on their hosts. The third type of spider-ant association discussed is myrmecophagy—or predatory specialization on ants. A table of known spider myrmecophages is provided as is information on their biology and hunting strategies. Myrmecophagy provides these predators with an essentially unlimited food supply and may even confer other protections to the spiders.

1. Introduction

The majority of spiders are solitary generalist predators of insects [1]. Most spiders, as with most arthropod predators, are averse to ant predation because ants are generally aggressive, some are venomous, and most are simply noxious for a variety of reasons [2]. Nevertheless, hundreds of arthropod species live in some level of proximity or association with ants [3–5]. The present paper supplements a review I published in 1997 [5] identifying and describing the biology of spiders that are found in association with ants. In the earlier article, I summarized what was then known about the biology and identities of ant-mimicking, or myrmecomorphic, spiders as well as spiders living in close proximity to or living within ant colonies, known as myrmecophiles. That review included tables listing known spider myrmecomorphs and myrmecophiles. The purpose of the present paper is not to replicate information contained in the 1997 article but, instead, to provide a summary of the extensive theoretical and empirical work that has been carried out in recent years testing the adaptational significance of the various spider-ant associations. Additionally, I summarize instances of a

different kind of spider-ant association—that of predator-prey relationships, or myrmecophagy—and provide a table of known species of spiders that feed on or specialize on ants.

2. Spider Myrmecomorphy

2.1. Morphological and Behavioral Adaptations. Morphological adaptations conferring mimetic resemblance to ants include color pattern similarities as well as more dramatic morphological changes such as abdominal constrictions and/or constriction of the cephalothorax, both of which give the illusion that the spider has more than two body parts [5–7] (Figures 1(a) and 1(b)). One recent paper demonstrated that some of these morphological adaptations may be synapomorphic for lineages [8], suggesting that at least some of the morphological adaptations associated with myrmecomorphy may be under phylogenetic constraint. Additional morphological adaptations seen in some spider myrmecomorphs include enlargement of the chelicerae or enlargement or other adaptations associated with the pedipalps or first legs. For example, males of some species of salticids in

FIGURE 1: Myrmecomorphy in spiders. (a) The model ant *Pseudomyrmex simplex* (Smith) and its mimic, (b) *Synemosyna petrunkevitchi* (Chapin) (Salticidae). Photos © Lyn Atherton, used by permission. (c) *Myrmarachne formicaria* (De Geer) (Salticidae) showing the enlarged chelicerae of the male. Photo © Jay Cossey/PhotographsFromNature.com, used by permission. Scale bars = 1 mm.

the genus *Myrmarachne* have greatly enlarged chelicerae that extend anteriorly [9] (Figure 1(c)). These large chelicerae are thought to have evolved via sexual selection [10]. Recent research demonstrated that male *Myrmarachne* with enlarged chelicerae mimic encumbered ants (worker ants carrying items in their mandibles) [11, 12]. In the myrmecomorphic species in the family Corinnidae, *Pranburia mahannopi* Deeleman-Reinhold, the first pair of legs of males has a thick brush of setae around the distal part of the femora. When the spider is disturbed or alarmed, it brings the femora together and the brushes give the illusion of an ant head (i.e., the spider behaviorally and morphologically acquires a third body part [13]).

Spider myrmecomorphs resemble the model ants to varying degrees of accuracy. Some myrmecomorphs are, at least to the human observer, nearly perfect mimics; others generally resemble ants but no specific model species in the vicinity of the spider can be identified. The latter are termed "imperfect" or "inaccurate" mimics [14, 15]. Some species of myrmecomorphic spiders are polymorphic mimics, mimicking multiple species of ants found in the habitat (see [5, Table 1], and [9, 16–20]). One species of jumping spider (Salticidae), *Myrmarachne bakeri* Banks, is polymorphic in color patterns and individual spiders can even change patterns during the course their lives, even after molting to maturity [20]. Individuals can change their patterns even under constant environmental conditions and feeding regime [20]. Other myrmecomorphs are transformational mimics, mimicking different species of ants during their different developmental stages (see [5, Table 1], and [9, 16–18, 21]).

In addition to morphological resemblance to ants, most spider myrmecomorphs are also behavioral mimics (see citations in [5]). This behavioral mimicry includes erratic movement, much more akin to the movement of ants than the movement of spiders, and lifting the first or second pair of legs when moving through the environment as an antennal illusion [22]. Myrmecomorphic salticid spiders also hunt

their prey by lunging at and sometimes tapping the prey rather than by leaping on it as is common in most non-mimetic salticids [11, 22–24]. In other words, these spiders maintain their resemblance to ants even when hunting.

2.2. General Adaptive Significance of Myrmecomorphy. Myrmecomorphy has long been hypothesized to be an example of Batesian mimicry, conferring an adaptive advantage to the mimics against visually hunting arthropod predators that have either an innate or learned aversion to ants. Several studies have provided strong support for this hypothesis, demonstrating that myrmecomorphic spiders are less likely to be chosen as prey by visually hunting predators that would otherwise readily accept spiders [25–32]. In several of these studies, the predators used are naïve and have never encountered ants before, demonstrating that aversion to ants, at least in some arthropod predators, is innate rather than learned [27, 28, 30]. In order for myrmecomorphy to provide an adaptive advantage to the mimics, the mimics must live in close proximity to the models [33–38]. In addition, mimics should be rarer than models [15, 34, 36, 39, 40].

However, myrmecomorphic spiders, particularly those in the salticid genus *Myrmarachne*, often live in high concentrations within a given area. For example, *Myrmarachne melanotarsa* Wesolowska and Salm lives in aggregated groups in which their silken nest complexes are in close association with nests of their model ant, *Crematogaster* sp. [24]. Since ants live in often very large colonies, it has been hypothesized that aggregations of myrmecomorphs are an example of "collective mimicry" in which the myrmecomorphic spiders are, by living in aggregated groups, mimicking the colonial aspects of the models. Groups of mimics may be perceived by predators as more aversive than single individuals found in the habitat [24, 29]. A counter to this hypothesis is that the mimic may therefore outnumber the model in small areas of the habitat, making it more likely that predators will sample and learn the patterns of the palatable mimics and making Batesian mimicry less effective [34]. In some visually hunting spider predators, such as the wasp *Pison xanthopus* (Brulle) (Sphecidae), individuals can develop search images for myrmecomorphic spiders and stock proportionally more mimics in their mud cells than would be expected if the wasp was randomly hunting spiders in the environment [41]. Therefore, some predators are capable of learning to search for myrmecomorphs. However, in a study of the mud-dauber *Sceliphron spirifex* (L.), Jocqué found no myrmecomorphic spiders among almost 600 spiders removed from mud nests, despite *Myrmarachne* species being common in the habitat suggesting that, at least for this wasp, ant mimicry does provide protection from visually hunting predators [42].

Yet, it has been pointed out that mimics can still confer protection against predators even when they are more abundant than the model if certain conditions exist: (1) if the model is very noxious, then the predators will avoid good mimics regardless of the relative proportions of models and mimics; (2) if the mimic has low nutritional value and is, therefore, not worth pursuing; (3) if very profitable alternative prey are present in which case the predator will avoid both model and mimic regardless of the relative abundance of each; or (4) if the relative perception of abundance is different, for example, if the predator perceives the model as more abundant than the mimic (perhaps because of the higher activity levels of the models) [37].

2.3. Evolution of Polymorphic Mimicry. In recent years, researchers have explored the adaptive basis and the conditions under which polymorphic mimicry might arise. Theoretically, a mimic species should converge on mimetic resemblance of the single model species found in that habitat, particularly for predators that learn to avoid the model [37]. Yet many instances of polymorphic mimicry among spider myrmecomorphs have been documented (see citations in Section 2.1). Several hypotheses have been proposed to explain the existence of polymorphic mimics. For example, Ceccarelli and Crozier [43] suggested that the evolutionary rates between different morphs of the salticid *Myrmarachne* and their presumed models differ [43]. These authors demonstrated that morphs of the mimics radiated rapidly leading to higher degrees of polymorphism and provided evidence of possible sympatric speciation. *Myrmarachne plataleoides* (O. P.-Cambridge) mimics the weaver ant *Oecophylla smaragdina* (Fabricius). Borges et al. [19] showed that the different color morphs of *M. plataleoides* may mimic different models in the habitat besides *O. smaragdina*. Males of each color morph showed greatest interest in the silk retreats of females of their own color morph. Disruptive selection may be maintaining the polymorphism in this population [19]. In addition, it has been proposed that polymorphic mimicry, in essence, provides a "moving target" for template learning among visually hunting predators that learn to avoid aversive prey [44]. Nelson [44] proposed that polymorphism in a myrmecomorphic species reduces the apparent number of mimics per model. Therefore, predators cannot easily distinguish palatable mimics from the unpalatable models

because the characteristics of the prey are continuously changing. The new mimetic form will be advantageous since it is rare, but if this morph increases too much in frequency within the habitat, it may lose its mimetic protection and be selected against [37]. This selective process itself may generate selection for new morphs [37].

Sexual dimorphism can be considered a type of polymorphism. In many cases of sexually dimorphic spider myrmecomorphs, the male is more mimetic than the female, such as in species of the Corinnidae genus *Castianeira* and the Oonopidae genus *Antoonops* [13, 45]. Such sexual dimorphism may be adaptive if the sexes are different in ecology and are thus exposed to different predation pressures and selective forces [46]. Joron [46] provides a model supporting this mode of evolution and selection for sexual dimorphism among mimetic species. Although mimics gain protection from the resemblance to noxious species, they are often more conspicuous in their color markings than related species that have evolved cryptic coloration. Thus conspicuousness can be considered a cost of Batesian mimicry [47]. A palatable species may be evolutionarily maximizing its level of protection for the smallest cost (in terms of conspicuousness) and this evolutionary balancing act may lead to sexual dimorphism in which the more active sex (which, in spiders, is typically the males) evolves mimetic resemblance to noxious models whereas the other sex remains relatively more concealed and camouflaged behaviorally and morphologically [47].

In some species of the salticid genus *Myrmarachne*, the males and females are both mimetic but the males have extraordinarily long chelicerae. This sex mimics ants carrying an object in their mandibles [11, 12]. The large chelicerae of males are thought to have evolved via sexual selection [10]. These large chelicerae are an encumbrance to males during prey capture; however, they make males much more efficient than females in breaking into other spiders' silken retreats and feeding on eggs or juveniles [10]. Consequently, in this case of sexual dimorphism, both sexes have maintained mimetic resemblance to the models, although the male is mimicking a slightly different type of model ant (an encumbered ant). Any costs incurred from the dimorphism may be outweighed by benefits in opening up a different trophic niche for the males (oophagy).

2.4. Evolution of Imperfect or Inaccurate Mimicry. It is well documented that many mimics are imperfect in their mimetic resemblance to the model. These species generally resemble the putative models but are not accurate mimics [14, 31, 37, 47, 48]. Some authors propose that poor mimics are just on an evolutionary trajectory towards perfection. This hypothesis is discussed by Edmunds [49] and Gilbert [37]. Gilbert [37] refutes this hypothesis saying, "*In my view it is better to assume that poor mimetic patterns have evolved to an equilibrium state, rather than being in the process of being perfected by constant directional selection*" since there is no experimental or theoretical support for the hypothesis that imperfect mimics are just mimics on their way towards perfection.

Recently, authors have instead proposed various evolutionary scenarios that may select for imperfect or inaccurate mimicry rather than explain this phenomenon away as "evolution in progress." Many papers point out that if a model is extremely unpalatable, noxious, or difficult to capture, then even imperfect mimics will gain strong selective advantage from a general resemblance to this model and there may be no selective advantage or pressure for more accurate mimetic resemblance [34, 37, 39, 47, 50]. In fact, the fitness costs of close morphological resemblance (see Section 2.5) may select against accurate mimicry and may select for imperfect mimicry if either confers approximately the same selective advantage in terms of escape from predation. In a study by Duncan and Sheppard [50], the authors experimentally demonstrate that, when the model is very noxious, even imperfect mimics gain protection. However, when the model is only moderately distasteful, selection favors more accurate mimics. They showed that when the cost of making a mistake, attacking a distasteful model because it is mistaken for a palatable mimic, is high, the predator rejects a greater proportion of mimics and there is little selection for more accurate mimicry. When the penalty for making a mistake is low, tiny improvements in mimetic resemblance confer a selective advantage to the mimics, leading to more accurate mimicry [50]. In a study by Speed and Ruxton [47], the authors propose that if generalization by the selective agents (the predators) is narrow, selection towards accurate mimicry is predicted. If generalization by predators is relatively wide (e.g., in the case of a particularly noxious model), variations in mimetic forms may be selected for with both accurate and inaccurate mimics. Finally if generalization by predators is intermediate, then the rate of evolution selecting for accurate mimicry will be slow and polymorphic mimetic forms will be stable.

In situations in which the model either becomes rare or is weakly aversive and the incentive to attack and sample the models (by predators) is high, then close mimics may in fact be selected against. Kin selection among the mimetic population would select for less accurate mimics that diverge in their mimetic resemblance to the weakly defended model [15, 34, 37]. Inaccurate mimicry can also be favored in species with limited dispersal and high local abundance in which neighboring mimics are related (i.e., kin selection) [15].

A study by Kikuchi and Pfennig [39] provided experimental support for the hypothesis that evolution of accurate mimicry is a gradual process and depends on the relative abundance of the model. In this study, the authors found that in areas where the model was abundant, predators attacked cryptic (or camouflaged) prey, accurate mimics, and intermediate (or imperfect) mimics with the same low frequency. In other words, in areas where the model was abundant, predators generalize and imperfect mimics gain the same relative protection as more accurate mimics. In habitats where the model population was low, camouflaged species and mimics attained greater protection than imperfect mimics. Thus the authors showed that Batesian mimicry can evolve through gradual steps towards more accurate mimicry depending on conditions and context (particularly the abundance of models in the habitat) [39]. This study also suggests that

mimics may have evolved from cryptic or camouflaged ancestors.

Accuracy of the mimetic resemblance may depend largely on the visual acuity of the selective agent. If predators with keen vision serve as the primary selective agents, then these predators may select for more accurate mimicry [34]. Then again, mimicry may be in the eyes of the beholder. Arthropods that humans view as poor mimics were perceived by pigeons, in an experimental test, as very good mimics [14]. Dittrich et al. [14] also showed that slight changes in the morphology of the mimic led to sometimes dramatic improvements, from the perspective of the selective agent, in perceived mimetic resemblance. They further pointed out that discrimination between a good and a poor mimic occurs via multiple features (e.g., color, form, size), not a single characteristic [14]. Other authors have also suggested that selection for increasingly better mimetic resemblance can, in fact, be a gradual process through directional selection [50, 51].

Related to the hypothesis that mimetic accuracy is dependent on the visual acuity of the selective agent is the multi-predator hypothesis, which proposes that inaccurate Batesian mimics evolved as a result of selective forces from a suite of predators [52]. For example, model averse predators select for more accurate morphological mimics in a given habitat while specialist predators on the model (e.g., ant predators or myrmecophages) select for inaccurate mimicry or for secondary defenses in the mimic [52]. Secondary defenses may include fast evasive movements by the mimics (quickly dropping all pretense of behavioral mimicry) or signaling the predator in such a way as to communicate its true identity [52]. If both kinds of predators are present in a habitat, there may be selection for inaccurate mimics or for polymorphic mimicry [52].

One hypothesis explaining imperfect Batesian mimicry that has gained some momentum in recent years is the multi-model hypothesis. If many potential model species live in a given habitat (e.g., many different species of ants), then it may be adaptive for the mimetic species to evolve a general, imperfect resemblance—a gestalt resemblance—to all of them than to evolve a specific morphological resemblance to a particular model [33, 37, 49]. For example, a general ant-mimicking spider in such a habitat can then have a much greater range than a spider that resembles only one of the potential models. If it is an accurate mimic, then its range is limited to the range of that one species in order to be an effective Batesian mimic. In one study, the authors found that some species of accurate ant mimics were found in association with a single model (measured as the closest ant collected where the spider was found). Some imperfect mimics (by human standards) were collected in proximity to more than one species of ant, conferring some support for the multi-model hypothesis [33]. However, in this same study, the author also found habitats in which accurate and inaccurate mimics did not associate with the models as predicted.

2.5. Trade-Offs Affecting the Evolution of Myrmecomorphy. A close morphological resemblance to ants makes myrmecomorphs more attractive to ant predators or myrmecophages. Thus myrmecomorphs are faced with an evolutionary trade-off: they gain protection from general arthropod predators but risk predation from a completely different suite of predators ([11, 12, 53] and citations above under discussion of multi-predator hypothesis). Many spider myrmecomorphs confront a threat from a myrmecophage by completely dropping their behavioral mimicry. These spiders will stop their erratic ant-like movement and run away, drop on a silk thread, signal to the predator in a spider-specific manner, or otherwise communicate their true identity to the predator [11, 52, 54]. This strategy is effective in allowing the spider to escape from the myrmecophage (or from ants that may confront it directly) [11, 24, 54].

Myrmecomorphs face other costs that may affect their fitness, including (1) constraint of the circadian rhythm of the mimic since it must be active at the same time of day as the model for the resemblance to be adaptive; (2) an imposed limit to the myrmecomorph's trophic niche because it would only have access to prey that lived in the same habitat as the model; (3) a possible detrimental or costly effect on mating or reproduction since many myrmecomorphs must mate in a sheltered location, where their non-ant-like behavior will not "give the game away" or may mate for a shorter duration than non-mimetic relatives for the same reason; (4) a lowering of fecundity with the abdominal narrowing or constrictions often associated with myrmecomorphy and the resultant decrease in the number of eggs a female can produce [37, 55]. It has been documented that narrower abdomens in female spiders limit the number of eggs that can be produced in comparison to non-mimetic relatives [9, 18, 56–61]. In addition, there may be a cost associated with alteration in the prey capture behaviors, such as those seen in myrmecomorphic salticids that lunge rather than jump upon their prey, which may be a much less effective prey capture strategy.

Nevertheless, if the primary predators demonstrate an innate, rather than learned, aversion to ants, the circadian rhythm of the myrmecomorphs may not be greatly affected and they can be active at any time of day. The limitation of trophic niches may not apply to general ant mimics since these spiders can exist, according to the multi-model hypothesis, across a potentially broad range of habitats. It does seem though that most spider myrmecomorphs do share the same habitat as their models and are active at the same time of day. It has even been pointed out that no species of wolf spider (family Lycosidae) has been reported to be an ant mimic because most lycosids are nocturnal and not active when visually hunting arthropod predators are most active [4]. Researchers investigating the inaccurate myrmecomorphs *Liophrurillus flavitarsis* (Lucas), *Phrurolithus festivus* (C. L. Koch) (both in the family Corinnidae), and *Micaria sociabilis* Kulczynski (Gnaphosidae) found that, in comparison to these species' closest relatives, the trophic niche of each was constrained by their resemblance to ants because they were limited to catching only small invertebrates found in the same habitat as the models. The circadian rhythms of these myrmecomorphs were also constrained because the myrmecomorphs were all diurnal (as were the models) but the closest relatives were nocturnal. However, the reproductive traits were not constrained since the fecundity of the inaccurate mimics was about the same as the non-mimetic relatives and the myrmecomorphs mated out in the open on bark, not dropping their behavioral mimicry when copulating [55].

The evolution of close morphological and behavioral mimicry of ants is costly and these costs should be measured as fitness components [37]. In addition, more studies should attempt to identify the operators or selective agents selecting for mimetic resemblance since the visual acuity of these selective agents (if they can be identified) may affect the accuracy of the resemblance. All these costs, trade-offs, and constraints should be taken into account when testing or modeling the adaptive significance of myrmecomorphy. The relative measures of the costs and benefits of mimetic resemblance may have a significant impact on the accuracy of the resemblance. If, for a particular species, the fitness costs of close mimetic resemblance due to lower fecundity greatly outweigh the benefits, then imperfect or inaccurate mimicry may be selected for. For example, in a habitat where the primary selective agent is a predator with low visual acuity, increased mimetic accuracy may impose a higher cost in terms of fecundity than is gained in terms of escape from predation. In small species of spiders in which greater mimetic resemblance would lead to dramatically lower fecundity due to a narrowing of the female's abdomen, dimorphic mimicry may be selected for and males may show greater mimetic resemblance than females. Too few models take into account fitness costs of mimetic resemblance and the relative effect such trade-offs may have on the evolution of imperfect, polymorphic, transformational, and dimorphic mimicry.

3. Spider Myrmecophily

3.1. Additional Records of Spider Myrmecophiles. Myrmecophiles are defined as ant guests, arthropods that have evolved close associations with ant species, often living alongside the ants or within the ant colonies [2, 3, 5, 62]. Some, but not many, of these myrmecophiles are also myrmecomorphs. Recent work (cited below) has found that, among spider myrmecophiles, some are also myrmecophages.

An extensive table of spider myrmecophiles was presented by Cushing [5]. Table 1 supplements this earlier table and provides records of spider myrmecophiles not included in the previous table. Not as much work has been carried out exploring the natural history, adaptations, or evolutionary significance of spider myrmecophiles as has been done with spider myrmecomorphs and myrmecophages. Nevertheless, some significant research has been conducted recently that expands our understanding of the biology of these interesting ant associates and how this unique lifestyle may have evolved in a group of arthropods that otherwise includes primarily free-living, solitary predators.

3.2. Adaptive Significance of Myrmecophily. An ant colony, as pointed out by Hölldobler and Wilson [2], can be considered

TABLE 1: Spider myrmecophiles found in association with or inside ant nests. This table is meant to supplement the table of Araneae myrmecophiles found in Cushing [5]. Spider taxonomy according to Platnick [63]; ant taxonomy according to http://antbase.org/.

Spider myrmecophile	Ant host	Notes on biology	References
Linyphiidae			
Diastanillus pecuarius (Simon)	Formica cf. fusca L. and F. lemani Bondroit	Found under stone near ants.	[64, 65]
Pseudomaro aenigmaticus Denis	Lasius flavus (Fabricius)	Associated with nests.	[65]
Syedra myrmicarum (Kulczynski)	Manica rubida (Latreille) and Formica sp.	Found under stone near ants.	[64, 65]
Oonopidae			
Dysderina principalis (Keyserling)	Labidus praedator (Smith) (publ. as Eciton praedator)	Found inside nests.	[66]
Gamasomorpha maschwitzi Wunderlich	Leptogenys processionalis distinguenda (Emery) (publ. as L. distinguenda)	Found inside nests. Chemical mimic. Phoretic. Follows emigration trails of hosts. Builds webs inside nest.	[65, 67–69]
Gamasomorpha wasmanniae Mello-Leitão	Eciton sp.	Found inside nests.	[70]
Xestaspis loricata (L. Koch) (publ. as G. loricata)	Myrmecia dispar (Clark)	Found inside nests.	[71]
Salticidae			
Cosmophasis bitaeniata (Keyserling)	Oecophylla smaragdina (Fabricius)	Lives inside nest. Is chemical mimic of ant. Feeds on ant larvae by using tactile mimicry.	[72–76]
Phintella piatensis Barrion and Litsinger	O. smaragdina	Lives in proximity to ants.	[77]
Theridiidae			
Eidmannella pallida (Emerton)	Atta sexdens (L.)	Lives in old fungus chambers of nest.	[78]

an isolated ecosystem. Arthropods symbiotic with ant hosts typically experience a stable microclimate, plentiful food (either in the form of other symbionts, the hosts themselves, or other resources brought into the colony by the hosts), and protection from their own predators and parasites [5, 68, 77]. The degree of integration into the colonies varies greatly from species with just a loose affiliation or association with the ant nests to symbionts that spend their entire lives within the ant nests and fail to thrive when removed from this habitat [5, 79]. These symbionts can have a neutral, a positive, or a negative influence on the host colonies depending on their natural history. If the effect of the myrmecophile on the host is costly enough, there should be selection for the host to recognize and attack or remove these guests from the nest [69]. For example, the myrmecophile Masoncus pogonophilus Cushing (Linyphiidae) feeds on collembolans and other symbionts found in the colonies of the harvester ant, Pogonomyrmex badius (Latreille) [80] (Figure 2). Therefore, this spider may have a slightly negative effect on the colonies of these ants since the primary prey of the spiders, collembolans, graze fungal spores found inside the nest chambers, particularly the seed storage chambers [80], and thus keep fungal infestations low. However, populations of these spiders are so small within any given colony that their net effect on the host's success is probably negligible [79, 80]. Some evidence suggests that hosts can recognize and will attack these symbionts, particularly those introduced from

a neighboring nest [81, Cushing pers. obs.]. The myrmecophilic spider Gamasomorpha maschwitzi (Wunderlich) (Oonopidae) is found inside the nests and bivouacs of the army ant, Leptogenys distinguenda (Emery), where it apparently feeds on insects captured by the hosts. Therefore, this myrmecophile has a negative impact on host fitness as a kleptoparasite on the host's prey. However, as with M. pogonophilus, the abundance of spiders within any given colony is so low that its negative impact is likely negligible and these spider guests are either ignored or treated with only very low levels of aggression [67, 68]. Sometimes spiders are even groomed by the host ants [69]. The spider Attacobius attarum (Roewer) (Corinnidae) (originally published as the clubionid Myrmecques attarum) lives with Atta sexdens (L.) where it feeds on ant larvae and pupae [82] and thus also has a negative impact on host colonies. The hosts are known to antennate the spiders but do not show any aggression towards these myrmecophiles [82].

It has been noted that certain types of ant colonies are more open to invasion by myrmecophiles than others. Characteristics of host colonies that are most open to invasion by myrmecophiles include: colonies with multiple queens (polygynous colonies), colonies with multiple nest sites (polydomous colonies, which are often also polygynous), and very large colonies [83]. These societies tend to be more "loose, flexible, and dynamic" than monogynous colonies and tend to have less social cohesion leading to increased

(a)

(b)

FIGURE 2: Myrmecophily in spiders. (a) The host ant *Pogonomyrmex badius* (Latreille) at the nest entrance. (b) The myrmecophilic spider, *Masoncus pogonophilus* Cushing on the surface, walking along the emigration trail of the host ant. Scale bar in (a) = 8 mm, scale bar in (b) = 1 mm. Photos © author.

vulnerability to invasion by myrmecophiles [83]. In general, myrmecophile populations tend to occur in one of the following distinct patterns: (1) a myrmecophilic species is found in many colonies at certain locations throughout a host species' range but not at other locations (i.e., high infestation but low transmission), (2) a myrmecophilic species is found throughout the host's range but only within a few colonies at any given locality (i.e., low infestation but high transmission), or (3) the myrmecophile is found in only a few colonies at any one locality and not throughout the host's range (i.e., low infestation and low transmission) [83]. Population size of myrmecophiles is often quite low within a colony, but this depends on the type of myrmecophile. Spider myrmecophiles that have been studied in any depth, in general, tend to have small populations within a colony [67, 68, 79]. Intraspecific aggression between spider myrmecophiles within a colony has been reported [69] and may be one factor in keeping populations small.

3.3. General Adaptations Facilitating Integration into Colonies.
Close integration within ant colonies seems to be more common in certain families, such as the Linyphiidae and Oonopidae [5]. These spiders have several characteristics (morphological and behavioral) that may serve as preadaptations to a symbiotic lifestyle inside ant nests [67]. For example, both families include very small spiders (typically less than 5 mm); the species are often found in moist, humid microhabitats such as leaf litter, under rocks or logs, or under bark; and many species in these families (particularly oonopids) have morphological adaptations such as hard sclerotized scuta covering their abdomens that may provide some protection against attacks by host ants. Witte and colleagues point out that some species of oonopids may scavenge insect remains in the webs of other spiders [67]. All these behavioral and ecological characteristics may preadapt spiders to a myrmecophilic lifestyle within ant colonies. Smaller body sizes allow them to "sneak" inside the nests and become integrated. Protective scuta (and small sizes) may provide some protection against attacks from the hosts. A scavenger lifestyle may be considered a preadaptation to stealing food (insects

or ant brood) from workers. The constant temperature and humidity of an underground ant nest may be an attractive environment to species otherwise restricted to similar temperature and humidity regimes.

Once integrated into colonies, spider myrmecophiles certainly have evolved dramatic host-specific adaptations allowing them to become even more integrated into various aspects of the host's life cycle. These adaptations, in turn, place severe constraints on the geographical distribution of these inquilines or ant guests; the symbionts are restricted to the range of that host species [83]. This may explain why such inquilines are very localized or rare and may be subject to frequent extinctions [83]. Adaptations common to myrmecophiles include evasive devices such as behaviors, morphological structures, or chemical signals used to appease hosts or to mimic hosts; protective morphological structures such as sclerotized cuticular "shields" or plates; mechanisms to communicate with hosts via chemical cues, tactile cues, or even auditory cues [83].

3.4. Chemical Mimicry.
Among spider myrmecophiles, besides the preadaptations mentioned above, many have evolved the capacity to absorb, biosynthesize, or otherwise mimic the host ant's cuticular hydrocarbon colony odor. To survive inside the host colony, the guest must be considered a nest mate by the hosts and should, therefore, have somehow acquired the chemical odor of the hosts via either biosynthesis of the key compounds or by passively acquiring the chemical cues [84]. Thus far, no research has definitively documented glandular secretions that spider myrmecophiles might use to biosynthesize the compounds. If such glands are documented, then it is likely that the association between the host and the myrmecophile is an ancient association and the myrmecophile and host coevolved [85, 86]. However, biosynthesis may evolve rapidly in myrmecophile populations if the compounds biosynthesized can be easily manufactured by co-opting an already existing chemical pathway or if the guest can re-purpose an already existing compound [86].

The chemical signature of ant colonies may change over time [2]. Thus intruders (guests) into colonies must be able to update their profiles constantly in order to avoid detection and attack. If the myrmecophile's chemical profile does not match the host's closely enough then it will become more difficult for the guest to approach the host in order to update its profile, making social integration into the colonies a "well-balanced and potentially fragile system" [69]. Myrmecophiles can acquire colony odors by rubbing against the host ants, associating with nest materials, or by eating the ant's brood (larvae or pupae) [84]. All these mechanisms are seen in spider myrmecophiles. It may be that these myrmecophiles do not need to acquire an exact chemical match to the host's hydrocarbon profile, but need only one or two key constituents that are biologically most important in nest recognition and acceptance by the hosts [86].

For example, the oonopid, *G. maschwitzi*, found with the army ant, *L. distinguenda*, has a cuticular hydrocarbon profile that includes only compounds also seen in the host ant's profile but not all the compounds seen in the host's profile [69]. These spiders crawl on top of workers, moving their legs actively over the cuticle of the host, perhaps as an adaptation to acquire the host's chemical odor [68, 69, 78]. The hydrocarbon profile of the myrmecophilic spider matches that of the host's to a high degree; however, colony-specific matching was not evident [69]. Nevertheless, ants of *L. distinguenda* from different colonies did not show high levels of intercolony aggression; therefore, it may not matter that the myrmecophile's profile lacks these colony-specific compounds but just generally matches the gestalt odor of the species (i.e., has key chemical constituents that identify it as an ant and a member of the same species) [68, 69]. Research has also demonstrated that the phoresy displayed by *G. maschwitzi* may also function as a behavioral mechanism for the spider to acquire food (ant larvae, pupae, or insects being carried by the workers) via kleptoparasitism [68]. The spider riding on the back of the ant snatches the food item directly from the host's mandibles. In fact, these spiders have not been observed to hunt prey on their own [67] so this kleptoparasitic lifestyle may be another example of extreme adaptation related to their symbiotic life with these ants.

The salticid *Cosmophasis bitaeniata* (Keyserling) lives inside the colonies of the weaver ant, *Oecophylla smaragdina* (Fabricius), where it feeds on the larvae of the host ant [72–75]. The spider is more often found in and around older nests that have lots of larvae [72]. The spider touches the antennae and head of minor workers with its front legs, stimulating the workers to release the larva that the worker is carrying [72]. The spider otherwise avoids direct contact with the worker ants [72, 75]. The spider is a chemical mimic of the host [73–76]. It has been shown that the spider acquires the colony specific hydrocarbon profile by handling and eating the ant larvae [74, 76]. The hydrocarbon profile of the spider is colony specific but does not match the profile of the major workers [75]. Larvae from different colonies do not elicit aggressive responses from the host; thus spiders that mimic the hydrocarbon profile of the larvae rather than the workers may be more easily accepted by both their own hosts as well as those of neighboring colonies [76].

The spider *Attacobius attarum* that lives inside the nests of the leaf cutter ant, *Atta sexdens* (L.) rides on the dorsa of

workers and alates [78, 82, 87]. The spiders may disperse to new colonies via the alates [78, 82, 87]. *Attacobius attarum*, like *G. maschwitzi* and *C. bitaeniata*, is a kleptoparasite; the spider feeds on ant larvae and pupae and can steal the brood directly from the mandibles of workers [82]. The ants antennate the spiders and the spiders reciprocate by "antennating" the ants with their front legs, possibly providing mimetic tactile cues [82]. No aggression towards these kleptoparasites has been reported [82].

The theridiid spider, *Eidmannella pallida* (Emerton) (published as *Eidmannella attae*), also lives with *A. sexdens* where it is found in unused fungus chambers that the ants use to store refuse and dead ants [78]. Likewise, the linyphiid, *M. pogonophilus*, lives in seed chambers and empty chambers of the seed harvester ant, *P. badius* [79, 80]. Both these spider myrmecophiles may acquire host colony odor passively via the nest materials. Neither has been reported as phoretic, as kleptoparasitic, or as a predator of the hosts or their brood. Thus passive integration and acquisition of colony odor is likely for these symbionts.

3.5. Ability to Follow Chemical Cues of the Hosts. *Cosmophasis bitaeniata* can distinguish between nestmate and non-nestmate major workers and shows less tendency to try and escape when confined with nestmates, demonstrating that these myrmecophiles are not only chemical mimics but are also able to interpret chemical cues provided by the hosts [74]. Data suggests that the ability to interpret chemical signals of the hosts may be a general characteristic of spider myrmecophiles that are closely integrated into ant colonies. Research on *M. pogonophilus* and *G. maschwitzi* showed that spiders are able to follow trail pheromones laid by the ants [67, 68, 79, 80]. In controlled tests, Witte et al. found that *G. maschwitzi* is sensitive to high concentrations of naturally laid ant trail pheromones [67]. I found *M. pogonophilus* in the emigration trails of *P. badius* when the hosts emigrated to new nest sites [79, 80] (Figure 2(b)).

Spider myrmecophiles may use ant trail pheromones as a means of dispersing to new colonies. In a given habitat, it is not uncommon to find spider myrmecophiles in all or nearly all the nests of a given host, even if the host is not polygynous or polydomous [68, 79]. Thus in at least these instances, dispersal to new colonies must be occurring. Only one study has attempted to examine the population structure of a myrmecophilic spider, *M. pogonophilus*, which was found in nearly all colonies of *P. badius* in a given habitat (i.e., 10 colonies out of 12 that were excavated) [79]. *Pogonomyrmex badius* colonies are established by single inseminated queens [88] that can live for at least 15 years [89]. I hypothesized that spider populations might be considered metapopulations [90], made up of isolated demes, or local populations, with very low per-generation migration between populations resulting in low genetic diversity between individuals within populations (i.e., myrmecophiles within an ant nest) and higher genetic heterogeneity between populations (i.e., between populations of spiders found in different colonies) due to genetic drift [79]. Instead, I found that genetic diversity among individual spiders within populations (within a colony) was greater than the genetic diversity between populations from neighboring ant nests suggesting that spiders do disperse to new nests frequently enough to maintain high intra-population differentiation and low inter-population differentiation [79]. Although tests of the spiders' ability to follow trail pheromones (naturally laid and artificial trails) were inconclusive, I further hypothesized that spiders were able to locate new nests by following trail pheromones. They were found to emigrate with their hosts to new nest sites (see above), thus they may, during emigration, get "side-tracked" onto the foraging trail of a neighboring *P. badius* colony [79].

3.6. Life Cycle of Spider Myrmecophiles. Very little is known about the life cycle of any spider myrmecophile. Even for one of the best studied species, *G. maschwitzi*, no spiderlings have ever been detected in the emigration trails nor inside the nests [67, and Volker Witte, pers. communication]. *Masoncus pogonophilus* builds prey capture webs inside nest chambers and females deposit small silken egg sacs each containing up to seven eggs in depressions in the walls of the chambers [80]. The salticid, *C. bitaeniata* also deposits its egg sacs within the nest chambers of *O. smaragdina* [72]. A *G. maschwitzi* female was collected with one large egg in the abdomen and another with five smaller eggs [67]. Both *M. pogonophilus* and *C. bitaeniata* have female biased sex ratios [72, 80].

3.7. Future Directions. A great deal more research needs to be done to understand the basic biology of spider myrmecophiles. Questions and directions for future research include the following.

(i) How closely integrated are spider myrmecophiles with their host ants?

(ii) How do these spiders reproduce inside the ant colonies or does reproduction occur outside the nests?

(iii) How do they disperse to colonize other nests?

(iv) Is chemical integration a widespread phenomenon among spider myrmecophiles?

(v) Can any spider symbiont biosynthesize chemical compounds that act to appease or mimic the hosts?

(vi) Are spider myrmecophiles generally able to interpret the chemical signals of their hosts?

(vii) Is there evidence of a co-evolutionary relationship between symbionts and hosts?

(viii) How closely related are spider myrmecophiles within a colony and do these patterns of relatedness explain the female-biased sex ratios seen in some species?

4. Spider Myrmecophagy

4.1. Species Records. Spiders, like other arthropod predators, generally avoid preying upon ants. However, ants have been documented as part of the diet for well over 100 species of spiders (Table 1). Fossil evidence of spider myrmecophagy dates back 30–50 mya in Baltic amber specimens including one containing an inclusion of spider silk with an ant that had been fed upon as well as another showing a spider with an ant in its chelicerae [91]. Myrmecophagic spiders exist on a continuum from euryphagous to stenophagous predators [92]. Huseynov et al. [92] propose five categories of spider myrmecophages: (1) non-acceptors of ants (the majority of spider species); (2) reluctant acceptors that do prey on ants but prefer other prey; (3) indifferent acceptors that feed indiscriminately on ants and other prey; (4) facultative ant choosers that prefer ants to other prey; (5) obligatory ant choosers that feed exclusively on ants (unless severely food deprived). In Table 2, the various spider myrmecophages that have been documented from the literature are categorized as (R) Reluctant acceptors, (I) Indifferent acceptors, (F) Facultative ant choosers, or (O) Obligatory ant choosers based upon information about their biology provided in the literature. If researchers have only documented that the particular species eats ants but provide no other information about the hunting behavior or prey preference of the spiders, the species is categorized as (Unk) Unknown. However, these spiders are likely to turn out to be either reluctant or indifferent acceptors of ants in the diet. Details of the predatory biology of spider myrmecophages are also included in the table.

4.2. Evolutionary Costs and Benefits of Myrmecophagy. Spider myrmecophagy is a high risk hunting strategy. Risks for myrmecophages include being attacked by the prey, living in close proximity to dangerous prey, being attacked when mating, having the prey attack and destroy one's eggs if nesting and oviposition occur close to the ant nests [58, 143, 175]. However, a spider that evolves strategies for overcoming an ant's defenses and aggression faces relatively little competition for a nearly unlimited food resource [114, 143] (Figure 3(a)).

One study demonstrated that myrmecophagic spiders may actually derive protection against attacks from their own prey: when myrmecophagic, myrmecophilic, myrmecomorphic, and non-ant associating salticids were trapped with ants, the myrmecophagic spiders showed the highest survival rate followed by the myrmecomorphs and myrmecophiles, suggesting that ant associates may signal the ants in such a way that the ants show little aggression towards these spiders [176]. Thus not only are myrmecophagic spiders obtaining a nutrient rich, unlimited food supply through their specialized diet, but they may also be deriving protection from the ants, just as myrmecophilic and myrmecomorphic spiders do.

Although it has not been suggested that spider myrmecophages are chemical mimics of ants, as has been demonstrated for spider myrmecophiles, there is some evidence that certain species of myrmecophages may either be releasing chemical compounds that appease their potential prey or may be able to "read" chemical cues released by ants. For example, Lubin suggested that the thomisid, *Tmarus stoltzmanni* Keyserling, may use its 1st and 2nd pairs of legs to detect chemical or tactile cues from the ants [148]. *Habronestes bradleyi* (O. P.-Cambridge) (Zodariidae) waves its front legs around when hunting and, when the legs are amputated, the spider has a difficult time locating ant prey

Table 2: Spider myrmecophages. *Categories (defined in text) include R: Reluctant myrmecophage; I: Indifferent acceptor; F: Facultative ant predator; O: Obligatory ant predator; Unk: cannot be determined from information about their biology presented in the literature (these are most likely R or I myrmecophages). Spider taxonomy according to Platnick [63]; ant taxonomy according to http://antbase.org/.

Spider myrmecophage	Category of myrmecophage*	Notes on biology	References
Araneidae			
Metepeira gosoga Chamberlin and Ivie	Unk	Author suggests that spiders may feed on ants found only on cholla where spider is also found.	[93]
Metepeira sp.	Unk	Reported feeding on *Crematogaster opuntiae* Buren.	[93]
Deinopidae			
Deinopis sp.	Probably I	Throws web over ants passing below.	[94]
Eresidae			
Seothyra sp.	F	Lives in silk lined burrows. Mouth of burrow covered by prey capture web. Captures mostly ants. Male spider runs on ground during day and is myrmecomorph and behavioral mimic of *Camponotus* sp. and mutillid wasps (dimorphic mimicry).	[95]
Gnaphosidae			
Callilepis nocturna (L.)	May be F	Feeds on *Formica* spp. and *Lasius* spp. Actively searches for ants and may enter nests to hunt workers. Approaches ant and bites on base of antenna. Antennae seem to act as stimulus to trigger attack.	[96–98]
Linyphiidae			
Frontinella communis (Hentz)	I	Occasionally preys on ants.	[99]
Oecobiidae			
Oecobius annulipes Lucas	O	Main food is *Plagiolepis pygmaea* (Latreille) but other ants (e.g., *Lasius flavus* (Fabricius)) accepted in lab. Bites at base of antenna. Swaths ant in silk and encircles it. Sometimes uses last pair of legs as well as spinnerets to direct silk over prey. Reduced chelicerae and enlarged gnathocoxae may be adaptations to myrmecophagic lifestyle.	[100]
O. cellariorum (Dugès)	O	Feeds on *Plagiolepis pygmaea* (Latreille). Bites at base of antenna.	[100]
O. templi O. P.-Cambridge	O		[100]
Oonopidae			
Triaeris stenaspis Simon (publ. as *T. patellaris*)	Unk	Reported attacking *Cyphomyrmex costatus* Mann.	[101]
Oxyopidae			
Oxyopes apollo Brady	Unk	Eats ants.	[102]
O. globifer Simon	I/F	Ants constitute large % of prey.	[99, 102]
O. licenti Schenkel	Unk	Eats ants.	[102]
O. salticus Hentz	Unk	Eats ants.	[102]
O. scalaris Hentz	I	Occasionally eats ants.	[99, 102]
O. sertatus L. Koch	Unk	Eats ants.	[102]
Peucetia viridans (Hentz)	Unk	Eats ants.	[103]
Pholcidae			
Crossopriza lyoni (Blackwall) (publ. as *Crossopriza stridulans*)	Unk	Feeds on fire ants, *Solenopsis invicta* Buren.	[104]
Salticidae			
Aelurillus aeruginosus (Simon), *A. cognatus* (O. P.-Cambridge), and *A. kochi* Roewer	F	Prefer ants over other prey. Innately recognize ants even if ants are not moving. Attack from front unless ant is passing (then switch to rear attack). Use different hunting behavior for ants than for other prey. If hungry, show no preference for ants over other prey.	[105]

TABLE 2: Continued.

Spider myrmecophage	Category of myrmecophage*	Notes on biology	References
Aelurillus m-nigrum Kulczyński	F	Prefers ants over other prey; 85% of diet in field consists of ants. Uses different hunting behaviors for ants than for other prey: lunges, attacks from front, bites, releases, bites again.	[92]
Aelurillus spp.	F	Species in genus prefer ants over other prey. Use different hunting behaviors for ants than for other prey.	[106]
Anasaitis canosa (Walckenaer) (publ. as *Corythalia canosa* or as *Stoidis aurata*)	F	Prefers ants over other prey. Uses different hunting behaviors for ants than for other prey: attacks from front, holds forelegs away from struggling ant. Also stilts body off ground.	[107, 108]
Anasaitis spp.	F	Species in genus prefer ants over other prey. Use different hunting behaviors for ants than for other prey.	[106]
Chalcotropis spp.	F	Use different hunting behaviors for ants than for other prey: some attack from rear, some head-on, then lunge, bite, release, and wait.	[106, 109]
Chrysilla lauta Thorell	F	Prefers ants. Uses different hunting behaviors for ants than for other prey: attacks from rear, bites gaster (not appendages), retreats and waits, may lunge and strike several times. When ant quiescent, spider approaches, bites again, and carries it away.	[110]
Chrysilla spp.	F	Species in genus prefer ants over other prey. Use different hunting behaviors for ants than for other prey.	[106]
Cosmophasis sp.	Unk	Feeds on ants and is myrmecomorph.	[59]
Euophyrs spp.	F	Use different hunting behaviors for ants than for other prey: some attack from rear, some attack head-on, then lunge, bite, release, and wait.	[106]
Evarcha albaria (L. Koch)	I/F	Robs ants of their prey and of their brood (eggs and larvae) that workers carry (kleptoparasites).	[111]
Habrocestum pulex (Hentz)	Some F Some I	Some individuals prefer ants over other prey; some prefer other prey over ants. Myrmecophagic individuals use different behaviors for ants than for other prey: lunge or leap onto petiole or thorax, bite, release, repeat (up to 6 times). Keep front legs off ground away from ant. Reported preying on *Crematogaster* spp.	[112–114]
Habrocestum spp.	F	Species in genus prefer ants over other prey. Use different hunting behaviors for ants than other prey.	[106]
Hasarius adansoni (Audouin)	Probably I	Will feed on ants.	[115]
Hentzia palmarum (Hentz) (publ. as *Eris marginata*)	Unk	Reported feeding on workers of *Myrmica* sp.	[113]
Icius sp.	Unk	Reported feeding on small brown ants.	[113]
Menemerus fulvus (L. Koch) (publ. as *Menemerus confuses*)	I/F	Robs ants of their prey and of their brood (eggs and larvae) that workers carry (kleptoparasites).	[111]
Myrmarachne foenisex Simon	F	Regularly feeds on weaver ant (*Oecophylla*) larvae. Also mimics weaver ants.	[59]
Natta horizontalis Karsch (publ. as *Cyllobelus rufopictus*)	F	Prefer ants. Uses different hunting behaviors for ants than for other prey: attacks from rear, bites gaster (not appendages), retreats, and waits, may lunge and strike several times. When ant quiescent, spider approaches, bites again, and carries it away.	[110]
Natta spp.	F	Species in genus generally prefer ants. Use different hunting behaviors for ants than for other prey: attack from rear, bite gaster (not appendages), retreat and wait, may lunge and strike several times. When ant quiescent, spider approaches, bites again, and carries it away.	[106, 110]

TABLE 2: Continued.

Spider myrmecophage	Category of myrmecophage*	Notes on biology	References
Phidippus johnsoni (Peckham and Peckham)	I	Occasionally eats ants.	[99, 116]
Plexippus setipes Karsch	I/F	Robs ants of their prey and of their brood (eggs and larvae) that workers carry (kleptoparasites).	[111]
Siler cupreus Simon (publ. as *Silerella vittata*)	F/O	Eats ants. Spider population increases in areas infested with Argentine ants, *Linepithema humile* (Mayr). Also robs worker ants of brood including eggs, larvae, and pupae being carried by workers (kleptoparasitism).	[117–120]
Siler semiglaucus (Simon)	F	Prefer ants. Uses different hunting behaviors for ants than for other prey; bites gaster (not appendages), retreats and waits, may lunge and strike several times. When ant quiescent, spider approaches, bites again, and carries it away.	[110]
Siler spp.	F	Use different hunting behaviors for ants than for other prey: some attack from rear, some from head-on, lunge, bite, release and wait.	[106, 109]
Tutelina formicaria (Emerton)	F	Also myrmecomorph. Preys on red and black ants.	[121]
Tutelina similis (Banks)	F	Preys primarily on ants and is also a myrmecomorph. Uses different hunting behaviors for ants than for other prey: bites quickly, releases, retreats, carries paralyzed prey to safe area.	[99, 113]
Tutelina spp.	F	Other species of *Tutelina* found on mound of *Pogonomyrmex salinus* Olsen (publ. as *P. owyheei*) feeding on worker ants.	[113]
Xenocytaea spp.	F	Species in genus prefer ants over other prey. Use different hunting behaviors for ants than other prey.	[106]
Zenodorus durvillei (Walckenaer), *Z. metallescens* (L. Koch), and *Z. orbiculatus* (Keyserling)	F	Prefer ants over other prey. Feed on ants caught in other spider's webs—but only if spiders can approach safely without getting caught. Ambush ants; hang upside down and lunge at ant while releasing dragline. Repeatedly bite larger ants. Do not hold onto injured ant.	[106, 108]
Zenodorus spp.	F	Species in genus prefer ants over other prey. Use different hunting behaviors for ants than other prey.	[106]
Scytodidae			
Scytodes sp.	Unk	Feeds on fire ants, *Solenopsis invicta* Buren.	[104]
Theridiidae			
Achaearanea spp.	Unk	Feed on "carpenter ants." Ants become entangled in gum footed sticky thread attached to substrate. Movement of ant causes thread to snap and ant is lifted off ground.	[93]
Argyrodes sp.	Unk	Reported feeding on *Pogonomyrmex rugosus* Emery.	[93]
Asagena fulva (Keyserling) (publ. as *Steatoda fulva*) and *A. pulcher* (Keyserling) (publ. as *S. pulcher*)	Unk	Feed on *Pogonomyrmex badius* (Latreille) and *P. subnitidus* Emery. When ant workers captured in webs, major workers (patrollers) may attempt to free them but become caught in webs themselves.	[93, 122]
Cryptachaea riparia (Blackwall) (publ. as *Theridion saxatile* and as *Acaeoraneae riparia*)	F	Captures ants with above-ground web that has sticky threads attached to substrate. Webs built in areas of high ant activity or traffic. Greater than 88% of diet made up of ants (mostly *Formica* spp.). Ant gets tangled in sticky silk, struggling causes line to snap, ant is suspended, spider responds to vibrations, bites ant several times in legs and antennae while wrapping in silk, cuts paralyzed ant, and carries it to sand-covered tube retreat.	[123, 124]
Dipoena punctisparsa Yaginuma	Unk	Feeds on small ants in genus *Lasius*.	[125]

TABLE 2: Continued.

Spider myrmecophage	Category of myrmecophage*	Notes on biology	References
Enoplognatha ovata (Clerck) (publ. as *Theridion lineatum* or *T. lineamentum*)	Unk	Feeds on *Pogonomyrmex barbatus* (Smith). Builds webs in grass near colony. Ants crawling up into grass or passing below get entangled.	[126]
Euryopis californica Banks	I/F	Reported feeding on *Pogonomyrmex rugosus* Emery.	[93]
Euryopis coki Levi	I/F	Preys on *Pogonomyrmex salinus* Olsen (publ. as *P. owyheei*). Spider captures ant on the mound by trapping ant against ground with sticky silk. Bites on leg. Ant swings off ground on thread. When paralyzed, spider drags it away using a web sling attached to the ant and to the spinnerets.	[127]
Euryopis episinoides (Walckenaer) (publ. as *E. acuminata*)	I/F	Feeds on ants. Attacks *Crematogaster* ants and transports each attached to spinnerets.	[128]
Euryopis formosa Banks	I/F	Captures and carries workers of *Pogonomyrmex salinus* Olsen. Carries ant across ground. One attack described: spider bit gaster, released ant, moved to front and waited, reapproached paralyzed ant, climbed onto ant and began dragging across ant nest using web sling.	[129]
Euryopis funebris (Hentz)	F/O	Reported feeding on *Camponotus castaneus* (Latreille). Throws adhesive silk over ant passing by on tree trunk and fastens it to tree. Encircles ant, throwing silk. Bites leg. Cuts paralyzed ant free and carries it to crack or crevice or drops on line to feed.	[130, 131]
Euryopis scriptipes Banks	I/F	Feeds on ants.	[132]
Euryopis texana Banks	I/F	Female reported preying upon moving line of small ants.	[133]
Other *Euryopis* spp.	I/F	Prey on ants. Throw adhesive silk over ants and fasten to trees.	[131–133]
Latrodectus corallinus Abalos	Unk		[93, 134]
Latrodectus hesperus Chamberlin and Ivie	Probably I	Feeds on *Pogonomyrmex rugosus* Emery. Builds web on colony mound over foraging trail. Spider throws silk on ant that gets caught in gum threads. Spider approaches ant from above, bites posterior femur, retreats, returns after ant paralyzed, and pulls ant to retreat or to hidden part of web. Also feeds on other species of ants.	[93]
Latrodectus mactans (Fabricius)	I/F	75% of prey in cotton fields in Texas made up of fire ants, *Solenopsis invicta* Buren. Also reported feeding on *Pogonomyrmex badius* (Latreille) and *P. barbatus*.	[89, 126, 135]
Latrodectus mirabilis (Holmberg)	Unk	Feeds on *Acromyrmex* spp. and *Camponotus* spp. Builds webs over colony entrances.	[93, 134]
Latrodectus pallidus O. P.-Cambridge	F	Primary prey are ants. Feeds on *Monomorium semirufus* (*nomen dubium*, but probably *Messor semirufus* (André)). Females build webs over foraging trails. Capture ants from above with trip line attached to substrate and pull prey into retreat. Spiders can also descend to ground and catch ants running on trails.	[136–138]
L. quartus Abalos	Unk	Feeds on *Acromyrmex* spp. and *Camponotus* spp. Builds webs over colony entrances.	[93, 134]
Latrodectus revivensis Shulov	Unk	Remains of *Messor* sp. found in webs.	[136]
Latrodectus tredecimguttatus (Rossi)	Unk	Remains of *Messor* sp. found in webs.	[136, 137]
Latrodectus spp.	Unk	Members of genus may generally be myrmecophages. Reported feeding on *Monomorium* sp. and *Messor semirufus* (André).	[136–138]
Parasteatoda tepidariorum (C. L. Koch) (publ. as *Achaearanea tepidariorum*)	Unk	Feeds on fire ants, *Solenopsis invicta* Buren.	[107]

TABLE 2: Continued.

Spider myrmecophage	Category of myrmecophage*	Notes on biology	References
Phycosoma mustelinum (Simon) (publ. as *Dipoena mustelina*)	Unk	Captures various species of ants of wide range of sizes.	[125]
Steatoda albomaculata (De Geer)	I	Feeds on ants; ant remains found in webs.	[139]
Steatoda fulva (Keyserling)	I/F	Reported building webs near nest entrance of colonies of *Pogonomyrmex badius* (Latreille).	[122]
S. triangulosa (Walckenaer)	I	Feeds on fire ants, *Solenopsis invicta* Buren.	[104]
Yaginumena castrata (Bösenberg and Strand) (publ. as *Dipoena castrata*)	Unk	Mostly feeds upon *Camponotus* sp. and *Lasius* sp. and most individual spiders feed upon single type of prey. The larger the spider, the larger the ant it can attack.	[125]
Thomisidae			
Amyciaea albomaculata (O. P.-Cambridge)	O	Myrmecomorph of *Oecophylla smaragdina* (Fabricius) (publ. as *O. virescens*). Adult spiders with eye spots on abdomen. Juvs. yellow and mimic other species of yellow ants (transformational mimics). Spider waits near foraging trail of ant, attacks from behind, bites back of body, drags paralyzed ant to edge of vegetation, drops down to feed.	[140]
Aphantochilus rogersi O. P.-Cambridge (publ. as *Cryptoceroides cryptocerophagum*)	O	Also a myrmecomorph of *Cephalotes pusillus* (Klug) (publ. as *Zacryptocerus pusillus*). Attacks from behind. Holds dead ant as "protective shield." Females oviposit near ant nest and defend egg sacs against worker ants.	[141–143]
Aphantochilus spp.	Unk	Feed on cephalotine ants.	[57, 141–143]
Bucranium spp.	Unk	Feed on cephalotine ants. Hold dead ants as protective shield against attacks from other ants.	[57, 141–143]
Mecaphesa californica (Banks) (publ. as *Misumenops californicus*)	Unk	Feeds on *Pogonomyrmex rugosus* in vegetation near ant nests.	[93]
Mecaphesa coloradensis (Gertsch) (publ. as *Misumenops coloradensis*)	Unk	Feeds on alate females of *Pogonomyrmex maricopa* Wheeler and *P. desertorum* Wheeler after they have removed their wings and while resting on bushes waiting for temperatures to drop in order to dig new nest chambers.	[144]
Mecaphesa lepida (Thorell) (publ. as *Misumenops lepidus*)	I	Occasionally feeds on ants.	[99]
Misumenops argenteus (Rinaldi)	Probably I	17% of prey are ants; mostly ants that get caught in trichomes of plant *Trichogoniopsis adenantha* (OC), where spider spends most of its time.	[145]
Runcinioides argenteus Mello-Leitão (publ. as *Misumenops argenteus*)	Unk	Includes ants in diet.	[146]
Saccodomus formivorus Rainbow	May be F or O	Builds a basket-like web that appears to attract wandering *Iridomyrmex* ants. Spider also uses behavioral tactics-tapping ant with its own legs before attacking.	[4, 147]
Thomisus onustus Walckenaer	I	42.8% of diet consists of ants.	[147]
Tmarus stoltzmanni Keyserling	O	Feeds exclusively on ants; but only those without stings such as dolichoderine and formicine ants. Uses frontal attacks. May have sensory structures on 1st or 2nd pair of legs to detect chemical or tactile cues from ants.	[148]
Other *Tmarus* sp. (from Australia)	Unk	Includes ants in diet.	[148, 149]
Xysticus californicus Keyserling	Unk	Attacks harvester ants in California (cites unpubl. work of Snelling).	[148, 149]
X. loeffleri Roewer	R	Ants comprise only a minor part of diet.	[150]
Other *Xysticus* spp.	I/F	30–35% of diet of some spp. of *Xysticus* comprised of ants. One spider seen preying on *Pogonomyrmex salinus* Olsen. Spider seen on back of ant where it rode around, biting ant until paralyzed. Spider bit at base of petiole.	[129, 150]

TABLE 2: Continued.

Spider myrmecophage	Category of myrmecophage*	Notes on biology	References
Zodariidae			
Diores spp.	Probably F or O	Feed on ants.	[151]
Habronestes bradleyi (O. P.-Cambridge)	O	Spider also myrmecomorph. Waves front legs around when hunting ants. When legs are amputated, spider has difficult time locating prey (*Iridomyrmex purpureus* (Smith)).	[152–154]
Lachesana insensibilis Jocqué	I	Polyphagous but will eat ants smaller than themselves. Uses different hunting behaviors for ants than for other prey: bites, releases, re-approaches, bites again.	[155]
Lachesana tarabaevi Zonstein *and Ovtchinnikov*	F	Preys mostly on harvester ants in genus *Messor* and on isopods.	[156]
Pax islamita (Simon)	I	Polyphagous but will eat ants smaller than themselves. Uses different hunting behaviors for ants than for other prey: bites, releases, re-approaches, bites again.	[155]
Trygetus sexoculatus (O. P.-Cambridge)	O	Paralysis latency longer for male and juvenile attacks than for female attacks.	[157]
Trygetus spp.	O	Paralysis latency longer for male and juvenile attacks than for female attacks.	[155, 157]
Zodariellum asiaticum (Tyschchenko)	O	Specializes on formicine ants. Attacks other kinds of ants readily but there is shorter paralysis latency for formicine ants suggesting biochemical specificity of venom for certain kinds of ants.	[155]
Zodariellum spp.	Probably all O	Feed on ants.	[155]
Zodarion cyrenaicum Denis	O	Shows cooperative foraging behavior. But some individuals steal prey from others (kleptoparasitism). Paralysis latency longer for male and juvenile attacks than for female attacks.	[157–159]
Zodarion frenatum (Simon)	O	Feeds on *Cataglyphis bicolor* (Fabricius). Locates nests at night (maybe via odor cues?). Sometimes builds retreats near nest. Digs open closed nest entrances, which triggers ants to come out and repair. Spider sometimes enters nest. Bites ant's legs and carries paralyzed ant away from nest. Also kills ants in morning when they emerge from nest.	[158, 160, 161]
Zodarion germanicum (C. L. Koch)	O	Myrmecomorph as well as myrmecophage. Waves 1st legs as antennal illusion. Holds dead ant in chelicerae and presents dead ant to approaching live ant while "antennating" live ant with its own forelegs. Presumably presenting both odor and tactile cues to living ant to deceive it and avoid attack. Attacks *Cataglyphis bicolor* (Fabricius).	[162, 163]
Zodarion jozefienae Bosmans	O	Females and juveniles actively hunt ants. Mature males are kleptoparasites on females' prey (spend energy on mate searching, not prey capture). Sexual size dimorphism (females larger).	[161, 164, 165]
Zodarion lutipes (O. P.-Cambridge)	O	Paralysis latency longer for male and juvenile attacks than for female attacks.	[157]
Zodarion nitidum (Audouin)	O	Paralysis latency longer for male and juvenile attacks than for female attacks.	[157]
Zodarion rubidum Simon	O	Myrmecomorph as well as myrmecophage. Waves 1st legs as antennal illusion. Holds dead ant in chelicerae and presents dead ant to approaching live ant while "antennating" live ant with its own forelegs. Presumably presenting both odor and tactile cues to living ant to deceive it and avoid attack.	[163, 166–168]

TABLE 2: Continued.

Spider myrmecophage	Category of myrmecophage*	Notes on biology	References
Zodarion spp.	O	All species obligate myrmecophages. Species also imperfect myrmecomorphs. Documented hunting various species. Do not survive well on non-ant diet. Seem to be behaviorally adapted to hunt ants and seem to have evolved nutritional limitations (non-ant prey do not provide required nutrients). Attack from rear, bite legs, retreat, may repeat, re-approach, pick up, and carry away paralyzed ants. Move front legs while hunting. Have femoral organ that may secrete chemical involved in prey capture.	[49, 98, 151, 157, 158, 160, 161, 166, 168–174]

(a)   (b)

FIGURE 3: Myrmecophagy in spiders. (a) *Zodarion rubidum* Simon eating an ant. (b) Femoral organ on *Z. rubidum*. Note the pore openings in the chitin between the two specialized setae of the femoral organ. Scale bar in (a) = 1 mm, scale bar in (b) = 10 μm. Photo of spider © author, SEM of femoral organ © Catherine Tuell, used by permission.

suggesting that the spider may have organs on its front legs that pick up chemical cues from ants [152, 153]. When these spiders detect chemical cues left by ants, they adopt prey capture posture and behavior [153]. Zodariid spiders in the genus *Zodarion* have a structure on the dorsolateral distal tip of the first femora called the femoral organ (Figure 3(b)). The organ consists of pores surrounded by specialized setae with secretory cells beneath the cuticle [171]. It is hypothesized that the femoral organ may release chemicals that somehow subdue the ants upon which the spiders prey (the setae may facilitate dispersion of the secretion) [171]. *Zodarion rubidum* Simon (and other species in the genus) move their front legs around while moving through the environment, similar to the antennal illusion of myrmecomorphs. The spiders seem to use the legs (perhaps via the femoral organs) to pick up cues about ants and conspecifics the spiders may encounter [166]. Recent work by Pekár and Jiroš [177] tested whether various species of myrmecomorphs including one myrmecophage, *Zodarion alacre* (Simon), were also chemical mimics of ants. They found little overlap in the chemical signature of the spiders and ants. Only a weak similarity in profiles was seen for the myrmecophage. The authors hypothesized that the femoral organ of *Zodarion* may be used to synthesize the compounds responsible for the similarity.

The family that includes the most specialized (stenophagous) myrmecophages is the Zodariidae (Table 2). Plesio-morphic representatives of this family, *Lachesana insensibilis* Jocqué and *Pax islamita* (Simon), are polyphagous but will eat ants and hunt ants differently from other prey [155]. Thus these plesiomorphic representatives of zodariids have behavioral preadaptations for hunting ants [155]. Pekár hypothesized that obligatory myrmecophagy may be a derived behavior because, within the Zodariidae, it is only seen in more recent taxa; primitive representatives of the family seem to be polyphagous [98].

4.3. Specialized Hunting Behaviors of Spider Myrmecophages. The majority of reluctant or indifferent myrmecophages will accept ants in the diet but typically show no specialized hunting behavior for these potentially dangerous predators, whereas the majority of facultative and obligatory myrmecophages have evolved specialized hunting strategies to subdue ants with minimum risk to themselves. It has been pointed out that "*when predators evolve prey-specific capture behaviour for use against dangerous prey, they also tend to evolve distinct preferences for these dangerous prey*" [114, 178]. Hunting dangerous but abundant and/or high quality prey seems to select for behavioral plasticity in hunting behavior [105]. Such behavioral flexibility, or using different hunting strategies depending on the identity of the prey and on the circumstances, is common to both myrmecophagic and araneophagic spiders [11, 105, 179].

Many myrmecophagic spiders, particularly facultative or obligatory predators, live in close proximity to ant colonies, often building their webs directly over nest entrances or foraging trails or establishing retreats close to or adjacent to nest mounds [44, 93, 122, 124, 126, 127, 130, 131, 134, 136–138, 140, 143, 147, 158, 160, 161, 180]. In addition to living in close proximity to their prey, these spiders also show specialized hunting behaviors as predicted for stenophagous predators hunting dangerous prey. Web-building myrmecophages (largely in the family Theridiidae, see Table 2) often build webs directly over ant foraging trails where they extend sticky silk strands down to the substrate. When an ant contacts the sticky strand, the ant is catapulted into the air and into the aboveground portion of the web where the spider waits [4, 93, 124, 127, 136–138, 140]. The spider then typically bites the ant one or more times and, each time, the spider retreats until the ant is paralyzed or moribund [93, 130, 131]. The spider then typically carries the ant to a secluded retreat to feed or may even drop on a line to feed [93, 124, 130, 131, 136–138] (possibly to avoid detection from worker ants that may be attracted to alarm pheromones released by the captured ant). When catching non-ant prey, theridiids and other web building spiders do not typically retreat after biting the prey and may or may not carry the paralyzed prey to a different part of the web.

Non-web-building spiders, such as zodariids and salticids (the other families with large numbers of myrmecophagic species), show similar specialized hunting behaviors when attacking ants. For example, zodariids typically attack quickly from the rear of an ant, bite a leg, retreat, and may repeat this sequence several times until the ant is paralyzed. The spider then lifts the moribund ant and carries it to a secluded place to feed (Table 2 and [98, 158, 162, 167, 168]). It has been suggested that the paralyzed ant is used as a shield and a decoy to protect the zodariid from attacks by living ants; the paralyzed ant provides pheromone cues to a curious worker ant that passes by and may provide tactile cues as well [163, 166, 167]. Additional tactile cues are provided by the zodariid, which holds and waves its first pair of legs in front of its body like antennae [163]. The crab spider, *Aphantochilus rogersi* O. P.-Cambridge (Thomisidae), also uses the paralyzed ant as a shield, presumably protecting it from attacks by living ants [142, 143].

Many salticids lunge, rather than jump, at ant prey, then quickly bite, release, and bite again, each time retreating. Even nonmyrmecomorphic ant-eating salticids hunt ants by lunging. This is quite different from the usual stalk and pounce behavior shown to non-ant prey. Myrmecophagic salticids are much more cautious in their approach of ants and much more deliberate in where they bite the prey; some nearly always position themselves in front of the ant and bite the petiole or thorax [92, 105, 106, 108, 109, 114]. Others nearly always attack ants from the rear, lunging at the gaster (not the appendages), but always retreating and waiting until the ant is quiescent before carrying it away [99, 106, 109, 110]. Many salticids keep their front legs extended off the ground when attacking an ant, away from the ant's mandibles [108, 113]. The salticids do not show these behaviors when hunting non-ant prey. The salticids, *Zenodorus durvillei*

(Walckenaer), *Z. metallescens* (L. Koch), and *Z. orbiculatus* (Keyserling), are all facultative myrmecophages that feed on ants caught in other spiders' webs, but only if there is a safe way to capture these prey [106]. These species of *Zenodorus* will walk across a line of detritus to the captured ant or will even hang upside down above the ant and lunge at the prey caught in the web [106]. Some spider myrmecophages, particularly *Callilepis nocturna* (L) (Gnaphosidae), and species of *Oecobius* (Oecobiidae) aim for the ant's antenna when hunting then retreat and wait as is seen in nearly all other species of myrmecophages [96, 97, 100].

4.4. Nutritional Costs of Myrmecophagy and a Stenophagous Diet. It has recently been demonstrated that at least some obligatory myrmecophages do not survive well on an ant-poor diet; some even starve rather than hunt non-ant prey [173]. Thus obligatory myrmecophages show both behavioral limitations (i.e., spiders are reluctant to hunt non-ant prey) and nutritional limitations (i.e., non-ant prey do not provide required nutrients for survival) [173]. In fact, in order to obtain the necessary nutrients for survival, these spiders selectively consume particular parts of the bodies of their ant prey suggesting that "*specialist predators can use a behavioral strategy to balance nutrient intake by selective exploitation of different prey body parts*" [174]. These authors found, for example, that *Zodarion rubidum* preferentially fed on the foreparts of the ant body, which were richer in proteins, than on the gaster, which is higher in lipids but also contains possible toxins such as formic acid. These obligatory myrmecophages may take their specialization a step further by feeding primarily on one or two types of their preferred prey. For example, *Zodarion* species possess more effective venoms against particular groups of ants, such as formicine ants rather than myrmicine ants [151, 157, 170]. *Zodarion germanicum* (C. L. Koch) does better, in terms of growth and survival, on a diet that includes the preferred formicine ants than on a diet restricted to myrmicine ants [172].

5. Discussion

Research on spider myrmecomorphs has demonstrated, unequivocally, that these spiders are Batesian mimics and that the mimicry confers strong adaptive advantages to their survival. Some research has also tested how and why polymorphic and imperfect mimicry evolved. Future research on myrmecomorphic spiders should focus on the costs, trade-offs, and constraints inherent in the evolution of close morphological (and behavioral) resemblance to ants. These factors may have a significant impact on the accuracy of the resemblance. It is also important to identify the selective agents involved in this type of mimetic resemblance since the characteristics of the selective agents (e.g., the visual acuity of the selective agents and whether there is more than one actor in the drama) may explain the phenomena of polymorphic and imperfect mimicry.

Research on spider myrmecophiles has not been extensive in the years since the first review article. Nevertheless, the research that has been carried out, particularly on the species

Gamasomorpha maschwitzi and *Cosmophasis bitaeniata*, is fascinating and demonstrates that the biology of these symbiotic spiders is closely linked to the lifestyle and biology of the host ants. From my earlier review article [5], and from Table 1, it is clear that many more species of myrmecophilic spiders can be studied and details of their biology explored. In the section on spider myrmecophiles, I suggest additional directions for future research such as: What adaptations are involved in colony integration? How do myrmecophiles disperse to neighboring colonies? Do all spider myrmecophiles mimic colony odors? To what extent can myrmecophiles interpret the chemical cues released by the hosts? What is the population structure of spider myrmecophiles (i.e., is the spider population within a single nest made up of close relatives)?

I also provide a summary of what is known about spider myrmecophages and present an extensive table listing all (I hope) records of spider myrmecophages from the literature. Recent research on these specialist predators has revealed the evolutionary costs and benefits of this stenophagous diet. It has also highlighted the extraordinary morphological and behavioral adaptations that have evolved enabling spiders to specialize on such dangerous prey.

Although spiders and ants seem unlikely co-evolutionary partners given ants' territorial aggressiveness and spiders' solitary lifestyles, it is clear that hundreds of species of spiders have evolved close relationships with ants. The information on spider myrmecomorphs, myrmecophiles, and myrmecophages included herein supplements information presented in the 1997 review [5]. The present paper includes the first comprehensive summary of the extensive research on myrmecophagic spiders. In addition, it presents an overview of the research carried out since 1997 that examines the evolutionary costs and benefits of the various spider-ant associations. One of my primary goals has been to provide ideas for new or expanded avenues of research on these fascinating arthropod relationships.

Acknowledgments

Sincere thanks to Kathy Honda for her extraordinary efforts and talent at tracking down citations for this paper. This paper would not have been possible without her assistance. Thanks are also due to Julie Whitman-Zai and two anonymous reviewers for helpful suggestions for improving the paper.

References

[1] R. F. Foelix, *Biology of Spiders*, Oxford University Press, 3rd edition, 2010.

[2] B. Hölldobler and E. O. Wilson, *The Ants*, Harvard University Press, Cambridge, Mass, USA, 1990.

[3] J. D. McIver and G. Stonedahl, "Myrmecomorphy: morphological and behavioral mimicry of ants," *Annual Review of Entomology*, vol. 38, pp. 351–379, 1993.

[4] M. A. Elgar, "Inter-specific associations involving spiders: kleptoparasitism, mimicry and mutualism," *Memoirs. Queensland Museum*, vol. 33, no. 2, pp. 411–430, 1993.

[5] P. E. Cushing, "Myrmecomorphy and myrmecophily in spiders: a review," *Florida Entomologist*, vol. 80, no. 2, pp. 165–193, 1997.

[6] J. Reiskind, "Morphological adaptation for ant-mimicry in spiders," in *Proceedings of the 5th International Congress of Arachnology*, vol. 1971, pp. 221–226, Brno, Czech Republic, 1972.

[7] J. Reiskind, "Ant-mimicry in Panamanian clubionid and salticid spiders (Araneae: Clubionidae, Salticidae)," *Biotropica*, vol. 9, no. 1, pp. 1–8, 1977.

[8] S. P. Benjamin, "Taxonomic revision and phylogenetic hypothesis for the jumping spider subfamily Ballinae (Araneae, Salticidae)," *Zoological Journal of the Linnean Society*, vol. 142, no. 1, pp. 1–82, 2004.

[9] F. S. Ceccarelli, "New species of ant-mimicking jumping spiders of the genus *Myrmarachne* MacLeay, 1839 (Araneae: Salticidae) from north Queensland, Australia," *Australian Journal of Entomology*, vol. 49, pp. 245–255, 2010.

[10] R. R. Jackson and M. B. Willey, "The comparative study of the predatory behaviour of *Myrmarachne*, ant-like jumping spiders (Araneae: Salticidae)," *Zoological Journal of the Linnean Society*, vol. 110, no. 1, pp. 77–102, 1994.

[11] R. R. Jackson and S. D. Pollard, "Predatory behavior of jumping spiders," *Annual Review of Entomology*, vol. 41, pp. 287–308, 1996.

[12] X. J. Nelson and R. R. Jackson, "Compound mimicry and trading predators by the males of sexually dimorphic Batesian mimics," *Proceedings of the Royal Society B*, vol. 273, no. 1584, pp. 367–372, 2006.

[13] C. L. Deeleman-Reinhold, "A new spider genus from Thailand with a unique ant mimicking device," *The Natural History Bulletin of the Siam Society*, vol. 40, no. 2, pp. 167–184, 1993.

[14] W. Dittrich, F. Gilbert, P. Green, P. Mcgregor, and D. Grewcock, "Imperfect mimicry: a pigeon's perspective," *Proceedings of the Royal Society B*, vol. 251, pp. 195–200, 1993.

[15] R. A. Johnstone, "The evolution of inaccurate mimics," *Nature*, vol. 418, pp. 524–526, 2002.

[16] A. Collart, "Quelques notes sur les *Myrmarachne* araignées oecophylliformes," *Bulletin du Cercle Zoologique Congolais*, vol. 5, pp. 117–118, 1929.

[17] A. Collart, "Quelques observations sur une araignée mimétique," *Revue de Zoologie et de Botanique Africaines*, vol. 18, pp. 147–161, 1929.

[18] A. Collart, "Notes Complémentares sur *Myrmarachne foenisex* Simon araignée myrmécomorphe du Congo Belge," *Bulletin du Musée Royal d'Histoire Naturelle de Belgique*, vol. 17, pp. 1–11, 1941.

[19] R. M. Borges, S. Ahmed, and C. V. Prabhu, "Male ant-mimicking salticid spiders discriminate between retreat silks of sympatric females: implications for pre-mating reproductive isolation," *Journal of Insect Behavior*, vol. 20, no. 4, pp. 389–402, 2007.

[20] X. J. Nelson, "Visual cues used by ant-like jumping spiders to distinguish conspecifics from their models," *Journal of Arachnology*, vol. 38, no. 1, pp. 27–34, 2010.

[21] M. G. Rix, "A new genus and species of ant-mimicking jumping spider (Araneae: Salticidae) from Southeast Queensland, with notes on its biology," *Memoirs of the Queensland Museum*, vol. 43, no. 2, pp. 827–832, 1999.

[22] F. S. Ceccarelli, "Behavioral mimicry in *Myrmarachne* species (Araneae, Salticidae) from North Queensland, Australia," *Journal of Arachnology*, vol. 36, no. 2, pp. 344–351, 2008.

[23] R. R. Jackson, "The biology of ant-like jumping spiders (Araneae, Salticidae): prey and predatory behaviour of *Myrmarachne* with particular attention to *M. lupata* from Queensland," *Zoological Journal of the Linnean Society*, vol. 88, no. 2, pp. 179–190, 1986.

[24] R. R. Jackson, X. J. Nelson, and K. Salm, "The natural history of *Myrmarachne melanotarsa*, a social ant-mimicking jumping spider," *New Zealand Journal of Zoology*, vol. 35, no. 3, pp. 225–235, 2008.

[25] W. Engelhardt, "Gestalt und Lebensweise der "Ameisenspinne" *Synageles venator* (Lucas) Zugleich ein Beitrag zur Ameisenmimikryforschung," *Zoologischer Anzeiger*, vol. 185, no. 5-6, pp. 317–334, 1970.

[26] B. Cutler, "Reduced predation on the antlike jumping spider *Synageles occidentalis* (Araneae: Salticidae)," *Journal of Insect Behavior*, vol. 4, no. 3, pp. 401–407, 1991.

[27] X. J. Nelson and R. R. Jackson, "Vision-based innate aversion to ants and ant mimics," *Behavioral Ecology*, vol. 17, no. 4, pp. 676–681, 2006.

[28] X. J. Nelson, R. R. Jackson, D. Li, A. T. Barrion, and G. B. Edwards, "Innate aversion to ants (Hymenoptera: Formicidae) and ant mimics: experimental findings from mantises (Mantodea)," *Biological Journal of the Linnean Society*, vol. 88, no. 1, pp. 23–32, 2006.

[29] X. J. Nelson and R. R. Jackson, "Collective Batesian mimicry of ant groups by aggregating spiders," *Animal Behaviour*, vol. 78, pp. 123–129, 2009.

[30] J.-N. Huang, R.-C. Cheng, D. Li, and I.-M. Tso, "Salticid predation as one potential driving force of ant mimicry in jumping spiders," *Proceedings of the Royal Society B*, vol. 278, pp. 1356–1364, 2011.

[31] C. A. Durkee, M. R. Weiss, and D. B. Uma, "Ant mimicry lessens predation on a North American jumping spider by larger salticid spiders," *Environmental Entomology*, vol. 40, no. 5, pp. 1223–1231, 2011.

[32] X. J. Nelson, "A predator's perspective of the accuracy of ant mimicry in spiders," *Psyche*, vol. 2012, Article ID 168549, 5 pages, 2012.

[33] M. Edmunds, "Do Malaysian *Myrmarachne* associate with particular species of ant?" *Biological Journal of the Linnean Society*, vol. 88, no. 4, pp. 645–653, 2006.

[34] O. H. Holen and R. A. Johnstone, "The evolution of mimicry under constraints," *American Naturalist*, vol. 164, no. 5, pp. 598–613, 2004.

[35] O. H. Holen and R. A. Johnstone, "Context-dependent discrimination and the evolution of mimicry," *American Naturalist*, vol. 167, no. 3, pp. 377–389, 2006.

[36] X. J. Nelson, R. R. Jackson, G. B. Edwards, and A. T. Barrion, "Living with the enemy: jumping spiders that mimic weaver ants," *Journal of Arachnology*, vol. 33, no. 3, pp. 813–819, 2005.

[37] F. Gilbert, "The evolution of imperfect mimicry," in *Insect Evolutionary Ecology*, M. D. E. Fellowes, G. J. Holloway, and J. Rolff, Eds., pp. 231–288, CABI, Wallingford, Wash, USA, 2005.

[38] G. B. Edwards and S. P. Benjamin, "A first look at the phylogeny of the Myrmarachninae, with rediscovery and redescription of the type species of *Myrmarachne* (Araneae: Salticidae)," *Zootaxa*, vol. 2309, pp. 1–29, 2009.

[39] D. W. Kikuchi and D. W. Pfennig, "High-model abundance may permit the gradual evolution of Batesian mimicry: an experimental test," *Proceedings of the Royal Society B*, vol. 277, pp. 1041–1048, 2010.

[40] S. Pekár and M. Jarab, "Assessment of color and behavioral resemblance to models by inaccurate myrmecomorphic spiders (Araneae)," *Invertebrate Biology*, vol. 130, no. 1, pp. 83–90, 2011.

[41] M. Edmunds, "Does mimicry of ants reduce predation by wasps on salticid spiders?" *Memoirs - Queensland Museum*, vol. 33, no. 2, pp. 507–512, 1993.

[42] R. Jocqué, "The prey of the mud-dauber wasp, *Scelephron spirifex* (Linnaeus), in Central Africa," *Newsletter of the British Arachnological Society*, vol. 51, p. 7, 1988.

[43] F. S. Ceccarelli and R. H. Crozier, "Dynamics of the evolution of Batesian mimicry: molecular phylogenetic analysis of ant-mimicking *Myrmarachne* (Araneae: Salticidae) species and their ant models," *Journal of Evolutionary Biology*, vol. 20, no. 1, pp. 286–295, 2007.

[44] X. J. Nelson, "Polymorphism in an ant mimicking jumping spider," *Journal of Arachnology*, vol. 38, no. 1, pp. 139–141, 2010.

[45] W. Fannes and R. Jocqué, "Ultrastructure of *Antoonops*, a new ant-mimicking genus of afrotropical Oonopidae (Araneae) with complex internal genitalia," *American Museum Novitates*, vol. 3614, pp. 1–30, 2008.

[46] M. Joron, "Polymorphic mimicry, microhabitat use, and sex-specific behaviour," *Journal of Evolutionary Biology*, vol. 18, pp. 547–556, 2005.

[47] M. P. Speed and G. D. Ruxton, "Imperfect Batesian mimicry and the conspicuousness costs of mimetic resemblance," *American Naturalist*, vol. 176, no. 1, pp. E1–E14, 2010.

[48] T. N. Sherratt, "The evolution of imperfect mimicry," *Behavioral Ecology*, vol. 13, no. 6, pp. 821–826, 2002.

[49] M. Edmunds, "Why are there good and poor mimics?" *Biological Journal of the Linnean Society*, vol. 70, no. 3, pp. 459–466, 2000.

[50] C. J. Duncan and P. M. Sheppard, "Sensory discrimination and its role in the evolution of Batesian mimicry," *Behaviour*, vol. 24, no. 3, pp. 269–282, 1965.

[51] J. Mappes and R. V. Alatalo, "Batesian mimicry and signal accuracy," *Evolution*, vol. 51, no. 6, pp. 2050–2053, 1997.

[52] S. Pekár, M. Jarab, L. Fromhage, and M. E. Herberstein, "Is the evolution of inaccurate mimicry a result of selection by a suite of predators? A case study using myrmecomorphic spiders," *American Naturalist*, vol. 178, no. 1, pp. 124–134, 2011.

[53] X. J. Nelson, D. Li, and R. R. Jackson, "Out of the frying pan and into the fire: a novel trade-off for Batesian mimics," *Ethology*, vol. 112, no. 3, pp. 270–277, 2006.

[54] X. J. Nelson, R. R. Jackson, and D. Li, "Conditional use of honest signaling by a Batesian mimic," *Behavioral Ecology*, vol. 17, no. 4, pp. 575–580, 2006.

[55] S. Pekár and M. Jarab, "Life-history constraints in inaccurate Batesian myrmecomorphic spiders (Araneae: Corinnidae, Gnaphosidae)," *European Journal of Entomology*, vol. 108, no. 2, pp. 255–260, 2011.

[56] W. S. Bristowe, *The Comity of Spiders I*, Ray Society, London, UK, 1939.

[57] W. S. Bristowe, *The Comity of Spiders II*, Ray Society, London, UK, 1941.

[58] M. Edmunds, "On the association between *Myrmarachne* spp. (Salticidae) and ants," *Bulletin of the British Arachnological Society*, vol. 4, pp. 149–160, 1978.

[59] F. R. Wanless, "A revision of the spider genera *Belippo* and *Myrmarachne* (Araneae: Salticidae) in the Ethiopian region," *Bulletin of the British Museum (Natural History), Zoological Series*, vol. 33, pp. 1–139, 1978.

[60] B. L. Bradoo, "A new ant-like spider of the genus *Myrmarachne* (Salticidae) from India," *Current Science*, vol. 49, no. 10, pp. 387–388, 1980.

[61] J.-L. Boevé, "Association of some spiders with ants," *Revue Suisse de Zoologie*, vol. 99, pp. 81–85, 1992.

[62] H. Donisthorpe, *The Guests of British Ants, Their Habits and Life Histories*, Routledge and Sons, London, UK, 1927.

[63] N. Platnick, "The world spider catalog, version 12.0," American Museum of Natural History, http://research.amnh.org/iz/spiders/catalog.

[64] K. Thaler, A. Kofler, and E. Meyer, "Fragmenta faunistica Tirolensia–IX," *Berichte des Naturwissenschaftlich-Medizinischen Vereins in Innsbruck*, vol. 77, pp. 225–243, 1990.

[65] J. Wunderlich, "Beschreibung bisher unbekannter Spinnenarten und Gattungen aus Malaysia und Indonesien (Arachnida: Araneae: Oonopidae, Tetrablemidae, Telemidae und Dictynidae)," *Beitraege zur Araneologie*, vol. 4, pp. 559–580, 1994.

[66] L. Fage, "Quelques Arachnides provenant de fourmilières ou de termitières du Costa Rica," *Bulletin Museum National d'Histoire Naturelle*, vol. 4, pp. 369–376, 1938.

[67] V. Witte, H. Hänel, A. Weissflog, H. Rosli, and U. Maschwitz, "Social integration of the myrmecophilic spider *Gamasomorpha maschwitzi* (Araneae: Oonopidae) in colonies of the South East Asian army ant, *Leptogenys distinguenda* (Formicidae: Ponerinae)," *Sociobiology*, vol. 34, pp. 145–159, 1999.

[68] V. Witte, A. Leingärtner, L. Sabab, R. Hashim, and S. Foitzik, "Symbiont microcosm in an ant society and the diversity of interspecific interactions," *Animal Behaviour*, vol. 76, pp. 1477–1486, 2008.

[69] V. Witte, S. Foitzik, R. Hashim, U. Maschwitz, and S. Schulz, "Fine tuning of social integration by two myrmecophiles of the ponerine army ant, *Leptogenys distinguenda*," *Journal of Chemical Ecology*, vol. 35, no. 3, pp. 355–367, 2009.

[70] M. Birabén, "Nuevas "Gamasomorphinae" de la Argentina," *Notas del Museo, Universidad Nacional de Eva Perón*, vol. 17, no. 152, pp. 181–212, 1954.

[71] B. Gray, "Notes on the biology of the ant species *Myrmecia dispar* (Clark) (Hymenoptera: Formicidæ)," *Insectes Sociaux*, vol. 18, no. 2, pp. 71–109, 1971.

[72] R. A. Allan and M. A. Elgar, "Exploitation of the green tree ant, *Oecophylla smaragdina*, by the salticid spider *Cosmophasis bitaeniata*," *Australian Journal of Zoology*, vol. 49, no. 2, pp. 129–137, 2001.

[73] R. A. Allan, R. J. Capon, W. V. Brown, and M. A. Elgar, "Mimicry of host cuticular hydrocarbons by salticid spider *Cosmophasis bitaeniata* that preys on larvae of tree ants *Oecophylla smaragdina*," *Journal of Chemical Ecology*, vol. 28, no. 4, pp. 835–848, 2002.

[74] M. A. Elgar and R. A. Allan, "Predatory spider mimics acquire colony-specific cuticular hydrocarbons from their ant model prey," *Naturwissenschaften*, vol. 91, no. 3, pp. 143–147, 2004.

[75] M. A. Elgar and R. A. Allan, "Chemical mimicry of the ant *Oecophylla smaragdina* by the myrmecophilous spider *Cosmophasis bitaeniata* is it colony-specific?" *Journal of Ethology*, vol. 24, no. 3, pp. 239–246, 2006.

[76] R. H. Crozier, P. S. Newey, E. A. Schlüns, and S. K. A. Robson, "A masterpiece of evolution—*Oecophylla* weaver ants (Hymenoptera: Formicidae)," *Myrmecological News*, vol. 13, pp. 57–71, 2010.

[77] X. J. Nelson and R. R. Jackson, "The influence of ants on the mating strategy of a myrmecophilic jumping spider (Araneae, Salticidae)," *Journal of Natural History*, vol. 43, no. 11-12, pp. 713–735, 2009.

[78] C. F. Roewer, "Zwei myrmecophile Spinnen-Arten Brasiliens," *Veröffentlichungen Deutschen Kolon, Uebersee Museum, Bremen*, vol. 1, pp. 193–197, 1935.

[79] P. E. Cushing, "Population structure of the ant nest symbiont *Masoncus pogonophilus* (Araneae: Linyphiidae)," *Annals of the Entomological Society of America*, vol. 91, no. 5, pp. 626–631, 1998.

[80] P. E. Cushing, "Description of the spider *Masoncus pogonophilus* (Araneae, Linyphiidae), a harvester ant myrmecophile," *Journal of Arachnology*, vol. 23, pp. 55–59, 1995.

[81] S. D. Porter, "*Masoncus* spider: a miniature predator of Collembola in harvester ant colonies," *Psyche*, vol. 92, no. 1, pp. 145–150, 1985.

[82] M. Erthal Jr. and A. Tonhasca Jr., "*Attacobius attarum* spiders (Corinnidae): myrmecophilous predators of immature forms of the leaf-cutting ant *Atta sexdens* (Formicidae)," *Biotropica*, vol. 33, no. 2, pp. 374–376, 2001.

[83] J. A. Thomas, K. Schönrogge, and G. W. Elmes, "17. Specializations and host associations of social parasites of ants," in *Insect Evolutionary Ecology*, M. D. E. Fellowes, G. J. Holloway, and J. Rolff, Eds., pp. 479–518, Royal Entomological Society, CABI Publishing, UK, 2005.

[84] A. Lenoir, P. D'Ettorre, C. Errard, and A. Hefetz, "Chemical ecology and social parasitism in ants," *Annual Review of Entomology*, vol. 46, pp. 573–599, 2001.

[85] R. W. Howard, R. D. Akre, and W. B. Garnett, "Chemical mimicry of an obligate predator of carpenter ants (Hymenoptera: Formicidae)," *Annals of the Entomological Society of America*, vol. 83, pp. 607–616, 1990.

[86] K. Dettner and C. Liepert, "Chemical mimicry and camouflage," *Annual Review of Entomology*, vol. 39, pp. 129–154, 1994.

[87] N. I. Platnick and R. L. C. Baptista, "On the spider genus *Attacobius* (Araneae, Dionycha)," *American Museum Novitates*, vol. 3120, pp. 1–9, 1995.

[88] A. C. Cole Jr., *Pogonomyrmex harvester ants: a study of the Genus in North America*, The University of Tennessee Press, Knoxville, Tenn, USA, 1968.

[89] J. B. Gentry, "Response to predation by colonies of the Florida harvester ant, *Pogonomyrmex badius*," *Ecology*, vol. 55, pp. 1328–1338, 1974.

[90] J. Antonovics, "Ecological genetics of metapopulations: the *Silene-Ustilago* plant-pathogen system," in *Ecological Genetics*, L. A. Real, Ed., pp. 3–17, Princeton University Press, Princeton, NJ, USA, 1994.

[91] J. Wunderlich, "Ober "Ameisenspinnen" in Mitteleuropa (Arachnida: Araneae)," *Beitrage Araneologie*, vol. 4, pp. 447–470, 1994.

[92] E. F. Huseynov, R. R. Jackson, and F. R. Cross, "The meaning of predatory specialization as illustrated by *Aelurillus m-nigrum*, an ant-eating jumping spider (Araneae: Salticidae) from Azerbaijan," *Behavioural Processes*, vol. 77, no. 3, pp. 389–399, 2008.

[93] W. P. MacKay, "The effect of predation of western widow spiders (Araneae: Theridiidae) on harvester ants (Hymenoptera: Formicidae)," *Oecologia*, vol. 53, pp. 406–411, 1982.

[94] A. D. Austin and A. D. Blest, "The biology of two Australian species of dinopid spider," *Journal of Zoology*, vol. 189, pp. 145–156, 1979.

[95] A. S. Dippenaar-Schoeman, "A revision of the African genus *Seothyra* Purcell (Araneae: Eresidae)," *Cymbebasia*, vol. 12, pp. 35–160, 1991.

[96] G. Heller, *Zur Biologie der ameisenfressenden Spinne Callilepis nocturna Linnaeus 1758 (Araneae, Drassodidae)*, Dissertation, Johannes Gutenberg-Universität, Mainz, Germany, 1974.

[97] G. Heller, "Zum Beutefangverhalten der ameisenfressenden Spinne *Callilepis nocturna* (Arachnida: Araneae: Drassodidae)," *Entomolgische Germanica*, vol. 3, pp. 100–103, 1976.

[98] S. Pekár, "Predatory behavior of two European ant-eating spiders (Araneae, Zodariidae)," *Journal of Arachnology*, vol. 32, no. 1, pp. 31–41, 2004.

[99] K. Wing, "*Tutelina similis* (Araneae: Salticidae): an ant mimic that feeds on ants," *Journal of the Kansas Entomological Society*, vol. 56, no. 1, pp. 55–58, 1983.

[100] L. Glatz, "Zur Biologie und Morphologie von *Oecobius annulipes* Lucas (Araneae, Oecobiidae)," *Zoomorphology*, vol. 61, no. 2, pp. 185–214, 1967.

[101] N. A. Weber, "Fungus-growing ants and their fungi: *Cyphomyrmex costatus*," *Ecology*, vol. 38, pp. 480–494, 1957.

[102] E. F. O. Huseynov, "The prey of the lynx spider *Oxyopes globifer* (Araneae, Oxyopidae) associated with a semidesert dwarf shrub in Azerbaijan," *Journal of Arachnology*, vol. 34, no. 2, pp. 422–426, 2006.

[103] M. Nyffeler, D. A. Dean, and W. L. Sterling, "Diets, feeding specialization, and predatory role of two lynx spiders, *Oxyopes salticus and Peucetia viridans* (Araneae: Oxyopidae) , in a Texas cotton agroecosystem," *Environmental Entomology*, vol. 21, pp. 1457–1465, 1992.

[104] W. P. MacKay and S. B. Vinson, "Evaluation of the spider *Steatoda triangulosa* (Araneae: Theridiidae) as a predator of the red imported fire ant (Hymenoptera: Formicidae)," *Journal of the New York Entomological Society*, vol. 97, no. 2, pp. 232–233, 1989.

[105] D. Li, R. R. Jackson, and D. P. Harland, "Prey-capture techniques and prey preferences of *Aelurillus aeruginosus*, *A. cognatus*, and *A. kochi*, ant-eating jumping spiders (Araneae: Salticidae) from Israel," *Israel Journal of Zoology*, vol. 45, pp. 341–359, 1999.

[106] R. R. Jackson and D. Li, "Prey-capture techniques and prey preferences of *Zenodorus durvillei*, *Z. metallescens* and *Z. orbiculatus*, tropical ant-eating jumping spiders (Araneae: Salticidae) from Australia," *New Zealand Journal of Zoology*, vol. 28, pp. 299–341, 2001.

[107] G. B. Edwards, J. F. Carroll, and W. H. Whitcomb, "*Stoidis aurata* (Araneae: Salticidae), a spider predator of ants," *The Florida Entomologist*, vol. 57, no. 4, pp. 337–346, 1975.

[108] R. R. Jackson and A. van Olphen, "Prey-capture techniques and prey preferences of *Corythalia canosa* and *Pystira orbiculata*, ant-eating jumping spiders (Araneae, Salticidae)," *Journal of Zoology, London*, vol. 223, pp. 577–591, 1991.

[109] R. R. Jackson, D. Li, A. T. Barron, and G. B. Edwards, "Prey-capture techniques and prey preferences of nine species of ant-eating jumping spiders (Araneae: Salticidae) from the Philippines," *New Zealand Journal of Zoology*, vol. 25, pp. 249–272, 1998.

[110] R. R. Jackson and A. van Olphen, "Prey-capture techniques and prey preferences of *Chrysilla*, *Natta* and *Siler*, ant-eating jumping spiders (Araneae, Salticidae) from Kenya and Sri Lanka," *Journal of Zoology, London*, vol. 227, pp. 163–170, 1992.

[111] N. Ii, "Observations on a strange plundering behaviour in salticid spiders," *Acta Arachnologica*, vol. 27, pp. 209–212, 1977 (Japanese).

[112] H. S. Fitch, "Spiders of the University of Kansas Natural History Reservation and Rockefeller Experimental Tract," *Miscellaneous Publications of the University of Kansas Museum of Natural History*, vol. 33, pp. 1–202, 1963.

[113] B. Cutler, "Ant predation by *Habrocestum pulex* (Hentz) (Araneae: Salticidae)," *Zoologischer Anzeiger*, vol. 204, no. 1-2, pp. 97–101, 1980.

[114] D. Li, R. R. Jackson, and B. Cutler, "Prey-capture techniques and prey preferences of *Habrocestum pulex*, an ant-eating jumping spider (Araneae, Salticidae) from North America," *Journal of Zoology, London*, vol. 240, pp. 551–562, 1996.

[115] J. L. Cloudsley-Thompson, "Notes on Arachnida 12. Mating habits of *Hasarius adansoni* Say," *Entomologist's Monthly Magazine*, vol. 85, pp. 211–212, 1949.

[116] R. R. Jackson, "Prey of the jumping spider, *Phidippus johnsoni* (Araneae: Salticidae)," *Journal of Arachnology*, vol. 5, no. 2, pp. 145–149, 1977.

[117] K. Nakahira, "Observations on a jumping spider *Silerella vittata* that attacks ants (2)," *Atypus*, vol. 9, p. 1, 1955 (Japanese).

[118] S. Jo, "Observations on jumping spider that carries off ant larva," *Atypus*, vol. 32, p. 11, 1964 (Japanese).

[119] K. Nakahira, "Ants on which jumping spider *Silerella vittata* preys," *Atypus*, vol. 61, p. 15, 1973 (Japanese).

[120] Y. Touyama, Y. Ihara, and F. Ito, "Argentine ant infestation affects the abundance of the native myrmecophagic jumping spider *Siler cupreus* Simon in Japan," *Insectes Sociaux*, vol. 55, no. 2, pp. 144–146, 2008.

[121] B. J. Kaston, "Spiders of Connecticut," *State of Connecticut Geological and Natural History Survey Bulletin*, vol. 70, pp. 1–374, 1981.

[122] B. Hölldobler, "*Steatoda fulva* (Theridiidae), a spider that feeds on harvester ants," *Psyche*, vol. 77, no. 2, pp. 202–208, 1970.

[123] J. T. Moggridge, *Harvesting ants and trapdoor spiders with notes and observations on their habits and dwellings*, London, UK, 1873.

[124] E. Nørgaard, "Environment and behavior of *Theridion saxatile*," *Oikos*, vol. 7, pp. 159–192, 1956.

[125] Y. Umeda, A. Shinkai, and T. Miyashita, "Prey composition of three *Dipoena* spp. (Araneae: Theridiidae) specializing on ants," *Acta Arachnologica*, vol. 45, no. 1, pp. 95–99, 1996.

[126] H. C. McCook, *The Natural History of the Agricultural Ant of Texas. A Monograph of the Habits, Architecture and Structure of Pogonomyrmex barbatus*, Lippincott and Co., Philadelphia, PA, USA, 1880.

[127] S. D. Porter and D. A. Eastmond, "*Euryopis coki* (Theridiidae), a spider that preys on *Pogonomyrmex* ants," *Journal of Arachnology*, vol. 10, no. 3, pp. 275–277, 1982.

[128] L. Berland, "Contribution a l'étude de la biologie des arachnides," *Archives de Zoologie Expérimentale et Générale*, vol. 76, no. 1, pp. 1–23, 1933.

[129] W. H. Clark and P.E. Blom, "Notes on spider (Theridiidae, Salticidae) predation of the harvester ant, *Pogonomyrmex salinus* Olsen (Hymenoptera: Formicidae: Myrmicinae), and a possible parasitoid fly (Chloropidae)," *Great Basin Naturalist*, vol. 52, pp. 385–386, 1992.

[130] A. F. Archer, "The Theridiidae or comb-footed spiders of Alabama," *Museum Papers of the Alabama Museum of Natural History*, vol. 22, pp. 5–67, 1946.

[131] J. E. Carico, "Predatory behavior of *Euryopis funeris* (Hentz) (Araneae: Theridiidae) and the evolutionary significance of web reduction," *Symposium of the Zoological Society of London*, vol. 42, pp. 51–58, 1978.

[132] H. W. Levi, "Spiders of the genus *Euryopis* from North and Central America," *American Museum Novitates*, vol. 1666, pp. 1–48, 1954.

[133] W. J. Gertsch, *American Spiders*, Van Nostrand Reinhold, New York, NY, USA, 2nd edition, 1979.

[134] J. W. Abalos, "Las arañas del género Latrodectus en la Argentina," *Obra del Centenario del Museo de La Plata*, vol. 6, pp. 29–51, 1980.

[135] M. Nyffeler, D. A. Dean, and W. L. Sterling, "The southern black widow spider *Latrodectus mactans* (Araneae, Theridiidae), as a predator of the red imported fire ant, *Solenopsis invicta* (Hymenoptera, Formicidae), in Texas cotton fields," *Journal of Applied Entomology*, vol. 106, pp. 52–57, 1988.

[136] A. Shulov and A. Weissman, "Notes on the life habitats and potency of the venom of the three *Latrodectus* spider species in Israel," *Ecology*, vol. 40, pp. 515–518, 1939.

[137] A. Shulov, "On the biology of two *Latrodectus* spiders in Palestine," in *Proceedings of the Linnean Society, London*, vol. 152, pp. 309–328, 1940.

[138] A. Shulov, "Biology and ecology of venomous animals in Israel," *Memórias do Instituto Butantan*, vol. 33, pp. 93–99, 1966.

[139] H. W. Levi, "The spider genera *Crustulina* and *Steatoda* in North America, Central America, and the West Indies (Araneae, Theridiidae)," *Bulletin of the Museum of Comparative Zoology*, vol. 117, no. 3, pp. 367–424, 1957.

[140] D. Cooper, A. Williamson, and C. Williamson, "Deadly deception," *Geo: Australia's Geographical Magazine*, vol. 12, no. 1, pp. 86–95, 1990.

[141] S. de T. Piza, "Novas espécies de aranhas myrmecomorphas do Brazil e considerações sobre o seu mimetismo," *Revista do Museu Paulista*, vol. 23, pp. 307–319, 1937.

[142] P. S. Oliveira and I. Sazima, "The adaptive bases of ant-mimicry in a neotropical aphantochilid spider (Araneae: Aphantochilidae)," *Biological Journal of the Linnean Society*, vol. 22, pp. 145–155, 1984.

[143] L. M. Castanho and P. S. Oliveira, "Biology and behavior of the neotropical ant-mimicking spider *Aphantochilus rogersi* (Araneae: Aphantochilidae): nesting, maternal care and ontogeny of ant-hunting techniques," *Journal of Zoology, London*, vol. 242, pp. 643–650, 1997.

[144] B. Hölldobler, "The behavioral ecology of mating in harvester ants (Hymenoptera: Formicidae: *Pogonomyrmex*)," *Behavioral Ecology and Sociobiology*, vol. 1, pp. 405–423, 1976.

[145] Q. R. Romero and J. Vasconcellos-Neto, "Natural history of *Misumenops argenteus* (Thomisidae): seasonality and diet on *Trichogoniopsis adenantha* (Asteraceae)," *Journal of Arachnology*, vol. 31, no. 2, pp. 297–304, 2003.

[146] E. F. Huseynov, "Natural prey of the crab spider *Thomisus onustus* (Araneae: Thomisidae), an extremely powerful predator of insects," *Journal of Natural History*, vol. 41, no. 37-40, pp. 2341–2349, 2007.

[147] K. C. McKeown, *Australian Spiders: Their Lives and Habits*, Angus & Robertson, Sydney, Australia, 1952.

[148] Y. D. Lubin, "An ant eating crab spider from the Galapagos," *Noticias de Galapagos*, vol. 37, pp. 18–19, 1983.

[149] R. Mascord, *Spiders of Australia*, A. H. and A. W. Reed Pty. Ltd., Sydney, Australia, 1980.

[150] E. F.O. Guseinov, "The prey of a lithophilous crab spider *Xysticus loeffleri* (Araneae, Thomisidae)," *Journal of Arachnology*, vol. 34, no. 1, pp. 37–45, 2006.

[151] S. Pekár, "Capture efficiency of an ant-eating spider, *Zodariellum asiaticum* (Araneae: Zodariidae), from Kazakhstan," *Journal of Arachnology*, vol. 37, no. 3, pp. 388–391, 2009.

[152] R. A. Allan, M. A. Elgar, and R. J. Capon, "Exploitation of an ant chemical alarm signal by the zodariid spider *Habronestes bradleyi* Walckenaer," *Proceedings of the Royal Society B*, vol. 263, no. 1366, pp. 69–73, 1996.

[153] R. J. Clark, R. R. Jackson, and B. Cutler, "Chemical cues from ants influence predatory behavior in *Habrocestum pulex*, an ant-eating jumping spider (Araneae, Salticidae)," *Journal of Arachnology*, vol. 28, no. 3, pp. 309–318, 2000.

[154] H. Gibb, "Dominant meat ants affect only their specialist predator in an epigaeic arthropod community," *Oecologia*, vol. 136, no. 4, pp. 609–615, 2003.

[155] S. Pekár and Y. Lubin, "Prey and predatory behavior of two zodariid species (Araneae, Zodariidae)," *Journal of Arachnology*, vol. 37, no. 1, pp. 118–121, 2009.

[156] S. L. Zonstein and S. V. Ovtschinnikov, "A new Central Asian species of the spider genus *Lachesana* Strand, 1932 (Araneae, Zodariidae: Lachesaninae)," *TETHYS Entomological Research*, vol. 1, pp. 59–62, 1999.

[157] S. Pekár, J. Král, and Y. Lubin, "Natural history and karyotype of some ant-eating zodariid spiders (Araneae, Zodariidae) from Israel," *Journal of Arachnology*, vol. 33, no. 1, pp. 50–62, 2005.

[158] R. D. Harkness, "Further observations on the relation between an ant, *Cataglyphis bicolor* (F.) (Hymenoptera, Formicidae) and a spider, *Zodarium frenatum* (Simon) (Araneae, Zodariidae)," *Entomologist's Monthly Magazine*, vol. 112, pp. 111–121, 1977.

[159] S. Pekár, M. Hrušková, and Y. Lubin, "Can solitary spiders (Araneae) cooperate in prey capture?" *Journal of Animal Ecology*, vol. 74, no. 1, pp. 63–70, 2005.

[160] R. D. Harkness, "The relation between an ant, *Cataglyphis bicolor* (F.) (Hymenoptera, Formicidae) and a spider, *Zodarium frenatum* (Simon) (Araneae, Zodariidae)," *Entomologist's Monthly Magazine*, vol. 111, pp. 141–146, 1976.

[161] M. L. R. Harkness and R. D. Harkness, "Predation of an ant (*Cataglyphis bicolor* (F.) Hymenoptera, Formicidae) by a spider (*Zodarion frenatum* (Simon) Araneae, Zodariidae) in Greece," *Entomologist's Monthly Magazine*, vol. 128, pp. 147–156, 1992.

[162] P. Schneider, "Ameisenjagende Spinnen (Zodariidae) an *Cataglyphis*—Nestern in Afghanistan," *Zoologischer Anzeiger, Leipzig*, vol. 187, no. 3-4, pp. 199–201, 1971.

[163] S. Pekár and J. Král, "Mimicry complex in two central European zodariid spiders (Araneae: Zodariidae): how *Zodarion* deceives ants," *Biological Journal of the Linnean Society*, vol. 75, no. 4, pp. 517–532, 2002.

[164] M. Martišová, T. Bilde, and S. Pekár, "Sex-specific kleptoparasitic foraging in ant-eating spiders," *Animal Behaviour*, vol. 78, no. 5, pp. 1115–1118, 2009.

[165] S. Pekár, M. Martišová, and T. Bilde, "Intersexual trophic niche partitioning in an ant-eating spider (Araneae: Zodariidae)," *PLoS ONE*, vol. 6, no. 1, pp. 1–7, 2011.

[166] J. M. Couvreur, "Le comportement d'presentation d'un leurre' chez *Zodarion rubidium* (Araneae, Zodariidae)," in *Proceedings of the 12th European Colloquium of Arachnology, Paris*, M. L. Célérier, J. Heurtault, and C. Rollard, Eds., vol. 1, pp. 75–79, Bulletin de la Sociéte Européene Arachnologique, Paris, France, 1990.

[167] J. M. Couvreur, "Quelques aspects de la biologie de *Zodarion rubidium*, Simon, 1914," *Nieuwsbrief van de Belgische Arachnologische Vereniging*, vol. 7, pp. 7–15, 1990.

[168] P. E. Cushing and R. G. Santangelo, "Notes on the natural history and hunting behavior of an ant eating zodariid spider (Arachnida, Araneae) in Colorado," *Journal of Arachnology*, vol. 30, no. 3, pp. 618–621, 2002.

[169] R. Jocqué and J. Billen, "The femoral organ of the Zodariinae (Araneae, Zodariidae)," *Revue de Zoologie Africaine*, vol. 101, pp. 165–170, 1987.

[170] S. Pekár, J. Král, A. Malten, and C. Komposch, "Comparison of natural histories and karyotypes of two closely related ant-eating spiders, *Zodarion hamatum* and *Z. italicum* (Araneae, Zodariidae)," *Journal of Natural History*, vol. 39, no. 19, pp. 1583–1596, 2005.

[171] S. Pekár and J. Šobotník, "Comparative study of the femoral organ in *Zodarion* spiders (Araneae: Zodariidae)," *Arthropod Structure and Development*, vol. 36, pp. 105–112, 2007.

[172] S. Pekár, S. Toft, M. Hrušková, and D. Mayntz, "Dietary and prey-capture adaptations by which *Zodarion germanicum*, an ant-eating spider (Araneae: Zodariidae), specializes on the Formicinae," *Naturwissenschaften*, vol. 95, no. 3, pp. 233–239, 2008.

[173] S. Pekár and S. Toft, "Can ant-eating *Zodarion* spiders (Araneae: Zodariidae) develop on a diet optimal for euryphagous arthropod predators?" *Physiological Entomology*, vol. 34, pp. 195–201, 2009.

[174] S. Pekár, D. Mayntz, T. Ribeiro, and M. E. Herberstein, "Specialist ant-eating spiders selectively feed on different body parts to balance nutrient intake," *Animal Behaviour*, vol. 79, no. 6, pp. 1301–1306, 2010.

[175] A. P. Mathew, "Observations on the habits of the two spider mimics of the red ant *Oecophylla smaragdina* (Fabr.)," *Journal of the Bombay Natural History Society,*, vol. 52, pp. 249–263, 1954.

[176] X. J. Nelson, R. R. Jackson, S. D. Pollard, G. B. Edwards, and A. T. Barrion, "Predation by ants on jumping spiders (Araneae: Salticidae) in the Philippines," *New Zealand Journal of Zoology*, vol. 31, no. 1, pp. 45–56, 2004.

[177] S. Pekár and P. Jiroš, "Do ant mimics imitate cuticular hydrocarbons of their models?" *Animal Behaviour* , vol. 82, no. 5, pp. 1193–1199, 2011.

[178] D. Li and R. R. Jackson, "Prey-specific capture behaviour and prey preferences of myrmecophagic and araneophagic jumping spiders," *Revue Suisse de Zoologie Hors Serie*, pp. 423–436, 1996.

[179] D. B. Richman and R. R. Jackson, "A review of the ethology of jumping spiders (Araneae: Salticidae)," *Bulletin of the British Arachnological Society*, vol. 9, pp. 33–37, 1992.

[180] W. H. Clark, "Predation on the harvester ant, *Pogonomyrmex tenuispina* Forel (Hymenoptera: Formicidae), by the spider, *Steatoda fulva* (Keyserling) (Araneae: Theridiidae) in Baja California Sur, México," *Southwestern Entomologist, Scientific Note*, vol. 21, no. 2, pp. 213–217, 1996.

16

Division of Labor in *Pachycondyla striata* Fr. Smith, 1858 (Hymenoptera: Formicidae: Ponerinae)

Adolfo da Silva-Melo and Edilberto Giannotti

Departamento de Zoologia, Instituto de Biociências, Universidade Estadual Paulista (UNESP), Campus Rio Claro SP, Caixa Postal 199, 13506-900, Rio Claro, SP, Brazil

Correspondence should be addressed to Adolfo da Silva-Melo, adolfoants@yahoo.com.br

Academic Editor: Jacques H. C. Delabie

Four colonies of the ant *Pachycondyla striata* were used to analyze the specie behavioral repertoire. Forty-six behavioral acts were recorded in laboratory. Here, we present the record the division of labor between the castes and the temporal polyethism of monomorphic workers. The queens carried out many of the behavioral traits recorded in this work however; they performed them less frequently compared to the worker. The workers activity involved chasing and feeding on fresh insects and usingthem to nourish larvae besides laying eggs in the C-posture, an activity also performed by queens, which is similar to that of wasps of the subfamily *Stenogastrinae*. The young workers were involved in activities of brood care, sexuate care, and nest maintenance, and the older workers were involved in defense, exploration, and foraging.

1. Introduction

The evolution of social behavior may be defined as the combination of care for young individuals by adults, overlapping generations, and division of labor in the reproductive and nonreproductive castes [1–4]. The ants are eusocial, and their behavior differs from that one of other social insects in three respects: (a) they have a varied diet, (b) nest building retains characteristics unique to this group, parental care in galleries, and workers performing tasks according to their age or size, and (c) adults remaining long time with their brood [5].

Among the aspects covered in ethologic studies of ants, division of labor (when individuals within a group perform different roles) or polyethism comprehends a widely explored subject and may present two divisions: (a) physical polyethism, when individuals show distinct morphological characteristics to perform specific tasks and (b) temporal polyethism, when the variation of tasks occurs according to age [1, 2, 4, 6]. Therefore, temporal polyethism may occur both in populations of monomorphic workers and in polymorphic workers [7, 8]. The ants of the genus *Pachycondyla* have a wide pantropical distribution with about 270 species

being described [9]. The *Pachycondyla species* are diverse in their morphology and their behavior [10].

Pachycondyla striata Smith 1858 [11], classified into the subfamily Ponerinae [12], presents relatively large individuals (13.2–16.7 mm long). The castes are slightly different. The workers are different from the queens by the absence of ocelli and wing scars. This species is distributed through northern Argentina, Paraguay, Uruguay, and Brazil [13–15].

The aim of this study was to verify whether there is division of labor among castes and age polyethism in *P. striata*. The results will contribute to better understanding and interpretation of its social organization and allow comparison with other species of the family Formicidae.

2. Materials and Methods

Four colonies were collected on the campus of the University UNESP—Universidade Estadual Paulista, Rio Claro (22°32′40″S/47°32′44″W), São Paulo State. The ethological analysis began two days after the collection. Observations were done in the foraging area and plaster nest.

TABLE 1: Composition of the colonies of *Pachycondyla striata*.

Colony	Number of individuals							Date of collection
	eggs	larvae	pupae	workers	winged females	males	queens	
N. 2	—	—	—	20	5	—	—	04/13/2006
N. 3	—	—	37	178	38	33	—	04/14/2006
N. 7	264	65	—	382	7	8	—	08/06/2006
N. 8	30	231	240	384	—	—	1	11/17/2006

The colonies selected in field contained queens and/or winged females. The latter were regarded as queens after wing loss. The colonies were transferred to a laboratory and placed in plastic containers (width: 30.0 cm; length: 48.0 cm; height: 12.0 cm). In each container, there was a plaster nest consisting of three chambers in different sizes, interconnected by tunnels of 1.0 cm in width and 3.0 cm in depth, covered with glass to avoid disturbance and red cellophane paper to prevent the passage of the full spectrum of light.

The diet of the ants consisted of sugar and water in a ratio of 1:1 (offered in test tubes, with cotton wool in the opening), termites, worms, cockroaches, larvae of Coleoptera (*Tenebrio molitor*), flies, and papaya seeds.

Previous observation was performed for 20 hours to obtain behavioral data, with the aim of identifying queens and workers. The ants were differentiated by covering their thorax with quick-drying paint for model airplanes (Revel), allowing the identification of the individuals by age group just after their emergence. Young workers are known for having a paler color in relation to older ones. Later, the scan sampling method described by Altmann [16] was used to qualify the acts.

The quantitative observation of the behavioral acts of the individuals in each colony was performed for five minutes, with one-minute intervals. The observation time was one hour a day, four times a week, during six months, for a total of 94 hours. A comparative ethogram for the individuals was developed. Sample coverage was defined by the formula $\varnothing = 1 - (\mathbf{N1/i})$, where $\mathbf{N1}$ = number of behavioral acts observed once and $\mathbf{i}$ = total number of behavioral acts, the more this value approaches to 1, the more complete the sample [17]. The behavioral catalog was divided into ten categories and used to build histograms and a dendrogram with clustering method (UPGMA) of Euclidean distance [18] (Table 1).

3. Results

3.1. Division of Labor. When introduced in laboratory, the individuals of *P. striata* immediately occupied the artificial nest. The ants carried the immature from the foraging area and accommodated them in the first and minor chamber for 12 hours. Only after this, they carried them to the last and bigger chamber. In the nest seven, the workers distributed randomly the immature to the chambers and tunnels of the nest.

As previously announced for this study, we considered the existence of two castes morphologically and subtly differentiated, containing monomorphical workers. In Table 2 the different categories, are distributed and quantified and behavioral acts of queens, workers, winged females, and males of *P. striata are defined as well*.

The sample coverage value ($\varnothing$) was 0. 981 meeting the expectations of Fagen and Goldman [17]. The dissimilarity dendrogram informs a great ethological difference between the castes. (Figure 1).

The inactivity of the males into the nest suggests their action to be more prevalent in the mating season, but this was not verified in this study (Table 2).

The behavorial acts supposedly regarded as less derived have been identified in the castes, such as feeding larvae and adults on fresh insects, and laying eggs in the C-posture. Furthermore, the queens performed activities that are exclusively carried out by workers in other more derived species, such as brood care, exploring, foraging, and nest maintenance (Table 2).

The dominance behavior involved both individuals for recruiting and reproductive labor. The latter case, the interaction of dominance occurred between queen and worker and among workers. Some workers developed ovaries to lay eggs. However, this data were not quantified.

3.2. Temporal Polyethism. Some activities were preferably carried out by younger workers or older workers. This suggests division of labor by age (Figure 2).

The younger workers (7 to 56 days of age) stayed in the nest for approximately 27.03 ± 12.72 days (7–56, $N = 27$). For this time, took they care the pupae, larvae, eggs, males, and winged females (Figure 2). However, some newly hatched ants did not taken care for the young individuals. This might be related the presence of physiological problems, because they died within two or three days.

The older workers (those at more than 56 days of age) performed several categories, but they pointed in the activities out of the nest, as defense, foraging, and exploring (Figure 2). Furthermore, the dominance is a category that deserves attention. It may be linked to the maintenance of the colony, as a measure of protection from the nest and obtaining food, or reproduction.

The intermediate group (queens, virgin queens, and winged females), which is regarded as a caste, showed clear transition tasks. The quantitative results of the group are

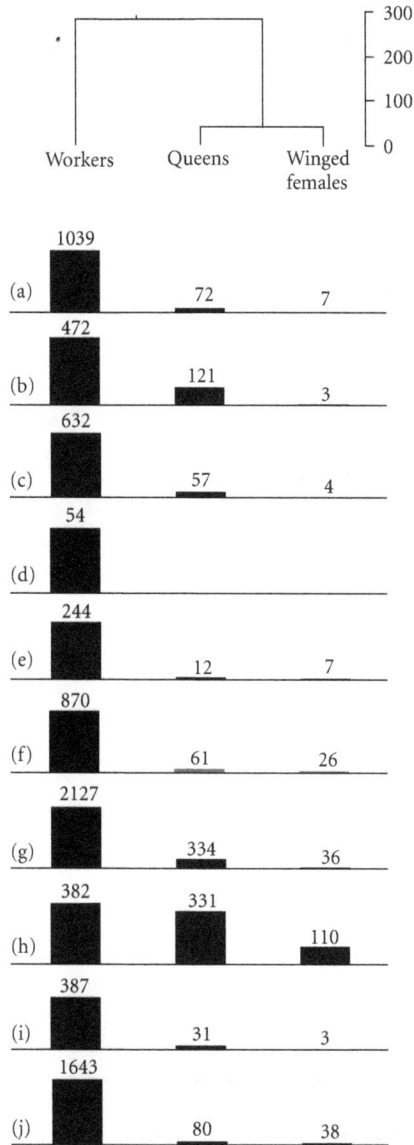

FIGURE 1: Dissimilarity dendrogram of individuals of *P. striata*. Behavioral categories: (a) feeding, (b) communication, (c) brood care, (d) sexuate care, (e) defense, (f) exploring and foraging, (g) grooming, (h) inactivity, (i) dominance, and (j) nest maintenance.

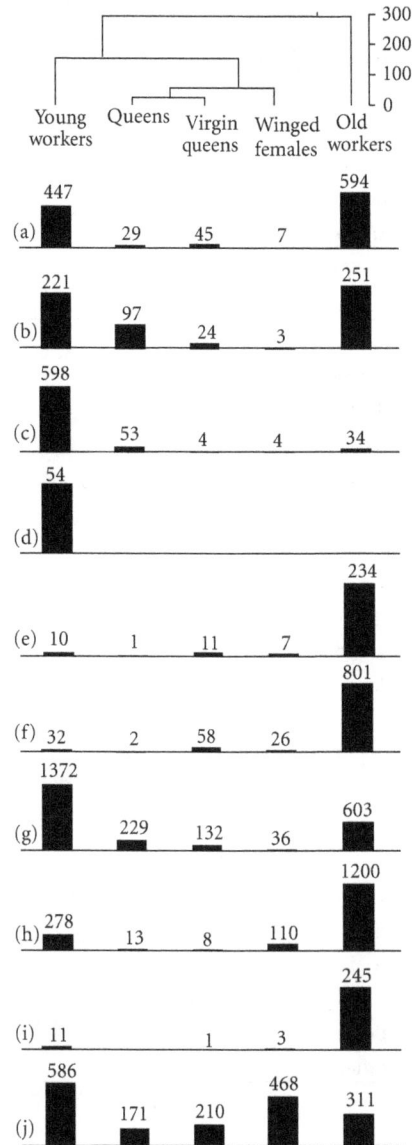

FIGURE 2: Dissimilarity dendrogram of individuals of *P. striata* showing the division of labor by age. Behavioral categories: (a) feeding, (b) communication, (c) brood care, (d) sexuate care, (e) defense, (f) exploring and foraging, (g) grooming, (h) inactivity, (i) dominance, and (j) nest maintenance.

smaller when compared to workers, and the activties have been concentrated within the nest.

4. Discussion

It is interesting to note that a small portion of behavioral acts is performed by queens within the nest. This type of occurrence is mentioned to the species of *P. (Neoponera) villosa*, *P. (Neoponera) apicalis*, and *P. (Neoponera) obscuricornis* [19]. The queens of *P. striata* presented more care for eggs than to the other immature individuals, while *P. (Neoponera) villosa* spends more energy caring for eggs and pupae, *P. (Neoponera) apicalis* and *P. (Neoponera)* *obscuricornis* invest more energy in caring for larvae and pupae [19]. The involvement of queens in brood care seems to be a little derived characteristic [20].

Feeding was a behavioral act frequently observed in the queens of *P. striata*, while the queen of *Nothomyrmecia macrops* was seen feeding once [21].

The queens and workers which to perform the laying eggs, retained the position in the form C. *Ectatomma planidens* [22, 23] and *Platythyrea punctata* [24], also acquired the same position.

This is characteristic of wasps of the genera *Listenogaster* [25] and *Eustenogaster* [26]. This condition may be an evidence of an attribute that might have been preserved.

TABLE 2: Behavioral catalog of *Pachycondyla striata*.

Category and behavioral acts	Queens	Workers	Winged females	Males
(A) Feeding				
01-Feeding on prey	0.0355	0.0700	0.0128	
02-Intake of liquids	0.0264	0.0587	0.0171	0.0116
44-Cannibalism	0.0030	0.0020	0.0059	
(B) Communication				
03-Antennate workers	0.0755	0.0549	0.0128	0.0516
04-Antennate queens	0.0345	0.0052		
(C) Brood care				
05-Antennate egg	0.0218	0.0129		
06-Antennate larvae	0.0010	0.003		
07-Antennate pupae		0.004		
08-Standing on eggs		0.0006		
09-Standing on larvae	0.0010	0.0014		
10-Standing on pupae		0.0009		0.0058
11-Handling eggs	0.0127	0.0045		
12-Handling larvae		0.0063	0.0043	
13-Handling pupae		0.0015		
15-Feeding larvae		0.0058		
16-Cleaning larvae		0.0085	0.0043	
18-Carrying eggs	0.0063	0.0126	0.0086	
19-Carrying pupae		0.0071		
20-Carrying larvae		0.0031		
22-Standing and holding an egg	0.0045	0.0061		
23-Standing and holding a pupa		0.0012		
24-Egg laying	0.0045	0.0004		
(D) Sexuate care				
14-Handling winged females		0.0003		
17-Cleaning males		0.0059		
21-Carrying males		0.0006		
(E) Defense				
25-Guarding the nest entrance	0.0110	0.0310	0.0210	
(F) Exploring and foraging				
26-Capturing prey	0.0118	0.0582		
27-Walking in the foraging arena	0.0427	0.0478	0.1070	0.0580
28-Tanden running	0.0010	0.0047	0.0043	
(G) Grooming				
29-Self-grooming their antennae	0.1164	0.082	0.059	0.0860
30-Self-grooming their 1st pair of legs	0.0582	0.03	0.0043	0.0660
31-Self-grooming their antennae and 1st pair of legs	0.0282	0.0252	0.0259	0.0233
32-Self-grooming their 2nd and 3rd pairs of legs	0.0591	0.0722	0.0212	0.0290
33-Self-grooming their anus	0.0054	0.0200		0.0260
34-Social grooming	0.0363	0.0062	0.0530	
(H) Inactivity				
38-Inactivity in the nest	0.3700	0.0323	0.4369	0.5000
39-Inactivity in the foraging arena	0.0219	0.0163	0.0350	0.0630
(I) Dominance				
35-Antennal boxing	0.0218	0.0405	0.0027	
36-Blocking	0.0054	0.0005		
37-Immobilization	0.0010	0.0081		0.0160

TABLE 2: Continued.

Category and behavioral acts	Queens	Workers	Winged females	Males
(J) Nest maintenance				
40-Carrying a dead ant	0.0010	0.1512		
41-Handling a dead ant		0.133		
42-Carrying garbage	0.0010	0.0103		
43-Handling garbage	0.0010	0.0085		
45-Exploring the plaster nest	0.0700	0.0160	0.1639	0.0643
46-Digging in the plaster nest		0.0080		
Total frequency	1	1	1	1
Total categories	9	10	9	8
Total behavioral acts	31	46	19	13

The agonistic behavioral acts were almost always related to reproduction or foraging activities. Antennal boxing occurred with winged females, queens, and workers. This behavior may be related to the recruitment of workers, as the measure was implemented in the nest, and a larger number of workers moved to the foraging arena. The same happens to *P. bertholudi* [27].

In nest 8, after the queen's death, one worker started laying eggs. Afterwards, agonistic encounters became frequent, and another worker that started laying eggs was mutilated. This suggests that *P. striata* presents a reproductive dominance, as does *P. crassinoda* [28]. Agonistic encounters were also reported for *P. (Neoponera) obscuricornis* [29, 30] and *P. bertholudi* [27].

Chagas and Vasconcelos [31] described the fighting behavior between workers of *P. striata* and *P. (Neoponera) obscuricornis* in the field. According to these researchers, this event occurred because *P. striata* invaded the foraging and/or life area of *P. obscuricornis*.

The agonistic behavioral acts observed in *P. striata* were also reported for *Dinoponera quadriceps* [32], *P. (Neoponera) apicalis* [33], *P. (Neoponera) obscuricornis* [29], *Rhytidoponera* sp. 12 [34], *P. inversa* [35], and *P. bertholudi* [27].

We checked that the workers ate larvae, pupae, other workers, and males. Some alive males had their abdominal region pulled off by workers. These behaviors may indicate stress or cannibalism. Wilson [1] reported that dead workers might be used as food or were discarded.

The eggs of *P. striata* collected from the natural environment and those laid by queens and workers in laboratory did not develop. They were predated by dominant individuals or by the whole group under stress. Egg predation was reported in *Ectatomma planidens* [22, 23] and *E. vizottoi* [36] although it has been absent or not observed in *Pachycondyla bertholudi* [37]. The eggs laid by workers are usually eaten by queens and larvae, which represents a stereotyped, conspicuous behavior pattern [1].

Oophagy is indispensable to the social Hymenoptera [1]. It is important because workers do not regurgitate food either for larvae or for queens, so they can use their own resources to produce immature oocytes [38]. This event seems restricted to some genera in the subfamily Ponerinae [38].

In the presence of a large number of eggs, the workers gathered them and stood still on them. They standing motionless on eggs, pupae, and larvae. This may suggest warming and protection of the immature individuals. When the number of eggs in the nest was small, the ants of this species kept the eggs clustered between their mandibles.

The behavioral act *tandem running* was carried out to recruit workers into the foraging arena. Medeiros and Oliveira [39] observed this as well. This behavior is common in several species such as *Pachycondyla (Brotoponera) tesserinoda* [40] and *Pachycondyla obscuricornis* [31].

The larvae of *P. striata* display a characteristic behavior to order food. They shake their necks and heads several times towards the ventral region of their body until a worker answers. This behavior is similar to that one of larvae of *Gnamptogenys striatula* [41]. The workers moved the larvae towards the prey. In some cases, the workers held the prey between their mandibles, while the larvae inserted their head into the sectioned part of the mealworm and fed on hemolymph. The workers feed preferentially larvae closer to them. Asking for food was a behavioral act observed more often in larvae in the last instar. The workers touched the buccal apparatus of the larvae with their mandibles open, but it was not possible to see the food transfer or the projection of the glossa of the workers. A similar behavioral act was described for *P. crassinoda* [28].

Small pieces of mealworm were placed in the ventral region of the larvae of ants by the workers. The larvae curved their necks and fed in the same manner as described for *Gnamptogenys horni* [42], *Ponera pennsylvannica* [43], and *Pachycondyla crassinoda* [43]. According to Wilson [1] and Traniello and Jayasuriya [44], feeding larvae on small fragments of prey is a less derived characteristic.

P. striata use their stinger to paralyze their prey. The sting might be stimulated by sudden movements of the prey, similar to way what happens to workers of *P. caffraria* [45]. According to Traniello and Jayasuriya [44], using the stinger to paralyze prey is a less derived characteristic.

The state of inactivity or deep sleep exhibited by *P. striata* is similar to one that described by Cassill et al. [46]. Many workers remained motionless in foraging area. This category may reflect the restricted space of the arena or, as Miguel and Del-Claro [47], the state, containment of spent energy. The inactivity behavior was observed in *Pachycondyla (Neoponera) villosa*, *P. (Neoponera) apicalis*, *P. (Neoponera) obscuricornis* [19], *P. crassinoda* [48], *Nothomyrmecia macrops* [21], *E. planidens* [22, 23], and *E. opaciventre* [47].

The monomorphic workers of *P. striata* present specialized task division, forming work groups to performing tasks linking to individuals with similar ages. Young individuals provide parental care, whereas older individuals carry out the activities of defense, exploration, and foraging.

Young workers stayed in the nest for 56 days, but some left earlier. They were recruited into the foraging area according to the necessity of food or to substitute the dead workers. In the first 45 days after emergence, *Ectatomma tuberculatum* performs tasks progressively according to the age of the individuals [49]. The same happens to workers of *Platythyrea lamellosa*, which after hatching (0–5 days of age) present association with pupae and later take care of eggs and larvae, performing specific tasks influenced by their age [50]. Unlike *P. striata*, newly hatched individuals of the species *Pachycondyla caffraria* (0–5 days of age) present four types of behavioral acts and are capable of foraging early at this age [51]. Each colony of this species has precise requirements as to carbohydrates and proteins, appropriate for labor division, which happens in relatively fixed proportions between hunting foragers and those which collect water with sugar [45]. Workers of *P. striata* were seen at the carbohydrate source in a very small frequency. This activity was included in the behavioral act of taking water in from the cotton wool. *P. striata* preferred to capture other insects to provide protein intake.

This research shows the profile of social organization of *P. striata*. We see that many behavioral acts are common for species of the subfamily Ponerinae. Although there is a narrow dimorphism in castes of *P. striata*, there is a great difference of division of labour between them. The age is a factor that controls the performance of tasks in workers.

References

[1] E. O. Wilson, *The Insect Societies*, Belknap Press, Cambridge, Mass, USA, 1971.

[2] E. O. Wilson, *Sociobiology The New Synthesis*, Belknap Press, Cambridge, Mass, USA, 1975.

[3] M. Andersson, "The evolution of eusociality," *Annual Review of Ecology and Systematics*, vol. 15, pp. 165–189, 1984.

[4] B. Hölldobler and E. O. Wilson, *The Ants*, Belknap Press, Cambridge, Mass, USA, 1990.

[5] H. D. Sudd, "Ants: foragim, nesting, brood behavior and polyethism," in *Social Insects*, H. R. Hermann, Ed., vol. 4, pp. 107–155, Academic Press, New York, NY, USA, 1982.

[6] E. O. Wilson, "The social biology of ants," *Annual Review of Entomology*, vol. 8, pp. 345–368, 1963.

[7] J. H. Sudd and N. R. Franks, *The Behavioural Ecology of Ants*, Chapman and Hall, New York, NY, USA, 1987.

[8] A. B. Sendova-Franks and N. R. Franks, "Self-assembly, self-organization and division of labour," *Philosophical Transactions of the Royal Society B*, vol. 354, no. 1388, pp. 1395–1405, 1999.

[9] B. Bolton, *A New General Catalogue of the Ants of the World*, Harvard University, Cambridge, Mass, USA, 1995.

[10] A. L. Wild, "The genus *Pachycondyla* (Hymenoptera: Formicidae) in Paraguay," *Boletin del Museo Nacional de Historia Natural del Paraguay*, vol. 14, pp. 1–18, 2002.

[11] Fr. Smith, *Catalogue of the Hymenopterous Insects in the Collection of the British Museum*, Part 6, British Museum, London, UK, 1858.

[12] B. Bolton, *Synopsis and Classification of Formicidae*, vol. 71 of *Memoirs of the American Entomological Institute*, American Entomological Institute, 2003.

[13] W. W. Kempf, "As formigas do gênero Pachycondyla Fr. Smith no Brasil (Hymenoptera: Formicinae)," *Revista Brasileira de Entomologia*, vol. 10, pp. 189–204, 1961.

[14] W. W. Kempf, "Catalogo abreviado das formigas da região neotropical (Hymenoptera: Formicidae)," *Studia Entomologica*, vol. 5, pp. 3–344, 1972.

[15] W. W. Kempf and K. Lenko, "Levantamento da formicifauna no litoral norte e ilhas adjacentes do Estado de São Paulo, Brasil.I. subfamilias dorylinae, Ponerinae e Pseudomyrmicinae (Hymenoptera: Formicidae)," *Studia Entomologica*, vol. 19, pp. 45–66, 1976.

[16] J. Altmann, "Observational study of behavior: sampling methods," *Behaviour*, vol. 49, no. 3-4, pp. 227–267, 1974.

[17] R. M. Fagen and R. Goldman, "Behavioural catalogue analysis methods," *Animal Behaviour*, vol. 25, no. 2, pp. 261–274, 1977.

[18] C. J. Krebs, *Ecological Methodology*, Benjamin Cummings, Redwood City, Calif, USA, 2nd edition, 1999.

[19] M. Pérez-Bautista, J. P. Lachaud, and D. Fresneau, "La division del trabajo em la hormiga primitiva *Neoponera villosa* (Hymenoptera: Formicidae)," *Folia Entomológica Mexicana*, vol. 65, pp. 119–130, 1985.

[20] D. Fresneau and P. Dupuy, "A study of polyethism in a ponerine ant: *Neoponera apicalis* (Hymenoptera, Formicidae)," *Animal Behaviour*, vol. 36, no. 5, pp. 1398–1399, 1988.

[21] P. Jaisson, D. Fresneau, and R. W. Taylor, "Social organization in some primitive australian ants. I. *Nothomyrmecia macrops* Clark," *Insectes Sociaux*, vol. 3, no. 4, pp. 425–438, 1992.

[22] W. F. Antonialli-Junior and E. Giannotti, "Division of labor in *Ectatomma edentatum* (Hymenoptera, Formicidae)," *Sociobiology*, vol. 39, no. 1, pp. 37–63, 2002.

[23] W. F. Antonialli, V. C. Tofolo, and E. Giannotti, "Population dynamics of *Ectatomma planidens* (Hymenoptera: Formicidae) under laboratory conditions," *Sociobiology*, vol. 50, no. 3, pp. 1005–1013, 2007.

[24] J. Korb and J. Heinze, "Multilevel selection and social evolution of insect societies," *Naturwissenschaften*, vol. 91, no. 6, pp. 291–304, 2004.

[25] S. Turillazzi, "The stenogastrinae," in *The Social Biology of Wasps*, K. G. Ross and R. W. Matthews, Eds., pp. 47–98, Cornell University, Ithaca, NY, USA, 1991.

[26] E. Francescato, A. Massolo, M. Landi, L. Gerace, R. Hashim, and S. Turillazzi, "Colony membership, division of labor, and genetic relatedness among females of colonies of *Eustenogaster fraterna* (Hymenoptera, Vespidae, Stenogastrinae)," *Journal of Insect Behavior*, vol. 15, no. 2, pp. 153–170, 2002.

[27] M. F. Sledge, C. Peeters, and M. R. Crewe, "Reproductive division of labour without dominance interactions in the queenless ponerine ant *Pachycondyla (Ophthalmopone) berthoudi*," *Insectes Sociaux*, vol. 48, no. 1, pp. 67–73, 2001.

[28] A. Henriques and P. R. S. Moutinho, "Algumas observações sobre a organização social de *Pachyconyla crassinoda* Latreille,1802 (Hymenoptera: Formicidae: Ponerinae)," *Revista Brasileira de Entomologica*, vol. 38, pp. 605–611, 1994.

[29] P. S. Oliveira and B. Hölldobler, "Agonistic interactions and reproductives dominance in *Pachycondyla obscuricornis* (Hymenoptera: Formicidae)," *Psyche*, vol. 98, pp. 215–225, 1991.

[30] B. Gobin, J. Heinze, M. Strätz, and F. Roces, "The energetic cost of reproductive conflicts in the ant *Pachycondyla obscuricornis*," *Journal of Insect Physiology*, vol. 49, no. 8, pp. 747–758, 2003.

[31] A. C. S. Chagas and V. O. Vasconcelos, "Comparação da freqüência da atividade forrageira da formiga *Pachycondyla obscuricornis* (Emery, 1890) (Hymenoptera: Formicidae) no verão e no inverno, em condições de campo," *Revista Brasileira de Zoociências*, vol. 4, p. 109, 2002.

[32] T. Monnin and C. Peeters, "Dominance hierarchy and reproductive conflicts among subordinates in a monogynous queenless ant," *Behavioral Ecology*, vol. 10, no. 3, pp. 323–332, 1999.

[33] P. S. Oliveira and B. Hölldobler, "Dominance orders in the ponerine ant *Pachycondyla apicalis* (Hymenoptera, Formicidae)," *Behavioral Ecology and Sociobiology*, vol. 27, no. 6, pp. 385–393, 1990.

[34] W. T. Tay and R. H. Crozier, "Nestmate interactions and egg-laying behaviour in the queenless ponerine ant *Rhytidoponera* sp. 12," *Insectes Sociaux*, vol. 47, no. 2, pp. 133–140, 2000.

[35] K. Kolmer and J. Heinze, "Rank orders and division of labour among unrelated cofounding ant queens," *Proceedings of the Royal Society B*, vol. 267, no. 1454, pp. 1729–1734, 2000.

[36] A. S. Vieira, W. D. Fernandes, and W. F. Antonialli-Junior, "Temporal polyethism, life expectancy, and entropy of workers of the ant *Ectatomma vizottoi* almeida, 1987 (Formicidae: Ectatomminae)," *Acta Ethologica*, vol. 13, no. 1, pp. 23–31, 2010.

[37] C. P. Peeters and R. M. Crewe, "Male biology in the queenless *Ophthalmopone berthoudi* (Formicidae: Ponerinae)," *Psyche*, vol. 93, pp. 227–283, 1986.

[38] C. Peeters, "Morphologically "primitive" ants; comparative review of social characters, and the importance of queen-worker dimorphism," in *The Evolution of Social Behavior in Insects and Arachnids*, J. Choe and B. Crespi, Eds., pp. 372–391, Cambridge University Press, Cambridge, UK, 1997.

[39] F. N. S. Medeiros and P. S. Oliveira, "Season-dependent foraging patterns case study of a neotropical Forest-dwelling ant (*Pachycondyla striata* Ponerinae)," in *Food Exploitation by Social Insects: Ecological, Behavioral, and Theoretical Approaches*, S. Jarau and M. Hrncir, Eds., chapter 4, pp. 81–95, CRC Press, 2009.

[40] V. U. Maschwitz, B. Hölldobler, and M. Möglich, "Tandelaufen als rekrutierungsverhalten bei *Brothroponera tesserinoda* forel (Formicidae: Ponerinae)," *Zei Tschrift Für Tierpsychologie*, vol. 35, pp. 113–123, 1974.

[41] N. Kaptein, J. Billen, and B. Gobin, "Larval begging for food enhances reproductive options in the ponerine ant *Gnamptogenys striatula*," *Animal Behaviour*, vol. 69, no. 2, pp. 263–299, 2005.

[42] S. C. Pratt, "Ecology and behavior of *Gnamptogenys horni* (Formicidae: Ponerinae)," *Insectes Sociaux*, vol. 41, no. 3, pp. 255–262, 1994.

[43] C. S. Pratt, F. N. Carlin, and P. Calabi, "Division of labor in *Ponera pennsylvannica* (Formicidae: Ponerinae)," *Insectes Sociaux*, vol. 41, no. 1, pp. 43–61, 1994.

[44] J. F. A. Traniello and A. K. Jayasuriya, "The biology of the primitive ant *Aneuretus simoni* (Emery) (Formicidae: Aneuretinae) II. the social ethogram and division of labor," *Insectes Sociaux*, vol. 32, no. 4, pp. 375–388, 1985.

[45] C. Agbogba and P. E. Howse, "Division of labor between foraging workers of the ponerine ant *Pachycondila caffraria* (Smith) (Hymenoptera: Formicidae)," *Insectes Sociaux*, vol. 39, pp. 455–458, 1992.

[46] D. L. Cassill, S. Brown, and D. Swick, "Polyphasic wake/sleep episodes in the fire ant, *Solenopsis invicta*," *Journal of Insect Behavior*, vol. 22, no. 4, pp. 313–323, 2009.

[47] T. B. Miguel and K. Del-Claro, "Polietismo etário e repertório comportamental de *Ectatomma opaciventre* roger, 1861 (Formicidae: Ponerinae)," *Revista Brasileira de Zoociências*, vol. 7, pp. 293–310, 2005.

[48] R. C. S. Brito, *Divisão De Trabalho: Aspectos Comportamentais Da Regulação Social Do Cuidado à Prole Em* Pachycondyla crassinoda *Latreille, 1802 (Hymenoptera: Formicidae: Ponerinae)*, thesis, USP-Univesidade de São Paulo, 1999.

[49] A. Champalbert and J. P. Lachaud, "Existence of a sensitive period during the ontogenesis of social behaviour in a primitive ant," *Animal Behaviour*, vol. 39, no. 5, pp. 850–859, 1990.

[50] M. H. Villet, "Social organization of *Platythyrea lamellosa* (Roger) (Hymenoptera: Formicidae): II division of labour," *Suid-Afrikaanse Tydskrif vir Plantkunde*, vol. 25, pp. 254–259, 1990.

[51] C. Agbogba, "Absence of temporal polyethism in the ponerine ant *Pachycondyla caffraria* (Smith) (Hymenoptera: Formicidae): early specialization of the foragers," *Behavioural Processes*, vol. 32, no. 1, pp. 47–52, 1994.

17

The Host Genera of Ant-Parasitic Lycaenidae Butterflies: A Review

Konrad Fiedler

Department of Tropical Ecology and Animal Biodiversity, Faculty of Life Sciences, University of Vienna, Rennweg 14, 1030 Vienna, Austria

Correspondence should be addressed to Konrad Fiedler, konrad.fiedler@univie.ac.at

Academic Editor: Volker Witte

Numerous butterfly species in the family Lycaenidae maintain myrmecophilous associations with trophobiotic ants, but only a minority of ant-associated butterflies are parasites of ants. *Camponotus*, *Crematogaster*, *Myrmica*, and *Oecophylla* are the most frequently parasitized ant genera. The distribution of ant-parasitic representatives of the Lycaenidae suggests that only *Camponotus* and *Crematogaster* have multiply been invaded as hosts by different independent butterfly lineages. A general linear model reveals that the number of associated nonparasitic lycaenid butterfly species is the single best predictor of the frequency of parasitic interactions to occur within an ant genus. Neither species richness of invaded ant genera nor their ecological prevalence or geographical distribution contributed significantly to that model. Some large and dominant ant genera, which comprise important visitors of ant-mutualistic lycaenids, have no (*Formica*, *Dolichoderus*) or very few ant-parasitic butterflies (*Lasius*, *Polyrhachis*) associated with them.

1. Introduction

Associations between ants and butterfly species in the families Lycaenidae and Riodinidae have attracted the interest of naturalists since more than 200 years. Building upon an ever-increasing number of field records and case studies (summarized in [1]) these interactions with their manifold variations and intricacies have developed into a paradigmatic example of the evolutionary ecology and dynamics of interspecific associations [2]. Interactions with ants are most well developed during the larval stages of myrmecophilous butterflies. To communicate with ants, myrmecophilous caterpillars possess a variety of glandular organs and often also use vibrational signals that may modulate ant behaviour [3, 4]. Essentially, interactions between myrmecophilous caterpillars and visiting ants comprise a trade of two commodities. The caterpillars produce secretions that contain carbohydrates and amino acids [5]. In turn, the ants harvest these secretions, do not attack myrmecophilous caterpillars and the presence of ant guards confers, at least in a statistical sense, protection against predators or parasitoids (reviewed in [2]). Thus, such interactions are basically mutualistic in

nature, even though the extent of benefits accruing to both partners may be asymmetric and manipulatory communication (by means of mimicking chemical or vibrational signals of ants) is not uncommon. In certain cases, especially if butterfly-ant associations are obligatory (from the butterfly's perspective) and involve specific host ants, interactions may extend into other life-cycle stages of the butterflies, such as pupae (if pupation occurs in ant nests or pavilions built by ants to protect their trophobiotic partners), adults (if egg-laying or nutrient acquisition occurs in company with ants), or eggs.

The vast majority of known butterfly-ant interactions are mutualistic or commensalic in nature. In the latter case the butterfly larvae benefit from their association with ants, while no costs accrue to the ants. Some few butterflies, however, have evolved into parasites of ants [6]. These unusual associations have served as models for host-parasite coevolution [7]. Ant parasitism requires very precise tailoring of the chemical and mechanical signals employed to achieve social integration into ant colonies. Accordingly, ant-parasitic lycaenid butterflies are highly specific with regard to their host ant use, which also renders them extraordinarily susceptible to the risk of coextinction [8]. Indeed, many ant-parasitic lycaenids are highly endangered species [9], and the well-studied Palaearctic genus *Maculinea* is now regarded as a prime example of insect conservation biology [10].

In this essay, I will focus on the ant genera that serve as hosts of parasitic butterflies. First, I summarize which ant genera in the world are known to be parasitized by butterflies. I then discuss whether this host ant use reflects the macroecological patterns seen in mutualistic butterfly-ant associations. Finally, I will explore if the observed host use patterns allow for generalizations and testable predictions, for example, with regard to expected host ant affiliations in underexplored faunas. Specifically, I expected that the number of associated parasitic lycaenids per host ant genera increases with their ecological prevalence, geographical distribution, and species richness.

2. What Constitutes an Ant-Parasitic Butterfly Species?

I here use a rather restrictive definition of ant parasitism. I regard a butterfly species as a parasite of its host ants only if (a) the butterfly caterpillars (at least from some developmental stage onwards) feed on ant brood inside ant nests ("predators") or (b) the caterpillars are being fed through trophallaxis by their host ants ("cuckoo-type" parasitism). Both these types of parasitism occur in *Maculinea* [11], but the extent of the nutrient flow from the ant colony to the caterpillars may vary across species. For example, in some lycaenid species feeding through trophallaxis apparently occurs only as a supplementary mode of nutrient acquisition. Yet include such cases here as parasites of ants, since the respective behavioural and communicative strategies are in place.

In contrast, I exclude two types of "indirect" parasitism. First, there are a few myrmecophilous lycaenid species that feed obligately on myrmecophytic ant plants. The best documented examples are certain SE Asian *Arhopala* species on ant-trees of the genus *Macaranga* [12, 13]. These caterpillars cause substantial feeding damage to the ant-trees and thereby likely inflict costs to the *Crematogaster* ants that inhibit these trees. *Arhopala* caterpillars on *Macaranga*, however, possess a nectar gland and secrete nectar at rates typical for ant-mutualistic lycaenids (K. Fiedler, unpublished observations). They are also not known to elicit trophallaxis or even to prey on ant brood. Accordingly, I did not score these associations as parasitic, but rather as competitors of ants for the same resource (namely, the ant-tree). Analogous cases are known, or suspected, to occur in other tropical lycaenid butterflies whose larvae feed on obligate myrmecophytes, such as various *Hypochrysops* species in Australia and New Guinea on *Myrmecodia* ant plants [14, 15].

Similarly, I do not include those lycaenid species (notably in the subfamily Miletinae) whose larvae prey upon ant-attended honeydew-producing homopterans and often also feed on homopteran honeydew [16–20]. In analogy to the case of myrmecophytes, these butterflies compete with ants for the same resource (here: trophobiotic homopterans), but

as a rule the caterpillars neither prey on ant brood nor elicit trophallaxis. Some species of the Miletinae, however, are known to supplement their diet through ant regurgitations, and these are included below since they show the behavioural traits considered here as essential for parasitism with ants.

Two further restrictions are (1) cases where trophallaxis or predation on ant brood have so far only been indirectly inferred, but not be confirmed through direct observational evidence, are largely excluded. This relates to a couple of tropical lycaenid species for which only old, or very incomplete or vague, information on their life cycles is available. In these cases, new data are needed, before any conclusions become feasible. (2) The butterfly family Riodinidae is also excluded. Ant-associations occur in at least two clades of Neotropical Riodinidae (tribes Eurybiini and Nymphidiini, see [21, 22] for many case studies and [23] for a tentative phylogeny). Circumstantial evidence exists that in at least one genus within the Nymphidiini (*Aricoris*) the larvae may feed on trophallaxis received from *Camponotus* host ants [21], but otherwise the existence of ant-parasitic life habits in the Riodinidae (though not unlikely to exist amongst Neotropical riodinids) must await confirmation.

3. Data Sources

Butterfly life-history data were compiled from a large variety of sources, ranging from faunal monographic treatments across hundreds of journal papers to databases in the Internet. The data tables in [1] formed the initial basis, and they have been continually extended and updated ever since [24, 25]. Here, I focus on that subset of sources where (a) the butterfly species qualifies as a parasite of ants according to the restrictions stated above and (b) the host ant has been reported at least at genus level. Three reasons justify the choice of the ant genus level for the subsequent comparisons. (1) For most ant genera, no modern revisions are available. Thus, proper species identifications are often impossible, especially in tropical realms. (2) Ant genus delimitations are quite stable and recognizable on a worldwide basis ([26], see also http://www.antweb.org). Accordingly, records (often reported by lepidopterists and not myrmecologists) should usually be reliable on this level. (3) Most ant-parasitic lycaenids are not bound to one single ant species, but are affiliated with a couple of congeneric ant species. For example novel *Myrmica* host ant species continue to be discovered in Eastern Europe for butterflies in the genus *Maculinea* [27, 28]. Therefore, I performed all analyses on the taxonomic level where the highest reliability can be achieved. Data on species richness of ant genera was extracted from the website antweb.org (as of 9 October 2011).

A complete bibliography of the evaluated literature would extend beyond the scope of this essay. For ant-parasitic Lycaenidae, many sources have been detailed in [6]. Full information on data sources is available upon request from the author.

4. Summary of Ant Genera That Are Confirmed as Hosts of Parasitic Lycaenid Butterflies

Of the 54 ant genera known to attend lycaenid larvae on a worldwide basis ([24], only *Liometopum* has been added to this list since) just 11 genera are for certain recorded as hosts of parasitic butterflies.

4.1. Subfamily Formicinae

4.1.1. Camponotus. This is one of the globally most prevalent ant genera in terms of species richness (>1050 described species) as well as ecological significance. It is also the numerically leading ant genus with regard to the number of as- sociated parasitic lycaenid species. At least 9 species of the large Afrotropical genus *Lepidochrysops* have been recorded from nests of either *Camponotus niveosetosus* or *C. maculatus*. *Lepidochrysops* larvae have a life cycle similar to the *Maculinea-Phengaris* clade. They initially feed on flowers of plants (mostly in the families Lamiaceae, but also Verbenaceae and Scrophulariaceae). At the onset of their third instar they are adopted by *Camponotus* workers into the ant colonies where they turn into predators of ant brood. There are more than 125 described Lepidochrysops species [29]. Many of them are microendemics of high conservation concern [30]. Presumably all *Lepidochrysops* species are parasites of *Camponotus* ants. The small South African genus *Orachrysops* is the closest relative of *Lepidochrysops*. *Orachrysops* larvae are not parasites of ants, but live in close association with *Camponotus* ants as leaf, and later root, herbivores of various Fabaceae plants [31]. *Orachrysops* species may therefore be seen as models for the evolutionary transition between "normal" phytophagous ant-mutualistic lycaenids and species that are parasites of ants.

The East Asian *Niphanda fusca* is an obligate cuckoo-type parasite of various *Camponotus* ants [32]. Unusual for ant-parasitic lycaenids, larvae of this species retain a fully functional nectar gland whose secretions are tuned towards the gustatory preferences of their host ants [33]. Life histories of other *Niphanda* species, that all occur in East and South-East Asia, are unknown. Within the genus *Ogyris* (13 species in New Guinea and Australia) most species maintain obligate mutualistic associations with ants, but two are reported to occur inside nests of *Camponotus* species, namely, *O. idmo* and *O. subterrestris* [15, 34, 35]. Finally, for at least two representatives of the aphytophagous African genus *Lachnocnema* (*L. bibulus*, *L. magna*) there is evidence that caterpillars supplement their diet by eliciting trophallaxis from *Camponotus* ants (in *L. bibulus* reportedly also from *Crematogaster* ants). The major nutrient source of *Lachnocnema* larvae, however, is preying on homopterans and drinking their honeydew excretions.

4.1.2. Oecophylla. The two species of weaver ants in the genus *Oecophylla* are extremely dominant insects in their habitats in tropical Africa, southern and south-eastern Asia, Australia, and New Guinea. Two lycaenid genera are specialist parasites of weaver ants. *Liphyra* (*L. brassolis*, *L. grandis*)

are predators of the brood of *Oecophylla smaragdina* in the Oriental region [15, 25], while African *Euliphyra* (*Eu. mirifica*, *Eu. leucyania*) are cuckoo-type parasites of *Oe. longinoda* by means of trophallaxis and also steal prey items of their host ants [36]. Many more lycaenid species are associated with weaver ants, including striking examples of obligate and specific interactions, but these all appear to be mutualistic associations.

4.1.3. Polyrhachis. Even though this large ant genus (>600 described species) ranks rather high in the visitors list of lycaenid caterpillars, only one of its reported associated 27 myrmecophilous butterfly species is a parasite. The rare *Arhopala wildei* in Australia and New Guinea preys on brood in nests of *Polyrhachis queenslandica* [37, 38].

4.1.4. Lasius. Ant species of this moderately rich genus (>100 species) are frequent visitors of lycaenid caterpillars, especially in the Palaearctic realm [25]. *Shirozua jonasi* from East Asia is the only ant-parasitic butterfly known to be affiliated with *Lasius* ants (*L. spathepus*, *L. fuliginosus*, and *L. morisitai*). The caterpillars apparently receive occasional trophallactic regurgitations, but their principle mode of feeding is to prey on a variety of homopterans and to drink their honeydew excretions [39].

4.1.5. Lepisiota. Butterflies of the South African genus *Aloeides* all have an obligate relationship to ants. *Lepisiota capensis* is their major host ant [40]. As far as known, most *Aloeides* species are phytophagous ant mutualists (host plants in the Fabaceae and more rarely the Malvaceae, Zygophyllaceae and Thymelaeaceae), but older larvae of *A. pallida* have been observed to feed on ant eggs and appear to be completely aphytophagous [40].

4.1.6. Anoplolepis. Another endemic South African butterfly genus is *Thestor*, with about 27 recognized species [41]. The life histories of these butterflies are still very incompletely known, but for sure they are essentially aphytophagous, as is the rule in the Miletinae to which this genus belongs. Younger larvae prey on various homopterans, and in at least 3 species (*Th. yildizae, rileyi*, and *basutus*) older larvae live inside ant nests where they feed on brood of the ant *Anoplolepis custodiens*. It is suspected that all *Thestor* species share this habit [41].

4.2. Subfamily Dolichoderinae

4.2.1. Papyrius. The small endemic Australian butterfly genus *Acrodipsas* can be divided into two clades [42]. Larvae of one of these, comprising the species *A. brisbanensis* and *A. myrmecophila*, are obligate parasites of *Papyrius nitidus* [35] from their first instar onwards, that is, without a phytophagous phase as in *Lepidochrysops* or the *Maculinea/Phengaris* clade. *Papyrius* species are highly dominant components of Australian ant assemblages and serve as mutualistic partners for some additional Australian lycaenids [34].

4.3. Subfamily Myrmicinae

4.3.1. Crematogaster.

This diverse ant genus (>450 described species) ranks second in terms of associated ant-parasitic lycaenid butterflies. In the lycaenid tribe Aphnaeini (about 260 species, of which >90% occur in Africa) caterpillar-ant associations are nearly always obligatory, and the predominant host ant genus is *Crematogaster*. Few Aphnaeini species, however, are well established to be parasites of *Crematogaster* ants. Only one of these is a brood predator (*Cigaritis acamas* [43]), whereas in other cases trophallactic feeding has been reported (e.g., *Aphnaeus adamsi, Chrysoritis (Oxychaeta) dicksoni, Spindasis takanonis*, and also *S. syama*; [40, 44]). Beyond the tribe Aphnaeini, parasitic relationships occur in the Australian *Acrodipsas* of which three species (*A. cuprea, illidgei*, and *aurata*) are predators of *Crematogaster* ants [35, 42, 45]. According to one old account caterpillars of the aphytophagous African *Lachnocnema bibulus* (which essentially prey on homopterans and drink their honeydew exudates, see above) also supplement their diet by trophallaxis obtained from *Crematogaster* ants [46].

4.3.2. Myrmica.

This genus is famous as being the host of the ant-parasitic *Maculinea* butterflies in temperate regions of Eurasia. *Maculinea* comprises about 10–15 species, depending on the status allocated to local forms and cryptic lineages detected through resent sequence analyses [47]. All *Maculinea* species are either brood predators or cuckoo-type parasites [11] of *Myrmica* ants. Host specificity was initially thought to be generally high [48], but research over the past two decades has revealed more complex, locally to regionally variable patterns of host specificity [27]. Especially in previously underexplored regions of central and east Europe many new local host associations have been elucidated through thorough field work [28]. Caterpillars of the closely related East Asian butterfly genus *Phengaris* also parasitize *Myrmica* species [44, 47].

4.3.3. Aphaenogaster.

There are two *Maculinea* species from East Asia (*M. arionides, M. teleius*) for which the use of *Aphaenogaster* ant species as hosts has been recorded. Both these butterfly species are known to parasitize mainly *Myrmica* host ants. It remains to be shown to what degree *Aphaenogaster* ants really qualify as valid hosts. Alternatively, these records might be based on misidentifications or represent rare affiliations that only occur under exceptional circumstances (see the discussion about primary and secondary hosts in [27]).

4.3.4. Rhoptromyrmex.

Representatives of this small Oriental ant genus have been observed to attend a range of lycaenid caterpillars in a mutualistic manner. Besides, trophallactic feeding does occur in one unusual case, the Miletinae species *Logania malayica*. *L. malayica* larvae prey essentially on homopterans and drink their honeydew exudates, but young larvae also elicit regurgitations from *Rh. wroughtonii* ants, with which the butterflies are closely and specifically associated over their entire life cycle [17, 49].

5. Macroecological Patterns of Host Ant Use among Ant-Parasitic Lycaenidae Butterflies

Myrmecophilous associations between lycaenid butterflies and ants are confined to that subset of ant genera which maintain trophobiotic interactions [24]. Trophobiotic ants form a highly significant fraction in terms of their ecological prevalence as well as species diversity. They essentially derive liquid nutrients from extrafloral plant nectar [50, 51] and from the excretions ("honeydew") of sap-sucking homopterans [52, 53]. Lycaenid and riodinid butterfly species that offer nectar-like secretions in exchange for protection largely "hitch-hike" on the behavioural and ecological syndromes which are associated with ant trophobiosis. Harvesting nutrient-rich liquids requires specialized anatomy [54] and behaviour in ants (e.g., trophallactic exchange of liquid food within the colony), with trophobiosis demanding a more complex suite of morphological and behavioural traits than licking-up plant nectar [55].

Ant-parasitic lycaenids form a very small subset of myrmecophilous ant-attended species in that butterfly family. Not surprisingly, the host ants parasitized by them constitute a small subset of ant genera known to visit and attend caterpillars in mutualistic associations. In two earlier studies the ecological prevalence and geographical distribution of ant genera were shown to be the best predictors for their representation in mutualistic lycaenid-ant associations [24, 25]. For parasitic interactions, this pattern changes according to a similar analysis. In analogy to [24], I constructed a multiple linear regression model, with the number of recorded ant-parasitic lycaenids as response variable and the species richness (log-transformed), representation in lycaenid-ant interactions (log-transformed), ecological prevalence, and geographical distribution of ant genera as predictors. Geographical distribution was scored on a rank scale (from 1 to 10) as the number of faunal regions from which an ant genus is known, using the following 10 regions: West Palaearctic region (Europe eastwards up to the Ural mountains, including Africa north of the Sahara, Asia Minor, and the Near East); East Palaearctic region (Asia east of the Ural mountains, including Japan and Taiwan); India; South East Asia (comprising Thailand, the Malay Peninsula, and the large islands of the Sunda shelf like Sumatra, Borneo, and Java); New Guinea; Australia; Central Africa (south of the Sahara to approx. 15° southern latitude); Southern Africa (mainly comprising South Africa, Namibia, Botswana, and Zimbabwe); North America (north of Mexico); Central and South America. Ecological prevalence (sensu [56]) was scored on a rank scale from 1 to 5 (Table 1).

The linear model revealed that only the number of associated lycaenid species had a significant and positive relationship with the number of recorded cases of lycaenid-ant parasitism in an ant genus (see Table 2 for full documentation). All three other potential predictors were far from having any significant effect. Inspection of residuals confirmed that the model assumptions were met with reasonable accuracy. Moreover, application of a Ridge correction (with $\lambda = 0.1$) to account for collinearity among predictors did not change the overall model outcome (data

TABLE 1: Classification of ant genera known to associate with Lycaenidae caterpillars into prevalence groups. Ant genera are classified into that group which corresponds to the dominance status of its most dominant component species involved in butterfly-ant associations. For example, *Formica* is scored as "top dominant" since many (but not all) *Formica* species are territorial key-stone ant species in their respective habitats and communities, adapted from [24].

Class	Score	Criteria	Genera
Top dominant	5	Dominant ants in habitat; defend territories and resources intra- as well as interspecifically; monopolize resources against all heterospecific competitors	Myrmicinae: *Pheidole*; Formicinae: *Formica, Oecophylla*; Dolichoderinae: *Anonychomyrma, Azteca, Forelius, Froggattella, Iridomyrmex, Papyrius*
Second-order dominant	4	Subordinate relative to top dominants, but may become dominant in the absence of these; monopolize resources[†]	Myrmicinae: *Crematogaster, Meranoplus, Monomorium, Myrmicaria, Solenopsis, Tetramorium*; Formicinae: *Anoplolepis, Camponotus, Polyrhachis, Lasius, Lepisiota, Myrmecocystus*; Dolichoderinae: *Dolichoderus, Linepithema, Liometopum, Ochetellus, Philidris*;
Submissive	3	Subordinate to both classes of dominants; usually opportunistic species with generalized feeding habits; rarely defend and monopolize resources against heterospecific ants	Myrmicinae: *Acanthomyrmex, Aphaenogaster, Myrmica, Rhoptromyrmex*; Formicinae: *Echinopla, Notoncus, Paratrechina, Prolasius*; Dolichoderinae: *Dorymyrmex, Tapinoma, Technomyrmex*; Ponerinae: *Ectatomma*
Solitary	2	Foraging individually; rarely monopolize resources	Myrmeciinae: *Myrmecia*; Myrmicinae: *Cataulacus*; Ponerinae: *Gnamptogenys, Odontomachus, Rhytidoponera* Pseudomyrmecinae: *Tetraponera, Pseudomyrmex*
Cryptic	1	Minute species foraging on the ground or in leaf litter; inferior to all other ants in direct confrontation	Myrmicinae: *Leptothorax*; Formicinae: *Brachymyrmex, Plagiolepis*; Dolichoderinae: *Bothriomyrmex*

[†] Includes many species that become dominant in disturbed habitats or when introduced as alien species into non-adapted ant communities.

not shown). In a stepwise forward model selection, again only the frequency of nonparasitic associations remained as significant predictor. Likewise, using Poisson-type (instead of Gaussian) error distributions did not affect the outcome of this analysis (data not shown).

Hence, it is not the ecological or geographical prevalence that is decisive for the establishment of parasitic relationships between lycaenid butterflies and ants. Rather, the more butterfly species do interact with a given ant clade, the more likely it is that some of these interactions may turn, in evolutionary time, into parasitic relationships.

This also becomes evident when the incidence of ant-parasitism is plotted against the rank the ant genera have in interactions with lycaenid caterpillar species (Figure 1). Instances of social parasitism are more likely amongst those ant genera that are numerically more important in lycaenid-ant associations in general, whereas again species richness of the respective ant genera had no significant influence (Table 3).

A number of ant genera (e.g., *Pheidole, Dolichoderus, Formica,* and *Iridomyrmex*) that are ecologically dominant in

TABLE 2: Results of general linear model relating the number of parasitic lycaenid species associated with an ant genus to its species richness (log-transformed), ecological prevalence, geographical distribution, and importance in nonparasitic lycaenid-ant associations (log-transformed). Given are standardized regression coefficients β, and the F and p scores for each variable. SS: sum of squares; MS: mean of squares. Overall model fit: $R = 0.5394$, $R^2_{korr} = 0.2332$, $F_{4;49} = 5.0288$; $P = 0.0018$.

	SS	df	MS	β	F	p
constant	12.78	1	12.78		2.607	0.113
dominance	0.35	1	0.35	−0.0370	0.072	0.789
associated lycaenid spp.	47.54	1	47.54	0.5051	9.698	0.003
species richness	3.44	1	3.44	0.1201	0.702	0.406
geographic regions	0.018	1	0.018	−0.0097	0.004	0.952
error	240.21	49	4.90217			

their habitats and serve as hosts for many well-integrated myrmecophilous ant parasites from other insect groups (e.g., [55]) are thus far completely missing in the host list of

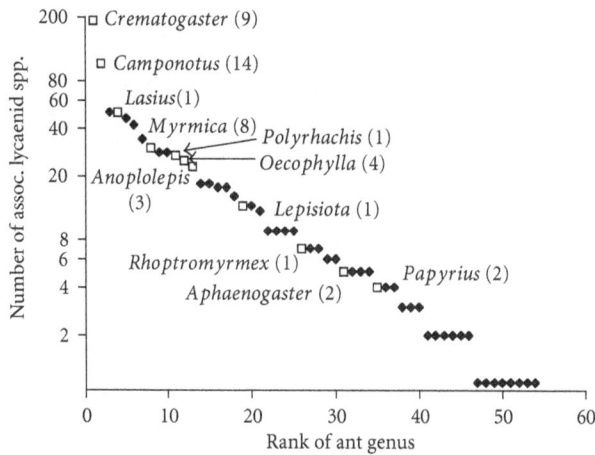

FIGURE 1: Rank-frequency plot of ant genera of the world involved in myrmecophilous associations of Lycaenidae butterflies, based on 927 record pairs of 497 butterfly species with 54 ant genera. Rank 1: ant genus with largest number of associated lycaenid species reported. Ranks 47 to 54: ant genera with only one associated lycaenid species known thus far. Filled diamonds: ant genera only known to be involved in mutualistic interactions with butterflies; open squares: ant genera that also serve as hosts for ant-parasitic Lycaenidae larvae (with genus name included; figure in parentheses: number of confirmed ant-parasitic lycaenid species). Note log-scale of y-axis.

TABLE 3: Results of a bivariate logistic regression, modelling the incidence of ant-parasitic associations within an ant genus (N = 54 ant genera), in relation to species richness and number of nonparasitic associations with lycaenid immatures (both log-transformed) per ant genus. Given are the regression coefficients b_i, their standard errors and corresponding t and p values. Overall model score: χ^2_{2df} = 6.577; P = 0.0373.

	$b_i \pm 1$ SE	t	p
Constant	-3.0864 ± 1.1160	2.766	0.008
Number of associated lycaenid species	0.9105 ± 0.3936	2.313	0.025
Species richness of genus	-0.1799 ± 0.2327	0.773	0.443

ant-parasitic lycaenids. Even considering that trophobiosis is an important evolutionary prerequisite for the establishment of lycaenid-ant interactions (thereby excluding nontrophobiotic ants such as army ants, leaf-cutter ants or harvester ants as potential hosts), the discrepancy in host use between ant-parasitic lycaenids, and other well-integrated myrmecophilous parasites remains striking.

Only two ant genera, *Camponotus* and *Crematogaster*, have been the target of multiple evolutionary trajectories towards parasitic life habits amongst the Lycaenidae. Even though complete phylogenetic analyses are still lacking for the family Lycaenidae, there can be no doubt that parasitism of *Camponotus* through the butterfly genera *Lepidochrysops*, *Niphanda*, *Ogyris*, and *Lachnocnema* has evolved independently—these four butterfly genera are far apart from each other in all systematic accounts of the family Lycaenidae, and they represent all three potential pathways

to ant-parasitism suggested earlier [6]. Likewise, parasitism of *Crematogaster* ants by *Acrodipsas* is certainly unrelated in phylogenetic terms to the multiple (and probably again: independent) occurrences amongst single species of Aphnaeini that all belong to larger genera where the majority of species is nonparasitic (*Cigaritis*, *Spindasis*, and *Aphnaeus*).

Overall, the scattered occurrence of ant-parasitism amongst the Lycaenidae gives evidence that such interactions have evolved multiple times, rather independently from another, and under quite different circumstances [6]. Only few such cases have given rise to moderate or even substantial radiations, most notably in the African genus *Lepidochrysops* (over 120 species) and in the Eurasiatic *Phengaris-Maculinea* clade (some 10–20 species). The host ant use of the latter remains a mystery in terms of its evolutionary and ecological roots. *Myrmica* ants are visitors of only a moderate number of ant-mutualistic lycaenids in the Holarctic region (recorded with 22 species thus far). Moreover, *Myrmica* ants usually neither form very large colonies nor are they territorial and ecologically dominant in most habitats where they occur today. Hence, they lack typical characters of other host ants of parasitic myrmecophiles. On the other hand recent phylogenetic evidence [47] strengthens the notion that evolution of parasitic associations with *Myrmica* ants occurred just once, at the base of the *Phengaris-Maculinea* clade. Similarly, the affiliation with *Camponotus* ants in parasitic *Lepidochrysops* as well as mutualistic *Orachrysops* suggests that specialization on *Camponotus* hosts predated the evolution of parasitism in that butterfly lineage.

Host shifts among ant-parasitic butterflies from one ant host genus to another have apparently rarely occurred in the Lycaenidae. One well-documented case is the Australian genus *Acrodipsas*, where some species parasitize *Papyrius* ants, but one clade subsequently shifted to *Crematogaster* hosts [42]. This rare case even implies a switching of hosts across ant subfamily boundaries. In contrast, the significance of *Aphaenogaster* recorded as host ants of some East Asiatic *Maculinea* needs to be rigorously addressed. In all likelihood, these are stray (or even erroneous) records rather than an indication of host shifts beyond ant genus boundaries.

6. Which Cases of Ant-Parasitism Might Await Detection amongst the Lycaenidae?

Starting from the patterns of host-ant use among ant-parasitic Lycaenidae, and in combination with other information on life-history traits of lycaenid butterflies, I here finally outline a few expectations in which butterfly clades and biomes further instances of parasitic interactions might most likely be uncovered. These expectations are amenable to testing by systematic assembly of further life history data or by evaluating earlier inconclusive reports.

One major group of lycaenid butterflies where a larger number of instances of trophallactic feeding by ants can be expected is the subfamily Miletinae. Miletinae larvae are essentially predators of homopterans. Since many homopterans are attended by ants and since quite a number of Miletinae larvae also drink honeydew, it would not come as a surprise to see more cases of trophallaxis with ants being documented in the future. Particularly likely candidates are those Miletinae species that are specifically adapted to spend their entire life cycle (including adult feeding on homopteran honeydew) with individual ant species. This is the case for *Logania malayica* with *Rhoptromyrmex wroughtonii*, and analogous candidate species occur in tropical SE Asia (*Miletus* spp. with *Dolichoderus* spp.; *Allotinus unicolor* with *Anoplolepis longipes*; [18, 20]). In two cases (*Allotinus apries* with *Myrmicaria lutea* [17]; *Logania hampsoni* with *Iridomyrmex* [15]) parasitic interactions have explicitly been suspected to exist, but until now these cases remain unsupported by direct observations of parasitic behaviours of the lycaenid caterpillars (A. Weissflog, personal communication for *A. apries*). As stated above, it is also quite likely that most, if not all *Thestor* species in South Africa will turn out to maintain parasitic relationships to *Anoplolepis custodiens* and allied ants [41]. Such cases of ant-parasitic relationships may also occasionally shift from the lower trophic level of cuckoo-feeding to the higher trophic level of brood predation (as in the genera *Liphyra* and *Euliphyra*). However, certain Miletinae do not interact intensively with ants that attend their homopteran prey [17, 57–59]. It is unlikely that traits required to entering into host-specific parasitic butterfly-ant interactions have evolved here. All further examples of ant-parasitism derived from predation on homopterans would obviously fall into the "Miletinae type" [6].

Another lycaenid clade where further cases of ant-parasitism can surely be expected to occur is the tribe Aphnaeini. Even though the few confirmed cases of ant-parasitism are rather isolated incidences nested within larger clades of ant-mutualists (e.g., *Chrysoritis dicksoni* in the genus *Chrysoritis* [60]), further species may show up to depend on nutrients derived from their close association with ants, as has been speculated many times in the literature (for critical reviews see [40, 61, 62]). Most additional instances of ant-parasitism in the Aphnaeini are expected to involve *Crematogaster* ants (the prevalent ant partner in mutualistic Aphnaeini species), but in *Aloeides* also further incidences of *Lepisiota*-parasitism may be found.

Other obvious candidates to furnish more ant-parasitic lycaenids are the genera *Lepidochrysops* (with *Camponotus*), *Maculinea*, and *Phengaris* (hitherto undescribed host associations in East Asia expected to refer to *Myrmica*), and *Niphanda* (probably with *Camponotus*). Beyond that, no valid extrapolations seem feasible at present. For example, the parasitic association between *Arhopala wildei* and *Polyrhachis queenslandica* does not seem "predictable" in a phylogenetic framework [13]. The most likely candidates for the discovery of novel ant-parasitic lycaenids of the "Aphnaeini type" are clades where a number of butterfly species show intimate host-specific mutualistic relationships towards specific host ants.

From the ant perspective, two genera which account for a very substantial fraction of records with lycaenids (namely, *Lasius* and *Formica*) score strikingly low as hosts of ant-parasitic butterflies. The only confirmed case with *Lasius* involves a species (*Shirozua jonasi*) whose larvae obtain most of their nutrient income from preying on homopterans and drinking their honeydew. This hairstreak species is ecologically similar to Miletinae butterflies and does not enter into *Lasius* nests to prey on ant brood. Possibly, the lack of brood being present in *Lasius* nests over winter poses a constraint in the evolution of ant-parasitism in temperate-zone climates. This would also explain why so far no case of ant-parasitism has been confirmed from the genus *Formica*. In East Asia, larvae of *Orthomiella rantaizana* have been found in *Formica* nests (Shen-Horn Yen, personal communication), but whether these are parasites, commensales, or mutualists of ants remains to be uncovered. Clearly, *Lasius* as well as *Formica* species serve as hosts for a large range of well-integrated myrmecophiles [55], but the majority of these parasites have evolved from detritivorous or predacious ancestors, and not from herbivores.

Two other sociobiological traits of ant colonies that have been suggested to be related to the evolution of parasitic myrmecophily are the level of polygyny or polyandry, and the brood cycle. With regard to the latter, as already noted above the absence of winter brood may have prevented the intrusion of Holarctic lycaenids as parasites into *Lasius* and *Formica* colonies. With regard to ants from the humid tropics, however, seasonal fluctuations in brood availability are less likely to constrain the evolution of lycaenid butterflies into parasites of ants, so that this factor (if valid at all) would have to be restricted to seasonal climates. Genetic intracolonial heterogeneity, which can result from the presence of multiple queens and/or the occurrence of multiple matings during their nuptial flight, may facilitate the intrusion of social parasites as well as of parasitic myrmecophiles [63]. It is presently impossible to rigorously test these two hypotheses, since data on the colony structure and population demography of many tropical and subtropical ants that are parasitized by lycaenids are too scant. Polygyny seems to be common among ants that serve as hosts [64], but in at least one instance (*Camponotus japonicus*, the host ant of *Niphanda fusca*) monogyny and claustral colony foundation have been confirmed [65].

7. Perspective

Ant-parasitic lycaenid butterflies are a bewildering evolutionary outcome: carnivores or cuckoo-type feeders in an otherwise phytophagous clade of insects. The communication modes required for integration into their host colonies, the phylogenetic roots, and population genetic consequences of their unusual interactions with ants, and their repercussions into conservation biology [66, 67] will continue to attract the interest of scientists. However, these parasitic interactions encompass only a small minority of myrmecophilous Lycaenidae butterfly species. Also the ant genera involved comprise but a small minority as compared to the range of trophobiotic ants that could potentially be parasitized. For sure, some further extensions can be expected, especially in hitherto underexplored tropical regions or in butterfly clades whose life histories are thus far very poorly documented. Most known ant-parasitic lycaenids occur in seasonally cold and/or dry regions [6], where both the butterfly and the ant faunas are comparatively well covered. It has even been suggested, though not yet rigorously tested, that avoidance of unfavourable seasons might have promoted the entering of ant nests as safe places for lycaenid caterpillars. The detection of additional cases of butterfly-ant parasitism in these regions in all likelihood will not radically turn the robust patterns described here upside down. For tropical faunas, some more unexpected incidences of ant-parasitism may await discovery, yet it does not seem likely that many instances of butterfly caterpillars living in brood chambers of ant nests would have gone undetected thus far. Rather, future progress will be made in uncovering the microevolutionary steps that drive host-parasite co-evolution [7]. It will also be rewarding to rigorously assess the macroevolutionary pathways leading to ant-parasitism in a phylogenetically controlled manner. To achieve this goal, besides elucidating the phylogenetic relationships of lycaenids and their ant hosts, more bionomic data on both of these players, but especially a better documentation of the sociobiology and ecology of the host ants (beyond the well-studied *Myrmica* case) will be essential.

Acknowledgments

The author is grateful to Jean-Paul Lachaud for the invitation to write this contribution. Volker Witte and an anonymous reviewer provided helpful comments that served to improve this paper. Shen-Horn Yen, Andreas Weissflog, and Alain Dejean contributed some records of ant hosts of parasitic lycaenid larvae. Alain Heath generously sent him copies of papers that are otherwise difficult to obtain. Special thanks are due to Ulrich Maschwitz (formerly University of Frankfurt, Germany), who initially stimulated the author's studies on myrmecophilous Lycaenidae, and to Bert Hölldobler (formerly University of Würzburg, Germany), who provided an exciting working atmosphere for years. He thanks Phil J. DeVries, the late John N. Eliot, Graham W. Elmes, David R. Nash, Naomi E. Pierce, and Jeremy A. Thomas for fruitful discussions of butterfly-ant interactions over many years.

References

[1] K. Fiedler, "Systematic, evolutionary, and ecological implications of myrmecophily within the Lycaenidae (Insecta: Lepidoptera: Papilionoidea)," *Bonner zoologische Monographien*, vol. 31, pp. 1–210, 1991.

[2] N. E. Pierce, M. F. Braby, A. Heath et al., "The ecology and evolution of ant association in the Lycaenidae (Lepidoptera)," *Annual Review of Entomology*, vol. 47, pp. 733–771, 2002.

[3] K. Fiedler, B. Hölldobler, and P. Seufert, "Butterflies and ants: the communicative domain," *Experientia*, vol. 52, no. 1, pp. 14–24, 1996.

[4] F. Barbero, J. A. Thomas, S. Bonelli, E. Balletto, and K. Schönrogge, "Queen ants make distinctive sounds that are mimicked by a butterfly social parasite," *Science*, vol. 323, no. 5915, pp. 782–785, 2009.

[5] H. Daniels, G. Gottsberger, and K. Fiedler, "Nutrient composition of larval nectar secretions from three species of myrmecophilous butterflies," *Journal of Chemical Ecology*, vol. 31, no. 12, pp. 2805–2821, 2005.

[6] K. Fiedler, "Lycaenid-ant interactions of the *Maculinea* type: tracing their historical roots in a comparative framework," *Journal of Insect Conservation*, vol. 2, no. 1, pp. 3–14, 1998.

[7] D. R. Nash, T. D. Als, R. Maile, G. R. Jones, and J. J. Boomsma, "A mosaic of chemical coevolution in a large blue butterfly," *Science*, vol. 319, no. 5859, pp. 88–90, 2008.

[8] L. P. Koh, R. R. Dunn, N. S. Sodhi, R. K. Colwell, H. C. Proctor, and V. S. Smith, "Species coextinctions and the biodiversity crisis," *Science*, vol. 305, no. 5690, pp. 1632–1634, 2004.

[9] T. R. New, Ed., *Conservation Biology of the Lycaenidae (Butterflies)*, Occasional Paper of the IUCN Species Survival Commission, no. 8, IUCN, Gland, Switzerland, 1993.

[10] J. A. Thomas, D. J. Simcox, and R. T. Clarke, "Successful conservation of a threatened *Maculinea* butterfly," *Science*, vol. 325, no. 5936, pp. 80–83, 2009.

[11] T. D. Als, R. Vila, N. P. Kandul et al., "The evolution of alternative parasitic life histories in large blue butterflies," *Nature*, vol. 432, no. 7015, pp. 386–390, 2004.

[12] U. Maschwitz, M. Schroth, H. Hänel, and T. Y. Pong, "Lycaenids parasitizing symbiotic plant-ant partnerships," *Oecologia*, vol. 64, no. 1, pp. 78–80, 1984.

[13] H.-J. Megens, R. de Jong, and K. Fiedler, "Phylogenetic patterns in larval host plant and ant association of Indo-Australian Arhopalini butterflies (Lycaenidae: Theclinae)," *Biological Journal of the Linnean Society*, vol. 84, no. 2, pp. 225–241, 2005.

[14] D. P. A. Sands, "A revision of the genus *Hypochrysops* C. & R. Felder (Lepidoptera: Lycaenidae)," *Entomonograph*, vol. 7, pp. 1–116, 1986.

[15] M. Parsons, *The Butterflies of Papua New Guinea, Their Systematics and Biology*, Academic Press, San Diego, Calif, USA, 1999.

[16] U. Maschwitz, K. Dumpert, and P. Sebastian, "Morphological and behavioural adaptations of homopterophagous blues (Lepidoptera: Lycaenidae)," *Entomologia Generalis*, vol. 11, no. 1-2, pp. 85–90, 1985.

[17] U. Maschwitz, W. A. Nässig, K. Dumpert, and K. Fiedler, "Larval carnivory and myrmecoxeny, and imaginal myrmecophily in Miletine lycaenids (Lepidoptera: Lycaenidae) on the Malay Peninsula," *Tŷo to Ga*, vol. 39, no. 3, pp. 167–181, 1988.

[18] K. Fiedler and U. Maschwitz, "Adult myrmecophily in butterflies: the role of the ant *Anoplolepis longipes* in the feeding and oviposition behaviour of *Allotinus unicolor*," *Tŷo to Ga*, vol. 40, no. 4, pp. 241–251, 1989.

[19] N. E. Pierce, "Predatory and parasitic Lepidoptera: carnivores living on plants," *Journal of the Lepidopterists' Society*, vol. 49, no. 4, pp. 412–453, 1995.

[20] D. J. Lohman and V. U. Samarita, "The biology of carnivorous butterfly larvae (Lepidoptera: Lycaenidae: Miletinae: Miletini) and their ant-tended hemipteran prey in Thailand and the Philippines," *Journal of Natural History*, vol. 43, no. 9-10, pp. 569–581, 2009.

[21] P. J. DeVries, I. A. Chacon, and D. Murray, "Toward a better understanding of host use and biodiversity in riodinid butterflies (Lepidoptera)," *Journal of Research on the Lepidoptera*, vol. 31, no. 1-2, pp. 103–126, 1994.

[22] P. J. DeVries, *The Butterflies of Costa Rica and Their Natural History: Riodinidae*, vol. 2, Princeton University Press, Princeton, NJ, USA, 1997.

[23] D. L. Campbell, A. V. Z. Brower, and N. E. Pierce, "Molecular evolution of the Wingless gene and its implications for the phylogenetic placement of the butterfly family riodinidae (Lepidoptera: Papilionoidea)," *Molecular Biology and Evolution*, vol. 17, no. 5, pp. 684–696, 2000.

[24] K. Fiedler, "Ants that associate with Lycaeninae butterfly larvae: diversity, ecology and biogeography," *Diversity & Distributions*, vol. 7, no. 1-2, pp. 45–60, 2001.

[25] K. Fiedler, "Ant-associates of Palaearctic lycaenid butterfly larvae (Hymenoptera: Formicidae; Lepidoptera: Lycaenidae)—a review," *Myrmecologische Nachrichten*, vol. 9, pp. 77–87, 2006.

[26] B. Bolton, G. Alpert, P. S. Ward, and P. Naskrecki, *Bolton's Catalogue of Ants of the World: 1758–2005*, Harvard University Press, Cambridge, Mass, USA, 2007.

[27] J. Settele, E. Kühn, and J. A. Thomas, Eds., *Studies on the Ecology and Conservation of Butterflies in Europe: Species Ecology along a European Gradient: Maculinea Butterflies as a Model*, vol. 2, Pensoft, Sofia, Bulgaria, 2005.

[28] M. Sielezniew and A. M. Stankiewicz, "*Myrmica sabuleti* (Hymenoptera: Formicidae) not necessary for the survival of the population of *Phengaris (Maculinea) arion* (Lepidoptera: Lycaenidae) in Eastern Poland: lower host-ant specificity or evidence for geographical variation of an endangered social parasite?" *European Journal of Entomology*, vol. 105, no. 4, pp. 637–641, 2008.

[29] P. R. Ackery, C. R. Smith, and R. I. Vane-Wright, Eds., *Carcasson's African Butterflies*, CSIRO, Collingwood, Australia, 1995.

[30] S. F. Henning and G. A. Henning, *South African Red Data Book—Butterflies*, Foundation for Research and Development, Council for Scientific and Industrial Research, Pretoria, South Africa, 1989.

[31] D. A. Edge and H. van Hamburg, "Larval feeding behaviour and myrmecophily of the Brenton Blue, *Orachrysops niobe* (Trimen) (Lepidoptera: Lycaenidae)," *Journal of Research on the Lepidoptera*, vol. 42, pp. 21–33, 2010.

[32] M. K. Hojo, A. Wada-Katsumata, T. Akino, S. Yamaguchi, M. Ozaki, and R. Yamaoka, "Chemical disguise as particular caste of host ants in the ant inquiline parasite *Niphanda fusca* (Lepidoptera: Lycaenidae)," *Proceedings of the Royal Society B*, vol. 276, no. 1656, pp. 551–558, 2009.

[33] M. K. Hojo, A. Wada-Katsumata, S. Yamaguchi, M. Ozaki, and R. Yamaoka, "Gustatory synergism in ants mediates a species-specific symbiosis with lycaenid butterflies," *Journal of Comparative Physiology A*, vol. 194, no. 12, pp. 1043–1052, 2008.

[34] R. Eastwood and A. M. Fraser, "Associations between lycaenid butterflies and ants in Australia," *Australian Journal of Ecology*, vol. 24, no. 5, pp. 503–537, 1999.

[35] M. F. Braby, *Butterflies of Australia, Their Identification, Biology and Distribution*, CSIRO Publishing, Collingwood, Australia, 2000.

[36] A. Dejean and G. Beugnon, "Host-ant trail following by myrmecophilous larvae of Liphyrinae (Lepidoptera, Lycaenidae)," *Oecologia*, vol. 106, no. 1, pp. 57–62, 1996.

[37] A. J. King and L. R. Ring, "The life history of *Arhopala wildei wildei* Miskin (Lepidoptera: Lycaenidae)," *The Australian Entomologist*, vol. 23, pp. 117–120, 1996.

[38] R. Eastwood and A. J. King, "Observations of the biology of *Arhopala wildei* Miskin (Lepidoptera: Lycaenidae) and its host ant *Polyrhachis queenslandica* Emery (Hymenoptera: Formicidae)," *The Australian Entomologist*, vol. 25, pp. 1–6, 1998.

[39] H. Fukuda, E. Hama, T. Kuzuya et al., *The Life Histories of Butterflies in Japan*, vol. 3, Hoikusha, Osaka, Japan, 2nd edition, 1992.

[40] A. Heath, L. McLeod, Z. A. Kaliszewska, C. W. S. Fisher, and M. Cornwall, "Field notes including a summary of trophic and ant-associations for the butterfly genera *Chrysoritis* Butler, *Aloeides* Hübner and *Thestor* Hübner (Lepidoptera: Lycaeni-

dae) from South Africa," *Metamorphosis*, vol. 19, no. 3, pp. 127–148, 2008.

[41] A. Heath and E. L. Pringle, "A review of the Southern African genus *Thestor* Hübner (Lepidoptera: Lycaenidae: Miletinae)," *Metamorphosis*, vol. 15, no. 3, pp. 91–151, 2004.

[42] R. Eastwood and J. M. Hughes, "Molecular phylogeny and evolutionary biology of *Acrodipsas* (Lepidoptera: Lycaenidae)," *Molecular Phylogenetics and Evolution*, vol. 27, no. 1, pp. 93–102, 2003.

[43] M. Sanetra and K. Fiedler, "Behaviour and morphology of an aphytophagous lycaenid caterpillar: *Cigaritis (Apharitis) acamas* Klug, 1834 (Lepidoptera: Lycaenidae)," *Nota Lepidopterologica*, vol. 18, no. 1, pp. 57–76, 1995.

[44] S. Igarashi and H. Fukuda, *The Life Histories of Asian Butterflies*, vol. 2, Tokai University Press, Tokyo, Japan, 2000.

[45] R. Eastwood and J. M. Hughes, "Phylogeography of the rare myrmecophagous butterfly *Acrodipsas cuprea* (Lepidoptera: Lycaenidae) from pinned museum specimens," *Australian Journal of Zoology*, vol. 51, no. 4, pp. 331–340, 2003.

[46] C. B. Cottrell, "Aphytophagy in butterflies: its relationship to myrmecophily," *Zoological Journal of the Linnean Society*, vol. 80, no. 1, pp. 1–57, 1984.

[47] L. V. Ugelvig, R. Vila, N. E. Pierce, and D. R. Nash, "A phylogenetic revision of the *Glaucopsyche* section (Lepidoptera: Lycaenidae), with special focus on the *Phengaris-Maculinea* clade," *Molecular Phylogenetics and Evolution*, vol. 61, no. 1, pp. 237–243, 2011.

[48] J. A. Thomas, G. W. Elmes, J. C. Wardlaw, and M. Woyciechowski, "Host specificity among *Maculinea* butterflies in *Myrmica* ant nests," *Oecologia*, vol. 79, no. 4, pp. 452–457, 1989.

[49] K. Fiedler, "The remarkable biology of two Malaysian lycaenid butterflies," *Nature Malaysiana*, vol. 18, no. 2, pp. 35–42, 1993.

[50] S. Koptur, "Extrafloral nectary-mediated interactions between insects and plants," in *Insect-Plant Interactions*, Bernays E., Ed., vol. 4, pp. 81–129, CRC Press, Boca Raton, Fla, USA, 1992.

[51] J. L. Bronstein, "The contribution of ant-plant protection studies to our understanding of mutualism," *Biotropica*, vol. 30, no. 2, pp. 150–161, 1998.

[52] B. Stadler, P. Kindlmann, P. Šmilauer, and K. Fiedler, "A comparative analysis of morphological and ecological characters of European aphids and lycaenids in relation to ant attendance," *Oecologia*, vol. 135, no. 3, pp. 422–430, 2003.

[53] T. H. Oliver, S. R. Leather, and J. M. Cook, "Macroevolutionary patterns in the origin of mutualisms involving ants," *Journal of Evolutionary Biology*, vol. 21, no. 6, pp. 1597–1608, 2008.

[54] D. W. Davidson, S. C. Cook, and R. R. Snelling, "Liquid-feeding performances of ants (Formicidae): ecological and evolutionary implications," *Oecologia*, vol. 139, no. 2, pp. 255–266, 2004.

[55] B. Hölldobler and E. O. Wilson, *The Ants*, Harvard University Press, Cambridge, Mass, USA, 1990.

[56] E. O. Wilson, "Which are the most prevalent ant genera?" *Studia Entomologica*, vol. 19, pp. 187–200, 1976.

[57] M. G. Venkatesha, "Why is homopterophagous butterfly, *Spalgis epius* (Westwood) (Lepidoptera: Lycaenidae) amyrmecophilous?" *Current Science*, vol. 89, no. 2, pp. 245–246, 2005.

[58] E. Youngsteadt and P. J. DeVries, "The effects of ants on the entomophagous butterfly caterpillar *Feniseca tarquinius*, and the putative role of chemical camouflage in the *Feniseca* interaction," *Journal of Chemical Ecology*, vol. 31, no. 9, pp. 2091–2109, 2005.

[59] D. J. Lohman, Q. Liao, and N. E. Pierce, "Convergence of chemical mimicry in a guild of aphid predators," *Ecological Entomology*, vol. 31, no. 1, pp. 41–51, 2006.

[60] D. B. Rand, A. Heath, T. Suderman, and N. E. Pierce, "Phylogeny and life history evolution of the genus *Chrysoritis* within the Aphnaeini (Lepidoptera: Lycaenidae), inferred from mitochondrial cytochrome oxidase I sequences," *Molecular Phylogenetics and Evolution*, vol. 17, no. 1, pp. 85–96, 2000.

[61] A. Heath and A. J. M. Claassens, "Ant-association among Southern African Lycaenidae," *Journal of the Lepidopterists'Society*, vol. 57, no. 1, pp. 1–16, 2003.

[62] A. Heath and E. L. Pringle, "Biological observations and a taxonomic review based on morphological characters in the myrmecophilous genus *Chrysoritis* Butler (Lepidoptera: Lycaenidae: Aphnaeini)," *Metamorphosis*, vol. 18, no. 1, pp. 2–44, 2007.

[63] M. A. Fürst, M. Durey, and D. R. Nash, "Testing the adjustable threshold model for intruder recognition on *Myrmica* ants in the context of a social parasite," *Proceedings of the Royal Society B*, vol. 279, no. 1728, pp. 516–522, 2012.

[64] E. A. Schlüns, B. J. Wegener, H. Schlüns, N. Azuma, S. K. A. Robson, and R. H. Crozier, "Breeding system, colony and population structure in the weaver ant *Oecophylla smaragdina*," *Molecular Ecology*, vol. 18, no. 1, pp. 156–167, 2009.

[65] Z. Liu, S. Yamane, J. Kojima, Q. Wang, and S. Tanaka, "Flexibility of first brood production in a claustral ant, *Camponotus japonicus* (Hymenoptera: Formicidae)," *Journal of Ethology*, vol. 19, no. 2, pp. 87–91, 2001.

[66] N. Mouquet, V. Belrose, J. A. Thomas, G. W. Elmes, R. T. Clarke, and M. E. Hochberg, "Conserving community modules: a case study of the endangered lycaenid butterfly *Maculinea alcon*," *Ecology*, vol. 86, no. 12, pp. 3160–3173, 2005.

[67] J. Settele, O. Kudrna, A. Harpke et al., "Climatic risk atlas of European butterflies," *Biorisk*, vol. 1, pp. 1–719, 2008.

18

Carcass Fungistasis of the Burying Beetle *Nicrophorus nepalensis* Hope (Coleoptera: Silphidae)

Wenbe Hwang and Hsiu-Mei Lin

Department of Ecoscience and Ecotechnology, National University of Tainan, 33, Section 2, Shu-Lin Street, Tainan 70005, Taiwan

Correspondence should be addressed to Wenbe Hwang; wenbehwang@mail.nutn.edu.tw

Academic Editor: Vladimir Kostal

Our study investigated the fungistatic effects of the anal secretions of *Nicrophorus nepalensis* Hope on mouse carcasses. The diversity of fungi on carcasses was investigated in five different experimental conditions that corresponded to stages of the burial process. The inhibition of fungal growth on carcasses that were treated by mature beetles before burial was lost when identically treated carcasses were washed with distilled water. Compared with control carcasses, carcasses that were prepared, buried, and subsequently guarded by mature breeding pairs of beetles exhibited the greatest inhibition of fungal growth. No significant difference in fungistasis was observed between the 3.5 g and the 18 to 22 g guarded carcasses. We used the growth of the predominant species of fungi on the control carcasses, *Trichoderma* sp., as a biological indicator to examine differences in the fungistatic efficiency of anal secretions between sexually mature and immature adults and between genders. The anal secretions of sexually mature beetles inhibited the growth of *Trichoderma* sp., whereas the secretions of immature beetles did not. The secretions of sexually mature females displayed significantly greater inhibition of the growth of *Trichoderma* sp. than those of sexually mature males, possibly reflecting a division of labor in burying beetle reproduction.

1. Introduction

Burying beetles (*Nicrophorus* spp.) use small vertebrate carcasses as food for their larval broods by depositing their eggs around a buried carcass [1, 2]. Carcasses are nutritious yet rare resources [3, 4]. During the lifetime of a beetle, it may find only one carcass that is suitable for reproduction [5]. Competition for carcasses is intense [6–8], and burying beetles of the same or different species may fight to maintain occupancy of the carcass [1, 9–11].

Bacterial and fungal decomposers destroy carcasses, and scavenging animals have evolved behavioral and physiological counterstrategies to maintain food sources [12]. Before burying a carcass, the burying beetles remove the fur or feathers from the carcass, compact the carcass by rolling it repeatedly, and smear its surface with their anal secretions [1]. Carcasses used by beetles typically vary in size from 1 to 75 g [9, 10, 13] and are encountered in variable states of decay. Burying beetles exhibit adaptive strategies that enable them to manage the carrion resources in such diverse conditions, such as adjusting the number of eggs laid [13, 14] and practicing infanticide [15, 16], with the number of surviving larvae positively correlated with carcass size [9, 10, 17].

Although the loss of biomass resulting from microbial growth on a carcass is not large, microorganisms often produce toxins that can affect beetle-larvae survival [18–20]. The oral and anal secretions of various burying beetle species have bacteriostatic effects [21]. The oral secretions contain phospholipase A_2 that may disrupt the cell membranes of bacteria [22]. Fungal growth may also be inhibited following the preburial treatments by burying beetles [23]. The temperature and the composition of food materials can influence the antimicrobial activity of the oral secretions [24], and the antibacterial activity of the anal secretions has been shown to be upregulated following the discovery of carrion [25].

Burying beetles' preference for appropriately sized carcasses for reproduction may be related to their capacity to secrete antimicrobial substances [26]. Although burying beetles can feed more offspring on a larger carcass, the energy expenditure for the preburial preparation of larger carcasses is also higher. Scott [27] proposed that microorganisms are more serious competitors on larger carcasses because of the difficulties associated with preburial preparations. Scott [27] also reported that mold often renders substantial amounts of large carcasses unusable, and Hwang and Shiao [26] reported that large carcasses decay more rapidly than small carcasses, resulting in lower trophic efficiency for large carcasses. Therefore, communal breeding observed in some species of burying beetles may prevent the decay of a large carcass, contributing to better breeding efficiency [28–31].

Both uniparental and biparental breedings are commonly observed in burying beetles [8, 32]. Females reproducing without the assistance of a male do not display reduced reproductive success [33]. However, with the help of a male, the carcass can be better preserved [6, 34], and dipteran larvae and conspecific competitors can be more efficiently excluded [32, 35–38]. In addition, the male can also substitute for the female in brood care [39]. However, because the primary role of the male in brood care is providing protection against competitors, we propose that the female likely makes a greater antimicrobial contribution to the carcass.

In our current study, we assessed fungistasis in carcasses in the laboratory that were colonized by *Nicrophorus nepalensis*, a common burying beetle in southern Taiwan. We investigated whether the fungistatic efficiency of beetles correlated with the sexual maturity or the sex of the parent, and we examined whether fungistatic capacity of beetles was sex or age dependent.

2. Materials and Methods

2.1. Field Collection and Laboratory Rearing of Beetles. The *N. nepalensis* Hope beetles were collected using 15 hanging pitfall traps that were baited with 40 g of chicken meat each and placed at 100 m intervals along the Fengkang Forest Road (22°00′N, 120°41′E) in Kaohsiung City in southern Taiwan at altitudes of 1100 to 1600 m above sea level from January to July, 2009. All field-collected beetles were anesthetized with carbon dioxide, and any mites were removed under a microscope using forceps. To avoid the influence of parasites, only the laboratory-reared F1 and F2 offspring of field-collected beetles were used in our experiments. All beetle cages used in our study were 10.4 × 10.4 × 6 cm transparent plastic containers. All the beetles used in our study were reared at 20°C with 12 h light-dark cycling.

A breeding pair of field-collected beetles and a 20 g mouse carcass were added to a cage with 4 cm thick moist peat. Following oviposition, the eggs were removed and placed on wet toilet paper in an 8.5 cm Petri dish for hatching. The larvae and the parents were transferred to a new cage with 1 cm thick moist peat. When the larvae emerged from the burrow to pupate, up to 8 were placed in a new cage with 4 cm thick moist peat. Groups of up to 6 newly eclosed adults of the same sex were transferred to new cages with 3 cm thick moist peat. Prior to the fungistasis experiments, the laboratory-reared adult beetles were fed twice a week

with freshly decapitated *Tenebrio molitor* or cut sections of *Zophobas morio*. No beetles were exposed to carcasses before being used in the fungistasis experiments.

2.2. Experimental Design. To investigate whether preburial preparations affect fungal growth on carcasses, fungal growth on mouse (ICR strain) carcasses was assessed in five different conditions. All fresh frozen mice were purchased in CMLAC in National Taiwan University and thawed before the experiments. Untreated (control) carcasses, treated carcass balls, washed carcass balls, protected carcass balls, and large protected carcass balls were examined for visible fungal growth over the course of 14 days under the standard rearing conditions. The control carcasses (approximately 3.5 g) were not exposed to burying beetles and were placed on the surface of the moist peat in an otherwise empty cage. The treated carcass balls were obtained at 3 days following presentation of a carcass (approximately 3.5 g) to a mating pair of sexually mature adults by removing the carcass immediately after burial. During this stage, the fur removal, the carcass compaction, and the deposition of anal excretions had occurred prior to the removal of the carcass, but larval hatch had not yet occurred. The treated carcasses were each transferred to a new cage with moist peat and no beetles. The washed carcass balls were obtained using the same procedure as the treated carcass balls, with an additional step in which the carcass balls were rinsed with distilled water before being transferred to a new cage. The protected and large protected carcass balls were obtained using the same procedure as the treated carcass balls, except that the same mating pair of beetles was added after the carcass was transferred to a new cage, and 18 to 22 g mouse carcasses were used for the large protected carcass balls. To remove newly oviposited eggs, the carcass ball and the adult beetles were transferred to a new cage daily, with the transfers performed under red light to avoid disrupting the light-dark cycle.

2.3. Cultivation and Identification of Carcass Fungi. Any fungus that grew on a carcass within 14 days in any of the 5 experimental conditions was cultivated for identification. Solid malt extract agar (MEA) medium was prepared from 26 g of malt extract agar (Fluka) in 500 mL distilled water and sterilized at 121°C for 15 min. To prepare the solid cornmeal agar (CMA) medium, 10 g of cornmeal was boiled in distilled water. Following filtration, the volume of the cornmeal filtrate was adjusted to 500 mL, and 10 g of agar (Sigma-Aldrich, St. Louis, MO, USA) was added before sterilization at 121°C for 15 min. After cooling the media to 50°C, 40 ppm of Streptomycin sulfate and 40 ppm of Penicillin G were added to both the MEA and CMA media, and the media were poured into Petri dishes before solidification. A hypodermic needle was used to remove a specimen of the carcass skin containing the fungi, and the specimen was used to inoculate MEA plates, and the fungi were cultured for 7 days at 25°C. The fungi cultured on MEA plates were used to inoculate CMA plates that were subsequently cultured at 25°C. The cultured fungi were dyed with cotton blue in lactoglycerol, and the various taxa were identified to genus or species.

2.4. Fungistasis Quantification Assays. To examine differences in the fungistatic capacity of the anal secretions from males versus females and sexually mature adults versus sexually immature adults, the anal secretions were collected from each and used in fungistasis quantification assays. We observed that *Trichoderma* sp. was the predominant species on control carcasses and that *Trichoderma* sp. growth was inhibited on carcasses treated by sexually mature beetles. Therefore, we used *Trichoderma* sp. growth as a biological indicator of the fungistatic efficiency of burying beetle secretions. Green colonies of *Trichoderma* sp. formed on CMA plates after culturing for 7 days at 25°C. One colony was suspended in 300 μL of ultrapure water. Inoculums were prepared by mixing 30 μL of the fungal suspension with 30 μL of ultrapure water (control group) or 30 μL of anal secretions from male or female beetles that were taken at 6 days (sexually immature) or 35 days (sexually mature) after eclosion, and 2 μL of each inoculum was separately used to inoculate CMA plates that were subsequently cultured at 25°C. The secretory volume of each individual was different; 30 μL of anal secretions were collected from different individuals. The number of *Trichoderma* sp. colonies present on the CMA plates at 7 days after inoculation was recorded. The number of days at which the fungal growth reached confluency on the CMA plates was also recorded.

2.5. Statistical Analysis. We used the Fisher exact test to compare the fungal growth in the various experimental conditions, and we used a χ^2 test to compare the differences in the fungal species that were isolated in each set of experimental conditions. An independent sample Student's t test was used to examine the differences in fungistatic efficiency between sexually mature males and females. All statistical analyses were conducted using the SPSS version 17.0 computer software, with an alpha value of 0.05 as the accepted level of statistical significance.

3. Results

3.1. Inhibition of Fungal Growth on Carcasses. Compared with the control carcasses ($n = 19$), fungal growth was significantly inhibited on the treated and protected carcass balls (Fisher exact test: treated carcass balls: $P < 0.01$, $n = 18$; protected carcass balls: $P < 0.001$, $n = 16$). However, the fungistatic effect was significantly greater on protected carcass balls than on treated carcass balls (Fisher exact test: $P = 0.01$). The fungistasis on the washed carcass balls ($n = 18$) was not significantly different than that of the control carcasses (Fisher exact test: $P = 0.08$), and the fungistasis on the protected carcass balls ($n = 16$) was not significantly different than that of the large protected carcass balls ($n = 17$; Fisher exact test: $P = 0.68$; Figure 1).

3.2. Fungus Diversity on Carcasses. The following 12 fungal species were isolated from the mouse carcasses in the various experimental conditions: *Alternaria* sp., *Aspergillus fumigates*, *Cladosporium herbarum*, *Cladosporium* sp. 1, *Cladosporium* sp. 2, *Conidiobolus* sp., *Dactylaria* sp., *Graphium* sp.,

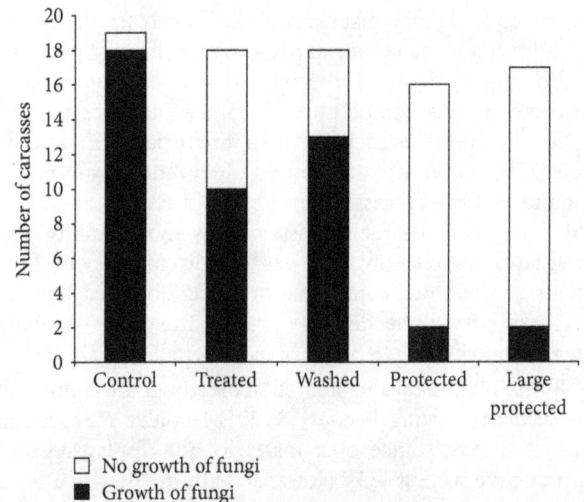

FIGURE 1: Number of control carcasses (untreated, $n = 19$), treated carcass balls ($n = 18$), washed carcass balls ($n = 18$), protected carcass balls (3.5 g, $n = 16$), and large protected carcass balls (18 to 22 g, $n = 17$) on which fungi grew within 14 d. The beetles had opportunity to come into contact with carcass on a protected carcass ball, while beetles were removed at 3 days after burial on a treated carcass ball.

Mucor sp., *Phoma* sp., *Trichoderma* sp., and *Verticillium* sp. On control carcasses, 23 fungal samples were acquired, which consisted of the following 7 species: *Aspergillus fumigates*, *Alternaria* sp., *Cladosporium* sp. 1, *Graphium* sp., *Mucor* sp., *Trichoderma* sp., and *Verticillium* sp. The *Trichoderma* sp. Had the highest frequency of occurrence on control carcasses, accounting for 72% of the acquired fungal samples. On the treated carcass balls, 10 fungal samples were acquired, which consisted of the following 4 species: *Aspergillus fumigates*, *Cladosporium herbarum*, *Mucor* sp., and *Phoma* sp. The *Mucor* sp. was the most predominant fungus on treated carcass balls, accounting for 70% of the acquired fungal samples. On the washed carcass balls, 14 fungal samples were acquired, which consisted of the following 8 species: *Aspergillus fumigates*, *Alternaria* sp., *Cladosporium* sp. 2, *Conidiobolus* sp., *Dactylaria* sp., *Graphium* sp., *Mucor* sp., and *Trichoderma* sp. The *Mucor* sp. was the predominant fungus on washed carcass balls, accounting for 62.5% of the acquired fungal samples (Figure 2). Only *Aspergillus fumigates* and *Cladosporium* sp. 1 were identified on the protected carcass balls, and only *Aspergillus fumigates* and *Mucor* sp. were identified on the large protected carcass balls.

The incidences of the various fungal species were not significantly different between the control carcasses and the washed carcass balls (χ^2 test: $P = 0.07$, control carcasses: $n = 23$, washed carcass balls: $n = 14$), or between the treated carcass balls and the washed carcass balls (χ^2 test: $P = 0.41$, treated carcass balls: $n = 10$). However, there were significant differences in the incidences of the various fungal species between the control carcasses and the treated carcass balls (χ^2 test: $P = 0.001$, treated carcass balls: $n = 10$).

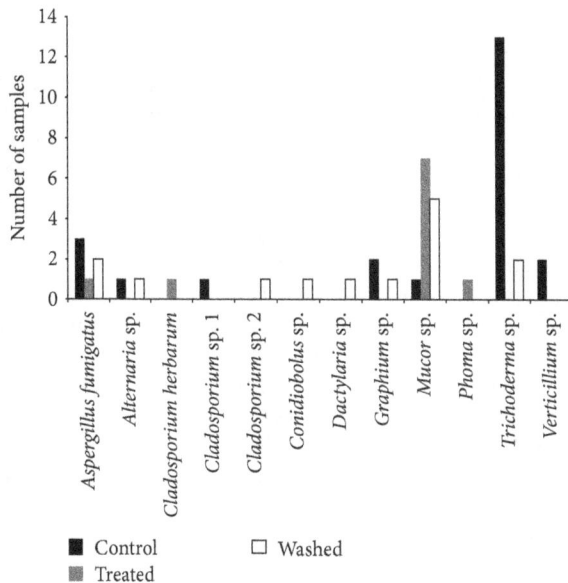

FIGURE 2: The sample number of the fungi species that were collected from control carcasses (n = 23), treated carcass balls (n = 10), and washed carcass balls (n = 14).

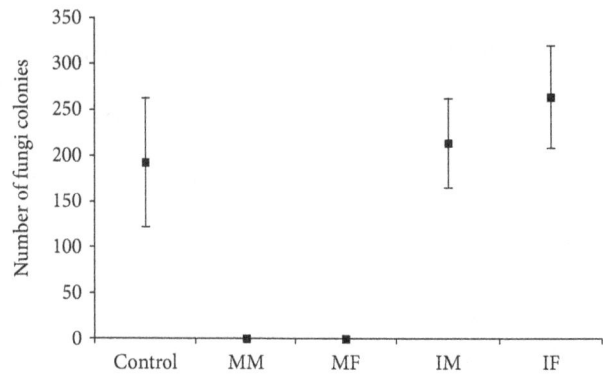

FIGURE 3: The number of *Trichoderma* sp. colonies (mean ± standard error) on CMA plates at 7 d after inoculation that were produced from the various inoculums. The inoculums were prepared from *Trichoderma* sp. suspended in ultrapure water (control group: n = 16) and the anal secretions of sexually mature males (MM: n = 16), sexually mature females (MF: n = 15), immature males (IM: n = 16), and immature females (IF: n = 16).

3.3. Effects of Beetle Sexual Maturity on Fungistasis. After cultivation for 7 days, 192.56 ± 70.41 colonies of *Trichoderma* sp. were present on the CMA plates that had been inoculated with the control inoculums (n = 16). The inoculums that contained the anal secretions of sexually mature beetles produced no colonies on the CMA plates. Compared to the control inoculums, the anal secretions from both sexually mature males and females could significantly inhibit fungal growth (independent sample Student's t test: P < 0.001; sexually mature male: n = 16, sexually mature female: n = 15). Inoculums containing the anal secretions of immature beetles produced fungal growth on the CMA plates, regardless of sex (immature male: 214.06 ± 48.86 colonies, n = 16; immature female: 264.38 ± 55.95 colonies, n = 16). The numbers of *Trichoderma* sp. colonies produced from the control inoculums were not significantly different than that produced from the immature male inoculums (independent sample Student's t test: P = 0.32). The inoculums that contained the anal secretions of sexually immature females produced a significantly greater number of colonies compared with the control inoculums (independent sample Student's t test: P = 0.003) and the inoculums that contained the anal secretions of sexually immature males (independent sample Student's t test: P = 0.011) (Figure 3).

The fungal growth reached confluency on the CMA plates in 3.0 ± 0.0 days using the control inoculums, 6.0 ± 0.0 days using the sexually mature male inoculums, and 9.0 ± 0.0 days using the sexually mature female inoculums. Thus, although the anal secretions from sexually mature beetles inhibited the growth of *Trichoderma* sp., the number of days required for fungal growth to reach confluency was significantly longer for inoculums containing mature female anal secretions than those produced from mature males (independent sample Student's t test: P < 0.001, sexually mature male: n = 16,

sexually mature female: n = 15), which is considered a better fungistasis from mature females.

4. Discussion

Vertebrate carcasses are a high-quality source of nutrition for many species, with insects, scavengers, and microbes competing for the food resources. Insects typically begin to consume carcasses before the arrival of the larger scavenging species, and microbes release toxins that may drive away competitors [40]. Burying beetles use a small vertebrate carcass as a source of nutrition of their larval broods [1, 14], putting them in direct competition with intraspecific or interspecific insects, bacteria, fungi, microorganisms, and so on [27].

The decomposition rate of a buried carcass is slower than that of an exposed carcass because subsequent access to the carcass may be hindered for many insects and other scavengers [40]. Thus, the burial behavior of burying beetles is an adaptation that reduces competition for food resources. Observed in our study, the adults of *N. nepalensis* often feed on the intestines of carcasses before removing the fur, which may reduce the rate of decay of the carcass by eliminating the bacteria that are normally present in intestines. However, soil is also rich in microorganisms, such as fungi, that may subsequently diminish the quality of a carcass after burial.

The efficiency of carcass preservation may thus directly affect the successful production of burying beetle offspring. The deposition of oral and anal secretions on a carcass is one burying beetle behavior that reduces the rate of decay [23, 27, 33, 35]. Before burying the whole carcass, *N. nepalensis* removes hair or feathers prior to coating the carcass with secretions first. Unlike other species in North America, burying beetles bury the whole carcass first [41]. In our study, fungistasis was most efficient when a breeding pair of beetles remained present with the carcass. Thus, it is likely that the anal secretions of beetles are continuously deposited on the carcass. Our findings support the claim that the activity of the antimicrobial chemicals in the secretions is maintained over time [23, 24] because the fungistasis was also observed on the treated carcass balls without attending adult beetles. But when a prepared carcass was given a rinse in water, the protection of fungistatic effect was absent in the washed carcass balls. However, the diversity of fungal species isolated from washed carcass balls was nonetheless influenced by the preexisting anal secretions because the dominant fungal species on both treated and protected carcass balls was *Mucor* sp., whereas *Trichoderma* sp. was the dominant species on control carcasses.

The preparation of large carcasses for burial requires more time and energy and often leads to reduced quality of maintenance by beetles [27]. A previous study showed that *N. nepalensis* was unable to efficiently use the resources of a 130 g carcass, resulting in lower offspring weight to carcass weight ratios, compared with that of smaller carcasses, because the larger carcasses decayed rapidly [26]. Compared with the 3.5 g protected carcass balls, the 18 to 22 g protected carcass balls exhibited no significant difference in fungistasis. The 2 sizes of carcasses that were used in our study are within the size range of carcasses typically used by *N. nepalensis*. Nonetheless, the fungistatic capacity of burying beetle behavior in the field may be limited by the size of a carcass because the maintenance of carcasses in the field may involve greater competition with microorganisms and other competitors than was replicated in our laboratory experiments. Therefore, the effects of carcass size on reproduction success should be further investigated in the field.

In our study, the 12 fungal species that were collected from the carcasses are common in the natural environment of the burying beetle, especially in the soil and the decaying organic matter of leaf litter [42]. However, whether the source of the fungi in our experiments was the mouse carcasses, the beetles, or the moist peat used in the cages was not determined. *Conidiobolus* sp. and *Mucor* sp. belong to the Zygomycotina, and the other 10 species that were identified in our study are members of Ascomycotina. Fungi of the Zygomycotina are commonly found in leaf litter and soil, and some species may parasitize insects [43, 44]. Members of Ascomycotina may cause disease in certain plants, and other members, such as *Cordyceps sinensis*, may parasitize insects [45]. The dominant species on the control carcasses in our study was *Trichoderma* sp., which is widespread in soil [46, 47].

The oral and anal secretions of burying beetles contain antimicrobial chemicals [48]. The antimicrobial and lytic activities of the anal secretions in *N. vespilloides* are upregulated following the discovery of a carcass [25]. Cotter and Kilner [25] suggested that the antimicrobial activity may be influenced by juvenile hormone. Our findings support the role of juvenile hormone in the fungistatic properties of anal secretions because the secretions from sexually immature adults did not inhibit fungal growth. Thus, it is doubtful that the fungistatic properties of secretions from sexually immature beetles are upregulated following contact with a carcass, despite the burying of carcasses by sexually immature beetles [26]. We suggest that the burying behavior of sexually

immature beetles may serve to protect the carcass for subsequent feeding or reproduction.

Burying beetles rear their offspring by biparentally caring for the brood [1]. The participation of the male in biparental care can significantly improve resistance to alien invaders, compared with uniparental care by a female [32, 37, 49]. However, no significant increase in larval weight or brood weight occurs with biparental care, compared with uniparental care by a female [33, 35, 36]. In biparental care, males spend more time protecting the larvae and carcass from invaders than females, whereas females spend more time feeding larvae than males [50–53]. The division of labor in reproduction among male and female burying beetles may extend to the deposition of oral and anal secretions on the carcass. Our results indicate that both sexually mature male and female burying beetles produce anal secretions that inhibit the growth of fungi. However, the degree of fungistasis conferred by the secretions was significantly different between the sexes.

Conflict of Interests

The authors have no conflict of interests with the SPSS used inside the paper.

Acknowledgments

The authors wish to thank Professor P.-J. Lee for her help with identifying the fungi. This research was supported by Grants from the National University of Tainan (99AB2-14 and AB100-319) and the National Science Council, Taiwan (NSC 97WFA0F00038).

References

[1] E. Pukowski, "Ökologische untersuchungen an *Necrophorus* F.," *Zeitschrift für Morphologie und Ökologie der Tiere*, vol. 27, no. 3, pp. 518–586, 1933.

[2] S. B. Peck, "Nicrophorus (Silphidae) can use large carcasses for reproduction (Coleoptera)," *The Coleopterists Bulletin*, pp. 40–44, 1986.

[3] E. O. Wilson, *The Insect Societies*, Belknap Press, Cambridge, UK, 1971.

[4] I. Hanski and Y. Cambefort, *Dung Beetle Ecology*, Princeton University Press, Princeton, NJ, USA, 1991.

[5] M. P. Scott and D. S. Gladstein, "Calculating males? An empirical and theoretical examination of the duration of paternal care in burying beetles," *Evolutionary Ecology*, vol. 7, no. 4, pp. 362–378, 1993.

[6] D. S. Wilson and J. Fudge, "Burying beetles: intraspecific interactions and reproductive success in the field," *Ecological Entomology*, vol. 9, no. 2, pp. 195–203, 1984.

[7] D. S. Wilson, W. G. Knollenberg, and J. Fudge, "Species packing and temperature dependent competition among burying beetles (Silphidae, Nicrophorus)," *Ecological Entomology*, vol. 9, no. 2, pp. 205–216, 1984.

[8] S. T. Trumbo, "Interference competition among burying beetles (Silphidae, Nicrophorus)," *Ecological Entomology*, vol. 15, no. 3, pp. 347–355, 1990.

[9] J. Bartlett and C. M. Ashworth, "Brood size and fitness in *Nicrophorus vespilloides* (Coleoptera: Silphidae)," *Behavioral Ecology and Sociobiology*, vol. 22, no. 6, pp. 429–434, 1988.

[10] M. Otronen, "The effect of body size on the outcome of fights in burying beetles (Nicrophorus)," *Annales Zoologici Fennici*, vol. 25, no. 2, pp. 191–201, 1988.

[11] J. K. Müller, A. K. Eggert, and E. Furlkröger, "Clutch size regulation in the burying beetle *Necrophorus vespilloides* Herbst (Coleoptera: Silphidae)," *Journal of Insect Behavior*, vol. 3, no. 2, pp. 265–270, 1990.

[12] D. H. Janzen, "Why fruits rot, seeds mold, and meat spoils," *The American Naturalist*, vol. 111, no. 980, pp. 691–713, 1977.

[13] J. K. Müller, A. K. Eggert, and J. Dressel, "Intraspecific brood parasitism in the burying beetle, *Necrophorus vespilloides* (Coleoptera: Silphidae)," *Animal Behaviour*, vol. 40, no. 3, pp. 491–499, 1990.

[14] J. K. Müller, "Replacement of a lost clutch: a strategy for optimal resource utilization in *Necrophorus vespilloides* (Coleoptera: Silphidae)," *Ethology*, vol. 76, pp. 74–80, 1987.

[15] J. Bartlett, "Filial cannibalism in burying beetles," *Behavioral Ecology and Sociobiology*, vol. 21, no. 3, pp. 179–183, 1987.

[16] J. K. Müller and A. K. Eggert, "Time-dependent shifts between infanticidal and parental behavior in female burying beetles a mechanism of indirect mother-offspring recognition," *Behavioral Ecology and Sociobiology*, vol. 27, no. 1, pp. 11–16, 1990.

[17] C. W. Beninger and S. B. Peck, "Temporal and spatial patterns of resource use among *Nicrophorus* carrion beetles (Coleoptera: Silphidae) in a Sphagnum bog and adjacent forest near Ottawa, Canada," *Canadian Entomologist*, vol. 124, no. 1, pp. 79–86, 1992.

[18] T. N. Sherratt, D. M. Wilkinson, and R. S. Bain, "Why fruits rot, seeds mold and meat spoils: a reappraisal," *Ecological Modelling*, vol. 192, no. 3-4, pp. 618–626, 2006.

[19] D. E. Rozen, D. J. P. Engelmoer, and P. T. Smiseth, "Antimicrobial strategies in burying beetles breeding on carrion," *Proceedings of the National Academy of Sciences of the United States of America*, vol. 105, no. 46, pp. 17890–17895, 2008.

[20] C. L. Hall, N. K. Wadsworth, D. R. Howard et al., "Inhibition of microorganisms on a carrion breeding resource: the antimicrobial peptide activity of burying beetle (Coleoptera: Silphidae) oral and anal secretions," *Environmental Entomology*, vol. 40, no. 3, pp. 669–678, 2011.

[21] W. W. Hoback, A. A. Bishop, J. Kroemer, J. Scalzitti, and J. J. Shaffer, "Differences among antimicrobial properties of carrion beetle secretions reflect phylogeny and ecology," *Journal of Chemical Ecology*, vol. 30, no. 4, pp. 719–729, 2004.

[22] R. L. Rana, W. Wyatt Hoback, N. A. A. Rahim, J. Bedick, and D. W. Stanley, "Pre-oral digestion: a phospholipase A_2 associated with oral secretions in adult burying beetles, *Nicrophorus marginatus*," *Comparative Biochemistry and Physiology B*, vol. 118, no. 2, pp. 375–380, 1997.

[23] S. Suzuki, "Suppression of fungal development on carcasses by the burying beetle *Nicrophorus quadripunctatus* (Coleoptera: Silphidae)," *Entomological Science*, vol. 4, no. 4, pp. 403–405, 2001.

[24] B. J. Jacques, S. Akahane, M. Abe, W. Middleton, W. W. Hoback, and J. J. Shaffer, "Temperature and food availability differentially affect the production of antimicrobial compounds in oral secretions produced by two species of burying beetle," *Journal of chemical ecology*, vol. 35, no. 8, pp. 871–877, 2009.

[25] S. C. Cotter and R. M. Kilner, "Sexual division of antibacterial resource defence in breeding burying beetles, *Nicrophorus*

vespilloides," *Journal of Animal Ecology*, vol. 79, no. 1, pp. 35–43, 2010.

[26] W. Hwang and S. F. Shiao, "Use of carcass by burying beetles *Nicrophorus nepalensis* Hope (Coleoptera: Silphidae)," *Formosan Entomologist*, vol. 28, no. 2, pp. 87–100, 2008.

[27] M. P. Scott, "The ecology and behavior of burying beetles," *Annual Review of Entomology*, vol. 43, pp. 595–618, 1998.

[28] A.-K. Eggert and J. K. Müller, "Joint breeding in female burying beetles," *Behavioral Ecology and Sociobiology*, vol. 31, no. 4, pp. 237–242, 1992.

[29] S. T. Trumbo, "Monogamy to communal breeding: exploitation of a broad resource base by burying beetles (Nicrophorus)," *Ecological Entomology*, vol. 17, no. 3, pp. 289–298, 1992.

[30] M. P. Scott, "Competition with flies promotes communal breeding in the burying beetle, *Nicrophorus tomentosus*," *Behavioral Ecology and Sociobiology*, vol. 34, no. 5, pp. 367–373, 1994.

[31] M. P. Scott, W. J. Lee, and E. D. Van Der Reijden, "The frequency and fitness consequences of communal breeding in a natural population of burying beetles: a test of reproductive skew," *Ecological Entomology*, vol. 32, no. 6, pp. 651–661, 2007.

[32] M. P. Scott, "The benefit of paternal assistance in intra- and interspecific competition for the burying beetle, *Nicrophorus defodiens*," *Ethology Ecology and Evolution*, vol. 6, no. 4, pp. 537–543, 1994.

[33] J. K. Müller, A. K. Eggert, and S. K. Sakaluk, "Carcass maintenance and biparental brood care in burying beetles: are males redundant?" *Ecological Entomology*, vol. 23, no. 2, pp. 195–200, 1998.

[34] G. S. Halffter, C. Huerta, and A. Huerta, "Nidification des *Nicrophorus* (Col. Silphidae)," *Bulletin de la Société Entomologique de France*, vol. 88, pp. 648–666, 1983.

[35] J. Bartlett, "Male mating success and paternal care in *Nicrophorus vespilloides* (Coleoptera: Silphidae)," *Behavioral Ecology and Sociobiology*, vol. 23, no. 5, pp. 297–303, 1988.

[36] S. T. Trumbo, "Reproductive benefits and the duration of paternal care in a biparental burying beetle, *Necrophorus orbicollis*," *Behaviour*, vol. 117, no. 1-2, pp. 82–105, 1991.

[37] I. C. Robertson, "Nest intrusions, infanticide, and parental care in the burying beetle, *Nicrophorus orbicollis* (Coleoptera: Silphidae)," *Journal of Zoology*, vol. 231, no. 4, pp. 583–593, 1993.

[38] S. T. Trumbo, "Interspecific competition, brood parasitism, and the evolution of biparental cooperation in burying beetles," *Oikos*, vol. 69, no. 2, pp. 241–249, 1994.

[39] I. A. Fetherston, M. P. Scott, and J. F. A. Traniello, "Behavioural compensation for mate loss in the burying beetle *Nicrophorus orbicollis*," *Animal Behaviour*, vol. 47, no. 4, pp. 777–785, 1994.

[40] D. O. Carter, D. Yellowlees, and M. Tibbett, "Cadaver decomposition in terrestrial ecosystems," *Naturwissenschaften*, vol. 94, no. 1, pp. 12–24, 2007.

[41] L. J. Milne and M. J. Milne, "The social behavior of the burying beetles," *Scientific American*, vol. 235, pp. 84–89, 1976.

[42] D. H. Larone, *Medically Important Fungi: A Guide to Identification*, ASM Press, Washington, DC, USA, 4th edition, 2002.

[43] C. M. S. Kumar, K. Jeyaram, and H. B. Singh, "First record of the entomopathogenic fungus Entomophaga aulicae on the Bihar hairy caterpillar Spilarctia obliqua in Manipur, India," *Phytoparasitica*, vol. 39, no. 1, pp. 67–71, 2011.

[44] C. Nielsen and A. E. Hajek, "Diurnal pattern of death and sporulation in Entomophaga maimaiga-infected Lymantria dispar," *Entomologia Experimentalis et Applicata*, vol. 118, no. 3, pp. 237–243, 2006.

[45] X. L. Wang and Y. J. Yao, "Host insect species of ophiocordyceps sinensis: a review," *ZooKeys*, vol. 127, pp. 43–59, 2011.

[46] A. Hagn, K. Pritsch, M. Schloter, and J. C. Munch, "Fungal diversity in agricultural soil under different farming management systems, with special reference to biocontrol strains of Trichoderma spp.," *Biology and Fertility of Soils*, vol. 38, no. 4, pp. 236–244, 2003.

[47] D. J. Roiger, S. N. Jeffers, and R. W. Caldwell, "Occurrence of Trichoderma species in apple orchard and woodland soils," *Soil Biology and Biochemistry*, vol. 23, no. 4, pp. 353–359, 1991.

[48] T. Degenkolb, R. A. Düring, and A. Vilcinskas, "Secondary metabolites released by the burying beetle *Nicrophorus vespilloides*: chemical analyses and possible ecological functions," *Journal of Chemical Ecology*, vol. 37, no. 7, pp. 724–735, 2011.

[49] S. T. Trumbo, "Defending young biparentally: female risk-taking with and without a male in the burying beetle, *Nicrophorus pustulatus*," *Behavioral Ecology and Sociobiology*, vol. 61, no. 11, pp. 1717–1723, 2007.

[50] I. A. Fetherston, M. P. Scott, and J. F. A. Traniello, "Parental care in burying beetles: the organization of male and female brood-care behavior," *Ethology*, vol. 95, no. 3, pp. 179–190, 1990.

[51] E. V. Jenkins, C. Morris, and S. Blackman, "Delayed benefits of paternal care in the burying beetle *Nicrophorus vespilloides*," *Animal Behaviour*, vol. 60, no. 4, pp. 443–451, 2000.

[52] P. T. Smiseth and A. J. Moore, "Behavioral dynamics between caring males and females in a beetle with facultative biparental care," *Behavioral Ecology*, vol. 15, no. 4, pp. 621–628, 2004.

[53] P. T. Smiseth, C. Dawson, E. Varley, and A. J. Moore, "How do caring parents respond to mate loss? Differential response by males and females," *Animal Behaviour*, vol. 69, no. 3, pp. 551–559, 2005.

19

Ecto- and Endoparasitic Fungi on Ants from the Holarctic Region

Xavier Espadaler[1] and Sergi Santamaria[2]

[1] Ecology Unit and CREAF, Autonomous University of Barcelona, 08193 Bellaterra, Spain
[2] Botany Unit, Autonomous University of Barcelona, 08193 Bellaterra, Spain

Correspondence should be addressed to Xavier Espadaler, xavier.espadaler@uab.es

Academic Editor: Alain Lenoir

The ant-specific fungi *Aegeritella*, *Laboulbenia*, *Rickia*, *Hormiscium*, and *Myrmicinosporidium* in the Holarctic region—nine species—are reviewed. Present knowledge is highly biased geographically, as shows the single record for Holarctic Asia, and this is to solve. The phylogenetic position of *Aegeritella*, *Hormiscium*, and *Myrmicinosporidium* is unknown. Hosts seem to be also skewed phylogenetically although this may be a true pattern.

1. Introduction

Extensive, massive mycoses are an extremely rare instance in ants [1] and involve individuals, rather than whole colonies. A fortiori, documented population level attacks are practically nonexistent. A case concerning *Tetramorium caespitum* [2, 3] seems to be an isolate within ant literature. Here we deal with ecto- and endoparasitic fungi, and we limit our survey to those that are ant specific. We differentiate parasitic fungi, that are not deadly to ants, and pathogenic fungi, which kill the host. Thus, generalist entomopathogenic fungi like *Beauveria* and *Metarhizium* or ant specifics like *Pandora myrmecophaga* (Figure 1) or *Telohannia solenopsae* are not included. Recent revisions of entomopathogens are those from Roy et al. [4], Kleespies et al. [5], Oi and Pereira [6] and, centred in social insects, in the seminal book by Schmid-Hempel [7]. We aim to review the knowledge of taxonomic and geographic distribution and, whenever possible, natural history and/or ecology of selected groups of fungi. The Holarctic is understood as comprising the nontropical parts of Europe and Asia, Africa north of the Sahara, and North America south to the Mexican desert region.

The fungi considered in this paper show a gradient of negative effects on the host. From a seemingly near absolute absence of any measurable—or measured—effect in some cases (*Aegeritella*, *Hormiscium*, and *Laboulbenia camponoti*), to a mild effect in other Laboulbeniales (reduced immunological response in *L. formicarum*; S. Cremer pers. comm.), or a possible strong negative effect in *Myrmicinosporidium*). This effect may concern exclusively infested individual ants (*Myrmicinosporidium*) although in some cases, because of the fungus life cycle and the social nature of ants, with many physical contacts between colony members outside of the nest and in the nest galleries, this may be multiplied and traduced directly to the colony level (Laboulbeniales, or *Aegeritella*). This general absence of strong negative effects indicates probably a very old interaction with ants.

An unfortunate circumstance is the completely unknown phylogenetic position of some of those specific ant fungi, and this is calling for a dedicated, focused study, using molecular techniques. We stress the necessity of enhanced attention from the part of myrmecologists and mycologists towards this interesting group of ectoparasitic fungi. Just remembering their existence, and with a little care and open mind, many more instances of Laboulbeniales, *Aegeritella*, *Myrmicinosporidium*, and pathogenic fungi on ants should surface in ample areas within the Holarctic region.

2. Material and Methods

Apart from our current files, we did a search in the ant data base FORMIS (version 2011) [9]. Search terms are as follow: ectoparasitic, endoparasitic, fungus, fungi, Laboulbeniales, *Laboulbenia*, *Rickia*, *Aegeritella*, *Myrmicinosporidium*, and filtered out a posteriori by geographical region

FIGURE 1: *Pandora myrmecophaga* having killed a worker *Formica rufa*, from The Netherlands, showing the characteristic attachment to the distal part of a grass leave caused by the summit disease [after [8]; Photo by H. Niesen; with permission).

(Holarctic). Within each fungus species, we give the country, ant species attacked, and reference. Taxonomical scheme and terminology follow Index Fungorum [10] (http://www.in-dexfungorum.org/).

3. Results

3.1. Ectoparasitic Fungi on Ants

3.1.1. Aegeritella Bałazy & J. Wiśn. Anamorphic Pezizomycotina. Those fungi were first noted by Wiśniewski in 1967 [11] although its fungal nature was not proven then. The fungi grow over the cuticle like dark protuberances (= bulbils). On a first sight, they look like dirt, and its form is usually a dome, rounded in perimeter, and up to 400 μm diameter (Figure 2). The number of bulbils may be from a single one to several hundreds. The distribution of bulbils on the body of ants is heterogeneous, being more abundant at the rear part [12–14]. The total number of bulbils is inversely related to ant size, with bigger ants having less bulbils than smaller ants [14]. Bulbils have been detected in workers and queens.

The ant-fungus relationship has not been properly ascertained although a reduced life duration or activity level has been suggested [15, 16]. In a similar vein, Bałazy et al. [17] note some workers with hundreds of bulbils, having immobilized bucal palps, all covered by hyphae. Nothing is known of the dynamics of infestation or transmission mechanisms of those enigmatic fungi, not even its phylogenetic position within the realm of Fungi.

(1) *Aegeritella superficialis* Bałazy & J. Wiś. 1974.

Europe

Czech Republic: *Formica sanguinea* Latreille, *Formica rufa* L., *Formica polyctena* Förster, *Formica pratensis* Retzius, *Formica truncorum* Fabricius, *Formica lugubris* Zetterstedt, *Formica exsecta* Nylander [18, 19].

Germany: *Formica polyctena* Förster [16].

Italy: *Formica lugubris* Zetterstedt [20].

Poland: *Formica polyctena* Förster, *Formica rufa* L., *Formica pratensis* Retzius; *Formica truncorum* Fabricius, *Formica fusca* L. [21–24]; *Formica sanguinea* Latreille [25].

Rumania: *Formica rufa* group [26].

Spain: *Formica decipiens* Bondroit [12].

Switzerland: *Formica rufa* L., *Formica polyctena* Förster, *Formica lugubris* Zetterstedt, *Formica sanguinea* Latreille [15].

(2) *Aegeritella tuberculata* Bałazy & J. Wiś. 1983.

Europe

Czech Republic: *Lasius distinguendus* Emery, *Lasius nitidigaster* Seifert (as *Lasius rabaudi*), *Lasius umbratus* (Nylander) [19].

Poland: *Lasius flavus* (Fabricius), *Formica fusca* L. [27].

Spain: *Lasius umbratus* (Nylander), *Lasius distinguendus* (Emery) [28], *Lasius umbratus* ([29], as *L. distinguendus*); *Formica pressilabris* Nylander [12]; *Formica rufa* L., *Formica rufibarbis* Fabr. [14]. Canary islands: Tenerife, *Lasius grandis* Forel [13].

North America

USA, Alaska: *Lasius pallitarsis* (Provancher) ([30], as *Lasius sitkaensis*).

(3) *Aegeritella roussillonensis* Bałazy, Lenoir & J. Wiś. 1986.

France. On *Cataglyphis cursor* (Fonscolombe) [17].

(4) *Aegeritella maroccana* Bałazy, Espad. & J. Wiś. 1990.

Morocco. On *Aphaenogaster baronii* Cagniant [31].

(5) An unidentified *Aegeritella* was noted on two workers

Polyergus breviceps Emery from Arizona [30].

3.1.2. Hormiscium Kunze, Incertae Sedis Pezizomycotina

(1) *Hormiscium myrmecophilum* Thaxter, 1914.

The species was described from an Amazonian *Pseudomyrmex* and remained elusive since its original description until it was found in Europe eighty years later. The filamentous, somewhat dichotomic thallus is undifferentiated and grows directly out of different parts of the ant body, without any apparent attaching structure. Mycelia have a maximum length of 163 μm and constant width of 10 μm. (Figure 3). Spores are unknown.

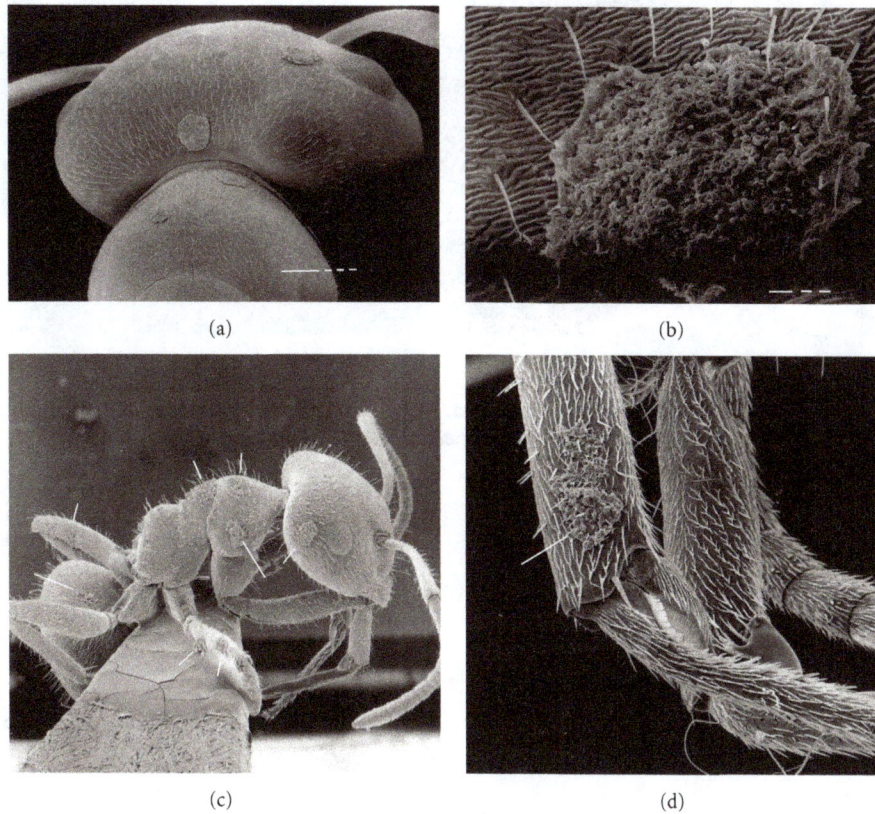

FIGURE 2: (a) *Aegeritella tuberculata* on *Formica pressilabris* (Spain). Two bulbils are in the pronotum, one at the back of head, (b) closeup of a bulbil; (c) *A. tuberculata* on *Lasius grandis* from Tenerife, Canary Islands; white arrows indicate bulbils; (d) closeup of bulbils in the first leg.

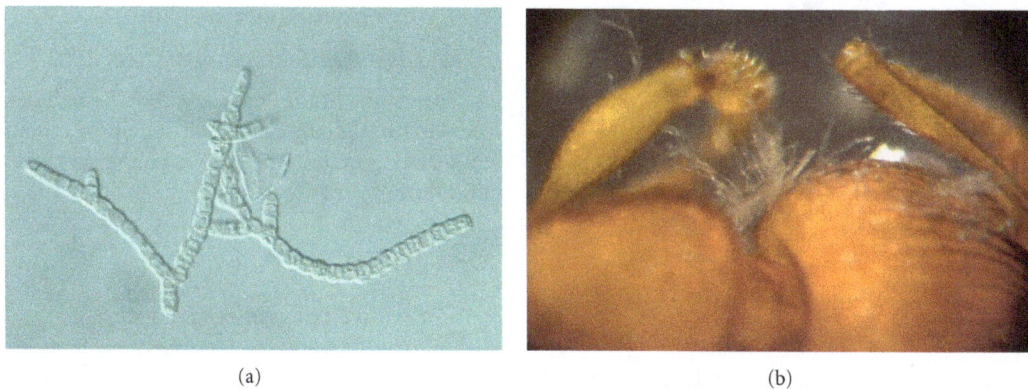

FIGURE 3: *Hormiscium myrmecophilum*. (a) hyphae on *Myrmica* sp.; (b) worker *Myrmica sabuleti* with hyphae on the head and lateral pronotum.

Europe

Portugal. On *Myrmica* sp. [32].

Spain. On *Myrmica sabuleti* Meinert (present paper).

3.1.3. Laboulbeniales (Ascomycota). Laboulbeniales are unusual among fungi because of their limited thallus with determinate growth. They are obligate external parasites of arthropods, especially insects. One key peculiarity is the ability to grow on their hosts without inflicting any noticeable injury. Ten orders of insects, in addition with millipedes and acari, may be affected although 80% of some 2000 species are recorded from beetles [33]. Only six are known to date infesting ants from the Holarctic region, and all castes are known to be susceptible to infestation.

(1) *Rickia wasmannii* Cavara, 1899.

The species is extremely characteristic in its microscopic morphological aspect (Figure 4) and is limited to several

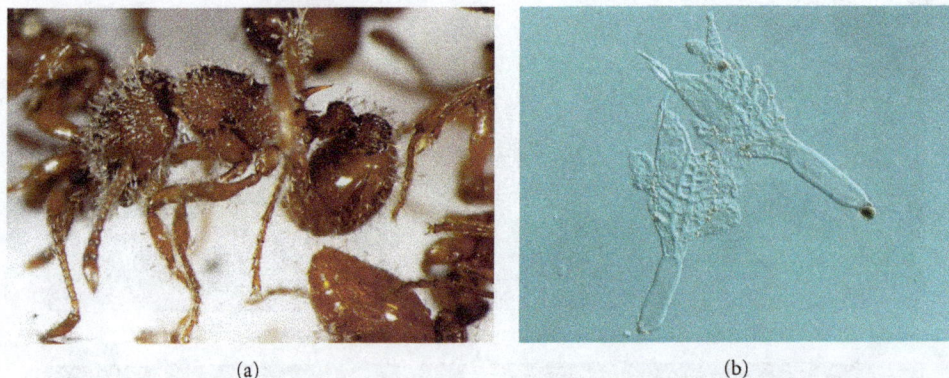

FIGURE 4: (a) *Rickia wasmannii* on *Myrmica scabrinodis* from Slovakia. Each "spatulate hair" is a thallus of *Rickia*. Photo by P. Bezděčka; with permission; (b) two mature thalli. Spores are oozing out of the perithecium on the specimen from the right.

species of *Myrmica*. Infested ants may harbour from a few thalli to several hundred thalli all over the body. Heavy infestations are visible to the naked eye and give a greyish shade, a pulverulent image to living individuals. Worker and queens may be infested.

Europe

Austria: *Myrmica rubra* (L.) [34].

Bulgaria: *Myrmica scabrinodis* Nylander [35].

Czech Republic: *Myrmica slovaca* Sadil, *Myrmica scabrinodis* Nylander [36].

France: *Myrmica scabrinodis* (Nylander) [37].

Germany: *Myrmica rubra* (L.) [38].

Hungary: *Myrmica slovaca* Sadil (as *M. salina*), *M. scabrinodis* Nylander, *M. specioides* Nylander, *M. vandeli* Bondroit [39].

Italy: *Myrmica scabrinodis* Nylander [40].

Luxembourg: *Myrmica rubra* L. [41].

Rumania: *Myrmica scabrinodis* Nylander [39].

Slovakia: *Myrmica scabrinodis* Nylander [42].

Slovenia: *Myrmica sabuleti* [41].

Spain: *Myrmica specioides* Bondroit [28, 43]; *Myrmica spinosior* Bondroit ([43], as *M. sabuleti*).

Switzerland: *Myrmica rubra* (L.) ([44], as *M. laevinodis*).

United Kingdom: *Myrmica sabuleti* Meinert [45, 46].

(2) *Rickia* sp.1.

Greece: On *Messor* (unpublished observation: description is pending).

(3) *Laboulbenia camponoti* S. W. T. Batra 1963.

Under the binocular, the thallus looks like a distorted ant hair (Figure 5) and is found all over the body, albeit more abundant in dorsal surfaces and external surface of legs.

Density is much lower than in other ant-specific Laboulbeniales. In the Holarctic, it has been detected exclusively in *Camponotus* species, all six from the subgenus *Tanaemyrmex*.

Asia

Turkey: *Camponotus baldaccii* Emery [47].

Europe

Bulgaria: *Camponotus aethiops* (Latreille), *Camponotus universitatis* Forel, *Camponotus* sp. (as *C. pilicornis*) [35].

Spain: *Camponotus pilicornis* (Roger) [48]; *Camponotus sylvaticus* (Olivier) [49].

(4) *Laboulbenia formicarum* Thaxt, 1902.

This is one of the smallest Laboulbeniales (up to 0.3 mm total length). Thalli can be extremely abundant on infested workers (Figure 6), which go foraging seemingly unaffected amid noninfested workers.

North America

Canada: *Lasius alienus* (Förster) [50].

USA: *Formica argentea* Wheeler [51]; *Formica aserva* Forel ([52], as *F. subnuda*); *Formica curiosa* Creighton ([53], as *F. parcipappa*); *Formica incerta* Buren [51]; *Formica lasioides* Emery [54]; *Formica montana* Wheeler ([54], as *F. neocinerea*); *Formica neogagates* Viereck [51, 55]; *Formica pallidefulva* Latreille ([54], as *F. nitidiventris*; [56], as *F. schaufussi*); *Formica puberula* Emery [52]; *Formica subintegra* Wheler [54]; *Formica subpolita* Mayr ([52], as *F. camponoticeps*); *Formica subsericea* Say [54]; *Formica vinculans* Wheeler [54]; *Lasius alienus* (Förster) ([55, 57], as *L. americanus*); *Lasius murphyi* Forel [58]; *Lasius neoniger* Emery [51, 59]; *Lasius pallitarsis* (Provancher) ([30], as *L. sitkaensis*); *Myrmecocystus mimicus* Wheeler [60]; *Polyergus breviceps* Emery [54]; *Polyergus lucidus* Mayr [54]; *Prenolepis impairs* (Say) [54].

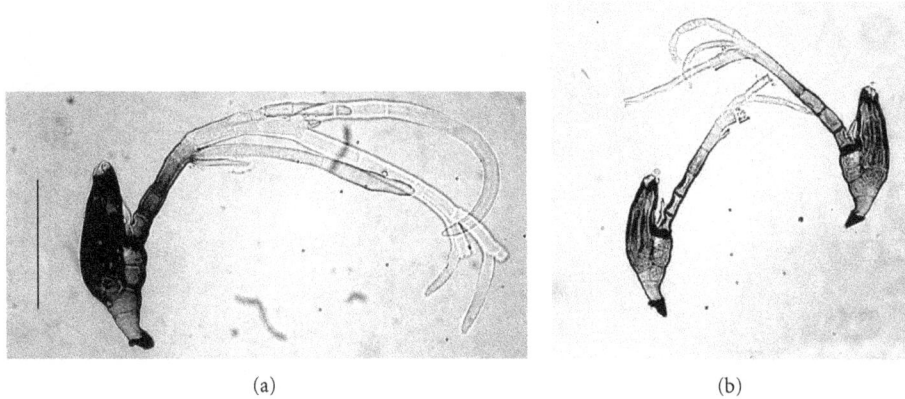

(a)   (b)

FIGURE 5: *Laboulbenia camponoti* from *Camponotus sylvaticus* (Spain); line: 1 mm. (a) A mature specimen; (b) two immature specimens.

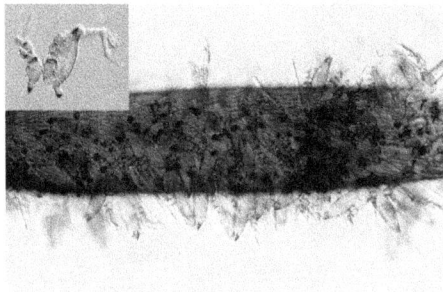

FIGURE 6: *Laboulbenia formicarum* on *Lasius grandis*. Worker tibia, showing full-grown thalli and dark spots which indicate attachment point of spores (more than 50 in the viewed side). Inset: one mature (right) and immature (left) specimens of *Laboulbenia formicarum*.

Europe

> France: *Lasius neglectus* Van Loon, Boomsma & Andrásfalvy [61].
>
> Portugal (Madeira): *Lasius grandis* Forel [62].
>
> Spain: *Lasius neglectus* Van Loon, Boomsma & Andrásfalvy [63].

3.2. Endoparasitic Fungi on Ants

3.2.1. Incertae Sedis

Myrmicinosporidium durum Hölldobler 1933. Those fungi were first noted by Hölldobler [64, 65] although they were formally described later, in 1933 [66]. Its phylogenetic position is still unknown, and their true fungal nature has been only proved recently [67]. Infested ants are usually well detected because the darker spores are visible through the integument (Figure 7); spores number may be very low, but usually they reach more than one hundred in a single ant. The caveat here is that the fungus may be much difficult to detect in ants having fuscous or black colouration. As a consequence, host range is probably biased. The usual aspect of concave spores, with a bow-like depression, is an artefact of fixation in alcohol [68].

Although the infested workers are almost certainly killed by the fungus when spores begin producing hyphae, life span seems not to be curtailed [67]. Infested workers seem scarcely affected in its normal behaviour [67, 69], and infested queens may participate in swarming flights [69] and show normal fertility [68]. Males have been found infected too [70]. Life cycle and mode on infestation are unknown although reports of *Myrmicinosporidium* from callow workers in *Pogonomyrmex badius* indicate that the infection is carried over from immature stages [71]. It is perhaps significant that the majority of diseased ants were collected in late summer and fall. After hibernation, those infected workers die [69]. Its geographical distribution is ample as is also the range of hosts.

Europe

Austria: *Plagiolepis vindobonensis* Lomnicki [67].

Croatia: *Temnothorax recedens* (Nylander), *Temnothorax affinis* (Mayr), *Temnothorax unifasciatus* (Latreille), *Plagiolepis pygmaea* (Latreille) [67].

France: *Solenopsis fugax* (Latreille), *Pheidole pallidula* (Nylander) [72]; *Temnothorax unifasciatus* (Latreille), *Temnothorax recedens* (Nylander) [68].

Germany: *Solenopsis fugax* (Latreille) [64, 65], *Temnothorax tuberum* (Fabricius) [66].

Hungary: *Solenopsis fugax* (Latreille), *Tetramorium caespitum* (L.), *Plagiolepis taurica* Santschi [73].

Italy: *Temnothorax unifasciatus* (Latreille) [67, 69], *Temnothorax albipennis* (Curtis) [67], *Temnothorax angustulus* (Nylander) [67], *Temnothorax exilis* (Emery) [67], *Temnothorax nylanderi* (Forster) [67], *Chalepoxenus muellerianus* (Finzi) [67].

Spain: *Pheidole pallidula* (Nylander), *Solenopsis* sp., *Strongylognathus caeciliae* Forel, *Tetramorium semilaeve* (André), *Plagiolepis pygmaea* (Latreille) [70], *Temnothorax lichtensteini* (Bondroit), *Temnothorax racovitzai* (Bondroit) [72].

Switzerland: *Solenopsis fugax* (Latreille) [68].

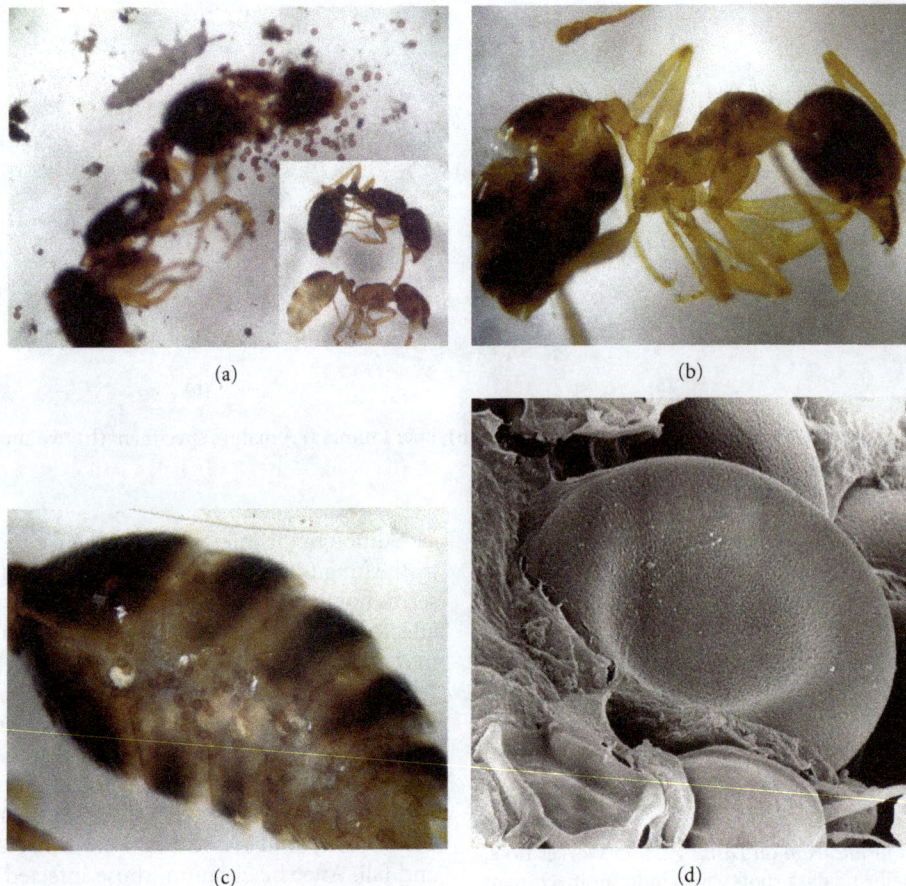

FIGURE 7: (a) *Myrmicinosporidium* mature spores inside workers *Tetramorium semilaeve* (inset: darker, infested worker, and normally coloured worker); (b) *Pheidole pallidula* with many spores on thorax, coxae, and gaster; (c) gaster of a male *Pheidole pallidula* with spores; (d) SEM image of a spore, showing the artifactual characteristic doughnut shaped form resulting from the alcohol fixation.

North America

USA: *Pogonomyrmex barbatus* (F. Smith) [67]; *Solenopsis carolinensis* Forel, *Solenopsis invicta* Buren, *Pheidole tysoni* Forel, *Pheidole bicarinata* Mayr, *Pyramica membranifera* (Emery), *Pogonomyrmex badius* (Latreille) [67]; *Nylanderia vividula* (Nylander) ([67], as *Paratrechina vividula*).

3.2.2. Dubious Cases. Across literature, two cases have been described but not identified. Although unproven, those are highly likely to belong in *Aegeritella* because of the macroscopic description given.

Bequaert ([56], page 74) wrote "*A number of so-called "imperfect fungi"—incompletely developed, conidia-bearing or sterile stages of various Ascomycetes—have been recorded from ants. A nest of* Formica rufa *Linné, at Potsdam, Germany, was heavily infested with fungous growths, about the size of a pinhead and attached mainly to the thorax, more rarely to other parts of the body. The ants were apparently but little hampered by their parasites. From cultures obtained with these fungi,* Bischoff *concluded that thy belonged to several species, among them a* Mucor, *a* Penicillium *and a yeast. Thaxter also found*

in the vicinity of Cambridge, Mass., a fungus forming blackish incrustations on various parts of ants and giving rise to a few short, colorless, erect branches; the exact nature of this plant has not been determined, nor is the name of its host mentioned."

Donisthorpe ([74], page 235 and Figure 86) commenting on *Lasius umbratus* var. *mixto-umbratus* Forel, [now *Lasius (Chthonolasius)* unrecognisable species] noted "*On August 11th, 1912, when at Weybridge in company with Professor Wheeler, we found two colonies of this variety, very many of the ants of both being infested with a curious dark brown warty growth in patches on parts of the body and legs—this Wheeler thought might be a fungus which was unknown to him. I kept a number of these ants in captivity, and added uninfected workers of* umbrata *from other localities; the growth however did not increase nor spread to the new ants, but rather seemed to decrease. I sent some of the infested ants alive and others in spirit, to Dr. Baylis Elliot, and she considered the patches were colonies of unicellular organisms growing on the outside of the ants; eventually she came to the conclusion that they were not fungoid growths, but probably colonies of an alga.*" Thus, albeit without a named host, *Aegeritella* is probably present too in the United Kingdom. A search with Donisthorpe's collection and/or in the vicinities of Weybridge could confirm this.

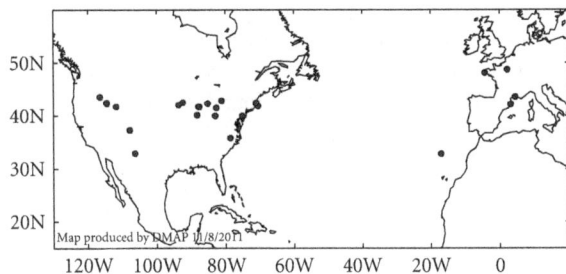

FIGURE 8: Distribution of *Laboulbenia formicarum*. North American records date from 1902 to 1979 and belong in 24 ant host species of five genera. European records date from 2003 to 2011 and imply two host species of *Lasius*.

4. Discussion

4.1. On Fungus Taxonomy. Laboulbeniales are taxonomically and nomenclaturally stable. There seems to be no major problem in morphological identification of the species involved. Perhaps, only, it would be worth examining the possibility of several species within *Laboulbenia formicarum* since its hosts belong in five genera, from three tribes—Formicini, Lasiini, and Plagiolepidini—in Formicinae.

Aegeritella is an especially difficult situation. Apart from its doubtful position within Fungi, bulbils are usually not in a perfect fruiting condition, and microscopic preparations are not easy to do since the bulbils are tightly attached to the ant's surface, anchored by the pubescence and hairs of the ant. The two most abundant species (*A. superficialis, A. tuberculata*) are well differentiated by the presence of hyphal elements in *A. superficialis* and by its absence in *A. tuberculata* [17].

Myrmicinosporidium is also an unsolved problem. All records but one are based simply on the presence of spores, which have a strikingly similar appearance across the two continents. Although they seem to be close to Chytridiomycetes [67], it remains to be studied where do those fungi belong within the phylogeny, and also the conspecificity of all so-called *M. durum* records. A similar situation is that of *Hormiscium*, from which only hyphae are known.

4.2. Host Phylogeny. A minimum of 13 subfamilies of ants are found in the Holarctic region. Only two (Myrmicinae and Formicinae) are noted with ecto- or endoparasitic fungi. Why should the distribution be so biased? If this is not a sampling artefact, it is noteworthy that the two subfamilies appear close together in the last comprehensive ant phylogenies [75, 76], thus indicating perhaps an ancestral susceptibility for both subfamilies.

Aegeritella is found on *Formica* and *Lasius*. *Laboulbenia* species infest exclusively ants from the subfamily Formicinae and *Rickia* infests Myrmicinae. This host specificity is not rare with Laboulbeniales [33]. Inasmuch *L. formicarum* is hosted by 24 ant species that belong in three tribes (Formicini, Lasiini, and Plagiolepidini), this calls for a dedicated evaluation (molecular and morphological) of the cospecificity of all populations of *L. formicarum*.

Myrmicinosporidium may be found in both ant subfamilies although the majority of cases belong in the Myrmicinae. We may speculate if the generic name is entirely appropriate or there is a detection bias of unknown origin towards Myrmicinae. Infested species belong in six tribes in Myrmicinae (Dacetini, Formicoxenini, Myrmicini, Pheidolini, Solenopsidini, and Tetramoriini), and one tribe in Formicinae (Plagiolepidini), widely scattered within ant phylogeny ([75], Figure 1; [76], Figure 1). Specificity is evidently not to uncritically assume in this fungus.

4.3. Geographical Distribution and Host Number. Knowledge is absolutely fragmentary and skewed. Asia in special, with a single record of ecto- and endoparasitic fungi, is a promising region to explore. The genus *Myrmica* with its many species should be searched for *Rickia*, and the genera *Formica* and *Lasius* for *Aegeritella*. Within Europe, countries such as Ireland, Belgium, The Netherlands, Denmark, Poland, or Portugal are obvious candidates for *Rickia*. The northernmost locale for *Rickia* seems to be Denbies Hillside, at $51°14'$N [45]. Some cases, such as *Laboulbenia formicarum* (Figure 8) or *Myrmicinosporidium durum* (Figure 9) agree with the usual worldwide or wide-ranging specific distribution of fungi although others are only known from its original description, from a single locality (*Aegeritella maroccana, Aegeritella roussillonensis*).

With host number, the situation seems to be dichotomous. Some fungi are known from a range of hosts: *A. superficialis* 9 hosts, *A. tuberculata* 10, *L. formicarum* 24, *L. camponoti* 7, *R. wasmannii* 8, and *Myrmicinosporidium* 27, while other fungi are known from single hosts, in parallel with geographical range, likely reflecting a sampling artefact. Horizontal transmission to slave-making ants is possible, as attested by *Aegeritella* [30] and *Laboulbenia formicarum* [54] on *Polyergus*, and by *Myrmicinosporidium* in *Chalepoxenus* [67] and *Strongylognathus* [70].

In the USA, three species (*Pheidole*, and 2 *Solenopsis*) from a single farm in Houston Co., Alabama [71] were noted as infested with *Myrmicinoporidium*. In southern Hungary, three genera (*Plagiolepis, Solenopsis*, and *Tetramorium*) [73] were noted as hosts in a single locality. A similar situation is that of an organic citrus field in Spain [70], in which up to four different genera (*Pheidole, Plagiolepis, Tetramorium*, and *Solenopsis*) have been detected as hosts during several years, their nests being at distances of 5–20 m. The disease may qualify as chronic in the three localities. In this last locality, *Aegeritella* on *Formica rufibarbis* and *Laboulbenia camponoti* on *Camponotus aethiops, C. pilicornis*, and *C. sylvaticus* exist too. The single circumstance we can suggest for this "abnormal" abundance of parasitic fungi in this last site is the intensity—monthly samples—and duration—since 2002 and ongoing—of ecological studies with abundant insect collection. This is suggestive of a general low-prevalence but ample geographic distribution. Thus, we cannot but expect a growth of information if proper attention is directed to those ecto- and endoparasitic fungi of ants. Myrmecologists, please, be aware!

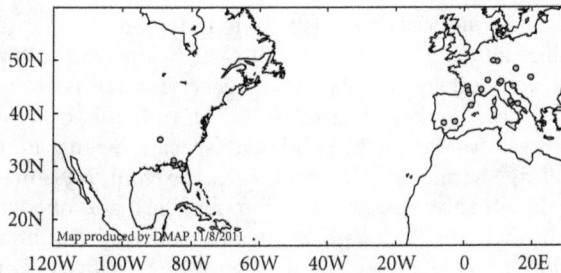

FIGURE 9: Distribution of *Myrmicinosporidium* sp. Eight ant host species are known from USA, and 19 from Europe.

Acknowledgments

The authors are grateful to L. Gallé and O. Kanizsai (Hungary) for help with references and unpublished information. They give their thanks to P. Bezděčka for allowing us to use the *Rickia wasmannii* image and to P. Boer and H. Niesen for the image *Formica rufa* infested with *Pandora myrmecophaga*. This work has been supported by Grants from MCYT-FEDER (CGL2004-05240-C02-01/BOS, CGL2007-64080-C02-01/BOS, and CGL2010-18182).

References

[1] B. Hölldobler and E. O. Wilson, *The Ants*, Springer, Berlin, Germany, 1990.

[2] P. I. Marikovsky, "On some features of behavior of the ants *Formica rufa* L. infected with fungous disease," *Insectes Sociaux*, vol. 9, no. 2, pp. 173–179, 1962.

[3] O. L. Rudakov, "Mikoz murav'ev (Predvant scobshchenie)," *Sb. Entomol. Rabot A kad. Nauk Kirghizsk. SSR, Kirghizsk. Otd. Vses. Entomol. Obshch*, vol. I, pp. 128–130, 1962.

[4] H. E. Roy, D. C. Steinkraus, J. Eilenberg, A. E. Hajek, and J. K. Pell, "Bizarre interactions and endgames: entomopathogenic fungi and their arthropod hosts," *Annual Review of Entomology*, vol. 51, pp. 331–357, 2006.

[5] R. G. Kleespies, A. M. Huger, and G. Zimmermann, "Diseases of insects and other arthropods: results of diagnostic research over 55 years," *Biocontrol Science and Technology*, vol. 18, no. 5, pp. 439–484, 2008.

[6] D. H. Oi and R. M. Pereira, "Ant behavior and microbial pathogens (Hymenoptera: Formicidae)," *Florida Entomologist*, vol. 76, pp. 63–74, 1993.

[7] P. Schmid-Hempel, *Parasites in Social Insects*, Princeton University Press, Princeton, NJ, USA, 1998.

[8] P. Boer, "Observations of summit disease in *Formica rufa* Linnaeus, 1761 (Hymenoptera: Formicidae)," *Myrmecological News*, vol. 11, pp. 63–66, 2008.

[9] D. P. Wojcik and S. D. Porter, "FORMIS: a master bibliography of ant literature," 2011, http://www.ars.usda.gov/saa/cmave/ifahi/formis.

[10] "Index Fungorum," 2011, http://www.indexfungorum.org/.

[11] J. Wiśniewski, "Narosla zaobserwowane na robotnicach *Formica polyctena* Forst. (Hym., Formicidae)," *Polskie Pismo Entomologiczne*, vol. 37, pp. 379–383, 1967.

[12] X. Espadaler and J. Wiśniewski, "*Aegeritella superficialis* Bał. et Wiś. and *A. tuberculata* Bał. et Wiś. (Deuteromycetes), epizoic fungi on two *Formica* (Hymenoptera, Formicidae) species

in the Iberian Peninsula," *Butlletí de l'Institució Catalana d' Història Natural*, vol. 54, pp. 31–35, 1987.

[13] X. Espadaler and P. Oromí, "*Aegeritella tuberculata* Bałazy et Wiśniewski (Deuteromycetes) found on *Lasius grandis* (Hymenoptera, Formicidae) in Tenerife, Canary Islands," *Vieraea*, vol. 26, pp. 93–98, 1998.

[14] X. Espadaler and S. Monteserín, "*Aegeritella* (Deuteromycetes) on *Formica* (Hymenoptera, Formicidae) in Spain," *Orsis*, vol. 18, pp. 13–17, 2003.

[15] D. Chérix, "Note sur la présence d'*Aegeritella superficialis* Bał. & Wiś. (Hyphomycetales, Blastosporae) sur des espèces du genre *Formica* (Hymenoptera, Formicidae) en Suisse," *Bulletin de la Societe Entomologique Suisse*, vol. 55, pp. 337–379, 1982.

[16] J. Wiśniewski and A. Buschinger, "*Aegeritella superficialis* Bał. et Wiś., ein epizootischer Pilz bei Waldameisen in der Bundesrepublik Deutschland," *Waldhygiene*, vol. 14, pp. 139–140, 1982.

[17] S. Bałazy, A. Lenoir, and J. Wiśniewski, "*Aegeritella roussillonensis* n. sp. (Hyphomycetales, Blastosporae) une espèce nouvelle de champignon epizoïque sur les fourmis *Cataglyphis cursor* (Fonscolombe) (Hymenoptera, Formicidae) en France," *Cryptogamie, Mycologie*, vol. 7, pp. 37–45, 1986.

[18] P. Bezděčka, "Epizootické houby rodu *Aegeritella* Bał. et Wiś. (Hyphomycetales, Blastosporae) na mravencích v Československu," *Česká Mykologie*, vol. 44, pp. 165–169, 1990.

[19] P. Bezděčka, "Parazitické houby na mravencích rodu *Formica*," *Formica, Zpravodaj Pro Aplikovaný Výzkum a Ochranu Lesních Mravenců (Liberec)*, vol. 2, pp. 71–75, 1999.

[20] J. Wiśniewski, "Occurrence of fungus *Aegeritella superficialis* Bał. & Wiś., 1974, on *Formica lugubris* Zett. in Italian Alps," *Bollettino della Società Entomologica Italiana*, vol. 109, pp. 83–84, 1977.

[21] S. Bałazy and J. Wiśniewski, "*Aegeritella superficialis* gen. et sp. nov., epifityczny grzyb na mrowkach z rodzaju *Formica* L.," *Prace Komisji Nauk Rolniczych i Komisji Nauk Lesnych, Poznanskie Towarzystwo Przyjaciol Nauk, Wydzial Nauk Rolniczych i Lesnych*, vol. 38, pp. 3–15, 1974.

[22] J. Wiśniewski, "Wystepowanie grzyba *Aegeritella superficialis* Bał. et Wiś. w Wielkopolskim Parku Narodowym," *Prace Poznan Tow Przyjaciol Nauk Wydz Nauk Roln Lesn*, vol. 42, pp. 41–45, 1976.

[23] S. Bałazy and J. Wiśniewski, "*Aegeritella superficialis* Bał. et Wiś. ein epizootischer Pilz auf Ameisen," *Die Waldameise*, vol. 2, pp. 49–50, 1989.

[24] J. Wiśniewski, "Aktueller Stand der Forschungen über Ameisen aus der *Formica* rufa-Gruppe (Hym., Formicidae) in Polen," *Bull. SROP / WPRS Bull. OILB*, vol. II-3, pp. 285–301, 1979.

[25] J. Wiśniewski and J. Sokolowski, "Nowe stanowiska grzybów *Aegeritiella superficialis* Bałazy et Wiśniewski i *Erynia myrmecophaga* (Turian et Wuest) Remaudière et Hennebert na mrówkach w Polsce," *Prace Komisji Nauk Rolniczych i Komisji Nauk Lesnych, Poznanskie Towarzystwo Przyjaciol Nauk, Wydzial Nauk Rolniczych i Lesnych*, vol. 56, pp. 137–144, 1983.

[26] V. D. Pascovici, "O noua entitate în microflora României: *Aegeritella superficialis* Bał. et Wiś., 1974 (Hiph., Blastosporae), parazita pe speciile din grupa *Formica* (Hym., Formicidae)," *Revista Padurilor Industria Lemnului, Celuloza si Hirtie; Celuloza si Hirtie*, pp. 148–149, 1983.

[27] S. Bałazy and J. Wiśniewski, "A new species of epizoic fungus on ants—*Aegeritella tuberculata* sp. nov.," *Bulletin de l'Academie Polonaise des Sciences Série des Sciences Biologiques*, vol. 30, pp. 85–88, 1982.

[28] X. Espadaler and D. Suñer, "Additions to iberian parasitic insect fungi," *Orsis*, vol. 4, pp. 145–149, 1989.

[29] F. García, X. Espadaler, P. Echave, and R. Vila, "Hormigas (Hymenoptera, Formicidae) de los acantilados de l'Avenc de Tavertet (Osona)," *Boletín de la Sociedad Entomológica Aragonesa*, vol. 47, pp. 363–367, 2010.

[30] X. Espadaler and X. Roig, "*Aegeritella* (Deuteromycetes) associated with ants in America North of Mexico," *Sociobiology*, vol. 23, pp. 39–43, 1993.

[31] S. Bałazy, X. Espadaler, and J. Wiśniewski, "A new myrmecophilic Hyphomycete, *Aegeritella maroccana* sp nov.," *Mycological Research*, vol. 94, pp. 273–275, 1990.

[32] S. Santamaria, "Sobre alguns fongs rars recol-lectats en insects vius," *Revista de la Societat Catalana de Micologia*, vol. 18, pp. 137–150, 1995.

[33] I. I. Tavares, "Laboulbeniales (Fungi, Ascomycetes)," *Mycological Memoirs*, vol. 9, pp. 1–627, 1985.

[34] J. Rick, "Zur Pilzkunde Vorarlbergs," *Österreichische Botanische Zeitschrift*, vol. 53, no. 4, pp. 159–164, 1903.

[35] A. Lapeva-Gjonova and S. Santamaria, "First record of Laboulbeniales (Ascomycota) on ants (Hymenoptera: Formicidae) in Bulgaria," *Zoonotes*, vol. 22, pp. 1–6, 2011.

[36] K. Bezděčková and P. Bezděčka, "First records of the myrmecophilous fungus *Rickia wasmannii* (Ascomycetes: Laboulbeniales) in the Czech Republic," *Acta Musei Moraviae, Sintiae Biologicae (Brno)*, vol. 96, pp. 193–197, 2011.

[37] S. Santamaria, "El orden Laboulbeniales (Fungi, Ascomycotina) en la Península Ibérica e Islas Baleares," *Edicions especials de la Societat Catalana de Micologia*, vol. 3, pp. 1–396, 1989.

[38] F. Cavara, "Di una nuova Laboulbeniacea, *Rickia wasmannii*, nov. gen. et nov. spec.," *Revue Mycologique*, vol. 22, pp. 155–156, 1899.

[39] A. Tartally, B. Szűcs, and J. R. Ebsen, "The first records of *Rickia wasmannii* Cavara, 1899, a myrmecophilous fungus, and its *Myrmica* Latreille, 1804 host ants in Hungary and Romania (Ascomycetes: Laboulbeniales; Formicidae)," *Myrmecological News*, vol. 10, p. 123, 2007.

[40] C. Spegazzini, "Primo contributo alla conoscenza delle Laboulbeniali italiani," *Redia*, vol. 10, pp. 21–75, 1914.

[41] L. Huldén, "Floristic notes on Palaearctic Laboulbeniales (Ascomycetes)," *Karstenia*, vol. 25, pp. 1–16, 1985.

[42] P. Bezděčka and K. Bezděčková, "First record of the myrmecophilous fungus *Rickia wasmannii* (Ascomycetes: Laboulbeniales) in Slovakia," *Folia Faunistica Slovaca*, vol. 16, pp. 77–78, 2011.

[43] X. Espadaler and D. Suñer, "Additional records of Iberian parasitic insect fungi: Laboulbeniales (Ascomycotina) and *Aegeritella* (Deuteromycotina)," *Orsis*, vol. 4, pp. 145–149, 1989.

[44] R. Baumgartner, "A propos de quelques Laboulbéniales (champignons sur insectes)," *Mitteilungen der Naturforschenden Gesellschaft in Bern*, vol. 1930, pp. 62–65, 1931.

[45] J. Pontin, *Ants of Surrey*, Surrey Wildlife Trust, United Kingdom, 2005.

[46] "Sifolinia's Ant Blog. *Rickia wasmannii* in the UK," 2011, http://sifolinia.blogspot.com/2009/10/rickia-wasmannii-in-uk.html.

[47] X. Espadaler and N. Lodos, "*Camponotus baldaccii* Emery (Hym., Formicidae) parasitized by *Laboulbenia camponoti* Batra (Ascomycetes) in Turkey," *Turkish Journal of Plant Protection*, vol. 7, pp. 217–219, 1983.

[48] J. Balazuc, X. Espadaler, and J. Girbal, "Laboulbenials (Ascomicets) ibèriques," *Collectanea Botanica*, vol. 13, pp. 403–421, 1982.

[49] X. Espadaler and J. Blasco, "*Laboulbenia camponoti* Batra, 1963 (Fungi, Ascomycotina) en Aragón," *Mallada*, vol. 2, pp. 75–80, 1991.

[50] W. W. Judd and R. K. Benjamin, "The ant *Lasius alienus* parasitized by the fungus *Laboulbenia formicarum* Thaxter at London, Ontario," *Canadian Entomologist*, vol. 90, p. 419, 1958.

[51] M. R. Smith, "Remarks concerning the distribution and hosts of the parasitic ant fungus, *Laboulbenia formicarum* Thaxter," *Bulletin of the Brooklyn Entomological Society*, vol. 23, pp. 104–106, 1928.

[52] A. C. Cole Jr., "*Laboulbenia formicarum* Thaxter, a fungus infesting some Idaho ants, and a list of its known North American hosts (Hym.: Formicidae)," *Entomological News*, vol. 46, p. 24, 1935.

[53] A. C. Cole Jr., "New ant hosts of the fungus *Laboulbenia formicarum* Thaxter," *Entomological News*, vol. 60, p. 17, 1949.

[54] M. R. Smith, "Ant hosts of the fungus, *Laboulbenia formicarum* Thaxter," *Proceedings of the Entomological Society of Washington*, vol. 48, pp. 29–31, 1946.

[55] R. Thaxter, "Preliminary diagnoses of new species of Laboulbeniaceae," *Proceedings of the American Academy of Arts and Sciences of Boston*, vol. 38, pp. 7–57, 1902.

[56] J. Bequaert, "A new host of *Laboulbenia formicarum* Thaxter, with remarks on the fungous parasites of ants," *Bulletin of the Brooklyn Entomological Society*, vol. 15, pp. 71–79, 1920.

[57] M. R. Smith, "An infestation of *Lasius niger* L. var. *americana* with *Laboulbenia formicarum* Thaxter," *Journal of Economic Entomology*, vol. 10, p. 447, 1917.

[58] T. P. Nuhn and C. G. Van Dyke, "*Laboulbenia formicarum* Thaxter (Ascomycotina: Laboulbeniales) on ants (Hymenoptera: Formicidae) in Raleigh, North Carolina with a new host record," *Proceedings of the Entomological Society of Washington*, vol. 81, pp. 101–104, 1979.

[59] W. M. Wheeler, "Colonies of ants (*Lasius neoniger* Emery) infested with *Laboulbenia formicarum* Thaxter," *Psyche (Cambridge)*, vol. 17, pp. 83–86, 1910.

[60] M. R. Smith, "Another ant genus host of the parasitic fungus *Laboulbenia* Robin (Hymenoptera: Formicidae)," *Proceedings of the Entomological Society of Washington*, vol. 63, p. 58, 1961.

[61] X. Espadaler, C. Lebas, J. Wagenknecht, and S. Tragust, "*Laboulbenia formicarum* (Ascomycota, Laboulbeniales) an exotic parasitic fungus, on an exotic ant in France," *Vie et Milieu*, vol. 61, pp. 41–44, 2011.

[62] X. Espadaler and S. Santamaria, "*Laboulbenia formicarum* crosses the Atlantic," *Orsis*, vol. 18, pp. 97–101, 2003.

[63] J. A. Herraiz and X. Espadaler, "*Laboulbenia formicarum* (Ascomycota, Laboulbeniales) reaches the Mediterranean," *Sociobiology*, vol. 50, no. 2, pp. 449–455, 2007.

[64] K. Hölldobler, "Über merkwürdige Parasiten von *Solenopsis fugax*," *Zoologischer Anzeiger*, vol. 70, pp. 333–334, 1927.

[65] K. Hölldobler, "über eine merkwürdige Parasitenerkrankung von ‚*Solenopsis fugax*‚" *Zeitschrift für Parasitenkunde*, vol. 2, no. 1, pp. 67–72, 1929.

[66] K. Hölldobler, "Weitere Mitteilungen Über Haplosporidien in Ameisen," *Zeitschrift für Parasitenkunde*, vol. 6, no. 1, pp. 91–100, 1933.

[67] S. R. Sanchez-Peña, A. Buschinger, and R. A. Humber, "*Myrmicinosporidium durum*, an enigmatic fungal parasite of ants,"

Journal of Invertebrate Pathology, vol. 61, no. 1, pp. 90–96, 1993.

[68] A. Buschinger and U. Winter, "*Myrmicinosporidium durum* Hölldobler 1933, Parasit bei Ameisen (Hym., Formicidae) in Frankreich, der Schweiz und Jugoslawien wieder aufgefunden," *Zoologischer Anzeiger*, vol. 210, pp. 393–398, 1983.

[69] A. Buschinger, J. Beibl, P. D'Ettorre, and W. Ehrhardt, "Recent records of *Myrmicinosporidium durum* Hölldobler, 1933, a fungal parasite of ants, with first record north of the Alps after 70 years," *Myrmecologische Nachrichten*, vol. 6, pp. 9–12, 2004.

[70] F. García and X. Espadaler, "Nuevos casos y hospedadores de *Myrmicinosporidium durum* Hölldobler, 1933 (Fungi)," *Iberomyrmex*, vol. 2, pp. 3–9, 2010.

[71] R. M. Pereira, "Occurrence of *Myrmicinosporidium durum* in red imported fire ant, Solenopsis invicta, and other new host ants in eastern United States," *Journal of Invertebrate Pathology*, vol. 86, no. 1-2, pp. 38–44, 2004.

[72] X. Espadaler, "*Myrmicinosporidium* sp., parasite interne des fourmis. Etude au MEB de la structure externe," in *La Communication chez les sociétés d'insectes*, A. De Haro and X. Espadaler, Eds., pp. 239–241, Colloque Internationale de l'Union Internationale pour l'Etude des Insectes Sociaux, Section française, Barcelona, Spain, 1982.

[73] O. Kanizsai, "*Myrmicinosporidium durum*, egy különös hangyapatogén," in *Proceedings of the 3rd Carpathian Basin Myrmecological Symposium*, p. 5, Senete, Romania, 2010.

[74] H. Donisthorpe, *British Ants, Their Life-History and Classification*, Brendon & Son, Plymouth, UK, 1915.

[75] C. S. Moreau, C. D. Bell, R. Vila, S. B. Archibald, and N. E. Pierce, "Phylogeny of the ants: Diversification in the age of angiosperms," *Science*, vol. 312, no. 5770, pp. 101–104, 2006.

[76] S. G. Brady, T. R. Schultz, B. L. Fisher, and P. S. Ward, "Evaluating alternative hypotheses for the early evolution and diversification of ants," *Proceedings of the National Academy of Sciences of the United States of America*, vol. 103, no. 48, pp. 18172–18177, 2006.

A Predator's Perspective of the Accuracy of Ant Mimicry in Spiders

Ximena J. Nelson

School of Biological Sciences, University of Canterbury, Private Bag 4800, Christchurch 8041, New Zealand

Correspondence should be addressed to Ximena J. Nelson, ximena.nelson@canterbury.ac.nz

Academic Editor: Jean Paul Lachaud

Among spiders, resemblance of ants (myrmecomorphy) usually involves the Batesian mimicry, in which the spider coopts the morphological and behavioural characteristics of ants to deceive ant-averse predators. Nevertheless, the degree of resemblance between mimics and ants varies considerably. I used *Portia fimbriata*, a jumping spider (Salticidae) with exceptional eyesight that specialises on preying on salticids, to test predator perception of the accuracy of ant mimicry. *Portia fimbriata*'s response to ants (*Oecophylla smaragdina*), accurate ant-like salticids (*Synageles occidentalis*), and inaccurate ant-like salticids (females of *Myrmarachne bakeri* and sexually dimorphic males of *M. bakeri*, which have enlarged chelicerae) was assessed. *Portia fimbriata* exhibited graded aversion in accordance with the accuracy of resemblance to ants (*O. smaragdina* > *S. occidentalis* > female *M. bakeri* > male *M. bakeri*). These results support the hypothesis that ant resemblance confers protection from visual predators, but to varying degrees depending on signal accuracy.

1. Introduction

Predator avoidance of dangerous prey is often exploited by deceptive prey species; the Batesian mimics are those that deceitfully advertise to potential predators that they also can induce the negative repercussions associated with this prey [1, 2], which often use warning (aposematic) signals to indicate their defences to would-be predators. The Batesian mimicry works solely to the advantage of the sender of the counterfeit signal, as both the receiver and the model are exploited. The receiver is cheated out of a source of food, and the model is less likely to benefit from its cues. The negative effect on models is due to frequency-dependent selection: if mimics exist in large numbers, the predators may take longer to learn an aversion or the potential for evolving innate fear of dangerous prey is lessened. Although studies of the Batesian mimicry have usually emphasised learning as a mechanism for the evolution of mimicry (e.g., [3]), both innate and learned fear of dangerous or distasteful prey can favour the evolution of the Batesian mimicry, as is clear from studies using naïve jumping spiders (Salticidae) as potential predators (e.g., [4]).

While we traditionally think of dangerous prey as one using bright, contrasting colours as aposematic signals, as in the case of poison dart frogs [3], not all dangerous species that are mimicked use aposematic signals. Correspondingly, deceitful use of aposematic signals appears to be an evolutionary strategy used by some Batesian mimics, but not others. Many spiders are the Batesian mimics of ants [5], animals which do not intuitively fit into the category of aposematic. Having a slender body, narrow waist, and an erratic style of locomotion, ants have a distinctive appearance, but this is unlikely to have evolved as an antipredator defence signal. Ants are, nevertheless, potentially harmful to predators through their ability to bite, sting, or spray formic acid. Being social, ants are all the more dangerous because they can mount communal attacks on potential predators [6]. Predators often respond to ant-like appearance as a cue for avoidance [4], and to disqualify ant mimicry as examples of the Batesian mimicry on the basis of hypotheses about the evolutionary origin of the ant's appearance places undue emphasis on a distinction that is irrelevant to the predator. In fact, ants appear to be particularly suitable as models for mimicry, especially among spiders. Illustrating how predation plays an important role in evolutionary diversification, ant mimicry (myrmecomorphy) has evolved in at least 43 spider genera within 13 families [5].

The 300 or so species of described myrmecomorphic spiders are typically characterised by a thin, elongated body, the creation of an antennal "illusion" by waving the forelegs, and an erratic style of locomotion [5, 7–9]. The vast majority of these species are Batesian mimics that are avoided by ant-averse arthropod predators [9–14], although the response of vertebrates is largely unknown. A few rough numbers may best express the efficacy of this deceptive signal. With over 5,300 described species, the Salticidae is the largest family of spiders [15]. The most speciose genus within the Salticidae, *Myrmarachne*, has over 200 described species—all of them ant mimics.

Theoretically the Batesian mimics are under selective pressure to closely resemble their models while the models are under pressure to distance themselves from the deceitful signalling of the mimics, so there should be an arms race in which mimics are expected to converge upon their models (e.g., [16]). Yet polymorphism can also be maintained in populations of the Batesian mimics [17], particularly when more than one model species is available [3]. It is especially noticeable that several species of ant mimics are polymorphic [18, 19]. As judged by humans, there is also considerable range in the accuracy of ant mimicry, with some being imprecise mimics, while others are remarkably similar in appearance to their model. Additionally, species in the large salticid genus *Myrmarachne* are sexually dimorphic as adults [20], with males seeming to be rather poor mimics due to their greatly enlarged chelicerae. Nevertheless, previous findings have suggested that males actually resemble ants carrying something in their mandibles [21]. In other words, they appear to be the Batesian mimics of a compound model (an ant plus the object it is carrying).

The exceptionally acute visual ability of salticids [22] enables them to identify motionless lures made from dead prey [23] and also enables them to escape some interactions with predators [11], such as ants. Although *Myrmarachne* can distinguish conspecifics and other mimics from ants [24–26], current evidence suggests that non-ant-like salticids are unable to make this distinction [4, 21]. The question of interest in this study is whether accuracy of ant mimicry, as judged by humans, is reflected in predator behaviour. The answer is of significance because most salticids will readily prey on each other [27], yet most salticids also appear to avoid ants [4], encounters with which are often lethal to salticids, including *Myrmarachne* [28, 29]. Clearly, it is also pertinent to determine how nonhuman animals classify objects and to determine the differences (or not) that may be found according to very different visual systems.

Here I tested *Portia fimbriata*, an Australian spider-eating (araneophagic) salticid that specialises on capturing other salticids as prey [30], with Asian weaver ants (*Oecophylla smaragdina*). I then compared whether their response toward ant-like salticids was similar to that elicited by *O. smaragdina* by testing *P. fimbriata* with males and females of *Myrmarachne bakeri* from the Philippines. This species is an imprecise ant mimic [19], and males are expected to be less precise than females due to their enlarged chelicerae. Finally, I tested *P. fimbriata* with an unrelated, but accurate, ant-like salticid from North America, *Synageles occidentalis*. In this study I address two specific questions: (1) does the non-ant-like salticid *P. fimbriata* avoid ants? (2) does *P. fimbriata* avoid or stalk ant-like salticids, and does this predators' behaviour differ depending on the accuracy of the mimic?

2. Materials and Methods

I collected *Myrmarachne bakeri* and *Oecophylla smaragdina* in the Philippines and conducted laboratory work at the University of Canterbury (Christchurch, New Zealand), where cultures of Australian *Portia fimbriata* and North American *Synageles occidentalis* were available. Sexually mature female *Portia fimbriata* (body length 8–10 mm) were tested with one of each of a variety of lures of four different types ($N = 15$ for each type), and the distance to which *P. fimbriata* approached lures was measured. Lures were made from dead ants (major workers of *O. smaragdina*, 8 mm in body length) and ant mimics (male and female *M. bakeri*, 8 and 6 mm in body length, respectively, and female *S. occidentalis*, 3.5 mm in body length). While *M. assimilis* is the accurate mimic of *O. smaragdina* [4], there were no longer any individuals of this species in the laboratory in New Zealand when this study was done. As we were unable to procure any more, tests were carried out using another excellent mimic, *S. occidentalis*, instead. No test spiders had any previous experience with ants or with ant mimics.

Spiders were maintained in individual plastic cages, cleaned weekly, with a cotton roll through the bottom that dangled in a small cup of water to provide humidity. Spiders were fed twice a week with house flies (*Musca domestica*). Testing was done between 0800 h and 1700 h (laboratory photoperiod 12L : 12D, lights on at 0800 h). A 200 W incandescent lamp, positioned *ca.* 600 mm overhead, lit the apparatus; fluorescent lamps provided additional, ambient lighting. Using standard protocol for experiments on predatory behaviour, spiders were fasted between 4 to 7 days prior to testing. No individual spider was tested more than once with a given type of lure.

The testing apparatus was a wooden ramp (see Figure 1 for dimensions) raised at a 20° angle, which was supported by a wooden pole, glued to a wooden base. The entire apparatus was painted with two coats of polyurethane and was wiped with 80% ethanol and allowed to dry for 30 min between each test to eliminate possible chemical traces from salticids in previous tests. The ramp was marked in a 5 mm grid to allow accurate distance measurements to be obtained. A thin piece of wood glued to the top end of the ramp served as a background against which the salticid saw the lure. The lure was placed 40 mm from the top end of the ramp, equidistant from both edges, and placed such that it was faced 45° away from the pit, enabling test spiders to view cues from both the body and the head or cephalothorax of the lure. Lures were made by immobilizing an arthropod with CO_2 and placing it in 80% ethanol. One day later, I mounted the arthropod in a life-like posture on the centre

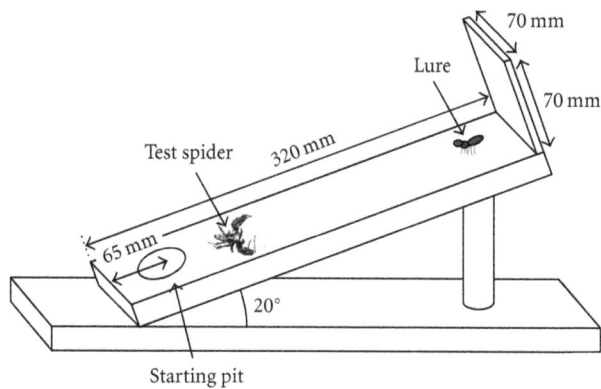

FIGURE 1: Ramp used for testing *Portia fimbriata* with lures of ants and ant mimics.

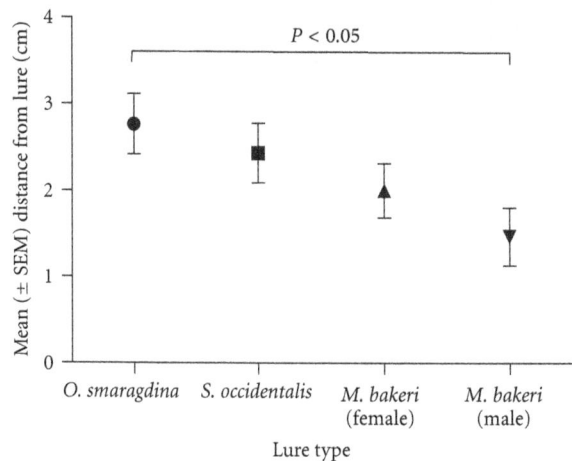

FIGURE 2: Mean (±SEM) approach distance by the spider-eating salticid *Portia fimbriata* to lures of ants (*Oecophylla smaragdina*) and ant mimics of varying degrees of accuracy of mimicry (*Synageles occidentalis* and male and female *Myrmarachne bakeri*).

of one side of a disc-shaped piece of cork (diameter *c.* 1.25 × the body length of the arthropod; thickness *ca.* 2 mm) using forceps to position the arthropod. Lures were then sprayed with a transparent aerosol plastic adhesive for preservation.

Before each test, *P. fimbriata* was placed in a 32 mm diameter "starting pit" drilled halfway through the thickness of the ramp 200 mm from the lure. The salticid was left in the pit to acclimate for 60 s before a piece of cardboard, which was placed over the pit, was removed, allowing the salticid to exit from the pit. A white paper screen running along three sides surrounded the apparatus, leaving one side open for observations. The ramp was positioned so that the salticid moved away from the observer during tests. Tests began when *P. fimbriata* walked out of the pit and on to the ramp and ended when *P. fimbriata* either attacked the lure or walked off the top end of the ramp. If the salticid jumped off the ramp at a point below the lure or if it stayed in the pit for more than 30 min (no spiders walked under the ramp), tests were aborted. After testing for normality (D'Agostino and Pearson omnibus test), data were analysed using ANOVA in Prism v.5.

3. Results

There was a significant overall effect of lure type on the distance to which *P. fimbriata* approached the lure (F_3 = 2.794, $P < 0.05$), although in general *P. fimbriata* showed an aversion to both ants and ant mimics. *P. fimbriata* avoided contact with lures by circling around the lure and then continuing up the ramp. Tukey's post hoc comparisons revealed no differences between responses to *O. smaragdina* and *S. occidentalis* or female *M. bakeri*, but male *M. bakeri* were approached significantly closer than *O. smaragdina* ($P < 0.05$). Overall *P. fimbriata* was kept furthest away from the ant (*O. smaragdina*), followed by *S. occidentalis*, then female *M. bakeri*, and lastly male *M. bakeri* (Figure 2). There were three instances of attacks towards lures, and all of these were aimed at lures of male *M. bakeri*.

4. Discussion

Portia fimbriata was unable to correctly classify the mimics as its preferred prey, salticids [30], and instead generally responded toward the mimics as it did toward ants. These results provide additional evidence that ant mimicry in spiders functions as Batesian mimicry, even with naïve predators. However, it appears that the degree of resemblance to ants may have repercussions when faced with predators with acute eyesight, such as salticids. *Synageles occidentalis* is thought to mimic *Lasius alienus* or *Myrmica americana*, with which it is associated [10]. The salticids we had in the laboratory bore an extremely accurate resemblance to the former ant species. Although *Myrmarachne bakeri* resemble ants, they do not have a specific model to which they render a faithful portrait [19]. *Portia fimbriata* apparently also classified the potential prey with which it was faced in a similar manner to the way in which humans classify these animals, which is by no means a given. Males of *M. bakeri* were significantly less effective at deterring *P. fimbriata* than ants and slightly less aversive than *M. bakeri* females and *S. occidentalis*. Nevertheless, it should be noted that in these experiments prey behaviour was not taken into account. It is known, for example, that some myrmecomorphs will actively display to ant-eating salticid predators, deterring potential attack through mistaken identity [31]. While there is currently no evidence supporting the idea that accurate ant-like spiders behave more like ants than poor mimics, it is conceivable that this might have exacerbated the results of the current study.

The only striking visible difference between the male and the other stimulus animals was the male's large chelicerae. The chelicerae of sexually mature *Myrmarachne* males, which can increase their body size by 30–50% [27], is believed to have evolved as a sexually selected trait [32]. To our eyes, *Myrmarachne* males resemble ants considerably less convincingly than *Myrmarachne* females and juveniles, suggesting that, along with impaired feeding mechanics [32], impaired predator deterrence through inaccurate mimicry has been a cost of sexual dimorphism for male *Myrmarachne*. Contrary to the other potential prey, lures of male *M. bakeri* were occasionally attacked. Nevertheless, *P. fimbriata* generally avoided lures of male *M. bakeri*, suggesting that mimicry among males, despite possessing some cost in terms of diminished efficacy of mimicry due to their enlarged chelicerae, is still effective at deterring visually based predators. This supports the idea that the shape of the chelicerae of male *Myrmarachne* is in keeping with its mimicry because it looks like an ant worker carrying something in its mandibles [21], as is commonly observed in worker ants [6].

In a study using hoverfly mimics of wasps as prey and pigeons as predators, Dittrich et al. [33] found that despite some species being poor mimics, they were still protected by their mimicry, perhaps due to some constraint in the birds' visual or learning systems. Here it is apparent that imprecise mimics, although not avoided to the same degree as accurate mimics, were nevertheless aversive to naïve predators, suggesting that learning is not essential for the same effects to be seen. A mutually compatible alternative explanation is simply that very numerous and very dangerous models may produce a wider "cone of protection," thus allowing for imprecise mimicry [34] because the payoff to a predator for attacking prey with a given resemblance to a numerous and highly noxious model is limited [35]. Furthermore, polymorphic mimics that do not resemble any particular ant species especially closely may gain other advantages. For example, imprecise ant mimics may not be restricted to the geographical area or microhabitat (e.g., arboreal ants) in which a specific model species is found. Ants are notorious for both their abundance and their formidable defences [6], and it may not be surprising to find that among ant mimics there is considerable variation in form, ranging from accurate to imprecise mimicry. What is unusual is that here we have an example of a mimic resembling one of its own predators [28, 29].

Acknowledgments

Work in the Philippines was generously assisted by the International Rice Research Institute. Robert Jackson provided helpful suggestions on early versions of the manuscript.

References

[1] M. E. Edmunds, *Defence in Animals: A Survey of Anti-Predator Defences*, Longman, London, UK, 1974.

[2] G. D. Ruxton, T. N. Sherratt, and M. P. Speed, *Avoiding Attack: The Evolutionary Ecology of Crypsis, Warning Signals and Mimicry*, Oxford University Press, Oxford, UK, 2004.

[3] C. R. Darst and M. E. Cummings, "Predator learning favours mimicry of a less-toxic model in poison frogs," *Nature*, vol. 440, no. 7081, pp. 208–211, 2006.

[4] X. J. Nelson and R. R. Jackson, "Vision-based innate aversion to ants and ant mimics," *Behavioral Ecology*, vol. 17, no. 4, pp. 676–681, 2006.

[5] P. E. Cushing, "Myrmecomorphy and myrmecophily in spiders: a review," *Florida Entomologist*, vol. 80, no. 2, pp. 165–193, 1997.

[6] B. Hölldobler and E. O. Wilson, *The Ants*, Springer, Heidelberg, Germany, 1990.

[7] J. Reiskind, "Ant-mimicry in Panamanian clubionid and salticid spiders (Araneae- Clubionidae, Salticidae)," *Biotropica*, vol. 9, pp. 1–8, 1977.

[8] P. S. Oliveira, "Ant-mimicry in some Brazilian salticid and clubionid spiders (Araneae: Salticidae, Clubionidae)," *Biological Journal of the Linnean Society*, vol. 33, no. 1, pp. 1–15, 1988.

[9] F. S. Ceccarelli, "Behavioral mimicry in *Myrmarachne* species (Araneae, Salticidae) from North Queensland, Australia," *Journal of Arachnology*, vol. 36, no. 2, pp. 344–351, 2008.

[10] B. Cutler, "Reduced predation on the antlike jumping spider *Synageles occidentalis* (Araneae: Salticidae)," *Journal of Insect Behavior*, vol. 4, no. 3, pp. 401–407, 1991.

[11] M. Edmunds, "Does mimicry of ants reduce predation by wasps on salticid spiders?" *Memoires of the Queensland Museum*, vol. 33, no. 2, pp. 507–512, 1993.

[12] M. Edmunds, "Do Malaysian *Myrmarachne* associate with particular species of ant?" *Biological Journal of the Linnean Society*, vol. 88, no. 4, pp. 645–653, 2006.

[13] X. J. Nelson and R. R. Jackson, "Collective Batesian mimicry of ant groups by aggregating spiders," *Animal Behaviour*, vol. 78, no. 1, pp. 123–129, 2009.

[14] X. J. Nelson, R. R. Jackson, D. Li, A. T. Barrion, and G. B. Edwards, "Innate aversion to ants (Hymenoptera: Formicidae) and ant mimics: experimental findings from mantises (Mantodea)," *Biological Journal of the Linnean Society*, vol. 88, no. 1, pp. 23–32, 2006.

[15] N. I. Platnick, "The world spider catalogue v.11.5," 2010, http://research.amnh.org/iz/spiders/catalog/INTRO1.html.

[16] J. Mappes and R. V. Alatalo, "Batesian mimicry and signal accuracy," *Evolution*, vol. 51, no. 6, pp. 2050–2053, 1997.

[17] M. P. Speed and G. D. Ruxton, "Imperfect Batesian mimicry and the conspicuousness costs of mimetic resemblance," *American Naturalist*, vol. 176, no. 1, pp. E1–E14, 2010.

[18] F. S. Ceccarelli and R. H. Crozier, "Dynamics of the evolution of Batesian mimicry: molecular phylogenetic analysis of ant-mimicking *Myrmarachne* (Araneae: Salticidae) species and their ant models," *Journal of Evolutionary Biology*, vol. 20, no. 1, pp. 286–295, 2007.

[19] X. J. Nelson, "Polymorphism in an ant mimicking jumping spider," *Journal of Arachnology*, vol. 38, no. 1, pp. 139–141, 2010.

[20] F. R. Wanless, "A revision of the spider genera *Belippo* and *Myrmarachne* (Araneae: Salticidae) in the Ethiopian region," *Bulletin of the British Museum of Natural History*, vol. 33, pp. 1–139, 1978.

[21] X. J. Nelson and R. R. Jackson, "Compound mimicry and trading predators by the males of sexually dimorphic Batesian mimics," *Proceedings of the Royal Society B*, vol. 273, no. 1584, pp. 367–372, 2006.

[22] M. Land, "The morphology and optics of spider eyes," in *Neurobiology of Arachnids*, F. G. Barth, Ed., pp. 53–78, Springer, New York, NY, USA, 1985.

[23] X. J. Nelson and R. R. Jackson, "Prey classification by an araneophagic ant-like jumping spider (Araneae: Salticidae)," *Journal of Zoology*, vol. 279, no. 2, pp. 173–179, 2009.

[24] X. J. Nelson and R. R. Jackson, "Vision-based ability of an ant-mimicking jumping spider to discriminate between models, conspecific individuals and prey," *Insectes Sociaux*, vol. 54, no. 1, pp. 1–4, 2007.

[25] F. S. Ceccarelli, "Ant-mimicking spider, *Myrmarachne* species (Araneae: Salticidae), distinguishes its model, the green ant, *Oecophylla smaragdina*, from a sympatric Batesian *O. smaragdina* mimic, Riptortus serripes (Hemiptera:Alydidae)," *Australian Journal of Zoology*, vol. 57, no. 5, pp. 305–309, 2009.

[26] X. J. Nelson, "Visual cues used by ant-like jumping spiders to distinguish conspecifics from their models," *Journal of Arachnology*, vol. 38, no. 1, pp. 27–34, 2010.

[27] R. R. Jackson and S. D. Pollard, "Predatory behavior of jumping spiders," *Annual Review of Entomology*, vol. 41, no. 1, pp. 287–308, 1996.

[28] X. J. Nelson, R. R. Jackson, G. B. Edwards, and A. T. Barrion, "Living with the enemy: jumping spiders that mimic weaver ants," *Journal of Arachnology*, vol. 33, no. 3, pp. 813–819, 2005.

[29] X. J. Nelson, R. R. Jackson, S. D. Pollard, G. B. Edwards, and A. T. Barrion, "Predation by ants on jumping spiders (Araneae: Salticidae) in the Philippines," *New Zealand Journal of Zoology*, vol. 31, no. 1, pp. 45–56, 2004.

[30] D. P. Harland and R. R. Jackson, "*Portia* Perceptions: the umwelt of an araneophagic jumping spider," in *Complex Worlds from Simpler Nervous Systems*, F. R. Prete, Ed., pp. 5–40, MIT Press, Cambridge, Mass, USA, 2004.

[31] X. J. Nelson, R. R. Jackson, and D. Li, "Conditional use of honest signaling by a Batesian mimic," *Behavioral Ecology*, vol. 17, no. 4, pp. 575–580, 2006.

[32] S. D. Pollard, "Consequences of sexual selection on feeding in male jumping spiders (Araneae: Salticidae)," *Journal of Zoology*, vol. 234, no. 2, pp. 203–208, 1994.

[33] W. Dittrich, F. Gilbert, P. Green, P. Mcgregor, and D. Grewcock, "Imperfect mimicry: a pigeon's perspective," *Proceedings of the Royal Society B*, vol. 251, no. 1332, pp. 195–200, 1993.

[34] D. W. Kikuchi and D. W. Pfennig, "High-model abundance may permit the gradual evolution of Batesian mimicry: an experimental test," *Proceedings of the Royal Society B*, vol. 277, no. 1684, pp. 1041–1048, 2010.

[35] T. N. Sherratt, "The evolution of imperfect mimicry," *Behavioral Ecology*, vol. 13, no. 6, pp. 821–826, 2002.

21

Cytogenetics of *Oryctes nasicornis* L. (Coleoptera: Scarabaeidae: Dynastinae) with Emphasis on Its Neochromosomes and Asynapsis Inducing Premature Bivalent and Chromosome Splits at Meiosis

B. Dutrillaux and A. M. Dutrillaux

UMR 7205, OSEB, CNRS/Muséum National d'Histoire Naturelle, 16, rue Buffon, CP 32, 75005 Paris, France

Correspondence should be addressed to B. Dutrillaux, bdutrill@mnhn.fr

Academic Editor: Howard Ginsberg

The chromosomes of specimens of *Oryctes nasicornis* from three locations in France and two locations in Greece were studied. All karyotypes have an X-Y-autosome translocation: 18, neoXY. Two male specimens from France (subspecies *nasicornis*) displayed an unusual behaviour of their meiotic chromosomes in 30–50% of spermatocytes, with asynapsis at pachynema, premature bivalent and chromosome split at metaphases I and II. The karyotypes remained balanced at metaphase I, but not at metaphase II. These particularities mimic the meiotic behaviour of B chromosomes and question about their existence, reported earlier in Spanish specimens. Due to the variable character of B chromosomes, complementary analyses are needed. To our knowledge, such meiotic particularities have not been described, beside cases of infertility. In specimens from Corsica (subspecies *laevigatus*) and Greece (subspecies *kuntzeni*), all spermatocytes I and II had a normal appearance. The meiotic particularity may thus be limited to male specimens from subspecies *nasicornis*.

1. Introduction

Beside pathological conditions such as malignancies or chromosome-instability syndromes, intraindividual variations of chromosomes are rare. Because of its usual stability, the karyotype of a limited number of cells is thought to represent that of a whole individual. This stability prevails for germ cells, so that parental and descendant karyotypes are similar. Consequently, the chromosome analysis of a limited number of cells from a limited number of individuals most frequently gives valuable information about the karyotype of their species. Exceptions exist, however, among which the presence of B chromosomes represents a major cause of numerical variation and polymorphism. B chromosomes have been described in plants and animals. They are characterized by a number of criteria among which is their particular meiotic behaviour: they do not pair like autosomes and tend to undergo premature centromere cleavage and non-disjunction at anaphase.

This leads to variations of their number from cell to cell and descendant to descendant [1].

Insect cytogenetics has essentially been developed through spontaneously dividing germ cells at diakinesis/metaphase I and metaphase II. At these stages, chromosome morphology is not optimal for analysis. Among several thousand of chromosome formulas reported in coleopterans, the presence of B chromosomes was noticed in about 40 instances [1–5]. *Oryctes nasicornis* L. 1758 (Coleoptera: Scarabaeidae: Dynastinae) is one of the very first insects in which dispensable supernumerary chromosomes were described [6] and later on considered as B chromosomes. This observation was quoted in reviews on both insect cytogenetics [5, 6] and B chromosomes [1].

Having analysed the mitotic chromosomes of a male specimen of *O. nasicornis* L. 1758, we were surprised to observe a karyotype different from its earlier descriptions. It had neither a Xyp (p for parachute, [6]) sex formula nor

Cytogenetics of Oryctes nasicornis L. (Coleoptera: Scarabaeidae: Dynastinae) with Emphasis on Its Neochromosomes and Asynapsis Inducing Premature Bivalent and Chromosome Splits at Meiosis

189

supernumerary B chromosomes, but neoXY as a conseq-uence of an X-Y-autosome translocation. B chromosomes being dispensable, we studied specimens from other localities and performed meiotic analyses to understand the causes of these discrepancies. We did not find B chromosomes, but, in two out of seven specimens, there were quite unexpected meiotic particularities. From pachytene to spermatocyte II stages, recurrent asynapsis, nonpairing, and premature cen-tromeric cleavages mimic the behaviour of B chromosomes. Checkpoints controlling meiotic chromosome behaviour have been identified, from yeast to mammals [7, 8]. They monitor elimination of spermatocytes with abnormal chro-mosome synapsis [9]. In some Oryctes nasicornis specimens, the anomalies at metaphase I and II, as consequences of pachytene asynapsis, suggest the low stringency of these checkpoints.

2. Material and Methods

Two male specimens (number 1 and 2) of O. nasicornis were obtained from the breeding developed at the Museum of Besançon (France). They were captured as larvae in the Besa-nçon area and are assumed to correspond to the nasicornis subspecies. They metamorphosed in June 2006. Two adult male specimens were captured in April 2007 (specimen num-ber 3) and September 2010 (specimen number 4), at Bois-le-Roi, at the Fontainebleau forest border (48°27′ N, 2°42′ E). They are assumed to belong also to the nasicornis subspecies. Another male (specimen number 5) was captured near Porto Vecchio, Corsica (41°36′ N, 9°11′ E), in June 2007. It is as-sumed to belong to the laevigatus Heer 1841 subspecies. Finally, two males were captured in Greece, one (specimen number 6) near Oros Kallidromo (38° 44′ N, 22°39′ E) in may 2010 and one (specimen number 7) near Kalambaka (39°47′ N, 21°55′ E) in June 2011. They are assumed to be-long to the kuntzeni Minck subspecies. Pachytene bivalent chromosome preparations were obtained following a long hypotonic shock and meiotic and mitotic metaphases after treatment with O.88 M KCL for 15 min. and another 15 min. in diluted calf serum (1 vol.) in distilled water (2 vol.) [10, 11]. Chromosomes at various mitotic and meiotic stages were studied after Giemsa and silver stainings and Q- and C-banding. Image capture was performed on a Zeiss Phomi 3 equipped with a high-resolution camera JAI M4+ and IKAROS (Metasystems) device or a Leica Aristoplan equip-ped with a JAI M300 camera and ISIS (Metasystems) device.

3. Results

Mitotic Karyotype (Figure 1). It is composed of 18 chromo-somes, including three sub-metacentric (number 1, 2 and 8) and five acrocentric (number 3–7) autosomal pairs. All of them carry large and variable heterochromatic segments around the centromeric region. The X chromosome is sub-metacentric and the Y acrocentric. Their size is much larger than that of gonosomes of most other Scarabaeid beetles. All heterochromatin is positively stained after C-banding and heterogeneously stained after Q-banding (not shown) which

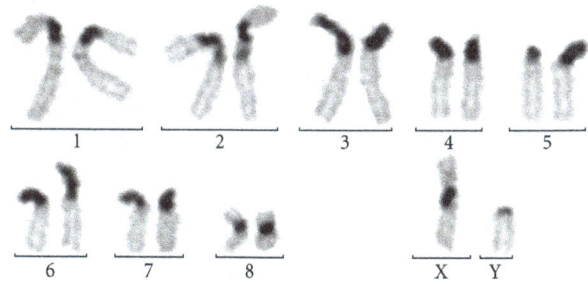

FIGURE 1: Mitotic karyotype of *Oryctes nasicornis* male (specimen number 1 from Besançon) after C-banding.

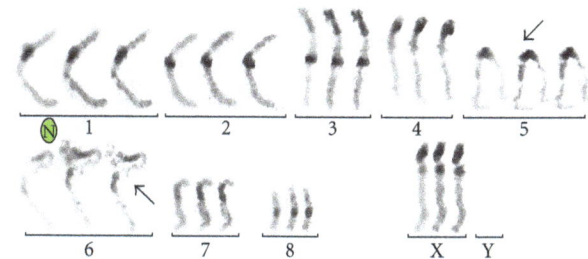

FIGURE 2: Karyotype from a spermatocyte at pachynema after the Giemsa staining (left), NOR staining displaying nucleoli (N) (cen-tre), and C-banding treatment (right). Acrocentric bivalents 5 and 6 are not synapsed, but associated by their heterochromatic short arms (arrows). Heterochromatin is more compact than on mitotic chromosomes. Specimen number 3 from Bois-le-Roi, as is the case in the next figures.

indicates its heterogeneous composition. Beside the varia-tions of the amounts of heterochromatin, all specimens had the same chromosome complement, as reported [4, 12].

Pachytene Chromosomes (Figures 2 and 3). As expected from the mitotic karyotype, nine bivalents were generally observ-ed. They could be identified by the amount and position of their heterochromatin, although heterochromatin was glob-ally more compact than in mitotic cells. The sex bivalent was quite characteristic. It had a large synapsed segment, similar-ly to autosomes, followed by juxtacentromeric heterochro-matin, and a compact segment. This was interpreted as the result of an X-Y-autosome translocation, the autosomal por-tion forming the long arm and the sex chromosomes forming the short arm. This translocation explains the low number of chromosomes (18 instead of 20 in most Scarabaeidae) and the large size of the sex chromosomes (the short arm relative length matches that of the X of other Dynastinae with a free X). Thus, the mitotic karyotype formula is 18, neoXY. Silver staining displayed a strong staining of all heterochromatin, as in most coleopterans. In addition, round nucleolar-like stru-ctures were recurrently associated with the short arm of a small acrocentric bivalent that we defined as number 6. Thus, according to previous studies [13], the Nucleolar Or-ganizer Region (NOR) is located on chromosome 6 short arm (Figure 2). The above description refers to observed spermatocytes. However, one or several bivalents displayed

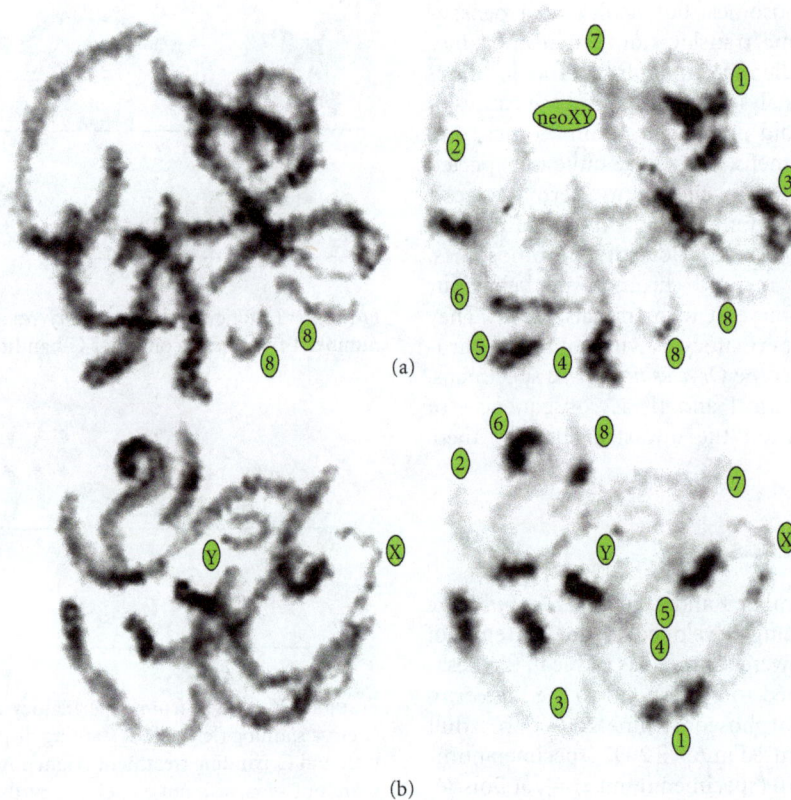

FIGURE 3: Spermatocytes at pachynema after the Giemsa staining (left) and C-banding (right) displaying asynapsis of chromosomes 8 (a) and sex chromosomes (b).

either asynapsis or incomplete synapsis in 29% and 41% of the spermatocytes from specimen number 2 and 3 from Besançon and Bois-le-Roi, respectively (Table 1). Smaller acrocentrics (numbers 6 and 7) were the most frequently involved, but all bivalents, including the sex bivalent (Figures 2, 3(a) and 3(b)), could be occasionally affected. In all instances, the non-synapsed autosomes were lying close to each other, suggesting either their premature desynapsis or deficient synapsis. The two homologues remained frequently at contact by their heterochromatic regions (Figure 2). Conversly the neoX and neoY chromosomes could be completely separated (Figure 3(b)). Specimen number 1 was immature and spermatocytes at pachynema of specimen number 5 could not accurately be studied and could not be considered as control. In specimen number 4 from Bois-le-Roi and specimens number 6 and 7 from Greece, the synapsis was strictly normal. We applied the same cytological techniques to specimens from more than other 100 species and observed such pachytene asynapsis only once and at a low frequency.

Diakinesis/Metaphase I (Figure 4). This stage was the most frequent in all the specimens studied: a total of 696 cells could be examined. Most of them displayed nine bivalents (biv), among which the sex bivalent could be identified by its asymmetrical constitution, as in other species with translocation-derived neoXY. No particularities were noticed in specimens number 4 to 7, whereas 43% and 34% of cells from the specimens number 2 and 3 (Table 1) displayed uni-

TABLE 1: Numbers and percentages of mitotic and meiotic cells analysed in specimen number 2 from Besançon and number 3 from Bois le Roi. Cells were scored as abnormal (abnl) when they displayed asynapsis (pachynema), univalents (diakinesis/metaphase I) or monochromatidic chromosomes (metaphase II), and normal (nl), when all chromosomes were in correct phase.

Cell stage	Besançon-image no. 2			Bois-le-Roi-image no. 3		
	nl	abnl	% abnl	nl	abnl	% abnl
Mitotic Metaphase	32	0	0	5	0	0
Pachynema	34	10	29	36	25	41
Diakinesis/ Metaphase I	41	31	43	195	99	34
Diakinesis/ Metaphase II	25	11	31	40	58	59

valents (univ), respectively. Their number was inversely proportional to that of bivalents: 9 biv + 0 univ; 8 biv + 2 univ; 7 biv + 4 univ; 6 biv + 6 univ, demonstrating that two univalents replaced one bivalent. The univalent occurrence, observed at both early diakinesis and late metaphase I, did not seem to depend on the progression towards anaphase. It preferentially involved smaller and sex chromosomes.

Metaphase II (Figures 5 and 6). No particularities were noticed among the 56, 48, 50 and 27 metaphases II analyzed

Cytogenetics of Oryctes nasicornis L. (Coleoptera: Scarabaeidae: Dynastinae) with Emphasis on Its Neochromosomes and Asynapsis Inducing Premature Bivalent and Chromosome Splits at Meiosis

191

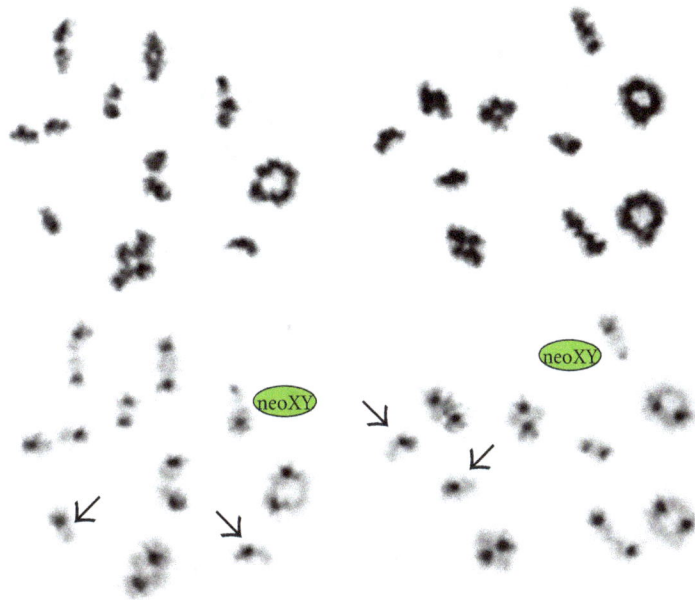

FIGURE 4: Spermatocytes at metaphase I after the Giemsa staining (top) and C-banding (bottom) with eight bivalents and two univalents (arrows).

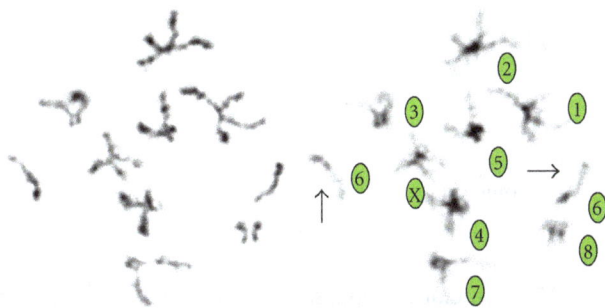

FIGURE 5: Spermatocyte at metaphase II displaying eight bi-chromatidic and two single-chromatid chromosomes, presumably number 6 (arrows).

from specimens 4, 5, 6, and 7, respectively. All were composed of 9 double-chromatid chromosomes. In specimens 2 and 3, 31% and 59% of metaphases II, respectively, comprised more than 9 chromosomes. C-banding allowed us to differentiate mono-chromatidic (monoc) and bi-chromatidic (bic) chromosomes. The number of monoc was roughly inversely proportional to that of bic: 9 bic + 0 monoc, 8 bic +2 monoc (Figure 5), 7 bic + 4 monoc, and 6 bic + 6 monoc. In a proportion of metaphases II, however, the ratio bic/monoc was different, indicating that aneuploidies occurred, as consequence of segregation errors at anaphase I (Figure 6). The premature centromeric split preferentially involved the small acrocentrics, the metacentric 8, and the sex chromosomes.

4. Discussion

The karyotypes of the specimens of *O. nasicornis* studied here obviously do not contain B chromosomes. *O. nasicornis* is

a widespread species in Western Europe, with eleven subspecies identified. The first mention of its karyotypic particularities was reported on Spanish specimens, which belong to the *grypus* Illiger 1803 subspecies [4]. The specimens from Besançon and Bois-le-Roi belong to the subspecies *nasicornis*. These two locations cover only a small part of the whole distribution area of the subspecies, but they are sufficiently distant (about 300 km) to assume that they do not constitute an isolate with abnormal gametogenesis. The specimens from Corsica and Greece, in which we failed to detect any meiotic particularity, belong to the subspecies *laevigatus* and *kuntzeni*, respectively, and there are no available data on the chromosomes of other specimens from this subspecies. Thus, the question of both the presence of B chromosomes and/or atypical meiosis, in relation with subspecies, remains open and needs further investigations.

The high recurrence of asynapsis and premature centromeric cleavage may be an artifact induced by hypotonic shock and spreading. However, the techniques used for pachynema and other meiotic stages were different, and we found fairly similar rates of aberrations at all stages. Furthermore, technical artifacts can hardly explain aneuploidies at metaphase II. We applied these techniques on meiotic chromosomes from many species of coleopterans without B chromosomes and observed such particularities only once at a low rate. Conversely, when B chromosomes were duly identified, they had a particular pairing leading to non-disjunctions at anaphase I, hence duplications and losses in spermatocytes II and variable numbers in descendants. It has no effect upon the phenotype, which indicates they carry no genes with major effect on the phenotype [1]. Here, all chromosomes can be involved in abnormal meiotic pairing. At metaphase II, 30–50% of spermatocytes displayed premature chromosome cleavage, which should induce a high rate of

FIGURE 6: Unbalanced sister metaphases II after the Giemsa staining (top) and C-banding (bottom). One acrocentric (presumably number 7) is single-chromatid on the left, while the complementary mono-chromatidic chromosome is in excess on the right (arrows).

unbalanced gametes. Indeed, aneuploid spermatocytes II were observed and a reduction of reproductive fitness should be expected, but we have no indication that it is the case. Furthermore, it is noteworthy that the rates of asynapsis at pachynema, premature bivalent cleavage at metaphase I, and premature centromere split at metaphase II are roughly similar at both intra- and interindividual levels. This suggests that metaphase I and II anomalies are direct consequences of pachytene asynapsis, and that there is both synapsis and checkpoint flaws at pachynema [8, 9]. It will be interesting to establish karyotypes of a series of eggs laid by parents with these meiotic particularities to know whether or not they induce a high rate of aneuploidies at early stages of development.

Another point of interest, in the karyotype of *O. nasicornis*, is the presence of neo-sex chromosomes. As described in the Scarabaeid beetles *Dynastes hercules* and *Jumnos ruckeri*, their meiotic behaviour, with an autosome-like synapsis of a long portion, indicates they originated from an X-Y-autosome translocation [13, 14]. As in these species too, the autosomal portion is separated from the original X component by the centromere, that is, constitutive heterochromatin. The insulating role of heterochromatin has been discussed for long in mammals, where it prevents inactivation spreading from the late replicating X to the attached autosome in female somatic

cells [15]. In meiotic prophase of the male, heterochromatin also isolates euchromatin from the inactivated sex chromosomes [16]. In *Drosophila*, the gene dosage compensation between males and females somatic cells is achieved by the overexpression of genes from the single X of the males [17]. This may also be the case of the beetle *Dynastes hercules*, but this was shown only for NOR expression [13]. In *Gryllotalpa fossor* (Orthoptera), the dosage compensation is of the mammalian type [18]. Finally, in *Musca domestica* (Diptera), no dosage compensation seems to exist [19]. These different situations demonstrate the existence of several regulatory mechanisms for X-linked gene expression in insect somatic cells. Whatever this mechanism, that is, over- or underexpression, there is an important character which is the existence of an epigenetic control spreading over large chromosome segments, if not whole chromosomes. We proposed that, in insects with overexpression of the X-linked genes in the male, as *Drosophila*, heterochromatin might play this insulating role [13]. This fits with the observation that in the few instances where an X-autosome translocation carrier *Drosophila* is fertile, the break point originating the translocation occurred within heterochromatin of the X ([20] and references herein). The presence of heterochromatin between gonosomal and autosomal components in the neo-sex chromosomes of *O. nasicornis* provides another example

Cytogenetics of Oryctes nasicornis L. (Coleoptera: Scarabaeidae: Dynastinae) with Emphasis on Its Neochromosomes
and Asynapsis Inducing Premature Bivalent and Chromosome Splits at Meiosis

193

suggesting the role of heterochromatin to avoid spreading of cis-acting epigenetic control elements.

In conclusion, this study shows that two chromosomal particularities exist in *O. nasicornis*. One is an X-Y-autosome translocation, frequently deleterious for reproduction, unless specific conditions prevent position effect, due to the different regulation of sex chromosomes and autosomes. Such translocations are not exceptional in Coleoptera, compared to other animals such as mammals. The other particularity is much more exceptional: two male specimens of *O. nasicornis nasicornis* display meiotic alterations usually considered as deleterious for fertility. These specimens were caught at two distant localities, which suggests these alterations are spread in the population and do not drastically prevent reproduction. Progress in the molecular biology of meiosis has shown the multiplicity of genes involved in synaptonemal complex formation and recombination [21, 22]. One of them may be altered in some specimens of *O. n. nasicornis* and maintained if associated with some hypothetical advantage. A third particularity, that is, the presence of B chromosomes, reported in specimens from Spain, may be an incorrect interpretation of the meiotic particularity described here and warrants further studies.

Acknowledgments

The authors are indebted to Jean-Yves Robert and Frédéric Maillot, Muséum de Besançon, France, and Laurent Dutrillaux, who provided us with the specimens from Besançon area and Corsica, respectively.

References

[1] R. N. Jones and H. Rees, *B Chromosomes*, Academic Press, London, UK, 1982.

[2] C. Juan and E. Petitpierre, "Chromosome numbers and sex determining systems in Tenebrionidae (Coleoptera)," in *Advances in Coleopterology*, M. Zunino, X. Belles, and M. Blas, Eds., pp. 167–176, AEC Press, Barcelona, Spain, 1991.

[3] E. Petitpierre, C. Segara, J. S. Yadav, and N. Virkki, "Chromosome numbers and meioformulae of chrysomelidae," in *Biology of Chrysomelidae*, P. Jolivet, E. Petitpierre, and T. H. Hsiao, Eds., pp. 161–186, Kluwer Academic Publishers, Dodrecht, The Netherlands, 1988.

[4] N. Virkki, "Akzessorische chromosomen bei zwei käfern, Epicometis hirta und Oryctes nasicornis L. (Scarabaeidae)," *Annales Academiæ Scientiarum Fennicæ*, vol. 26, pp. 1–19, 1954.

[5] J. S. Yadav, R. K. Pillai, and Kaaramjeet, "Chromosome numbers of Scarabaeidae (Polyphaga: Coleoptera)," *The Coleopterist Bulletin*, vol. 33, pp. 309–318, 1979.

[6] S. G. Smith and N. Virkki, *Insecta 5: Coleoptera*, vol. 3 of *Animal Cytogenetics*, Gebrüder Borntraeger, Berlin, Germany, 1978.

[7] N. Bhalla and A. F. Dernburg, "Cell biology: a conserved checkpoint monitors meiotic chromosome synapsis in Caenorhabditis elegans," *Science*, vol. 310, no. 5754, pp. 1683–1686, 2005.

[8] H. Y. Wu and S. M. Burgess, "Two distinct surveillance mechanisms monitor meiotic chromosome metabolism in budding yeast," *Current Biology*, vol. 16, no. 24, pp. 2473–2479, 2006.

[9] G. S. Roeder and J. M. Bailis, "The pachytene checkpoint," *Trends in Genetics*, vol. 16, no. 9, pp. 395–403, 2000.

[10] A. M. Dutrillaux, D. Pluot-Sigwalt, and B. Dutrillaux, "(Ovo-)-viviparity in the darkling beetle, Alegoria castelnaui (Tenebrioninae: Ulomini), from Guadeloupe," *European Journal of Entomology*, vol. 107, no. 4, pp. 481–485, 2010.

[11] A. M. Dutrillaux, S. Moulin, and B. Dutrillaux, "Use of meiotic pachytene stage of spermatocytes for karyotypic studies in insects," *Chromosome Research*, vol. 14, no. 5, pp. 549–557, 2006.

[12] A. M. Dutrillaux and B. Dutrillaux, "Sex chromosome rearrangements in polyphaga beetles," *Sexual Development*, vol. 3, no. 1, pp. 43–54, 2009.

[13] A. M. Dutrillaux, J. Mercier, and B. Dutrillaux, "X-Y-autosome translocation, chromosome compaction, NOR expression and heterochromatin insulation in the Scarabaeid beetle *Dynastes hercules hercules*," *Cytogenetic and Genome Research*, vol. 116, no. 4, pp. 305–310, 2007.

[14] N. Macaisne, A. M. Dutrillaux, and B. Dutrillaux, "Meiotic behaviour of a new complex X-Y-autosome translocation and amplified heterochromatin in Jumnos ruckeri (Saunders) (Coleoptera, Scarabaeidae, Cetoniinae)," *Chromosome Research*, vol. 14, no. 8, pp. 909–918, 2006.

[15] J. Couturier and B. Dutrillaux, "Replication studies and demonstration of position effect in rearrangements involving the human X chromosome," in *Cytogenetics of Human X Chromosome*, A. Sandberg and A. Liss, Eds., pp. 375–403, 1983.

[16] C. Ratomponirina, E. Viegas-Pequignot, and B. Dutrillaux, "Synaptonemal complexes in Gerbillidae: probable role of intercalated heterochromatin in gonosome-autosome translocations," *Cytogenetics and Cell Genetics*, vol. 43, no. 3-4, pp. 161–167, 1986.

[17] V. Gupta, M. Parisi, D. Sturgill et al., "Global analysis of X-chromosome dosage compensation," *Journal of Biology*, vol. 5, article 3, 2006.

[18] S. R. V. Rao and M. Padmaja, "Mammalian-type dosage compensation mechanism in an insect -Gryllotalpa fossor (Scudder)- Orthoptera," *Journal of Biosciences*, vol. 17, no. 3, pp. 253–273, 1992.

[19] A. Dübendorfer, M. Hediger, G. Burghardt, and D. Bopp, "*Musca domestica*, a window on the evolution of sex-determining mechanisms in insects," *International Journal of Developmental Biology*, vol. 46, no. 1, pp. 75–79, 2002.

[20] M. Ashburner, *Drosophila: A Laboratory Handbook*, Cold Spring Harbor Laboratory Press, Cold Spring Harbor, NY, USA, 1989.

[21] M. D. Champion and R. S. Hawley, "Playing for half the clock: the molecular biology of meiosis," *Nature Cell Biology*, vol. 4, pp. 50–56, 2002.

[22] P. E. Cohen, S. E. Pollack, and J. W. Pollard, "Genetic analysis of chromosome pairing, recombination, and cell cycle control during first meiotic prophase in mammals," *Endocrine Reviews*, vol. 27, no. 4, pp. 398–426, 2006.

22

Effect of Habitat Type on Parasitism of *Ectatomma ruidum* by Eucharitid Wasps

Aymer Andrés Vásquez-Ordóñez,[1, 2] Inge Armbrecht,[1] and Gabriela Pérez-Lachaud[3]

[1] *Departamento de Biología, Universidad del Valle, Calle 13 No. 100-00 Cali, Valle, Colombia*
[2] *Instituto de Ciencias Naturales, Universidad Nacional de Colombia, Apartado Aéreo 7495, Bogotá, Colombia*
[3] *El Colegio de la Frontera Sur, Entomología Tropical, Avenida Centenario km 5.5, 77014 Chetumal, QROO, Mexico*

Correspondence should be addressed to Gabriela Pérez-Lachaud, igperez@ecosur.mx

Academic Editor: Volker Witte

Eucharitidae are parasitoids that use immature stages of ants for their development. *Kapala* Cameron is the genus most frequently collected in the Neotropics, but little is known about the biology and behavior of any of the species of this genus. We aimed to evaluate the effect of habitat type on eucharitid parasitism and to contribute to the knowledge of the host-parasite relationship between *Kapala* sp. and the poneromorph ant *Ectatomma ruidum* (Roger) in Colombia. Twenty *E. ruidum* colonies were extracted from two different habitat types (woodland and grassland), and larvae and cocoons (pupae) were examined in search for parasitoids in different stages of development. Globally, 60% of the colonies were parasitized, with 1.3% of larvae and 4% of pupae parasitized. Planidia (first-instar larvae), pupae, and adults of the parasitoid were observed. All of the pupae and adult parasitoids belonged to *Kapala iridicolor* Cameron. All the colonies collected in the woodlands were parasitized and contained more parasitized larvae (2%) and parasitized cocoons (8%) than those collected in grasslands (4/12 parasitized colonies, 0.5% parasitized larvae, 0.8% parasitized cocoons). The relationship observed between habitat type and parasitism prevalence is a novel aspect of the study of eucharitid impact on ant host populations.

1. Introduction

Several dipteran, strepsipteran, and hymenopteran parasitoids are natural enemies of ants [1–9]. Among the hymenopterans, the Eucharitidae *sensu stricto* is the only monophyletic group, at the family level, where all of its members are parasitoids of ants. They are also one of the largest and most diverse groups attacking social insects [8].

Eucharitidae have a specialized life cycle that includes oviposition away from the host, on or into a host-plant [2]. Although there are more than 400 species of Eucharitidae already described [8], the hosts and host-plants of only a few species are known [10], and knowledge on the life history and ecology of these wasps is even scarcer. In the New World, detailed studies on selected species have only been carried out in a few localities in Mexico, Argentina and North America (e.g., [1, 11–17]). For Colombia, there is no detailed report on the biology of any species of this family.

The impact of eucharitids on their host populations has recently been explored in detail for some Mexican and South American ant populations [12, 17–19]. These, and earlier reports (e.g., [2, 11, 20]), signaled the aggregated nature of eucharitid populations. In fact, prevalence of parasitism by eucharitids varies greatly in time and space [2], with 100% of colonies parasitized at some sites, and other colonies escaping from parasitism (e.g., [18, 19]). Differences in local parasitism, in general, can be attributable to several different factors such as the presence of resources, other than hosts, necessary for maintaining high parasitoid populations locally (e.g., floral and extrafloral nectar, and refuge sites for adults), suitable host-plants, microclimatic differences, and/or dispersal capacity of adult parasitoids [21, 22]. In some cases, for example, parasitoids may be less effective at parasitizing hosts in sites with simpler vegetation [23]. In the case of eucharitids, an aspect not yet studied in detail is the effect of the habitat on the impact of these parasitoids on their ant-host populations, though preliminary results of a recent study suggest that differences in management in coffee agroecosystems (i.e., shade, pruning, weed management) might affect parasitism by eucharitids [24].

Ectatomma ruidum (Roger) (Hymenoptera: Formicidae: Ectatomminae) is a diurnal, earth-dwelling, Neotropical ant that nests in the soil. This ant is found from southern Mexico to Brazil, from sea level to an altitude of 1500–1600 m [25–27], and is dominant in several ecosystems such as forests [28], or economically important cultivated areas [29, 30]. Two species of *Kapala* (Eucharitidae) have been reported to parasitize this ant in Mexico [14, 31], and parasitism of *E. ruidum* by *Kapala* sp. is also known from Colombia (C. Santamaría and J. Herrera, unpub. data). The purpose of this study is to report observations of the host-parasite relationship between *Kapala* sp. and *E. ruidum* in Colombia and to compare the impact of this eucharitid on its ant host population in two different habitat types.

2. Materials and Methods

This study was carried out on the grounds of the Melendez Campus at the Universidad del Valle (3° 22′ N, 76° 32′ W), located at the south of the city of Cali, Department of Valle del Cauca, Colombia. The Campus has an area of approximately 100 ha, 8 ha of which are occupied by buildings, 44 ha by woodlands, 46 ha by grasslands, and 1 ha by two ponds. The average elevation is 970 m; mean annual temperature is 24.1°C and average relative humidity 73% [32]. Average annual rainfall is around 1500 mm, with two rainfall peaks, from March to May and from September to November (Instituto de Hidrobiología, Meteorología y Estudios Ambientales-IDEAM, unpublished data, cited by [32]). According to the Holdridge system, the study site is located in an area classified as Tropical Dry Forest (bs-T) [33].

Five sites on the campus were examined: 2 sites in grasslands dominated by Poaceae and other creeping plants and with no trees, and 3 sites located in woodlands. The sites in the latter habitat had a lesser amount of Poaceae among the creeping vegetation and had, in some cases, abundant litter. Common tree species in these sites were *Pithecellobium dulce* (Roxb.) Benth., *Samanea saman* (Jacq.) Merr., and *Calliandra pittieri* Standl. (Fabaceae); *Mangifera indica* L. (Anacardiaceae); *Ceiba pentandra* (L.) Gaertn. (Bombacaceae); *Ficus elastica* Roxb. (Moraceae); and *Tabebuia chrysantha* G. Nicholson, *T. rosea* (Bertol.) A. DC., and *Spathodea campanulata* P. Beauv. (Bignoniaceae) [32]. In each of the 5 sites chosen, we determined the number of *E. ruidum* nests in a plot of 8 × 8 m. One additional plot, placed 50 m from the closest grassland plot and comparable to the others, was censused for nest density evaluation in the grassland area, to get an even sample size. During April, May, and November 2009, 20 nests chosen at random were excavated (8 in woodlands and 12 in grasslands) and transported to the laboratory for examination.

Ant larvae were inspected for planidia (eucharitid first-instar larvae) attached to their cuticle by means of a stereoscopic-microscope (Nikon SZ645). Cocoons were kept

in petri dishes at room conditions for 5 days or more and were examined once daily to record emergence of adult eucharitids. At the end of the observation period, all of the cocoons were dissected to look for adults and pupae of dead, or not yet emerged, parasitoids, and to register the caste and sex of ants attacked by the parasitoids. Adult wasps were individually placed in vials covered with cloth mesh, and their survival time was evaluated. No food or water was provided. Pupae and adult eucharitids were identified with available keys [8, 34, 35], and their sex was determined, when possible, based on the dimorphism present in the antennae [8]. The material collected was measured using a stereomicroscope equipped with an ocular micrometer and preserved in 96% alcohol. Voucher specimens of both the ants and the parasitoids have been deposited in the Grupo de Investigación en Ecología de Agroecosistemas y Hábitats Naturales (GEAHNA) collection, at the Museo de Entomología of the Universidad del Valle, Colombia (MEUV), and at the Arthropod Collection of El Colegio de la Frontera Sur, Unidad Chetumal, Mexico (ECO-CH-AR).

A Fisher's exact test was carried out to establish whether there were significant differences between the proportions of parasitized colonies found in woodlands and in grasslands, and Z tests were used to search for differences in the number of parasitized larvae and parasitized pupae according to habitat. Nest density and colony size according to habitat (woodlands or grasslands), and colony size according to the presence or absence of parasitoids (both habitats), were compared using a Mann-Whitney test. Spearman nonparametric correlation was used to explore the relationship between the size of the colony (adults + brood) and total parasitized brood, between the number of larvae per colony and total parasitized larvae, and between the number of cocoons per colony and total parasitized cocoons. All statistics were calculated using STATISTICA 8.0 (StatSoft, Inc.) and R 2.13.1 (The Foundation for Statistical Computing) programs.

3. Results

Of the 20 *E. ruidum* colonies examined, 12 (60%) were parasitized (Table 1). The global rate of parasitism in the study area was 2.3% (parasitized brood per total ant brood, 27/1162), with 1.3% (9/714) of the larvae and 4.0% (18/448) of the pupae parasitized. In total, 29 eucharitid individuals or their remains were observed, with 2.4 ± 2.6 (mean ± standard deviation; n = 12 colonies; range: 1–10) parasitoids per parasitized colony. Parasitoids in 3 different stages of development were found: planidia in 7 colonies (1.6 ± 0.8 parasitized larvae per parasitized colony; range: 1–3), pupae in 3 colonies (3.3 ± 3.2 individuals; range: 1–7), and adults in 5 colonies (1.6 ± 0.9 individuals; range: 1–3). Pupae and adults were identified as belonging to *Kapala iridicolor* (Cameron).

All of the colonies collected in the woodlands were parasitized (n = 8), while in the grasslands only 33.3% (4/12) contained eucharitids (Table 1). Prevalence of parasitism and type of habitat were not independent (Fisher's two-tailed exact test: P = 0.0047), and there was a greater frequency of parasitized nests in the woodlands than in

TABLE 1: Composition of *Ectatomma ruidum* colonies in two different habitat types, percent parasitized brood, and number and stage of development of *Kapala iridicolor* individuals.

Nest Number	Habitat	Queen Number	Gynes	Males	Workers	Larvae	Pupae	Total	Parasitized larvae (%)	Parasitized pupae (%)	Planidia/ scar	Pupae	Adults
				Ectatomma ruidum								*Kapala iridicolor*	
1	Woodland	0	0	5	34	17	23	79	2 (11.8)	1 (4.4)	3	0	1(♂)
2	Woodland	0	0	0	9	8	3	20	1 (12.5)	0 (0)	1	0	0
3	Woodland	0	0	3	27	24	18	72	1 (4.2)	0 (0)	1	0	0
4	Woodland	0	0	0	5	20	58	83	0 (0)	10 (17.2)	0	7(5♂, 1♀, 1?)	3(1♂, 2♀)
5	Woodland	0	1	5	28	36	48	118	0 (0)	4 (8.3)	0	2(1♂, 1♀)	2(♀)
6	Woodland	0	0	2	47	121	46	216	0 (0)	1 (2.2)	0	0	1(♂)
7	Woodland	0	0	1	17	27	5	50	1 (3.7)	0 (0)	1	0	0
8	Woodland	0	0	14	18	91	0	123	2 (2.2)	—	2	0	0
9	Grassland	0	0	0	14	63	6	83	0 (0)	0 (0)	0	0	0
10	Grassland	0	0	0	9	18	22	49	0 (0)	0 (0)	0	0	0
11	Grassland	0	0	1	3	25	21	50	0 (0)	0 (0)	0	0	0
12	Grassland	0	0	0	15	25	32	72	0 (0)	0 (0)	0	0	0
13	Grassland	0	1	1	27	26	16	71	0 (0)	1 (6.3)	0	0	1(♂)
14	Grassland	0	0	1	38	31	23	93	0 (0)	0 (0)	0	0	0
15	Grassland	0	0	2	23	39	34	98	0 (0)	1 (2.9)	0	1(♂)	0
16	Grassland	0	0	2	20	1	19	42	0 (0)	0 (0)	0	0	0
17	Grassland	0	0	0	34	33	31	98	1 (3.0)	0 (0)	1	0	0
18	Grassland	1	0	0	39	42	20	102	0 (0)	0 (0)	0	0	0
19	Grassland	0	0	0	40	42	13	95	1 (2.4)	0 (0)	2	0	0
20	Grassland	0	0	0	3	25	10	38	0 (0)	0 (0)	0	0	0
	Total	1	2	37	450	714	**448**	1652	9 (1.3)	18 (4.0)	11	10	8

the grasslands (Fisher's one-tailed exact test: $P = 0.039$). Furthermore, the proportion of parasitized pupae differed between both habitats (Z-test, $Z = 2.9$; $P = 0.003$), with a greater number of parasitized pupae in the woodlands than in the grasslands (Figure 1). Although a greater proportion of parasitized larvae was also observed in the woodlands (Figure 1), there was no statistical difference according to habitat (Z-test, $Z = 1.79$, $P = 0.07$). The global rates of parasitism for the woodlands and the grasslands were 4.2% and 0.65%, respectively. The average number of parasitoids per parasitized colony was greater in the woodlands (3 ± 3.1 parasitoids; range: 1–10) than in the grasslands (1.25 ± 0.5 parasitoids; range: 1-2).

There was a significant, positive correlation between the number of parasitized pupae and the number of available pupae (Spearman correlation test, $r = 0.63$, $P = 0.004$, $n = 19$ colonies). However, no correlation was found between the following variables: (1) total parasitized brood and colony size (Spearman, $r = 0.39$, $P = 0.08$, $n = 20$ colonies), (2) total parasitized larvae and number of available host larvae (Spearman, $r = -0.04$, $P = 0.86$, $n = 20$ colonies), and (3) total parasitized brood and number of workers (Spearman, $r = 0.22$, $P = 0.36$, $n = 20$ colonies).

Workers and ant larvae were present in all of the *E. ruidum* colonies but there were no cocoons in one of them (Table 1). The global mean size of the colonies (queen, gynes, workers, males, pupae, and larvae) was 82.6 ± 41.7

individuals. Colony size was greater in colonies from the woodlands and in those parasitized (95.1 ± 59.2 and 93.8 ± 47.8 individuals, respectively) than in those from the grasslands (74.3 ± 24.0), or from nonparasitized colonies (66.1 ± 24.7). Nevertheless, there were no significant differences in colony size between parasitized and non-parasitized colonies (Mann-Whitney test, $U = 29$, $P = 0.14$, $n_1 = 12$, $n_2 = 8$), nor according to the habitat from which the colonies came (Mann-Whitney test, $U = 40$, $P = 0.62$, $n_1 = 12$, $n_2 = 8$). A significantly greater density of *E. ruidum* colonies was estimated for the grasslands (3281 colonies/ha), compared to that for the woodlands (1563 colonies/ha) (Mann-Whitney test, $U = 0$, $P = 0.049$, $n_1 = n_2 = 3$).

Six planidia attached to the cuticle of *E. ruidum* larvae (Figure 2(a)) were observed. Five were in the interior of a sclerotized ring (Figures 2(b) and 2(c)) while one was not (Figure 2(e)). In 5 cases, sclerotized rings with no planidia were observed (Figure 2(d)). In 2 host larvae from different colonies, 2 planidia (or empty scars) were observed (representing 22% of the parasitized larvae, 2/9). Very small host larvae were found parasitized. The length of parasitized larvae ranged from 2.77 to 10.10 mm ($n = 8$). Planidia were on average 0.086 ± 0.006 mm in length (range: 0.08–0.09 mm, $n = 3$), and the sclerotized rings had a diameter of 0.165 ± 0.072 mm (range: 0.07–0.27 mm, $n = 9$) and a thickness of 0.043 ± 0.026 mm (range: 0.08–0.09 mm, $n = 9$). More male (64.7%, 11/17) than female (35.3%, 6/17)

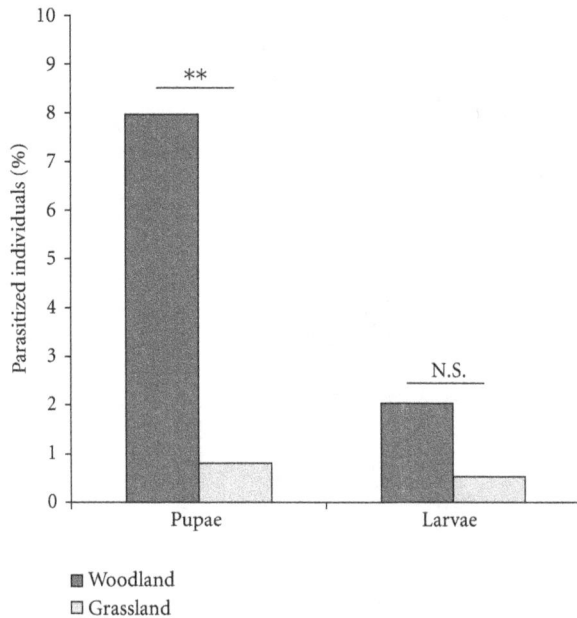

FIGURE 1: Global percentage of *Ectatomma ruidum* immature stages parasitized by *Kapala iridicolor* in two different habitat types (woodland versus grassland). **$P < 0.01$; N.S.: nonsignificant.

eucharitids were observed in six colonies with wasp pupae and adults. Adult *K. iridicolor* males, on the average, lived longer (5.3 ± 2.2 days, range: 2–7 days, $n = 4$) than females (4 ± 0.8 days, range: 3–5 days, $n = 4$) although no significant differences were found (Mann-Whitney test, $U = 4$, $P = 0.25$).

No cases of superparasitism were observed during dissection of cocoons. Empty cocoons from which eucharitid adults emerged had an operculum at the anterior end, opposite to the one with the ant's meconium, forming a regular circular cut (Figure 3(a)) which is made by the adult wasps with their falcate jaws. From the remains of the host ants found during dissection of the cocoons, it can be stated that male, queen, and worker pupae are attacked (Figure 3(b)). Queen pupae were not completely consumed by the developing wasp.

4. Discussion

In this study, a comparison of eucharitid parasitism was made between two contrasting habitats differing in tree cover and associated understory vegetation. This perspective had not previously been considered in detail in studies of the impact of eucharitids on their ant host populations. Although early works (e.g., [2, 11]) indicated variation of parasitism in space, no comparison of parasitism by eucharitids in different habitats involving grasslands is available. However, a recent study also found a greater parasitism of poneromorph ants in forest fragments, compared to the more disturbed areas of Mexican coffee plantations [24].

Because the plots in our study (woodlands and grasslands) were interspersed, a site effect is less possible than a "habitat" effect. Our results showed that parasitoids

were more prevalent in woodlands than in grasslands with respect to the number of parasitized colonies and parasitized host pupae, although the percentage of parasitized larvae did not differ statistically between both habitats. These findings suggest that the probability of an encounter between eucharitids and their host ant colonies is higher in more complex habitats such as those of the Valle del Cauca dry forest. Alternatively, the survival of eucharitids may be increased in shaded areas. The two habitats are different in vegetation (composition and structure) and environmental characteristics such as temperature and humidity [36], which may affect host and host-plant availability/distribution, and ant foraging strategies.

This study also represents the first detailed record of the interaction between *K. iridicolor* and *E. ruidum* in Colombia, which is the second locality for these species where aspects of the impact of parasitism and other information on the natural history of this eucharitid are known. *Kapala iridicolor* is known to parasitize several species of poneromorph ants in Mexico (*E. ruidum*, *Gnamptogenys regularis* Mayr, *G. sulcata* (Fr. Smith), *G. striatula* Mayr, and *Pachycondyla stigma* (F.) [14]), and it might probably interact with other ants in Colombia. On the Melendez Campus and in the City of Cali, *K. iridicolor* had been reported earlier [37], but its ant host was unknown. In the locality studied, other species of poneromorph ants have also been reported including *Odontomachus bauri* Emery, *O. erythrocephalus* Emery, and *Pachycondyla* sp. [37, 38].

The percentage of *E. ruidum* colonies with *K. iridicolor* parasitoids is very close to that observed for the interaction of *E. ruidum* with *K. iridicolor* and *K. izapa* Carmichael in Mexico [18]. The percentage of immature stages attacked was, however, low compared with the results of some studies that indicate over 16% of the brood parasitized (e.g., [11, 13, 20]). Nevertheless, these figures are within the range observed by Lachaud and Pérez-Lachaud [18] in their year-long study on *E. ruidum* in Mexico.

It is worth noting that a lower density of *E. ruidum* nests was found in the woodlands compared to the grasslands. These observations are consistent with those for the Departments of Valle del Cauca and Cauca [39], but not for Guajira, a drier region in the extreme north of Colombia, where a high abundance of *E. ruidum* nests was found in areas with higher presence of trees [40]. Furthermore, and although not studied in detail, we did note differences in the foraging hours of *E. ruidum* workers in the two habitats sampled. Foragers of this ant species displayed very low activity in the grasslands during the warmest hours (10:00 am to 16:00 pm), while ants were observed foraging during these hours in the woodlands. Nest distribution and nest density, as well as foraging times of ants, are factors that could also contribute to differences in parasitism by eucharitids. *Kapala iridicolor* is known to use a wide range of host plants for oviposition, including species of several plant families (Malvaceae, Boraginaceae, Asteraceae) [14, 34]. The plant(s) used by this eucharitid in our study site remain(s) unknown; however it is likely that differences in understory vegetation between the woodland and the grassland contribute also to the observed differences in

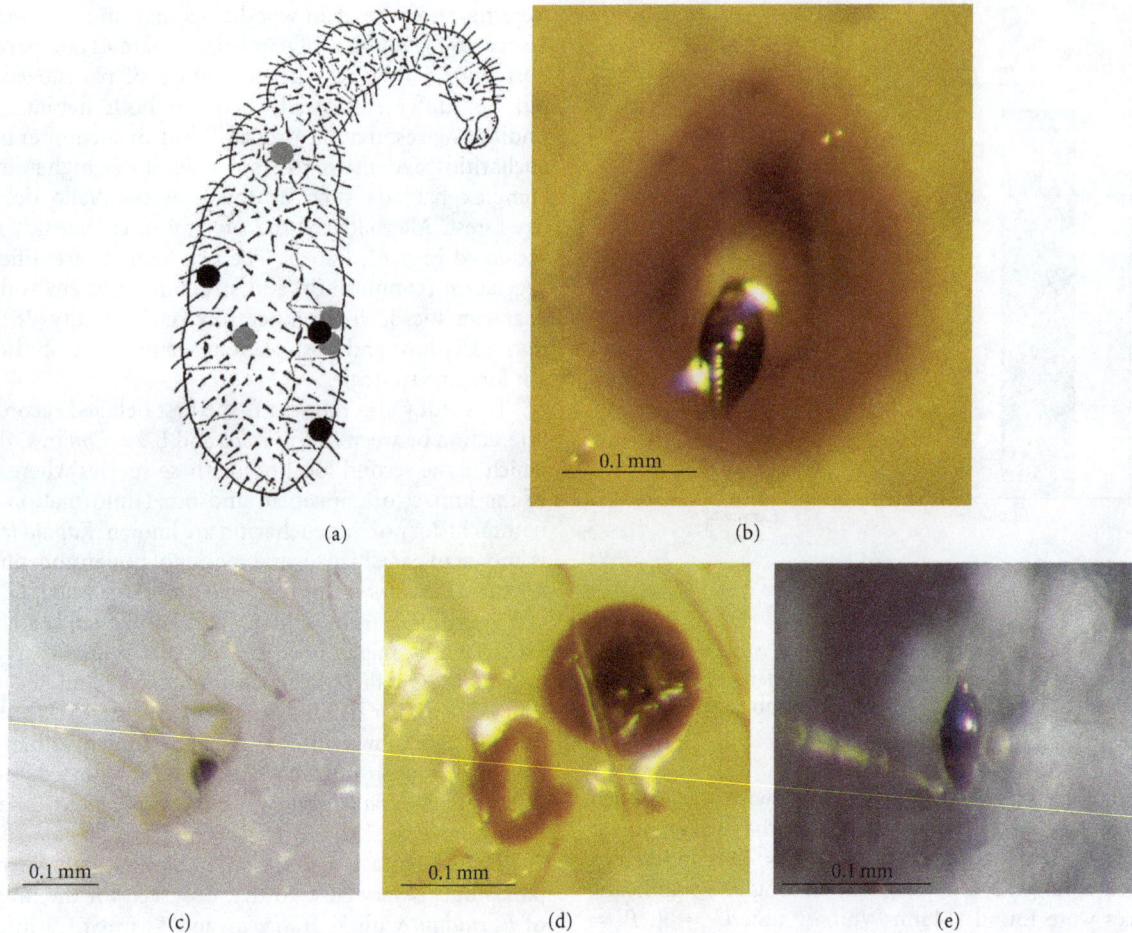

(a)   (b)

(c)   (d)   (e)

FIGURE 2: First larval stage (planidium) of *Kapala iridicolor* on its *Ectatomma ruidum* larval host. (a) Position of planidia (black dots) and scars (gray dots) on the host larva (modified from [43]). (b) Planidium with scar (sclerotized ring) around it. (c) Extreme back side view of a planidium on the interior of a partially sclerotized ring. (d) Scar without planidium. (e) Planidium joined to the host without formation of a sclerotized ring.

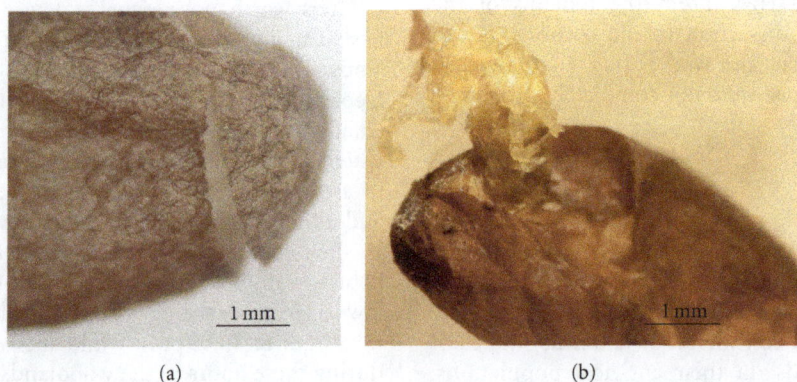

(a)   (b)

FIGURE 3: Operculum made by *Kapala iridicolor* wasp on emergence from the host cocoon. (a) View of the cocoon with operculum. (b) Host remains (only the ant cuticle is left).

parasitism of *E. ruidum* by *K. iridicolor*, as has been suggested for differences in eucharitid parasitism between coffee agro-ecosystems and the forest [24]. This issue deserves further study.

A positive correlation was established between the number of parasitized pupae and the number of pupae available in the colonies. This pattern was similar to that reported elsewhere [18, 19]. It was also found that parasitized ant larvae were quite variable in size and covered almost all sizes of larvae present in the colonies. This observation contrasts with previous reports [11–14, 41] where, in general, only late larval instars were found attacked by planidia, but very young host larvae with planidia have been reported in some cases [19, 42].

In summary, the results of this study showed an effect of habitat type (woodlands versus grasslands) on parasitism of *E. ruidum*, a widely distributed, dominant poneromorph ant, by *K. iridicolor*. It also records some aspects of the natural history of this parasitoid in Colombia.

Acknowledgments

The authors want to thank Julian Mendivil and Dagoberto Sinisterra for their help with field work, James Montoya and Carolina Giraldo for logistic support, the Grupo de Investigación de Biología, Ecología y Manejo de Hormigas (Universidad del Valle), and Grupo de Ecología de Agroecosistemas y Hábitats Naturales (GEAHNA) for providing space and equipment, and Mauricio Bermudez for statistical advice. Jean-Paul Lachaud (ECOSUR) and three anonymous reviewers provided useful comments on an earlier version of this work. Finally, the authors acknowledge the Universidad del Valle and The Basic Science program from Colciencias (Project Code Colciencias 1106 45 22 1048), for financial support.

References

[1] W. M. Wheeler, "The polymorphism of ants, with an account of some singular abnormalities due to parasitism," *Bulletin of the American Museum of Natural History*, vol. 23, pp. 1–93, 1907.

[2] C. P. Clausen, "The habits of the Eucharidae," *Psyche*, vol. 68, no. 2-3, pp. 57–69, 1941.

[3] J.-P. Lachaud and L. Passera, "Données sur la biologie de trois Diapriidae myrmécophiles: *Plagiopria passerai* Masner, *Solenopsia imitatrix* Wasmann et *Lepidopria pedestris* Kieffer," *Insectes Sociaux*, vol. 29, no. 4, pp. 561–568, 1982.

[4] M. S. Loiácono, "Un nuevo diáprido (Hymenoptera) parasitoide de larvas de *Acromyrmex ambiguus* (Emery) (Hymenoptera, Formicidae) en Uruguay," *Revista de la Sociedad Entomológica Argentina*, vol. 44, pp. 129–136, 1987.

[5] D. P. Wojcik, "Behavioral interactions between ants and their parasites," *Florida Entomologist*, vol. 72, no. 1, pp. 43–51, 1989.

[6] B. Hölldobler and E. O. Wilson, *The Ants*, The Belknap Press of Harvard University Press, Cambridge, Mass, USA, 1990.

[7] D. H. Feener Jr. and B. V. Brown, "Diptera as parasitoids," *Annual Review of Entomology*, vol. 42, pp. 73–97, 1997.

[8] J. M. Heraty, "A revision of the genera of Eucharitidae (Hymenoptera: Chalcidoidea) of the world," *Memoirs of the American Entomological Institute*, vol. 68, pp. 1–367, 2002.

[9] J. Kathirithamby, "Host-parasitoid associations in Strepsiptera," *Annual Review of Entomology*, vol. 54, pp. 227–249, 2009.

[10] J.-P. Lachaud, P. Cerdan, and G. Pérez-Lachaud, "Poneromorph ants associated with parasitoid wasps of the genus *Kapala* Cameron (Hymenoptera: Eucharitidae) in French Guiana," *Psyche*, vol. 2012, Article ID 393486, 6 pages, 2012.

[11] G. L. Ayre, "*Pseudometagea schwarzii* (Ashm.) (Eucharitidae: Hymenoptera), a parasite of *Lasius neoniger* Emery (Formicidae: Hymenoptera)," *The Canadian Journal of Zoology*, vol. 40, pp. 157–164, 1962.

[12] J. M. Heraty, D. P. Wojcik, and D. P. Jouvenaz, "Species of *Orasema* parasitic on the *Solenopsis saevissima*-complex in South America (Hymenoptera: Eucharitidae, Formicidae)," *Journal of Hymenoptera Research*, vol. 2, no. 1, pp. 169–182, 1993.

[13] J. M. Heraty, "Biology and importance of two eucharitid parasites of *Wasmannia* and *Solenopsis*," in *Exotic Ants. Biology, Impact, and Control of Introduced Species*, D. F. Williams, Ed., pp. 104–120, Westview Press, Boulder, Colo, USA, 1994.

[14] G. Pérez-Lachaud, J. M. Heraty, A. Carmichael, and J.-P. Lachaud, "Biology and behavior of *Kapala* (Hymenoptera: Eucharitidae) attacking *Ectatomma*, *Gnamptogenys*, and *Pachycondyla* (Formicidae: Ectatomminae and Ponerinae) in Chiapas, Mexico," *Annals of the Entomological Society of America*, vol. 99, no. 3, pp. 567–576, 2006.

[15] J. Torréns, J. M. Heraty, and P. Fidalgo, "Biology and description of a new species of *Laurella* Heraty (Hymenoptera: Eucharitidae) from Argentina," *Proceedings of the Entomological Society of Washington*, vol. 109, no. 1, pp. 45–51, 2007.

[16] J. Torrens, J. M. Heraty, and P. Fidalgo, "Biology and description of a new species of *Lophyrocera* Cameron (Hymenoptera: Eucharitidae) from Argentina," *Zootaxa*, no. 1871, pp. 56–62, 2008.

[17] L. Varone and J. Briano, "Bionomics of *Orasema simplex* (Hymenoptera: Eucharitidae), a parasitoid of *Solenopsis* fire ants (Hymenoptera: Formicidae) in Argentina," *Biological Control*, vol. 48, no. 2, pp. 204–209, 2009.

[18] J.-P. Lachaud and G. Pérez-Lachaud, "Impact of natural parasitism by two eucharitid wasps on a potential biocontrol agent ant in southeastern Mexico," *Biological Control*, vol. 48, no. 1, pp. 92–99, 2009.

[19] G. Pérez-Lachaud, J. A. López-Méndez, G. Beugnon, P. Winterton, and J.-P. Lachaud, "High prevalence but relatively low impact of two eucharitid parasitoids attacking the Neotropical ant *Ectatomma tuberculatum* (Olivier)," *Biological Control*, vol. 52, no. 2, pp. 131–139, 2010.

[20] C. P. Clausen, "The biology of *Schizaspidia tenuicornis* Ashm., a eucharid parasite of *Camponotus*," *Annals of the Entomological Society of America*, vol. 16, pp. 195–217, 1923.

[21] D. A. Landis, S. D. Wratten, and G. M. Gurr, "Habitat management to conserve natural enemies of arthropod pests in agriculture," *Annual Review of Entomology*, vol. 45, pp. 175–201, 2000.

[22] B. A. Shapiro and J. Pickering, "Rainfall and parasitic wasp (Hymenoptera: Ichneumonoidea) activity in successional forest stages at Barro Colorado Nature Monument, Panama, and La Selva Biological Station, Costa Rica," *Agricultural and Forest Entomology*, vol. 2, no. 1, pp. 39–47, 2000.

[23] A.-M. Klein, I. Steffan-Dewenter, and T. Tscharntke, "Rain forest promotes trophic interactions and diversity of trap-nesting Hymenoptera in adjacent agroforestry," *Journal of Animal Ecology*, vol. 75, no. 2, pp. 315–323, 2006.

[24] A. De La Mora and S. M. Philpott, "Wood-nesting ants and their parasites in forests and coffee agroecosystems," *Environmental Entomology*, vol. 39, no. 5, pp. 1473–1481, 2010.

[25] N. A. Weber, "Two common ponerine ants of possible economic significance, *Ectatomma tuberculatum* (Olivier) and *E. ruidum* Roger," *Proceedings of the Entomological Society of Washington*, vol. 48, no. 1, pp. 1–16, 1946.

[26] W. L. Brown Jr., "Contributions toward a reclassification of the Formicidae. II. Tribe Ectatommini (Hymenoptera)," *Bulletin of the Museum of Comparative Zoology, Harvard*, vol. 118, no. 5, pp. 173–362, 1958.

[27] C. Kugler and W. L. Brown, "Revisionary and other studies on the ant genus *Ectatomma*, including the descriptions of two new species," *Search: Agriculture, Ithaca*, no. 24, pp. 1–7, 1982.

[28] C. Kugler and M. C. Hincapié, "Ecology of the ant *Pogonomyrmex mayri*: distribution, abundance, nest structure, and diet," *Biotropica*, vol. 15, no. 3, pp. 190–198, 1983.

[29] J.-P. Lachaud, "Foraging activity and diet in some Neotropical ponerine ants. I. *Ectatomma ruidum* Roger (Hymenoptera, Formicidae)," *Folia Entomológica Mexicana*, vol. 78, pp. 241–256, 1990.

[30] I. Perfecto, "Ants (Hymenoptera: Formicidae) as natural control agents of pests in irrigated maize in Nicaragua," *Journal of Economic Entomology*, vol. 84, no. 1, pp. 65–70, 1991.

[31] J.-P. Lachaud and G. Pérez-Lachaud, "Fourmis ponérines associées aux parasitoïdes du genre *Kapala* Cameron (Hymenoptera, Eucharitidae)," *Actes des Colloques Insectes Sociaux*, vol. 14, pp. 101–105, 2001.

[32] M. C. Muñoz, K. Fierro-Calderón, and H. F. Rivera-Gutierrez, "Las aves del Campus de la Universidad del Valle, una isla verde urbana en Cali, Colombia," *Ornitología Colombiana*, no. 5, pp. 5–20, 2007.

[33] L. S. Espinal, *Apuntes Sobre Ecología Colombiana*, Universidad del Valle, Departamento de Biología, Biología, Colombia, 1967.

[34] J. M. Heraty and J. B. Woolley, "Separate species or polymorphism: a recurring problem in *Kapala* (Hymenoptera: Eucharitidae)," *Annals of the Entomological Society of America*, vol. 86, no. 5, pp. 517–531, 1993.

[35] J. M. Heraty, "Familia Eucharitidae," in *Introducción a los Hymenoptera de la Región Neotropical*, F. Fernández and M. J. Sharkey, Eds., pp. 709–715, Sociedad Colombiana de Entomología and Universidad Nacional de Colombia, Bogotá, Colombia, 2006.

[36] I. Armbrecht, "Comparación de la mirmecofauna en fragmentos boscosos del valle geográfico del río Cauca, Colombia," *Boletín del Museo de Entomología de la Universidad del Valle*, vol. 3, no. 2, pp. 1–14, 1995.

[37] A. A. Vásquez-Ordóñez, "Género *Kapala* Cameron (Hymenoptera: Eucharitidae) en el campus Meléndez de la Universidad del Valle (Cali-Colombia)," *Boletín del Museo de Entomología de la Universidad del Valle*, vol. 7, no. 2, pp. 17–18, 2006.

[38] P. Chacón de Ulloa, M. L. Baena, J. Bustos, R. C. Aldana, J. A. Aldana, and M.A. Gamboa, "Fauna de hormigas del Departamento del Valle del Cauca (Colombia)," in *Insectos de Colombia: Estudios Escogidos*, G. M. Andrade-C., G. Amat-García, and F. Fernández, Eds., pp. 413–451, Academia Colombiana de Ciencias Exactas, Físicas y Naturales, Centro Editorial Javeriano, Pontificia Universidad Javeriana, Bogotá, Colombia, 1996.

[39] C. Santamaría, I. Armbrecht, and J.-P. Lachaud, "Nest distribution and food preferences of *Ectatomma ruidum* (Hymenoptera: Formicidae) in shaded and open cattle pastures of Colombia," *Sociobiology*, vol. 53, no. 2, pp. 517–541, 2009.

[40] C. Santamaría, Y. Domínguez-Haydar, and I. Armbrecht, "Cambios en la distribución de nidos y abundancia de la hormiga *Ectatomma ruidum* (Roger 1861) en dos zonas de Colombia," *Boletín del Museo de Entomología de la Universidad del Valle*, vol. 10, no. 2, pp. 10–18, 2009.

[41] G. M. Das, "Preliminary studies on the biology of *Orasema assectator* Kerrich (Hym., Eucharitidae), parasitic on *Pheidole* and causing damage to leaves of tea in Assam," *Bulletin of Entomological Research*, vol. 54, pp. 373–378, 1963.

[42] G. C. Wheeler and E. W. Wheeler, "New hymenopterous parasites of ants (Chalcidoidea: Eucharidae)," *Annals of the Entomological Society of America*, vol. 30, pp. 163–175, 1937.

[43] G. C. Wheeler and J. Wheeler, "The ant larvae of the subfamily Ponerinae—part I," *The American Midland Naturalist*, vol. 48, no. 1, pp. 111–144, 1952.

Permissions

The contributors of this book come from diverse backgrounds, making this book a truly international effort. This book will bring forth new frontiers with its revolutionizing research information and detailed analysis of the nascent developments around the world.

We would like to thank all the contributing authors for lending their expertise to make the book truly unique. They have played a crucial role in the development of this book. Without their invaluable contributions this book wouldn't have been possible. They have made vital efforts to compile up to date information on the varied aspects of this subject to make this book a valuable addition to the collection of many professionals and students.

This book was conceptualized with the vision of imparting up-to-date information and advanced data in this field. To ensure the same, a matchless editorial board was set up. Every individual on the board went through rigorous rounds of assessment to prove their worth. After which they invested a large part of their time researching and compiling the most relevant data for our readers. Conferences and sessions were held from time to time between the editorial board and the contributing authors to present the data in the most comprehensible form. The editorial team has worked tirelessly to provide valuable and valid information to help people across the globe.

Every chapter published in this book has been scrutinized by our experts. Their significance has been extensively debated. The topics covered herein carry significant findings which will fuel the growth of the discipline. They may even be implemented as practical applications or may be referred to as a beginning point for another development. Chapters in this book were first published by Hindawi Publishing Corporation; hereby published with permission under the Creative Commons Attribution License or equivalent.

The editorial board has been involved in producing this book since its inception. They have spent rigorous hours researching and exploring the diverse topics which have resulted in the successful publishing of this book. They have passed on their knowledge of decades through this book. To expedite this challenging task, the publisher supported the team at every step. A small team of assistant editors was also appointed to further simplify the editing procedure and attain best results for the readers.

Our editorial team has been hand-picked from every corner of the world. Their multi-ethnicity adds dynamic inputs to the discussions which result in innovative outcomes. These outcomes are then further discussed with the researchers and contributors who give their valuable feedback and opinion regarding the same. The feedback is then collaborated with the researches and they are edited in a comprehensive manner to aid the understanding of the subject.

Apart from the editorial board, the designing team has also invested a significant amount of their time in understanding the subject and creating the most relevant covers. They scrutinized every image to scout for the most suitable representation of the subject and create an appropriate cover for the book.

The publishing team has been involved in this book since its early stages. They were actively engaged in every process, be it collecting the data, connecting with the contributors or procuring relevant information. The team has been an ardent support to the editorial, designing and production team. Their endless efforts to recruit the best for this project, has resulted in the accomplishment of this book. They are a veteran in the field of academics and their pool of knowledge is as vast as their experience in printing. Their expertise and guidance has proved useful at every step. Their uncompromising quality standards have made this book an exceptional effort. Their encouragement from time to time has been an inspiration for everyone.

The publisher and the editorial board hope that this book will prove to be a valuable piece of knowledge for researchers, students, practitioners and scholars across the globe.

List of Contributors

Cintia Lepesqueur
Programa de Ṕos-Graduac̨̃ao em Ecologia, Instituto de Ciˆencias Bioĺogicas, Universidade de Braśılia, 70910-900 Braśılia, DF, Brazil

Helena C. Morais
Departamento de Ecologia, Instituto de Ciˆencias Bioĺogicas, Universidade de Braśılia, 70910-900 Braśılia, DF, Brazil

Ivone Rezende Diniz
Departamento de Zoologia, Instituto de Ciˆencias Bioĺogicas, Universidade de Braśılia, 70910-900 Braśılia, DF, Brazil

W. D. Fernandes, M. V. Sant'Ana and J. Raizer
Faculdade de Ciˆencias Bioĺogicas e Ambientais, Universidade Federal da Grande Dourados (UFGD), MS 162, Km 12, 79804-970 Dourados, MS, Brazil

D. Lange
Laborat́orio de Ecologia Comportamental e de Interac̨̃oes, Universidade Federal de Uberlˆandia (UFU), P.O. Box 593, 38400-902 Uberlˆandia, MG, Brazil

V. K. Rahmathulla
P3 Basic Seed Farm, National Silkworm Seed Organization, Central Silk Board, Ring Road, Srirampura, Mysore, Karnataka 570 008, India

Arthur G. Appel and Xing Ping Hu
Department of Entomology and Plant Pathology, Auburn University, 301 Funchess Hall, Auburn, AL 36849-5413, USA

Jinxiang Zhou, Zhongqi Qin, Hongyan Zhu, Zhijing Wang, Xianqin Liu and Mingyan Liu
Fruit and Tea Institute, Hubei Academy of Agricultural Sciences, Wuhan 430209, China

Xiangqian Chang
Plant Protection and Fertilizer Institute, Hubei Academy of Agricultural Sciences, Wuhan 430070, China

Petros Damos and Matilda Savopoulou-Soultani
Laboratory of Applied Zoology and Parasitology, Department of Plant Protection, Faculty of Agriculture, Aristotle University of Thessaloniki, 54124 Thessaloniki, Greece

Pervin Erdogan and Aysegul Yildirim
Central Plant Protection Research Institute, Yenimahalle, 49.06172 Ankara, Turkey

Betul Sever
Faculty of Pharmacy, University of Ankara, Tandogan, 06100 Ankara, Turkey

Lucas A. Kaminski
Departamento de Biologia Animal, Instituto de Biologia, Universidade Estadual de Campinas, CP 6109, Campinas 13083-970, SP, Brazil

Fernando S. Carvalho-Filho
Laborat́orio de Ecologia de Invertebrados, Instituto de Biologia, Universidade Federal do Paŕa, Rua Augusto Corrˆea 01, Guaḿa, Beĺem 66075-110, PA, Brazil

Jean-Paul Lachaud
Departamento de Entomoloǵıa Tropical, El Colegio de la Frontera Sur, Avenida Centenario km 5.5, 77014 Chetumal, QRoo, Mexico
Centre de Recherches sur la Cognition Animale, CNRS-UMR 5169, Universit́e de Toulouse, UPS, 118 route de Narbonne, 31062 Toulouse Cedex 09, France

Gabriela Pérez-Lachaud
Departamento de Entomolog´ıa Tropical, El Colegio de la Frontera Sur, Avenida Centenario km 5.5, 77014 Chetumal, QRoo, Mexico

T. Delsinne
Biological Evaluation Section, Royal Belgian Institute of Natural Sciences, 29 rue Vautier, 1000 Brussels, Belgium

F. Fernández
Instituto de Ciencias Naturales, Universidad Nacional de Colombia, Apartado 7495, Bogot´a D.C., Colombia

Sergio López and Arturo Goldarazena
Neiker-Basque Institute of Agricultural Research and Development, Arkaute, 01080 Vitoria, Spain

Austin L. Hughes
Department of Biological Sciences, University of South Carolina, 715 Sumter Street, Coker Life Sciences Building, Columbia, SC 29208, USA

Martin N. Andersson
Department of Biology, Lund University, 223 62 Lund, Sweden

P. G. Tillman
USDA, ARS, Crop Protection and Management Research Laboratory, P.O. Box 748, Tifton, GA 31793, USA

T. E. Cottrell
USDA, ARS, Southeastern Fruit & Tree Nut Research Laboratory, 21 Dunbar Road, Byron, GA 31008, USA

Angsumarn Chandrapatya
Department of Entomology, Faculty of Agriculture, Kasetsart University, Bangkok 10900, Thailand

Prapassorn Bussaman, Chirayu Sa-uth and Paweena Rattanasena
Department of Biotechnology, Faculty of Technology, Mahasarakham University, Maha Sarakham 44150, Thailand

Paula E. Cushing
Department of Zoology, Denver Museum of Nature & Science, 2001 Colorado Boulevard, Denver, CO 80205, USA

Adolfo da Silva-Melo and Edilberto Giannotti
Departamento de Zoologia, Instituto de Bioci^encias, Universidade Estadual Paulista (UNESP), Campus Rio Claro SP, Caixa Postal 199, 13506-900, Rio Claro, SP, Brazil

Konrad Fiedler
Department of Tropical Ecology and Animal Biodiversity, Faculty of Life Sciences, University of Vienna, Rennweg 14, 1030 Vienna, Austria

Wenbe Hwang and Hsiu-Mei Lin
Department of Ecoscience and Ecotechnology, National University of Tainan, 33, Section 2, Shu-Lin Street, Tainan 70005, Taiwan

Xavier Espadaler
Ecology Unit and CREAF, Autonomous University of Barcelona, 08193 Bellaterra, Spain

Sergi Santamaria
Botany Unit, Autonomous University of Barcelona, 08193 Bellaterra, Spain

Ximena J. Nelson
School of Biological Sciences, University of Canterbury, Private Bag 4800, Christchurch 8041, New Zealand

B. Dutrillaux and A.M. Dutrillaux
UMR 7205, OSEB, CNRS/Mus´eum National d'Histoire Naturelle, 16, rue Buffon, CP 32, 75005 Paris, France

Aymer Andrés Vásquez-Ordóñez
Departamento de Biolog´ıa, Universidad del Valle, Calle 13 No. 100-00 Cali, Valle, Colombia
Instituto de Ciencias Naturales, Universidad Nacional de Colombia, Apartado A´ereo 7495, Bogot´a, Colombia

Inge Armbrecht
Departamento de Biolog´ıa, Universidad del Valle, Calle 13 No. 100-00 Cali, Valle, Colombia

Gabriela Pérez-Lachaud
El Colegio de la Frontera Sur, Entomolog´ıa Tropical, Avenida Centenario km 5.5, 77014 Chetumal, QROO, Mexico